Biological Learning and Control

Computational Neuroscience
Terence J. Sejnowski and Tomaso A. Poggio, editors

For a complete list of books in this series, see the back of the book and
http://mitpress.mit.edu/Computational_Neuroscience

Biological Learning and Control

How the Brain Builds Representations, Predicts Events, and Makes Decisions

Reza Shadmehr and Sandro Mussa-Ivaldi

The MIT Press
Cambridge, Massachusetts
London, England

This book was set in Syntax and Times Roman by Toppan Best-set Premedia Limited.

Library of Congress Cataloging-in-Publication Data

Shadmehr, Reza.
Biological learning and control : how the brain builds representations, predicts events, and makes decisions / Reza Shadmehr and Sandro Mussa-Ivaldi.
p. cm.—(Computational neuroscience)
Includes bibliographical references and index.
ISBN 978-0-262-01696-4 (hardcover : alk. paper)
ISBN 978-0-262-54955-4 (paperback)
1. Brain. 2. Neuropsychology. 3. Brain—Mathematical models. I. Mussa-Ivaldi, Sandro. II. Title.
QP376.S4373 2012
612.8'2—dc23

2011026392

Contents

Series Foreword

Computational neuroscience is an approach to understanding the development and function of nervous systems at many different structural scales, including the biophysical, the circuit, and the systems levels. Methods include theoretical analysis and modeling of neurons, networks, and brain systems and are complementary to empirical techniques in neuroscience. Areas and topics of particular interest to this book series include computational mechanisms in neurons, analysis of signal processing in neural circuits, representation of sensory information, systems models of sensorimotor integration, computational approaches to biological motor control, and models of learning and memory. Further topics of interest include the intersection of computational neuroscience with engineering, from representation and dynamics, to observation and control.

Terrence J. Sejnowski
Tomaso Poggio

Introduction

As you are reading this book, you are moving your eyes rapidly from one word to another. Each movement is a saccade. There is regularity in how you move your eyes in that the speed of your eye movements, the duration of each movement, and the places where you choose to move your eyes are similar to how another person reading this book would move their eyes. In fact, if you look closely at how we move our arms during reaching and how we move our legs during walking, there are also strong similarities across people. It is useful to notice regularity in nature because occasionally, regularity leads to discovery of fundamental mechanisms. Think of regularity in the motion of celestial objects: Kepler gave an approximate description of the orbits of the planets around the sun (the elliptical path, with the square of the orbital period being equal to the cube of the semimajor axis of the orbit), and Newton was able to account for this regularity by proposing that gravity produces a centripetal force inversely proportional to the square of the radius. In this book, we will focus on the regularity in how our brain perceives the world around us, the regularity in how our brain reacts to sensory stimuli, and most crucially, the regularity in how our brain controls our movements. Our goal is to first describe this regularity and then attempt to make sense of it using theory—in particular, theory of the kind that relies on mathematics.

Let us start with an example of regularity in perception. Children sometimes play a game in which a mischievous kid licks his *index* finger while a victim is watching and then runs over and wipes the *middle* finger on the arm of the second child. One of our students, Mollie Marko, told us of this story, and we had to try it. Indeed, the victim was certain that his arm felt wet. Why should our brain feel that the arm is wet when in fact it is perfectly dry? To account for this, we will build a theoretical framework in which perception is a combination of two streams of information: one due to prediction and the other due to observation. Prediction is due to internal models that describe what should happen. Measurements are due to reports from a sensory system that describes what actually did happen. Perception is a combination of the two, describing our belief about what happened.

But perhaps a more interesting question is: What is good about sensing the world this way? What do we gain by making predictions and then combining the predictions with our observations? By presenting the problem in a theoretical framework, we will see that our ability to sense the world is significantly improved by our ability to predict the world. In other words, by relying on predictions, our brain can dramatically improve upon the limitations of our sensors. Because of our brain's ability to predict, we can see what is around us better than what our eyes can tell us, and we can feel where our body is in space better than what our proprioception can tell us.

If our brain is going to rely on its own predictions to form beliefs about sensory observations, then a fundamental problem is how to maintain the accuracy of the predictions. To cope with this, the brain continuously learns from prediction errors. For example, in America soft drinks come in cans that are sold in colorful cardboard boxes. In a grocery store, if you were to pick up one of these boxes and it were empty, your arm would jerk upward, and you might lose your balance and fall backward. The reason is that your brain expects the box to be full and weigh a significant amount. The motor commands that you send to your arms (lift the box), and your legs (stabilize the body) are based on this expectation. If the box is empty, the result of the action is a prediction error, that is, a difference between the expected sensory consequences of the motor commands and the actual sensory consequences. This error produces a change in our prediction. If we were asked to lift another box, the brain would predict a smaller weight than before, and the motor commands would reflect this change in prediction.

Now consider the regularity in our movements. The next time that you open a web page on your computer, pay attention to where you look first. Say that the page has some text on it, and also some graphics, and one of the graphics contains a human face. It is quite likely that the first place that your eyes will seek is the face. Advertisers know this, and that is why they place graphics that contain faces in whatever they sell. Why should you look at the face first? Perhaps the reason is that every movement you make is a reflection of an economic decision: Among the potential things that I can look at, which is the most valuable? Attention is indeed a resource that comes in limited supply.

If you could measure your eye movements, you would note that the motor commands that your brain sends to your eyes to direct the gaze at a particular location are such that the resulting movement has a particular velocity and duration. For example, a 15-degree displacement of your eyes will take about 0.06 seconds and reach a peak velocity of 400 degrees per second. Among healthy people of a given age these numbers are fairly consistent. So there seems to be some regularity in where we tend to move our eyes, as well as in how our brain controls the motion of the eyes during that movement.

You might say that this regularity has a lot to do with the fact that across people, eyes are biomechanically very similar. Maybe the commands are similar because the biomechanics are similar. Biomechanics are indeed important, but consider the fact that our movements change from the teenage years to young adulthood, and they continue to change as we grow older. Our movements change because of maladies such as Parkinson's disease, schizophrenia, and depression. Our movements have changed through evolution: Movements made by monkeys are different than those made by humans. A clear example is how the brain moves our eyes during a saccade. When teenagers move their eyes, their saccades have very high velocities, and this speed declines at every decade of life. In Parkinson's disease, saccade speeds are abnormally slow, while in schizophrenia saccade speeds are abnormally fast. Monkeys have saccades that are nearly twice the speed of humans, despite nearly identical eye biomechanics. Why?

By presenting this puzzle in a theoretical framework, we will try to make sense of the facts. We will suggest that motor commands that move our body reflect an economic decision that our brain makes regarding reward and effort. In particular, if we view the goal of any movement as acquiring a more rewarding state for our body and further hypothesize that the brain discounts reward with the passage of time (we would rather receive reward sooner rather than later), then the theory predicts that as temporal discounting of reward changes in the brain (for example due to development, disease, or evolution), the shape of our movements will also change.

Temporal discounting of reward is typically measured in decision-making tasks by asking people and other animals to make decisions associated with reward and time. For example, given a choice between \$20 now and \$80 in one year, you may wait for a year to get \$80. We often assess impulsivity by quantifying the rate at which an individual discounts reward as a function of time. A person who is more impulsive will discount that \$80 more steeply as a function of time, and may therefore pick \$20 now. Diseases that affect the reward system can produce changes in impulsivity. Development produces changes in impulsivity, as has evolution. It is a peculiar fact that changes in impulsivity as measured in these decision-making tasks are sometimes correlated with changes in movement kinematics (Shadmehr et al., 2010). By attempting to build a theory that links movements with reward, we will build a story that has the potential to explain why development and aging alter the way we move, why diseases that affect the reward system alter the way we move, and why evolution that has brought about different primate species have also made them different in the way they move.

It may seem a bit odd that we are going to be using fairly sophisticated mathematics to study movements like saccades and reaching. These movements may not seem like much in a world where movies show robots fighting robots and enhancing human power as exoskeletons. Surely moving an eyeball from one place to another or an arm from here to there couldn't be that interesting. What's the big deal?

Well, consider that the motors in robots can produce the same force for a given input over and over again, whereas our muscles fatigue quickly and alter their responses from one movement to the next. The sensors that record the motion of a robot do so with far more precision than one finds in the response of our proprioceptive neurons. The transmission lines that connect a robot's motors and sensors to its controller move information at the speed of light, and the controller can process sensory information to issue commands in microseconds. In contrast, our transmission lines (axons) move information slower than the speed of sound, and neural computations often require tens of milliseconds. Indeed, our ability to produce a lifetime of accurate movements is not because we are born with an invariant set of actuators, precise set of sensors, or fast transmission lines. Once we see the problem in this way, we cannot help but be amazed by the consistency and fidelity of a lifetime of movements. There is really nothing ordinary about our most ordinary movements.

How do we maintain our ability to move accurately for our entire life? The answer seems to be that we are born with a nervous system that adapts to these limitations and continuously compensates for them. Internal models are representations that our nervous system builds of our own bodies and the world around us. Internal models allow the brain to predict the sensory consequences of motor commands. The brain uses these predictions to plan actions so that we can convert goals and desired rewarding states into motor commands that achieve them in some efficient way. Internal models change constantly because our body is changing from moment to moment and because the world around us provides us with innumerably rewarding states.

"Model" is a loaded word. Some models are abstract representations of the world. There are economic models of how prices fluctuate in free markets based on laws such as supply and demand. There are celestial models that allow us to know when and where a comet will appear in the sky. There are models of neuronal networks to explain how a collection of simple interconnected element can do something as difficult as recognizing a familiar face. All these models have in common the use of mathematics for understanding reality and, perhaps more important, for predicting the future. Predictions may take place over geologic eras or over fractions of a second. In all cases, the possibility to predict rests on the observation of some ongoing external influence, such as a gravitational force, combined with the knowledge of some relevant state variables that evolve through time. Another important meaning of the word "model" is the physical reproduction of an object, such as the model of ship or of a car. This is also another family of human-made artifacts. When we reproduce an object into a model, we capture some of the object's essential properties. These can be shape features, as in a doll, but also functional features, as in a paper airplane. The essence of all models may be less in what they represent than in what they omit. Modern physics was built upon ignoring large amounts of empirical facts to focus on those few that could bear a clear relation to ideas. This theoretical neglect is consistent with the way in which our brains perceive the external world. Our retinas have a tiny spot, called the fovea, where visual information is captured with high resolution. The rest is a blur. Yet, our visual perception is quite uniform. As we look around, we are constructing the very scene that we observe by combining a mosaic of fragments in our memories. You may be familiar with a video clip by Daniel Simmons in which students play by passing a ball to each other. At some point an actor dressed as a gorilla walks in the middle of the players and engages in gorilla-like gestures before leaving the scene. In Simmons's experiment, subjects look at the scene and count the number of times the ball changes hands. Almost invariably, when at the end they were asked if they noticed the gorilla, they were astonished. "What gorilla?" But when one sees the same clip after having been informed of the trick, then one is astonished for not having seen the gorilla the first time. Far from being a flaw, this kind of neglect is a foundation of our ability to operate in a world that presents us with a constant torrent of facts, among which only few are really important.

There is another type of model: the internal models that are formed by the brain as we act upon the world around us. Internal models are not human artifacts but products of biology. They are not made of clay or mathematical equations. However, they carry out many of the same functions of all other models: They constitute a form of fundamental knowledge of the physical world, of the space around us and of the consequences of actions. Most important, they provide the brain with one of the most essential skills for survival, the ability to predict based on past observations. In the following chapters we will present mathematical concepts that relate to the formation of internal models in movement control and perception. In a way, this book is about mathematical models of brain models. The mathematical notation is our own instrument for describing how representations are formed, what structure they may have, and how our own theories can be tested. We do not expect that the brain carry out computations in the way we do. But to the extent that we are capturing some truth about information processing in the brain, the behavior and the structure of our models have a degree of equivalence with the way in which biology operates.

Regularity in how people and other animals adapt to their environment provides us with clues as to how internal models are represented in the brain. We are going to be using primarily behavioral data from experiments in humans and other animals in order to describe the regularity, and then we will use mathematics to ask whether the behaviors make sense. Because we are not going to be talking about the "how" question with biology—for example, we are not going to be asking about how neurons in the brain might actually perform the computations that are implied by our mathematics—our approach is going to be fraught with danger. After all, it is quite possible that we cannot know why an organism does something until we know how it was built. That is, we may not be able to successfully theorize about the regularity in the way that our brain moves our body until we are much farther along in understanding the basic facts regarding the biology of our brain and its evolutionary history. Can we have any hope for success in using a purely theoretical approach?

Stephen Jay Gould (1995), the eminent paleontologist, considered this question in a broader sense and wrote the following regarding the role of theory:

> Geology, in the late 18th century, had been deluged with a rash of comprehensive, but mostly fatuous, "theories of the earth"—extended speculations about everything, generated largely from armchairs. When the Geological Society of London was inaugurated in the early 19th century, the founding members overreacted to this admitted blight by banning theoretical discussion from their proceedings. Geologists, they ruled, should first establish the facts of our planet's history by direct observation—and then, at some future time when the bulk of accumulated information becomes sufficiently dense, move to theories and explanations. . . . In mid career, in 1861, in a letter to Henry Fawcett, Darwin reflected on the false view of earlier geologists. In so doing, he outlined his own conception of proper scientific procedure in the best one-liner ever penned. "About thirty years ago there was much talk that geologists ought only to observe and not theorize; and I well remember someone saying that at this rate a man might as well go into a gravel pit and count the pebbles and describe the colors. How odd it is that anyone should not see that all observations must be for or against some view if it is to be of any service!"

Admittedly, theories can also act as straitjackets, focusing attention on observations that agree with the theory and blurring views that can challenge it. In telling our story, our goal will not to be to sell one view or another, but to point out that there is true wonder in the data, and that on occasion theory allows one to make sense of a good deal of it.

We would like to thank the colleagues who have read the manuscript and offered insightful comments and critical discussions. We are particularly grateful to Jean Jacques Orban de Xivry, Adrian Haith, Maura Casadio, Felix Huang, Alejandro Lopez Janez, Citlali Lopez Ortiz, Amir Karniel, John Krakauer, and Robert Scheidt. Special thanks to the many students that have endured our ramblings and contributed to the development of the ideas in this book. In particular we wish to thank Kurt Thoroughman, Maurice Smith, Opher Donchin, Haiyin Chen-Harris, Vincent Huang, Minnan Xu-Wilson, Jörn Diedrichsen, Jun Izawa, Sarah Criscimagna-Hemminger, Siavash Vaziri, Zachary Danziger, and Justin Horowitz.

1 Space in the Mammalian Brain

1.1 Where Am I?

You wake up one morning in a hotel room. You have just arrived in Cozumel, Mexico, for a short vacation. Coming out of a dream, you do not quite remember this fact. The room is dark; there is only a dim light coming from the shutter on the right side somewhere at a distance. You do not know where you are, or why you are there. It cannot be your home. Or is it? The bathroom would be immediately to the left of the door. But where is the door? This is a transient disorientation that many have occasionally experienced, puzzling and disturbing. But it does not last long. Our normal mode of being includes a persistent awareness of where we are in space, and this awareness seems to be tightly connected with our ability to remember facts—for example, to remember that we are on vacation in Cozumel.

The ability to form certain kinds of new memories as well as the ability to locate ourselves in space are both dependent on the hippocampus, a small yet very important seahorse-shaped structure inside the temporal lobes of the brain. The hippocampus (figure 1.1A) has attracted the attention of neuroanatomists since the beginning of modern neuroscience. The oldest drawings of the hippocampal network (figure 1.1B) are credited to the early giant of neuroanatomy, Santiago Ramón y Cajal, who was among the first to propose the "neuron doctrine." What is now common knowledge was then, at the end of the nineteenth century, the revolutionary idea that nerve cells form a communication network through a dense pattern of interconnections.

Damage to the hippocampus is known to cause anterograde amnesia, the inability to form new memories. This has been studied in the clinical case of HM (Milner, Corkin, and Teuber, 1968), a man who had the medial temporal lobes on both sides removed to treat a severe form of epilepsy. The intervention cured the epilepsy but had a very severe side effect: HM was left without the ability to form new memories of facts. Nevertheless, he maintained the ability to learn new motor skills. Brenda Milner trained HM to execute difficult manual tasks—for example,

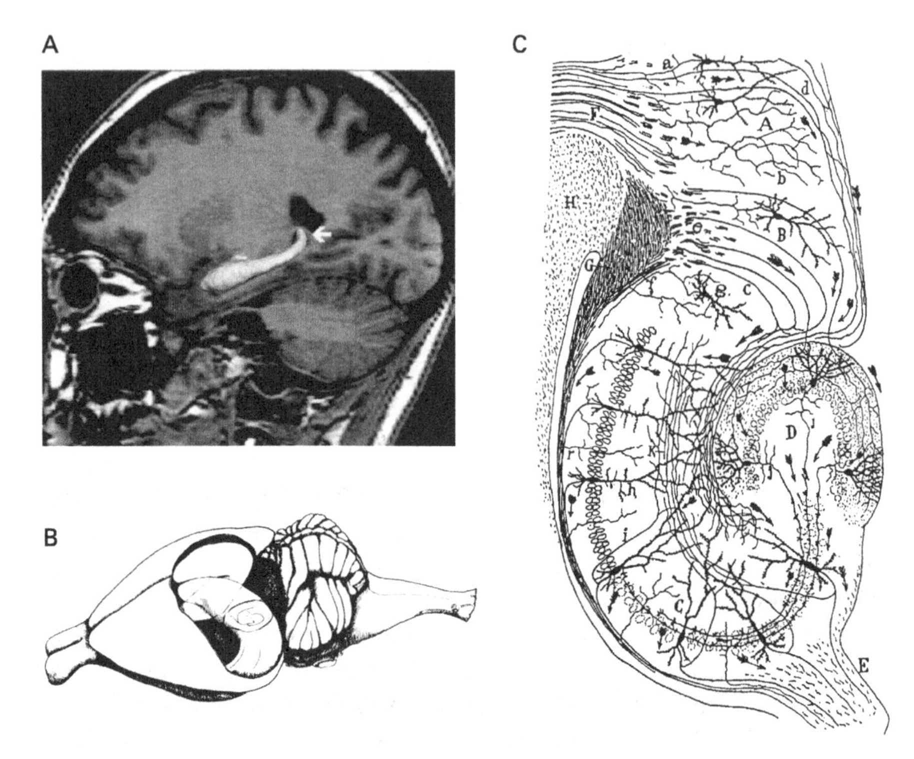

Figure 1.1
The hippocampus. (A) Sagittal MRI with medial aspect of human hippocampus. The arrow shows the alveus of the posterior part of the hippocampus (from Gardner and Hogan, 2005). (B) Rat hippocampus (from Teyler and DiScenna, 1984). (C) Rodent hippocampal circuitry drawn by S. Ramón y Cajal (1911).

tracing the shape of a star while looking at it through a mirror (Milner, 1962). He practiced for a few trials on one day, then was tested the next day. On the second day he retained the level of skill that he had acquired during the previous day. But he could not remember the fact that he had practiced that task earlier. In fact, he was quite surprised at his own unexpected dexterity in such a difficult activity.

Dramatically more common than the story of HM is Alzheimer disease, which causes a progressive loss of memory capacity and disorientation, associated with a severe damage to the hippocampus and the cerebral cortex.

1.2 Space Representations in the Mongolian Gerbil

The amnesia and disorientation following hippocampal damage are cues that the ability to locate ourselves in space and in time is strongly associated with the function of this brain region. This spatial skill is not merely human, for it is present in a multitude of animal species. A well-known example is a species of butterfly, the

Monarch, which migrates every year from central Mexico to Canada and then back. On more moderate geographical scales, rodents have notable skills to navigate across various terrains and in various lighting conditions.

Mongolian gerbils are small rodents that some people like to keep as pets. Besides being a cute domestic animal, the Mongolian gerbil is a skilled explorer and scavenger. The name derives from the hostile environment these gerbils come from, the Mongolian steppes. These are large, hostile, and arid territories, where food is scarce and long trips are necessary for the gerbils to find seeds and other nutrients and bring them back to their burrows for storage. These small rodents are among our best models of animal exploration.

In a set of now classic experiments, Collett, Cartwright, and Smith (1986) trained Mongolian gerbils to seek food inside a circular arena. They hid a sunflower seed under a layer of gravel scattered on the floor. The environment was carefully designed so that only the objects, or "landmarks," placed by the investigators could serve as spatial references for the gerbils. The arena had a diameter of 3.5 meters and was placed inside a room whose walls were painted black. A light bulb illuminated the floor and left the walls in the shadow. Somewhere in the center of the arena, the investigators placed a white cylinder that was clearly visible to the gerbil. The sunflower seed was hidden always at a distance of 50 cm and at fixed angle from the base of the cylinder (figure 1.2). Collett and colleagues placed the gerbil in the

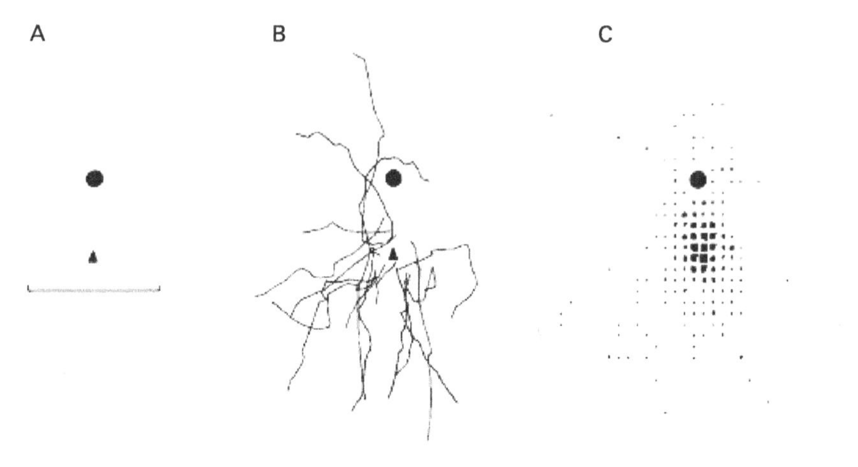

Figure 1.2
Search pattern of one gerbil. (A) Plan view of landmark (circle) and reward site (triangle). Calibration is 100 cm. Landmarks are not shown to scale. (B) Paths followed by the gerbil when released from different points within the arena. (C) Cumulative search-distribution resulting from twenty-one tests of 60 s duration. In this and subsequent figures, the animal's position in relation to the landmark array is given to within a cell 11 cm across and 13.3 cm high. The blacker the cell, the more time the gerbil spent there. Time spent in each cell is expressed as a percentage of that spent in the most visited cell. The latter is filled by twenty-five dots arranged in a square. (From Collett et al., 1986.)

arena and allowed it to find the seed for several training trials. On each trial, they changed the position of the landmark cylinder, but they maintained the position of the seed relative to the landmark. After a few trials, the gerbils learned to move straight toward the seed and retrieve it.

Once the investigators verified that the gerbils had learned this simple task, they removed the seed and watched as the gerbils searched for their reward. Figure 1.2C displays the performance of the gerbils in the test trials. The area of each black square is proportional to the total time spent by a gerbil searching within the corresponding region of space. This simple diagram demonstrates that the gerbil learned to predict the location of the seed in relation to the landmark. Importantly, the gerbil searched in the correct location even when starting different trials from different locations. That is, the gerbil had formed a spatial memory of the location of the goal in terms of its distance and orientation with respect to the landmark. And even if the investigators took care of removing the external visual cues, the ability to identify the correct direction with respect to the cue indicates that the gerbils had a sense of a fixed "north arrow" that could not have been supplied by the landmark alone, since this was uniform in color and cylindrical in shape.[1]

To gain a deeper knowledge about space representations in the gerbils, Collett and colleagues changed the pattern of landmarks once the gerbils had learned to retrieve the seed. Their objective was to investigate how the relation between the goal and the landmarks was understood and exploited by the gerbil's brain. In one experiment, the gerbils were first trained to retrieve the seed inside an arena with two identical landmarks (figure 1.3). The seed was placed in an equilateral triangle arrangement, at equal distance from the two landmarks. How would the trained gerbil react if the distance between the landmarks were unexpectedly doubled? We make the working hypothesis that the gerbil's brain forms a representation of the environment and that this representation, which we call an "internal model," is updated as the gerbil experiences new spatial relations among itself, the landmarks and the seed. We may think of two possible outcomes. In one case, the gerbil's internal model of the environment preserves the shape of the seed-landmark pattern and searches at one location that is at the same distance from the two landmarks. Since the distance between the landmarks has doubled, the distance of the seed from the line joining the landmarks must also increase so that the seed is at the vertex of an isosceles triangle. Alternatively, the internal model preserves the expected distance and orientation of the seed with respect to either landmark. Now, we would have not one but two distinct sites where the seed might be. This second outcome is what Collett and coworkers observed: the gerbils during the test trials searched for the seed at two distinct positions, each one corresponding to the position of the seed relative to the closest landmark.

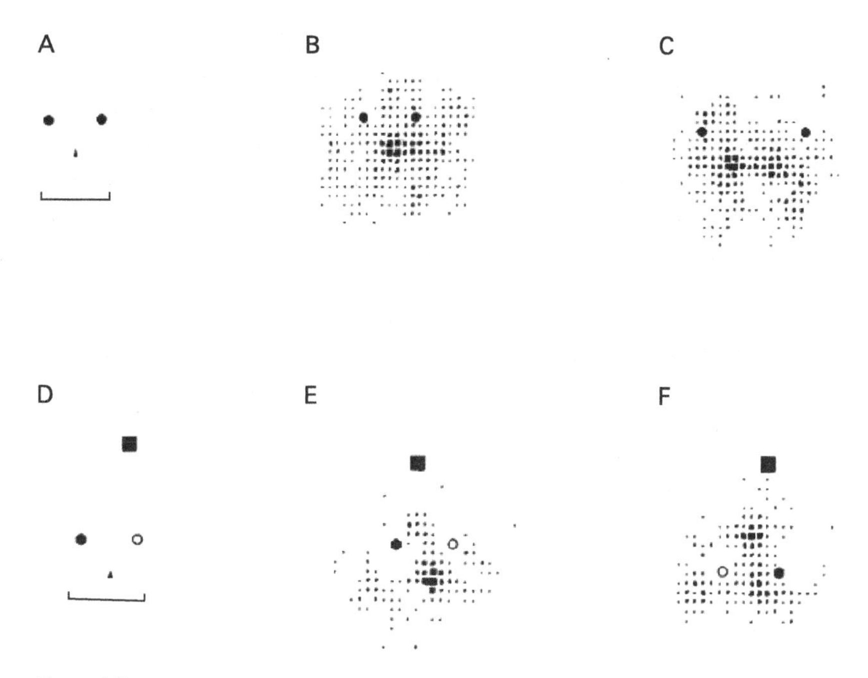

Figure 1.3
Euclidean versus scaling transformations. (A) Gerbils were trained to retrieve the seed placed at equal distance from two identical landmarks. (B) Search pattern after training. (C) When the distance between the landmarks is increased, the gerbils do not use a scaling rule. Instead, they look for the seed at two sites. Each searched site is at the same distance from the corresponding landmark in the trained configuration. This means that the gerbil understood that the environment had changed. The landmarks have been displaced, and the expected distance between the seed and each landmark has remained fixed. (D) In another experiment, gerbils were trained with two different landmarks: a dark aluminum cylinder on the left and a white cylinder on the right. The seed, again, was placed in an equilateral triangle configuration as in (A). The black square indicates the location from which the gerbils start searching. (E) Search distribution with the landmarks in the same configuration used for training. (F) Search distribution with the landmarks rotated by 180 degrees. In this case, they apply the same rotation to the expected site for the seed. (From Collett et al. 1986.)

Consider a new scenario (figure 1.3D) in which the gerbils are trained with two different landmarks, an aluminum cylinder and a white cylinder. Again, the seed is placed in a triangular arrangement, at equal distance from the two cylinders. After the gerbils have been trained to find the seed, they are tested with a 180-degree rotation of the landmark arrangement: their positions are now interchanged, but their distances are preserved. In this new environment, the gerbil searches the seed at a location that is consistent with this rigid rotation, that is, at a location that is reflected over the segment joining the two landmarks (figure 1.3F). These results suggest that not only the gerbils have a clear sense of distance, but that they also can distinguish between a rigid transformation that preserves the distances between the landmarks and can be interpreted as a rotation and a transformation in which the distances have been altered. In the latter case, the gerbils' brain deduces that the landmarks have moved. Accordingly, the seed is searched in relation to each landmark separately. Instead, in the former case, the gerbil's brain understands that distances have not changed. Therefore, the scene—landmarks and seed—has undergone either a rotation or a translation or both. Alternatively, the viewpoint has changed by a relative movement of the gerbil with respect to the fixed scene. The gerbil's brain, however, is not willing to accept the possibility that there has been a dilation of the environment. A dilation might occur through development, as the animal grows in size, but certainly not in the time scale of the experiment. As we will discuss in more detail, the two situations—rigid motions and scale changes—are representative of two types of affine transformations of space. These are all linear transformations, and only a subclass of them preserves the distance between points. This is the subclass of isometries or Euclidean transformations. As we move around in our environment, the points and objects around us remain unchanged. Therefore, our representation of these points and objects undergoes transformations of this Euclidean type.

1.3 Some General Properties of Space Maps in Psychology and Mathematics

There is an ancient debate between two views of animal intelligence. In the "reductionist" view, what we think of as intelligence is nothing but the apparent manifestation of automatic behaviors through which an organism seeks to acquire the largest amount possible of good stuff or, conversely, seeks to avoid bad stuff. Good and bad stuff are the less technical names of positive reward and negative reward. The reductionist viewpoint was once known as the stimulus-response or S-R theory. At its origin is the work of Clark Hull, who investigated learning as a consolidation of stimulus-response associations (Hull, 1930). This conceptual framework has had a revival in theories of optimal control (Todorov and Jordan, 2002) and reinforcement learning (Sutton and Barto, 1998), based on solid mathematical principles as well as on empirical observations.

In the "cognitive" view, organisms do not simply respond to internal and external stimuli, but they create knowledge. As they act in their environment, they acquire and maintain information that may not be immediately conducive to reward but may be later used to this purpose. Edward Tolman was an early champion of the cognitive view (Tolman, 1948). He vehemently opposed the Hull-stimulus-response approach. According to Tolman, when a rat moves inside a maze in search of food "something like a field map gets established in the rat's brain." The studies of Tolman as well as of other experimental psychologists of the time were mostly carried out on rats. The rats were placed within more or less complicated mazes, with corridors, curves, and a dead end. Tolman described one such experiment as being particularly supportive of the cognitive view. The rats entered a starting chamber at the base of a Y-shaped maze. They moved forward and were to choose between the right and left arms of the Y. At the end of each arm there was an exit. Near the right exit there was a bowl of water, and near the left exit there was a bowl of food. In an initial phase, the rats were satiated with food and water before entering the maze. They did not care about drinking or eating at the end of the maze. However, they wanted to find an exit quickly. Sometimes they took the right exit and sometimes the left with no particular preference. After a few repetitions of these initial trials, the rats were divided into two groups. One group was made hungry and the other was made thirsty before entering the maze. It turned out, Tolman reported, that the thirsty rats went immediately to the right, where there was water, and the hungry rats went immediately to the left, where there was food. He concluded that the rats during the first phase of the experiment learned where the food and the water were despite the fact that they did not receive any water or food reward. In the second phase, as they became hungry or thirsty, they went to the right place. The name for this kind of learning is "latent learning" because it is not associated to the delivery of reward. Tolman saw it as crucial evidence against S-R theory.

While there is a natural dialectic tension between reductionist and cognitivist views, these views are mutually incompatible only in their most extreme versions. In the initial part of the experiment, the rats entered the maze without interest for water or food. However, they were already endowed with the notion that food and water are important items from past stimulus-response associations. Then, the unexpected presence of food and water was registered as a salient event. One can say that the brain interprets such an event as an error over the expectation that the maze was merely an empty path. This type of unexpected event is known to trigger learning and the formation of new memories. Thus, the pairing between stimuli and responses or, in this case the association of motor commands with their potential to generate reward, does not have to be restricted to narrow temporal contiguity. In this text we are not considering the formation of maps in opposition to the mechanisms of reinforcement. It is abundantly evident that both are expressed in our brains. The challenge of reconciling stimulus-response mechanisms with the formation of cognitive maps may well lead to a deeper understanding of biological learning.

Before considering how maps are formed, we must ask what a space map is. In mathematics, a "mapping" establishes a correspondence between the elements of two sets. Take, for example, a typical street map of Paris. In this case we have a correspondence between the points on a sheet of paper, a small planar surface, and points in the French capital. The space of Paris is three-dimensional, whereas the space of the map is two-dimensional. Therefore, the street map involves a projection from 3D to 2D. The difference in dimensions imposes some restrictions on the way we go between the map and the space it represents. Each point in Paris has an image in each point of the planar surface of the map. But each point of the map corresponds to a whole line—a vertical line—in Paris. One can find different images for different buildings. But all the floors in a building have the same image on the planar map. We see here a first important point about topographical maps: They compress information so that not everything is represented but only what can be of some utility. The Argentinean writer Jorge Luis Borges wrote a short story about an emperor who asked a team of cartographers to create a map so detailed as to contain literally everything that was inside his empire: every grain of sand, every speck of wood, and so forth. The cartographers succeeded in their mission. But at the end, it turned out that their map, being so absolutely complete, was also perfectly useless.

On the street map of Paris, we find an index where we read something like "Eiffel Tower, E4." This means that the object Eiffel Tower is at the intersection of the row labeled *E* with the column labeled *4*. The map is organized in a grid, which divides the territory into a finite number of squares. Each square is identified by two symbols, corresponding to the rows and columns of the grid. In most street maps, one set of symbols is letters and the other set is integers. This is a trick to give a distinct identity to rows and columns. The letters, like the natural numbers follow an order: *A*, *B*, *C*, *D* . . . They only differ from numbers in that operations such as sums and subtractions are not explicitly defined. This is unimportant because the average tourist uses the map only to locate objects. However, if numbers were used in place of generic labels, then one could carry out arithmetic operations and place these operations in correspondence with objects in the real world. For example, one could add the displacements from landmark L1 to landmark L2 and from landmark L2 to landmark L3 to derive the displacement from landmark L1 to landmark L3. This is where a more rigorous mathematical definition of a mapping becomes most valuable.

The connection between numbers and "things"—for lack of a better term—is central to the development of topology and geometry. The prime example of this is the concept of the real line. A line is a geometrical entity, made of a continuum of points. The real line, indicated as $\mathbb{R}$, is a simple and fundamental correspondence between the real numbers—extending from $-\infty$ to $+\infty$—and the points of a straight line. The correspondence established by $\mathbb{R}$ is bijective: Each point on the line corresponds to one and only one real number, and each real number corresponds to one and only one point on the line. Then, real numbers and points on the line can be used interchangeably. What happens if instead of a real line we consider some other geometrical object? For example, consider a circle (figure 1.4). Because the real line is in some ways equivalent to the real numbers, we ask if the points on the real line can be placed in a bijective correspondence to the points on the circle by

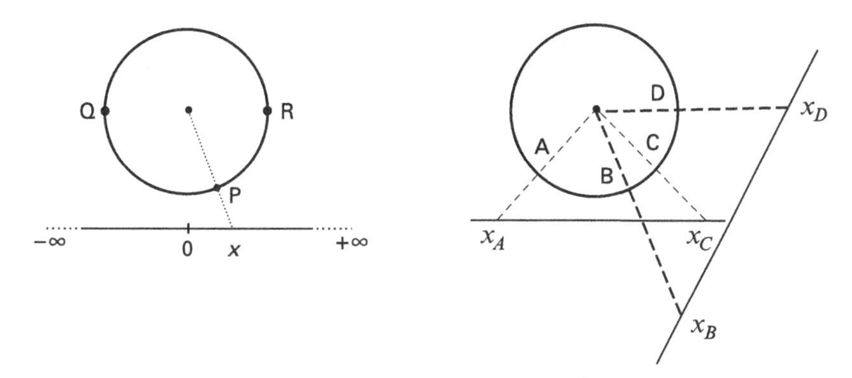

Figure 1.4
Mapping the circle. Left: The real line is placed in correspondence to the lower half-circle by drawing a line from the center of the circle, through the point P. The intersection with the real line is the number x. This establishes a one-to-one correspondence between the points in the lower half of the circle and the whole real line. Right: Two circular arcs, AC and BD, are mapped by two charts onto two real intervals, (x_A, x_C) and (x_B, x_D). Note that the two charts have different "gains," since the projecting distances between the real lines and the circle are different. They also have overlapping regions in their domains (the arc BC). A collection of charts that cover the whole circle is called an "atlas."

some type of projection. However, things are now more complicated. We see that, by projection, all the points in the lower half circle can be placed in correspondence with the entire real line, so that the two points on the "equator," Q and R, correspond to $-\infty$ and $+\infty$ respectively. The right panel of figure 1.4 illustrates how the entire circle is mapped over multiple segments on the real line. Each segment is placed in correspondence with a portion of the circle, that is, with an arc. The mapping on each segment is an example of what is called a "chart." By combining multiple charts, we may cover the entire circle and obtain what is called an "atlas." This mathematical terminology was borrowed from geography and from the ordinary concept of a world atlas, as a collection of charts that, put together, cover the entire globe. In building an atlas, it is of great importance to insure a precise correspondence in the regions where contiguous charts are overlapping. In a world atlas there are identical regions in different pages covering contiguous areas of the globe. Likewise, to make an atlas of the circle in figure 1.4, we need to rescale the two charts on the left of the figure so as to have a consistent mapping across charts.

The mathematical ideas of maps and charts have become increasingly relevant to describe the pattern of neural activities in the hippocampus. This is illustrated in figure 1.5, where the activities recorded from about 100 neurons in the hippocampus are represented over a region that corresponds to a 62 × 62 cm box in which a rat was free to move. These data come from an experiment of Matthew Wilson and Bruce McNaughton (1993) but were arranged in this particular representation in a subsequent article by Alexei Samsonovich and McNaughton (1997). Isolated

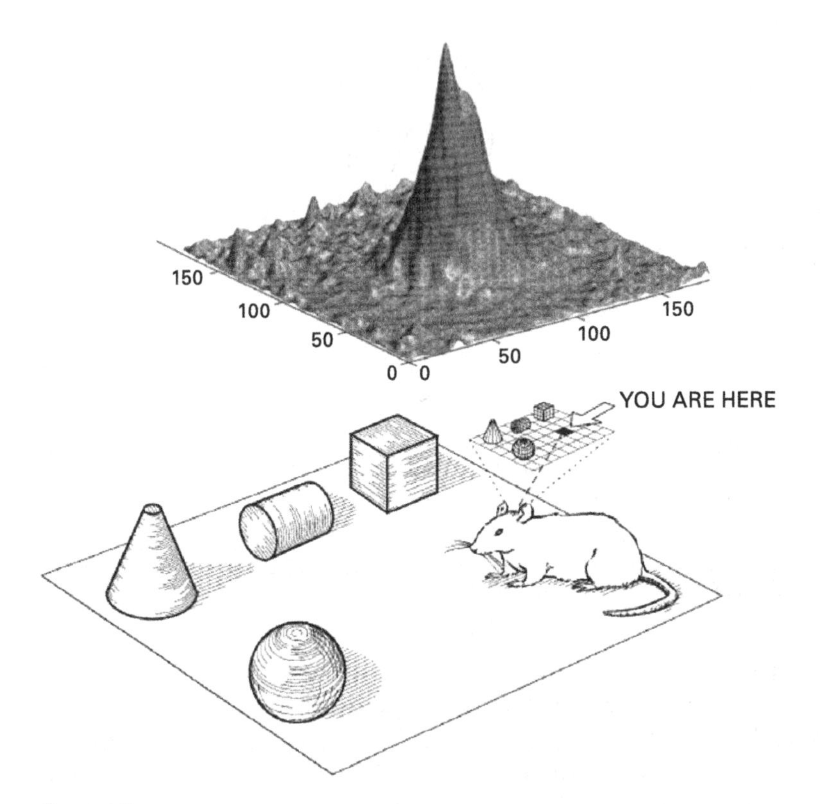

Figure 1.5
Neural charts. The model of Samsonovich and McNaughton depicts the activities of a collection of 100 hippocampal place neurons, by locating each neuron over a Cartesian x-y plane that represents the extrinsic space in which the rat is moving. The activity map is a snapshot of the activities over the entire set of recorded place cells when the rat was passing by the central location of the chart. (From Samsonovich and McNaughton, 1997; bottom portion from Eichenbaum et al., 1999.)

locations on the xy plane correspond to the activity of recorded hippocampal neurons. This correspondence establishes a chart that relates the locations of space explored by the rat to the activity of a family of neurons in the hippocampus that are called "place cells" and were discovered in the early 1970s.

1.4 Place Cells

The neurons studied by Samsonovich and McNaughton were discovered twenty years earlier by John O'Keefe and Jonathan Dostrovsky (1971). They were studying the activity of neurons in the rat's hippocampus and saw that a small fraction of these cells (about 10 percent) became active when the rat was placed in particular locations and was oriented in particular directions. In these early experiments at the

Department of Anatomy of the University College in London, the rats were placed and kept by hand in different sites. Being interested on the formation of space maps in the brain and being aware of the work of Tolman, O'Keefe and Dostrovsky immediately realized the importance of this relatively small sample of hippocampal neurons. In a later study, O'Keefe and David Conway (1976) had the rats moving within a maze populated with peculiar objects. This is well described by O'Keefe and Lynn Nadel (1978):

> The environment consisted of a 7 ft square set of black curtains within which was set a T-shaped maze. On the walls formed by the curtains were four stimuli: a low-wattage light bulb on the first wall, a white card on the second, and a buzzer and a fan on the third and fourth, respectively. Throughout the experiment the location of the goal arm of the T-maze and the four stimuli maintained the same spatial relationship to each other, but all other spatial relations were systematically varied. (205)

To record the neural activity from hippocampal cells while the rat was exploring the maze, and to place these activities in relation to the place where they occurred, the investigators developed an ingenious system. Remember, this is 1970, a time when computers were still in their infancy and video recorders were not yet on the scene:

> Rats were taught a place discrimination in this environment. They were made hungry and taught to go to the goal arm as defined by its relation to the four stimuli within the curtains in order to obtain food. After they had learned the task, place units were recorded. In order to relate the firing of these units to the animal's position in the environment, advantage was taken of the fact that these units have low spontaneous activity . . . outside the place field. Each action potential from the unit was used to trigger a voltage pulse which, when fed back to a light-emitting diode on the animal's head, produced a brief flash of light. A camera on the ceiling of the environment photographed the spots, recording directly the firing of the unit relative to the environment. (206)

The activity recorded from one hippocampal "place unit" is illustrated in figure 1.6. The dots indicate the location at which the activity was detected while the rats moved into the T-maze. The cell in this case became active only when the rat was near the buzzer and when the relative locations of the four landmarks were preserved. The cell was not sensitive to global rotations of the landmark systems, or to the particular physical arm of the maze that happened to be near the buzzer. These and subsequent studies confirmed that place cells became active when the rat was moving at specific locations and with specific heading directions with respect to the set of landmarks that collectively define a spatial frame of reference.

After the seminal work of O'Keefe and colleagues, the 1980s and 1990s saw a flourishing of studies on space representations in the hippocampus. While there is still much to understand about this neural code and its relation to episodic memory, there is no doubt that the place neurons are capable to represent single locations

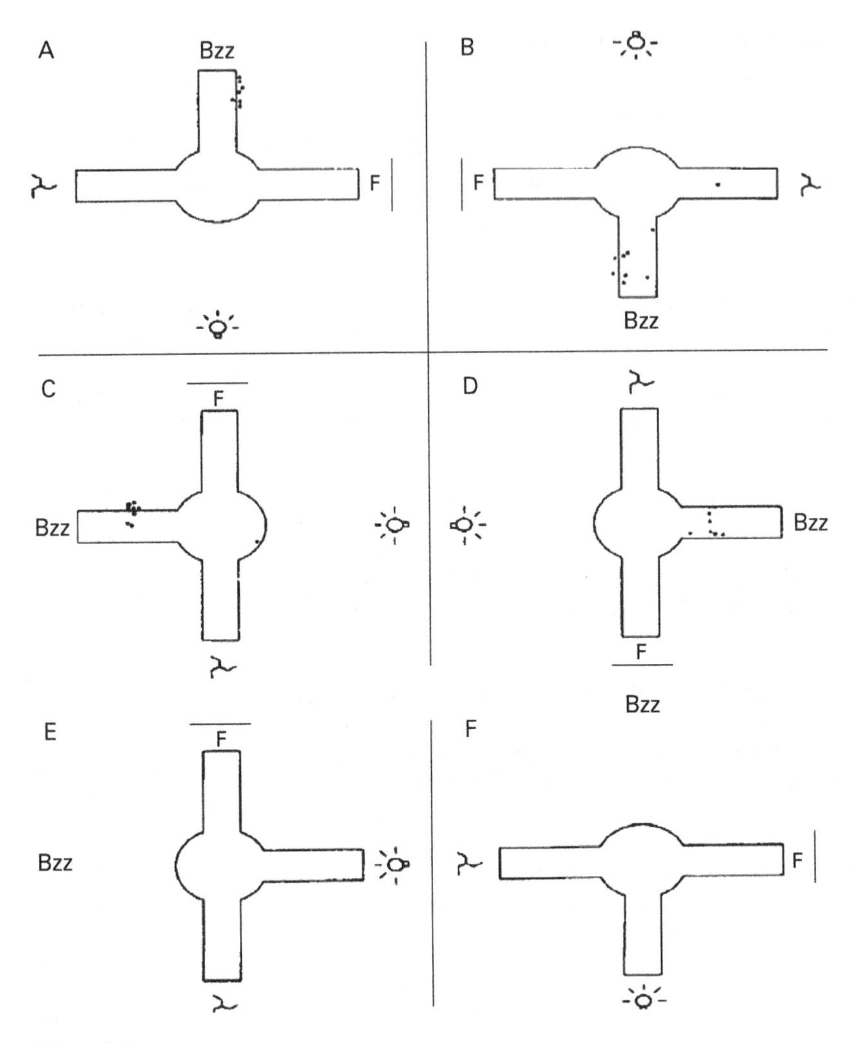

Figure 1.6
The firing of a place unit when a rat is on the T-shaped maze inside the cue-controlled enclosure. Each dot represents one action potential. Four ground trials are shown in A–D, in which the T-maze and the cues on the wall have four different orientations relative to the external world. The unit fires when the rat is in the start arm on the side close to the buzzer, regardless of the orientation relative to the external world. E and F show two ground trials with the start arm rotated 180° so that it is on the side close to the light. There is no unit firing in the start arm. (From O'Keefe and Nadel, 1978.)

in extrapersonal—or "allocentric" – space in a way that is consistent with the presence of a well-organized system of geometrical charts inside the brain of rodents. This rapidly led to functional imaging studies of the human hippocampus. Of particular significance is the work of Eleanor Maguire and colleagues, who studied the brains of London taxi drivers. One of the studies (Maguire et al., 2000) revealed that licensed taxi drivers have an expanded posterior hippocampus, compared to controls that do not drive taxis. This structural study was complemented by functional imaging that revealed significant differences in the activities of a collection of brain regions that includes the right hippocampus (Maguire, Frackowiak, and Frith, 1997). Functional imaging studies have less spatial and temporal resolution than electrophysiological recordings, which monitor the activities of isolated neurons on a millisecond scale. However, functional imaging offers the opportunity of looking at activity patterns across the whole brain. This has revealed that information processing associated with the representation of space spans a broad network of structures. The hippocampus is only one of them. But it is a very prominent one, as was recently demonstrated in a study by Demis Hassabis, Eleanor Maguire, and collaborators (Hassabis et al., 2009). The place cells investigated by O'Keefe and others with recording electrodes are relatively rare and there has not yet been evidence for any topographical organization over the hippocampus. Place cells representing two nearby locations of space may be relatively distant from each other. Conversely, two cells that are in nearby hippocampal sites may become active at two rather different and distant locations in allocentric space. Does this mean that there is no particular organization of neuronal populations in the hippocampus?

Hassabis and collaborators took a pragmatic approach to this question: If there is any functional organization in the activity of hippocampal neurons representing spatial locations, then it should be possible for a computer program to analyze hippocampal population activities and figure out where one is located in allocentric space. In a way, they took advantage of the poor resolution of functional MRI, where activities can only be discriminated to a limit of about $1mm^3$, the approximate size of a "voxel." While this may appear to be a small size, one cubic millimeter of hippocampal gray matter may actually contain between 10^4 and 10^5 neurons. Therefore, a speck of activity detected by fMRI originates from a rather large population of cells. If there were no particular organization in the distribution of place cells, then each voxel would contain more or less the same proportion of the same place cells. The resulting activity snapshot would look like a random blur. In that case, it would be virtually impossible to look at fMRI images of the hippocampus and make a good guess about what space region has "caused" the detected activity. There would be at most a diffuse pattern of signals indicating that the hippocampus is active in a spatial task, but with no possibility of extracting a space code. To test this possibility, Hassabis and colleagues asked subject to play a virtual reality game while lying in a MR scanner. Subjects were presented with the image of a room, with chairs, tables, and other objects that provided a spatial reference. Their task was simply to "move" within the room by pressing on arrow keys. Once at a target location, the view was switched down to the floor mat. The image was the same at all locations to avoid any visual cue about the place in the room. While in this location and without feedback, a functional image was acquired.

This operation was repeated at four different locations. A standard pattern classification algorithm was required to determine the locations based on the activities observed over a large region of the medial temporal lobe, which included the hippocampus. The algorithm was able to determine the location with high accuracy (> 80 percent) based only on the activity over the hippocampal region. This provides new evidence—albeit not conclusive—supporting the hypothesis that there is some topographical structure in the population activity over the hippocampus. In this case, activity charts, such as the one of figure 1.5, derived from neural recordings, might actually represent the internal model of the extrapersonal space. But remember that only a fraction of the hippocampal cells are place cells. The studies on amnesic patients, like H.M., had revealed that the hippocampus is critical to the formation of new memories. There is a close connection between the memory of an event and the location in space where the event occurred. We remember where we were when the twin towers of the World Trade Center in New York were hit on September 11, 2001. As stated by Hassabis and colleagues, this spatial representation may form "the scaffold upon which episodic memories are built." At the end of the next chapter, we will present a computational argument in support of this statement.

1.5 Grid Cells

What is the neural mechanism that leads to the formation of space cells in the hippocampus? While this question remains to be answered, an important clue has come from studies of May-Britt Moser, Edward Moser, Marianne Fyhn, and collaborators, who discovered in 2004 an intriguing pattern of activities in the entorhinal cortex of freely moving rats (Fyhn et al., 2004). The entorhinal cortex is part of the parahippocampal complex and is a major source of input to the hippocampus. Part of it can be seen in the top portion of Cajal's drawing, near the letter A (figure 1.1C). Some neurons in the entorhinal cortex express a very peculiar and impressive geometrical pattern. Like the place cell in the hippocampus, these neurons become active when the animal moves across certain regions of space. But unlike the place cell, the entorhinal neurons display a regular periodic structure (figure 1.7): they have distinct peaks of activity placed at the vertices of a grid of equilateral triangles!

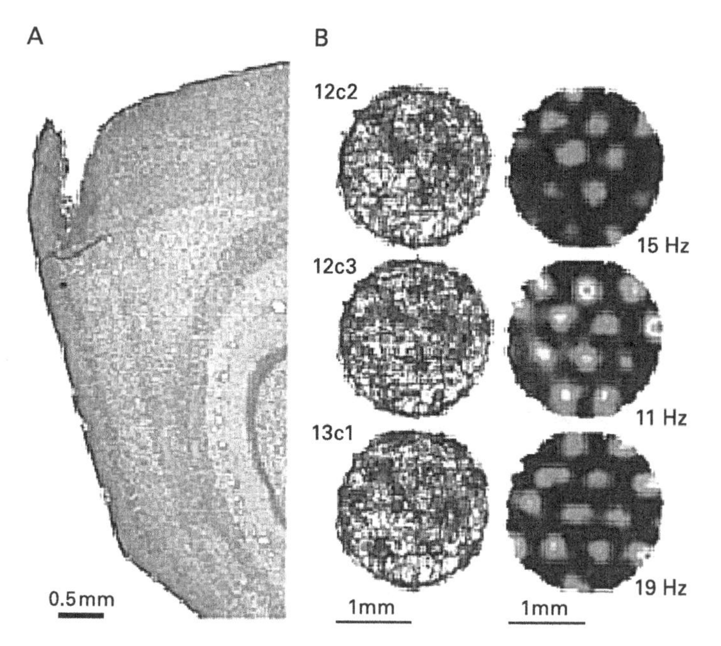

Figure 1.7
Firing fields of entorhinal grid cells. (A) Nissl-stained section indicating the recording location in layer II of the dorsomedial entorhinal cortex of a rat. (B) Firing fields of three simultaneously recorded cells as the rat moved within a large circular arena. Cell names refer to tetrode (*t*) and cell (*c*). The left column shows trajectories of the rat with superimposed firing locations (dark spots). The middle column is a grayscale map of the recorded activity (black: no activity). The peak rates are indicated on the side of each diagram. Note the distribution of activity peaks over the vertices of a grid of equilateral triangles. (From Hafting et al. 2005.)

A simple and elegant mathematical analysis by Trygve Solstad, Edward Moser, and Gaute Einevoll sheds some light on the significance—if not on the origin—of this pattern (Solstad et al., 2006). Let us begin by asking how a pattern of "peaks," similar to the activities of grid cells may come with a structure of equilateral triangles. Suppose we have a periodic function of space: a standing sine wave with wavelength λ. Three such waves are depicted in figure 1.8, with $\lambda = 3$ length units. These could be meters, inches, or centimeters—it does not matter for the present discussion. Instead, it matters that the three sine waves are oriented in three directions, 60 degrees apart from each other. Let us give a mathematical form for these sinusoids. They map each point on the plane, $\mathbf{r} = [x \quad y]^T$ into a number

$$F_i(\mathbf{r}) = \cos\left(\mathbf{k}_i^T \mathbf{r}\right) + 1 \qquad i = 1, 2, \text{ or } 3. \tag{1.1}$$

The three vectors $\mathbf{k}_i$ represent the wavefronts and are oriented in three directions 60 degrees apart from each other. Their amplitude is the spatial frequency of the wave in radians per unit lengths:

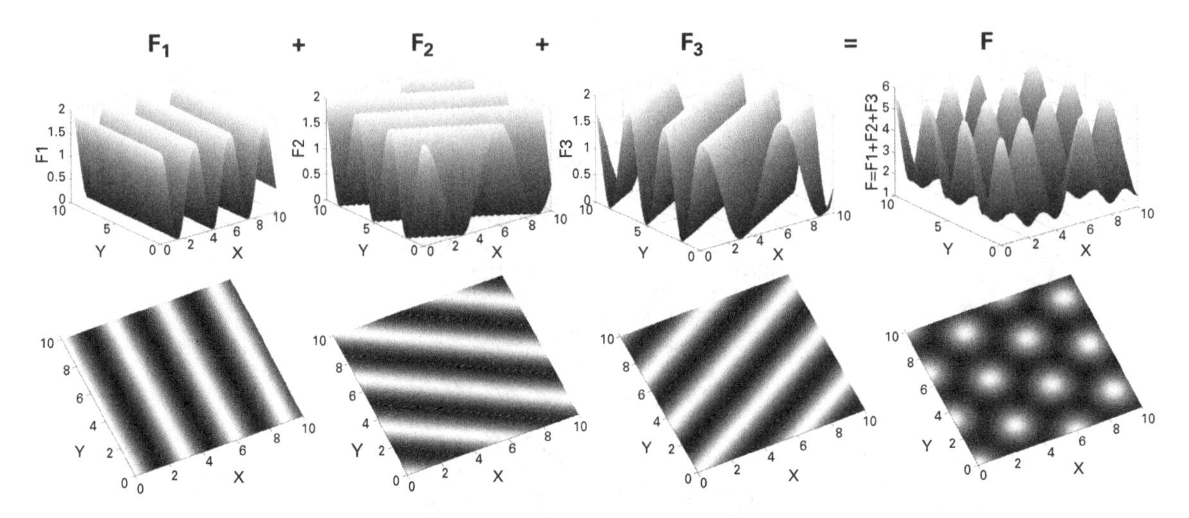

Figure 1.8
Summing three sinusoidal functions of space ($F_1 + F_2 + F_3$) with orientations that differ by 60 degrees results into a periodic distribution of equispaced peaks (right) in an equilateral triangle configuration. Note the similarity of this simple interference pattern with the activity patterns of the entorhinal grid cells (figure 1.7).

$$\begin{aligned}\mathbf{k}_1 &= \frac{2\pi}{\lambda}[\cos(\theta) \quad \sin(\theta)]^T \\ \mathbf{k}_2 &= \frac{2\pi}{\lambda}\left[\cos\left(\theta+\frac{\pi}{3}\right) \quad \sin\left(\theta+\frac{\pi}{3}\right)\right]^T \\ \mathbf{k}_3 &= \frac{2\pi}{\lambda}\left[\cos\left(\theta+2\frac{\pi}{3}\right) \quad \sin\left(\theta+2\frac{\pi}{3}\right)\right]^T\end{aligned} \quad . \tag{1.2}$$

The orientation of the first wavefront, $\mathbf{k}_1$, is the angle θ. In figure 1.8, $\theta = 0$ and the front is perpendicular to the x-axis. Adding 1 to the cosine functions insures that the range of each wave function remains positive ($0 \le F_i \le 2$). When the three functions are added together, they form the interference pattern shown on the right part of figure 1.8:

$$\begin{aligned}\phi(\mathbf{r}|\theta,\lambda) &= F_1(\mathbf{r}) + F_2(\mathbf{r}) + F_3(\mathbf{r}) \\ &= \cos(\mathbf{k}_1^T\mathbf{r}) + \cos(\mathbf{k}_2^T\mathbf{r}) + \cos(\mathbf{k}_3^T\mathbf{r}) + 3\end{aligned} \quad . \tag{1.3}$$

What we obtain by this simple summation of waves resembles the pattern of firing observed in the entorhinal grid cells (figure 1.7). The spacing, d, of the grid depends upon the wave length of the wavefronts, λ (figure 1.9):

$$d = \frac{2}{\sqrt{3}}\lambda. \tag{1.4}$$

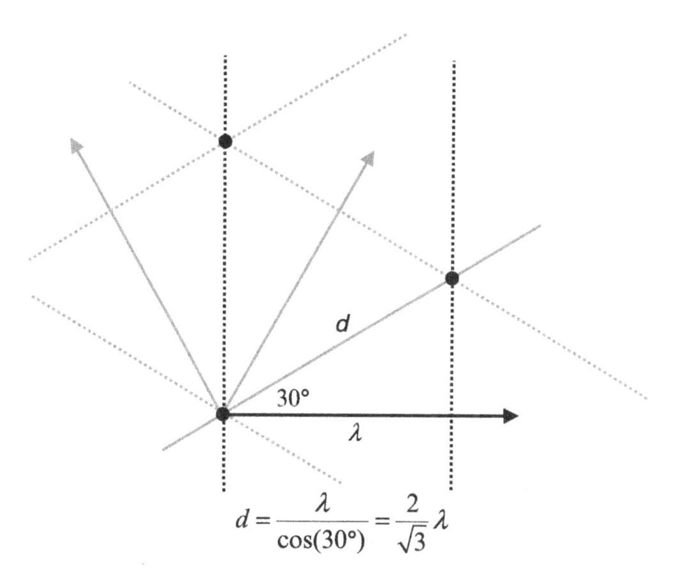

Figure 1.9
Interference pattern. The dotted lines represent the wavefronts of figure 1.8. These are lines at which each sine waves reach their maximum values. The arrows are the directions of the sine waves, that is, the directions of the vectors k_1, k_2, and k_3 (see main text). The intersections of three wavefronts are the points at which their sum reaches the maximum value. Simple trigonometry shows that the distance between two such peaks is slightly larger than the wavelength of each wave component (the distance between parallel dotted lines).

Each grid cell does not inform the rat's brain about the location at which the rat is. It only indicates a set of possible locations at which the rat could be. In that regard, we may see this as a particular coordinate system, like longitude and latitude. If we know our latitude we know a set of places where we may be. To know our position on the sphere, we need both the latitude and the longitude. As we shall see, the grid cells can be used as coordinates in a similar way. But many more than two coordinates are needed to specify a position.

1.6 Grid Cells to Place Cells: Functional Analysis

The French mathematician Jean Baptiste Joseph Fourier discovered two centuries ago the possibility of constructing arbitrary continuous functions by adding trigonometric functions with different frequencies. Fourier series are infinite sums that in the limit converge upon continuous functions. Most remarkably, any continuous function over a finite interval can be obtained as a Fourier series:

$$f(x) = \frac{1}{2}a_0 + \sum_{n=1}^{\infty} a_n \cos(nx) + \sum_{n=1}^{\infty} b_n \sin(nx). \quad (1.5)$$

Daniel Bernoulli first suggested such an infinite series in the late 1740s as he was working on a mathematical analysis of vibrating musical strings. However, Bernoulli was unable to solve for the coefficients of the series. Fourier's great accomplishment was determining the values of the coefficients. For example, for a function $f(x)$ over the interval $x = [-\pi, +\pi]$, Fourier demonstrated the following:

$$\begin{aligned} a_0 &= \frac{1}{\pi}\int_{-\pi}^{\pi} f(x)dx \\ a_n &= \frac{1}{\pi}\int_{-\pi}^{\pi} f(x)\cos(nx)dx. \\ b_n &= \frac{1}{\pi}\int_{-\pi}^{\pi} f(x)\sin(nx)dx \end{aligned} \tag{1.6}$$

Briefly, Fourier arrived at his solution by doing the following. First, he integrated the left and right sides of equation (1.5) over the range $[-\pi, +\pi]$ and then solved for a_0 by noting that the integrals of the trigonometric functions vanish over this range. Then, to derive the coefficients a_n and b_n, he used a more clever observation, which is fundamental to modern functional analysis. He noticed that

$$\int_{-\pi}^{\pi} \cos(nx)\sin(mx)dx = 0 \tag{1.7}$$

for all integer values of m and n. However, the integral

$$\int_{-\pi}^{\pi} \cos(nx)\cos(mx)dx \tag{1.8}$$

and the integrals

$$\int_{-\pi}^{\pi} \sin(nx)\sin(mx)dx \tag{1.9}$$

vanish only when $n \neq m$. Otherwise,

$$\int_{-\pi}^{\pi} \sin^2(nx)dx = \int_{-\pi}^{\pi} \cos^2(nx)dx = \pi. \tag{1.10}$$

To derive each coefficient a_n and b_n, Fourier multiplied both sides of equation (1.5) by the corresponding trigonometric function—i.e., by $\cos(nx)$ for a_n and by $\sin(nx)$ for b_n. For example, to derive a_3 one multiplies both sides of equation (1.5) by $\cos(3x)$ and calculates the integrals[2] in:

$$\int_{-\pi}^{\pi} f(x)\cos(3x)dx = \frac{1}{2}a_0 \int_{-\pi}^{\pi} \cos(3x)dx + \sum_{n=1}^{\infty} a_n \int_{-\pi}^{\pi} \cos(nx)\cos(3x)dx + \\ + \sum_{n=1}^{\infty} b_n \int_{-\pi}^{\pi} \sin(nx)\cos(3x)dx = \\ = a_3 \int_{-\pi}^{\pi} \cos^2(3x)dx = a_3 \cdot \pi. \tag{1.11}$$

The Fourier representation of a function as a sum of other functions has a powerful algebraic and geometric interpretation. The functions that appear in the sum of equation (1.5) are formally equivalent to vectors forming a basis in a vector space. While ordinary geometry is only three-dimensional, vector spaces can have an unlimited number of dimensions. What matters is that the elements that form a basis be mutually independent—like the three unit vectors pointing along the x, y, and z axes. But to be independent, vectors do not need to be mutually orthogonal. In ordinary 3D space, independence means that a vector that lies outside a plane cannot be obtained by adding vectors on that plane. Thus, one cannot obtain a vector sticking out of a plane by adding vectors on that plane. In symbols, if vectors φ_1, φ_2, and φ_3 are linearly independent, then we cannot write $\varphi_3 = a_1\varphi_1 + a_2\varphi_2$. One other way to say this is that

$$a_1\varphi_1 + a_2\varphi_2 + a_3\varphi_3 = 0 \tag{1.12}$$

can be true only if all the three coefficients, a_1, a_2 and a_3 are all zero. This is readily extended to an arbitrary number of vectors: N vectors $\varphi_1, \varphi_2, \ldots, \varphi_N$ are linearly independent if and only if

$$a_1\varphi_1 + a_2\varphi_2 + \ldots + a_N\varphi_N = \mathbf{0}. \tag{1.13}$$

This implies that all the a_i are zero. Extending this to an infinite number of independent vectors, we obtain the Fourier series, as in equation (1.5). And, going even further to a *continuum* of vectors, we have the Fourier transform. But let us limit this discussion to a finite number of independent vectors.

Now, suppose that the sum equation (1.13) is non-zero, that is:

$$a_1\varphi_1 + a_2\varphi_2 + \ldots + a_N\varphi_N = \mathbf{f} \neq \mathbf{0}. \tag{1.14}$$

The vector **f** in belongs to the n-dimensional vector space V_N spanned by the basis $\varphi_1, \varphi_2, \ldots, \varphi_N$. How can we use this basis to represent vectors in a higher-dimensional space? This can be achieved by approximation. In this case, the linear combination of the basis vectors cannot generate exactly the higher-dimensional vector. But it can get as close as possible to it. Consider a vector **g**, in a higher dimensional space, V_M, which includes V_N as a subspace ($M > N$). To gain an immediate intuition, one may think of $N = 2$ and $M = 3$. V_3 is the ordinary 3D space, with an associated Cartesian reference frame. V_2 is a planar surface, passing by the origin of V_3. The following discussion extends to spaces of higher dimension. We wish now to find the vector in V_N that is as close as possible to **g**. The intuitive solution to this problem is to look for the projection of **g** over V_N. Suppose that we have a basis for V_M which includes the basis in V_N, augmented by M-N vectors, $\boldsymbol{\varphi}_{N+1},\boldsymbol{\varphi}_{N+2},\ldots,\boldsymbol{\varphi}_M$, orthogonal to V_N. In this basis, the vector **g** has a representation

$$\mathbf{g} = b_1\boldsymbol{\varphi}_1 + b_2\boldsymbol{\varphi}_2 + \ldots + b_N\boldsymbol{\varphi}_N + b_{N+1}\boldsymbol{\varphi}_{N+1} + \ldots + b_M\boldsymbol{\varphi}_M. \tag{1.15}$$

Note that the Fourier expansion of equation (1.5) looks much like equation (1.15), with infinite terms. The first part of this representation,

$$\hat{\mathbf{g}} = b_1\boldsymbol{\varphi}_1 + b_2\boldsymbol{\varphi}_2 + \ldots + b_N\boldsymbol{\varphi}_N, \tag{1.16}$$

is the projection that we are looking for. We know the basis vectors, $\boldsymbol{\varphi}_1,\boldsymbol{\varphi}_2,\ldots,\boldsymbol{\varphi}_N$, but we do not know the coefficients $b_1,b_2,\ldots,b_N$. To find them we use the *inner product* operation and we exploit the fact that the inner product of the basis vectors in V_N with the vectors $\boldsymbol{\varphi}_{N+1},\boldsymbol{\varphi}_{N+2},\ldots,\boldsymbol{\varphi}_M$ is zero by hypothesis, because these vectors are orthogonal to V_N. Let us step back. We need to remember that the inner product of two vectors produces a number.[3] Here, we adopt the convention to use angled brackets to denote the inner product, as in $\langle\boldsymbol{\varphi}_1,\mathbf{g}\rangle$. In $\mathbb{R}^N$ we calculate the Euclidean inner product by multiplying component by component and by adding the results. In vector-matrix notation this is $\mathrm{u}^T \cdot \mathrm{v} = u_1v_1 + u_2v_2 + \ldots + u_Nv_N$.

However, here we use a more general notation that is not restricted to Euclidean spaces. To derive the coefficients $b_1,b_2,\ldots,b_N$, we begin by taking the inner product of both sides of equation (1.15) with each of the N basis vectors. This produces a system of N linear equations:

$$\begin{cases} \langle\boldsymbol{\varphi}_1,\mathbf{g}\rangle = \hat{g}_1 = \Phi_{1,1}b_1 + \Phi_{1,2}b_2 + \ldots + \Phi_{1,N}b_N \\ \langle\boldsymbol{\varphi}_2,\mathbf{g}\rangle = \hat{g}_2 = \Phi_{2,1}b_1 + \Phi_{2,2}b_2 + \ldots + \Phi_{2,N}b_N \\ \qquad\qquad\qquad \ldots \\ \langle\boldsymbol{\varphi}_N,\mathbf{g}\rangle = \hat{g}_N = \Phi_{N,1}b_1 + \Phi_{N,2}b_1 + \ldots + \Phi_{N,N}b_1 \end{cases} \tag{1.17}$$

with

$$\Phi_{i,j} = \langle\boldsymbol{\varphi}_i,\boldsymbol{\varphi}_j\rangle. \tag{1.18}$$

Note that while equation (1.15) contains vectors—**g** and the $\boldsymbol{\varphi}_i$'*s*—and numbers—the b_i'*s*—equation (1.17) contains only numbers. This is a system of N equations in N unknowns. A compact form for it is

$$\hat{\mathbf{g}} = \Phi\mathbf{b} \tag{1.19}$$

with

$$\hat{\mathbf{g}} = \begin{bmatrix} \hat{g}_1 \\ \hat{g}_2 \\ \cdots \\ \hat{g}_N \end{bmatrix} \quad \Phi = \begin{bmatrix} \varphi_{1,1} & \varphi_{1,2} & \cdots & \varphi_{1,N} \\ \varphi_{2,1} & \varphi_{2,2} & \cdots & \varphi_{2,N} \\ \cdots & \cdots & \cdots & \cdots \\ \varphi_{N,1} & \varphi_{N,2} & \cdots & \varphi_{N,N} \end{bmatrix} \quad \mathbf{b} = \begin{bmatrix} b_1 \\ b_2 \\ \cdots \\ b_N \end{bmatrix}.$$

In this notation, vectors are represented as matrices with a single column containing all the vector components. The matrix Φ is called the Gramian of the vectors $\varphi_1, \varphi_2, \ldots, \varphi_N$, after the Danish mathematician Jorgen P. Gram. Equation (1.19) provides us with a straightforward solution for the coefficients of equation (1.15):

$$\mathbf{b} = \Phi^{-1}\hat{\mathbf{g}}. \tag{1.20}$$

The only requirement for deriving **b** using the above expression is that the inverse of the matrix Φ exist or, equivalently that the determinant of Φ does not vanish. This condition is insured by the fact that the vectors $\varphi_1, \varphi_2, \ldots, \varphi_N$ in equation (1.15) form a basis for V_N.[4]

To summarize, so far we have shown that starting from a set of N independent vectors in V_N, it is possible (a) to represent all vectors in V_N and (b) to find the vector in V_N that lies closest to an arbitrary vector in a higher dimensional space, V_M. But what if the vectors are not all linearly independent? Then, they live in a space V_K of dimension K, lower than N. In this case, it is still possible to use the construct that led to equation (1.19). Now, however, the Gramian determinant is zero and the matrix cannot be inverted. We can still derive the projection of **g** over V_K by using the *pseudoinverse* of the Gramian. This is usually indicated by a + superscript, as in Φ^+ and the equation for **b** is quite similar to equation (1.20):

$$\mathbf{b} = \Phi^{+}\hat{\mathbf{g}}. \tag{1.21}$$

There are many ways to calculate the pseudoinverse of a matrix. Here, we limit ourselves to list its four defining properties.[5]

1. $\Phi\Phi^+\Phi = \Phi$.
2. $\Phi^+\Phi\Phi^+ = \Phi^+$.
3. $(\Phi\Phi^+)^T = \Phi\Phi^+$.
4. $(\Phi^+\Phi)^T = \Phi^+\Phi$.

Note that equation (1.21) is more general than equation (1.20), since the pseudoinverse of a matrix is equal to the standard inverse, whenever the latter exists. Once we have derived the coefficient vector using equation (1.21), we see that the vector $\hat{\mathbf{g}}$ of equation (1.16) is the projection of **g** over the smallest subspace of V_M, which contains the vectors $\varphi_1,\varphi_2,\ldots,\varphi_N$. In other words, $\hat{\mathbf{g}}$ is the closest approximation to $\mathbf{g}$ in this reduced subspace.

Next, we wish to see how all the above helps in understanding the function implemented by the grid cells in the entorhinal cortex, and their relation to the function implemented by the place cells in the hippocampus. In the previous discussion, we have assumed that certain quantities are vectors and others are numbers. The method of Fourier led to the idea that continuous functions are a type of vectors, although not of the kind we learned in our first courses of geometry. Mathematics seeks abstraction. In the case of vector calculus, the intuitive idea of a vector is extended by considering what its fundamental properties are. In this general sense, vectors are any objects that can be multiplied by a number and can be added to form other vectors. Thus, continuous functions form a *vector space* because the sum of any number of continuous functions generates another continuous function. But, most important, any continuous function can be obtained from the weighted sum of other continuous functions, such as sines and cosines. Because a Fourier series—as in equation (1.15)—has infinite independent terms, the vector space it spans has infinite dimensions. We conclude that the continuous functions are vectors in an infinite dimensional space spanned by an infinite number of "basis functions."

What happens if, instead of the infinite family of basis functions, one only considers a finite number of them? In this case all the previous discussion on vector spaces applies. With the available basis functions we use equation (1.20) for deriving the linear combination corresponding to a projection of a desired function over the space spanned by the basis functions. This is, in essence, one of the fundamental mechanisms to carry out function approximation by "least squares." Being a projection under the metric associated with the inner product, this approximation minimizes the square distance from the desired function.

So far, we have only presented the general principles in a rather informal way. Now, we need to clarify what is that we can call an inner product of two functions. All we need is a definition that satisfies the main general requirements for the inner product operation:

1. The inner product is a real number (but see note 3).
2. The inner product is symmetric. $<\varphi,\psi> = <\psi,\varphi>$.
3. The inner product is bilinear: $<a_1\varphi_1 + a_2\varphi_2,\psi> = a_1 <\varphi_1,\psi> + a_2 <\varphi_2,\psi>$ and $<\varphi,a_1\psi_1 + a_2\psi_2> = a_1 <\varphi,\psi_1> + a_2 <\varphi,\psi_2>$.
4. The square *norm* of a vector is the inner product of the vector with itself: $\|\varphi\|^2 \equiv \langle\varphi,\varphi\rangle \geq 0$. The norm is equal to zero if and only if the vector is the null vector.

The last requirement is perhaps the most important: the inner product defines what we mean by "size." Once we have an inner product, we are endowing a space with metric properties and the space becomes a *metric space*. The integral operation[6] offers a very simple definition of inner product. Of course, the requirement is that a function be integrable, or better, that the product of any two functions (or the square of a function) be integrable. Given two functions, $\varphi(x,y)$ and $\psi(x,y)$, both defined over a domain $D = \{x_{MIN} \le x \le x_{MAX},\quad y_{MIN} \le y \le y_{MAX}\}$, let us define their inner product as:

$$\langle \varphi, \psi \rangle \equiv \iint_D \varphi(x,y) \cdot \psi(x,y)\, dx\, dy. \tag{1.22}$$

We can readily verify that if both functions are integrable over D, then the preceding definition satisfies all four requirements. In practical calculations, the integrals are replaced by sums over the indices of the x and y variables. This is a convenient way of "extending" the natural definition of the inner product of two vectors, which is the simply the sum of the products between the corresponding components of each vector.

Let us go back to the physiology. We know that the entorhinal cortex supplies input signals to the hippocampus. Not the other way around. Based on this anatomical fact, Trygve Solstad, Edvard I. Moser, and Gaute T. Einevoll (2006) asked a simple question: Is it possible to obtain the firing pattern of a hippocampal place cell from the activities of multiple entorhinal grid? The answer is affirmative and derives directly from the previous discussion. In a first approximation, following Solstad and colleagues, we model the activity of a place cell as a Gaussian function (figure 1.10):

$$f = f_{MAX} \exp\left(-\frac{(x-x_0)^2 + (y-y_0)^2}{2\sigma^2} \right). \tag{1.23}$$

The place cell attains the maximum firing rate, f_{MAX}, at the location (x_0,y_0) of the arena where the rat is moving and the activity decays monotonically around this point. So, there is a "receptive field" of the place cell, with a "width" of σ^2. In contrast to the place cells, the grid cells in the entorhinal cortex do not specify the single location where the rat is at a given time. Each cell is firing whenever the rat is in one of several locations, as shown in figure 1.7. We have already shown that the superposition of three standing sine waves would reproduce this pattern (figure 1.8). The function that represents this grid cell has two parameters: the wavelength and the direction. Importantly, the grid cell functions, so reconstructed, contain trigonometric functions, which are known to provide a basis for representing other continuous functions, like the Gaussian of equation (1.23). Thus a linear combination of functions corresponding to grid cells can approximate the function corresponding to a place cell. This is shown in figure 1.10, where the combination of forty-nine grid functions generates a pattern that approximates the typical response of a Gaussian place cell. We derived this particular example following

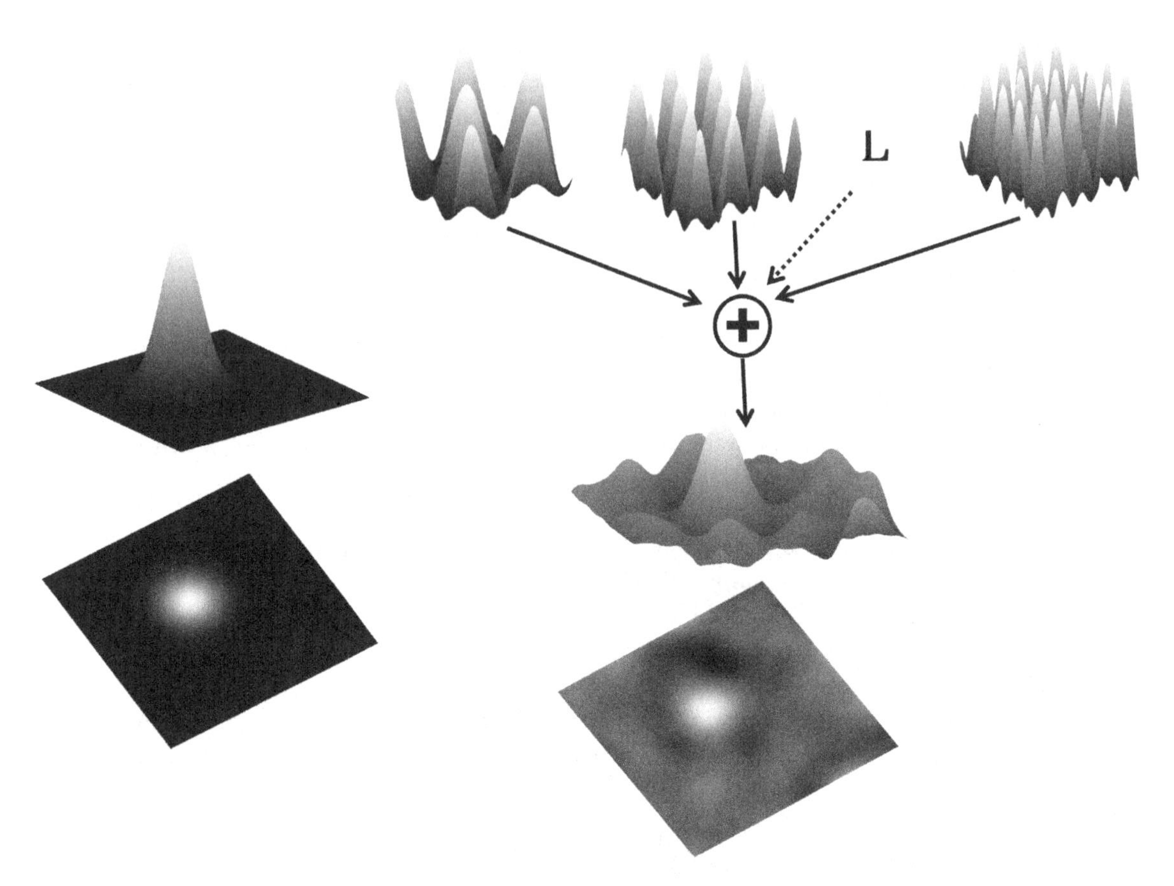

Figure 1.10
Place cells from grid cells. By a simple additive mechanism the firing patterns of multiple grid cells contribute to forming the firing pattern of a place cell (right). Each grid cell output is multiplied by a coefficient before being added to the other contributions. In this example, forty-nine simulated grid cells contributed to generate an activity pattern similar to a place cell. Each grid cell was obtained from the superposition of three standing sine waves, as shown in figures 1.8 and 1.9. The simulated space is a square with 1 meter side. The spacing of the peaks in the grid cells varied between 12 and 80 centimeters, and the direction of the central wavefront varied within 360 degrees. The multiplicative coefficients were obtained from the approximation of a Gaussian response profile (left) with a variance of 12 centimeters.

the least-squares approach, equation (1.21), with the inner product metric afforded by the definition (1.22). The weighted summation of the activities form relatively few grid cells—of the order of 10 to 50—can account for the responses of individual hippocampal place cells.

How are the hippocampal activities updated as the rat moves around? Figure 1.11 illustrates a simple approach. We tessellate the space with the contiguous receptive fields of place cells, following the logic of Samsonovich and McNaughton (figure 1.5). We are thus building a topographic chart, by associating each place cell with the location where its activity reaches a peak. This arrangement does not correspond to the actual distribution of place cells over the hippocampus. The anatomical distribution could be random (although this is disputed) and it would not matter for what one may call the "functional topography," which is the topography determined by what is being represented. For each place cell so arranged we derive the coefficient vector b using equation (1.21). Each element of the coefficient vector represents a "connection weight" that multiplies the input from the corresponding grid cell. All inputs are added, resulting in the net activity of the place cell. Of course, this is an oversimplified neural model to illustrate how a simple summation rule can produce a topographic map similar to that observed in the hippocampus.

If we partition a region of space in 20×20 place cells and if we have 50 grid cells feeding this system of place cells, we need to form a total of $20 \times 20 \times 50 = 20{,}000$ connections. While this is a large number of multiplications, they may be carried out simultaneously, in parallel, so that the total computational time of this whole charting operation may be as short as the time needed to carry out a single multiplication. As the fictional rat of our example moves within the environment, we see a wave of activity along a spatial map, as shown in figure 1.11. The peak of this wave tracks with good accuracy the instantaneous location of the rat.

This suggests how brain activities evolve between entorhinal cortex and hippocampus, as the rat moves in the environment. However, we have not yet addressed the most fundamental question: How, in the first place, does the rat's brain know where the rat is? How does the brain have an idea of the x and y coordinates that appear in the argument of the simulated grid cells? How can the brain have such basic information starting from sensory and motor data, supplied by the eyes and by the very movement instructions that the nervous system sends to the legs? We must say upfront that the answers to these questions are not yet available. We therefore cannot give them here. However, in the next chapter we will outline the computational problems that the brain must solve for creating and maintaining a representation of the extrapersonal space.

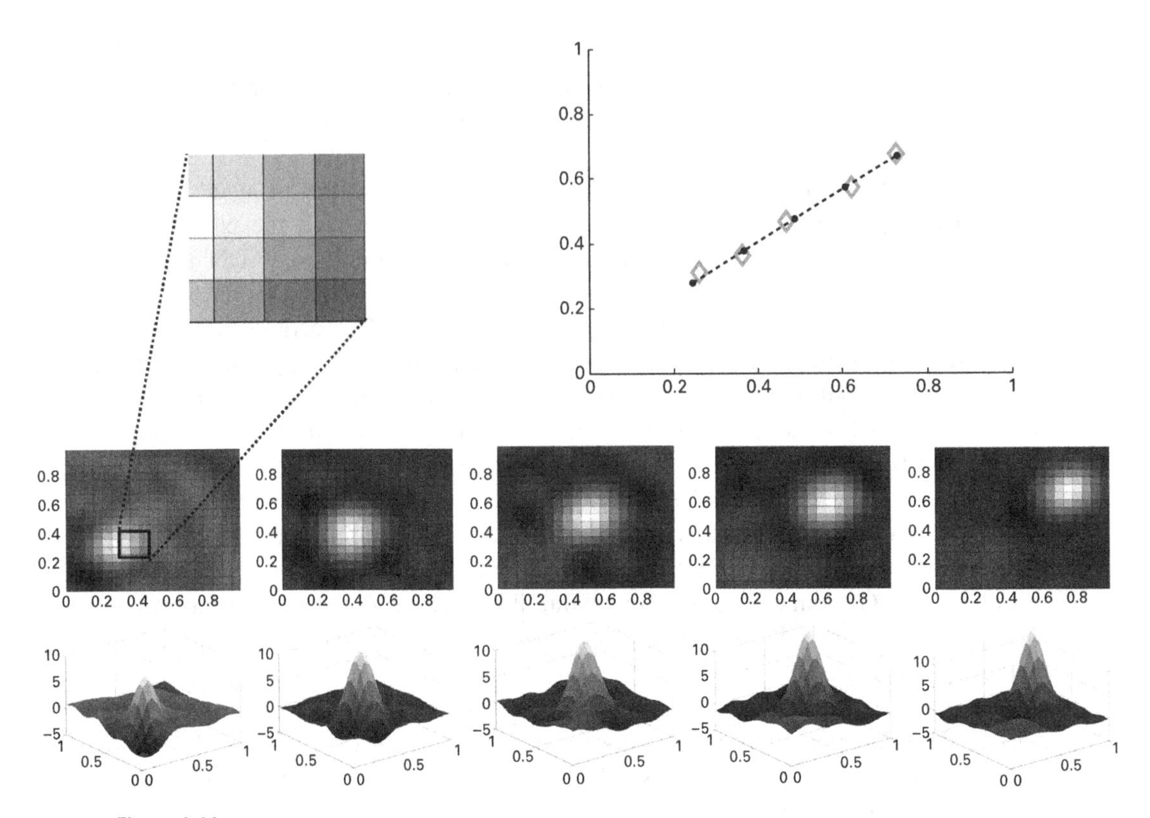

Figure 1.11
Hippocampal GPS. In this model, each simulated place cell receives inputs from 49 grid cells, as shown in figure 1.10. The space within which a fictional rat is moving is a 1 square meter region divided in 400 (20 × 20) small squares. The color of each small square represents the activity level of a simulated place cell, whose maximum of activity falls within that region. Thus, the place cells are distributed topographically to match the locations that they are coding for. Note that this is not intended to reproduce the spatial distribution of the cells within the hippocampus. We simply formed a chart in the style of Samsonovich and McNaughton (see figure 1.5). With 400 place cells and 50 grid cells, there is a total of 20,000 connections between the simulated grid and place system. As the rat moves along the dotted line (top right) the simulated activity on the hippocampal chart follows the pattern shown in the lower panels. The diamonds on the top right panels correspond to the place cells with maximal activity and tend to match closely the actual position of the rat.

Summary

Earlier studies on Mongolian gerbils demonstrated the ability of these rodents to form geometrical maps of the space in which they move. These maps represent the locations and the distances of objects in the environment. The way in which the gerbils use past experiences to search for food revealed their ability to represent the Euclidean properties of space: They have a sense of distance that is invariant by rotations and translations, but not by scaling.

The ability to locate ourselves in space is closely connected to our ability to form new memories of events. The relationship between memory and space maps has a physiological substrate in the mammalian hippocampus. Evidence that the hippocampus organizes a map of space camc with the first observations of "place cells" that encode single locations of extrapersonal space. A population of place cells in a rat's hippocampus forms a chart, where the instantaneous position of a rat in its environment is revealed as a moving hill of neural activity. The hippocampal place cell system is also studied in humans, where imaging studies suggest the existence of a topographical order.

Upstream from the hippocampus, cells in the entorhinal cortex appear to form a coordinate system, analogous to parallel and meridian lines on the earth's surface. Unlike place cells, the entorhinal "grid cells" become active at multiple places, disposed at the vertices of equilateral triangles over the surrounding environment. Fourier analysis can account for this pattern of activities as a superposition of three sinusoidal spatial waves along three directions, 60 degrees apart from each other. By applying Fourier analysis to a system of grid cells, we obtain a family of units with a single localized peak of activity, similar to the activity of the hippocampal place cells. Therefore, a local representation of the body in space, in the form of a topographic map with an isolated peak of activity, emerges from a linear superposition of elements with broad lines of activity implementing the representation of a coordinate system.

2 Building a Space Map

2.1 Ordinary Space

Philosophers have argued for over two thousand years on the nature of space. Some of the debate centered on whether space, as we think of it, really exists at all, or is it just a product of our minds. Perhaps a more approachable question is whether space exists independent of what fills it. Does "empty space" have any meaning at all?

Now add to this notion of space the word "ordinary." At first this may seem misleading, for "ordinary" is often used in a demeaning way to describe something as being trivial or uninteresting. In that sense, there is nothing ordinary about ordinary space.

From grade school we are exposed to notions like forces acting across a distance. We have learned from Isaac Newton that the motion of the earth around the sun can be accounted for by assuming that earth and sun pull on each other in direct proportion to their masses and in inverse proportion to the square of their distance. What is most intriguing is that the force that planets and stars exert on each other supposedly acts across vast regions of empty space. On a smaller scale, we all have experienced the force that a magnet exerts on another magnet across a distance. We are so used to these concepts that we do not question them. They seem natural and reasonable. Yet, these concepts were foreign to the very scientists that developed the foundations of modern physics. The concept of empty space was particularly hard to accept—so hard that Gottfried Leibniz, the mathematician who laid the foundations of infinitesimal calculus in the late seventeenth century, constructed the theory of a universe filled with special entities, the monads, without any space between them. The idea of a wave is so connected to the undulatory motion of a body of water or air that until recently physicists were convinced that light and other electromagnetic waves propagate within an invisible and mysterious substance called "ether." It took a long time and some crucial experiments to accept that a wave of pure energy may indeed travel across empty space.

Another critical concept is that of absolute space. Is there a point of view in which space can be considered as standing still? Modern physics teaches otherwise. When we sit on an airplane, the video screen mounted on the ceiling of the economy class section is fixed. The space around us in the cabin has many fixed points. However, to an observer on earth, these points are rapidly translating with the airplane. Over any interval, this observer sees these points as forming line segments. You can see one of the most compelling effects of changing viewpoint by placing a fixed camera on long pole so it looks down at a merry-go-round.

Many years ago, we left the lab that we shared at Cambridge, Massachusetts, and walked down the Charles River to a park that happened to have a merry-go-round. We brought a few tennis balls and sat across from each other on the merry-go-round. As the carousel spun, we threw balls at each other. Instead of moving straight, we saw, the ball curved opposite to the motion of the carousel. And yet, if we had a video camera, it would show that the ball moved along a perfectly rectilinear path. We encourage you to see examples of these movies and animations by searching the web for "Coriolis effect."

While we are interested in presenting some of the mathematical foundations of the concept of space, we will not dwell on the rich philosophical debate on the topic. The interested reader can find a concise summary of this debate, in relation to neuroscience, in the first chapter in O'Keefe and Nadel (1978). Instead, we will take a rather pragmatic approach by developing a detailed mathematical model of a simplified space and of a similarly simplified visual system.[1]

We want to represent some of the computational tasks from the perspective of a hypothetical organism moving inside an environment and receiving incomplete and distorted images of its surroundings. This is a simplification of the gerbil's viewpoint of the previous chapter. How can this organism's brain develop a sense of space? Or, in more modern terms, how does it develop an internal model of the space by combining sensor data with movement commands? We will address this question from the viewpoint of a Mongolian gerbil that we will call "G."

2.2 A Simple Model

The first challenge for G's brain as it moves inside an environment populated by various landmarks is to construct a representation of space from information captured by its eyes. The eye is a complex organ, where images are projected on a curved surface with an uneven distribution of neural elements that transform photons into electrical impulses. We do not want to develop a realistic model of this wonderful neural and optical machinery. We only wish to capture the idea that the information about space comes from projections on a curved element. So, let us begin by simplifying the dimensionality of the problem. Ordinary space is three-dimensional, and

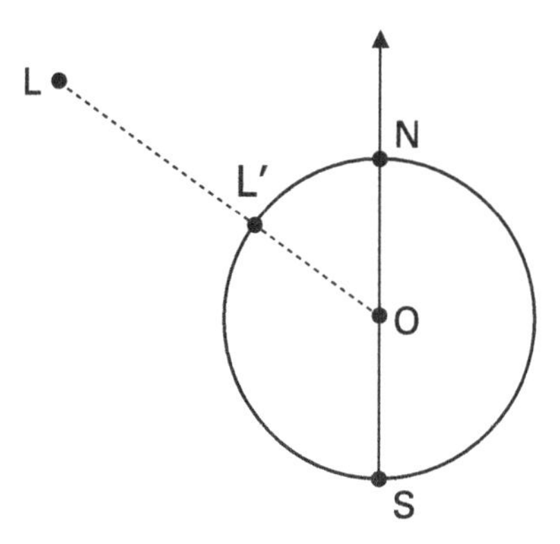

Figure 2.1
A simple eye. This is a projection model that maps points in the plane (L) into images over a one-dimensional "retina" in the shape of a circle centered at O. The heading direction is indicated by the arrow attached to N. The image of L on this retina is the arc $\widehat{NL'}$. Assuming that the radius ($\overline{OL'}$) has unit length, this arc measures in radians the angle $\widehat{NOL}$.

the surfaces of the eyes are two-dimensional. The math becomes manageable if we assume that the space is two-dimensional (that G is a flat gerbil) and, accordingly, that G's idealized eye is a circular, one-dimensional line. This geometrical expedient, which we will shamelessly call "eye," is depicted in figure 2.1.

In our model, the two-dimensional space is populated by "landmarks," that is, by points of particular significance. As G moves around, the landmarks are projected on the circumference of its eye. Now, we need to introduce something that makes this picture less "even," less symmetric. Ordinary space is isotropic: all points are identical and all directions are equivalent. However, our view is oriented because we have a body and we face the direction in which we move. That is, to us and to our eye, all directions are not equivalent. G's heading direction is indicated in the figure by an arrow that intersects the eye at a point N, which stands for "North." If L is a landmark in the external space, the projection of the landmark on the simplified one-dimensional "retina" (the circle) is obtained by tracing the segment $\overline{LO}$ joining the landmark to the center of the circle and by taking the intersection, L', of the segment with the circle. The point L' is for our purposes a perspective image of the landmark.

2.3 Points and Lines

So far, we have not introduced a metric notion, such as a measure or distance. At a very basic level, geometry, and particularly projective geometry, does not use distances. Projective geometry is only about objects such as points, lines, and surfaces. Much of its original raison d'être is the need to represent three-dimensional reality within the confines of two-dimensional paintings. Here, we consider perspective in the opposite (or "inverse") sense. We want to regenerate the reality of space from lower-dimensional pictures in our eyes. And we want to see this reconstruction as a combination of senses and motion. This is indeed the way in which the neurons in the hippocampus and in its main input structure, the entorhinal cortex, behave: They combine visual memory and self-motion information for generating a neural activity that code for the position of the body in the environment. But before developing a quantitative theory, we may ask what information can be extracted about the environment, without recourse to metric concepts. Is it possible for G's brain to understand that three or more distinct points are on the same straight line? The task would be easy if G could measure the distances between these points. G could take advantage of the well-known fact that the shortest distance between two points is measured along the straight segment that joins them. But what if one does not know how to measure distances? Then, G can make use of a simpler notion, the notion of "order." This is the intuitive idea of a point sitting "in between" two others. The order relation was formalized first by Moritz Pasch and subsequently by David Hilbert in a set of axioms known as Hilbert axioms. Can G's brain exploit the order of images on the sensor circle to infer something about the structure of the external space?

G can use the order relation, because it has a motor system that allows it to move around its environment. Let us start by accepting that given two points A and B, we can find a third point C such that C is between A and B. Then, the segment $\overline{AB}$ is simply the collection of all points that are between the two extremities, A and B. Points that belong to the same segment are said to be *collinear*. Consider now the situation depicted in figure 2.2. There are four collinear landmarks that are ordered as A, B, C, D, or equivalently as D, C, B, A. We say that A, B, C, D and D, C, B, A are equivalent orders because they only differ by the "reading direction" that is by a reflection. In this sense A, B, C, D and, say, A, C, B, D are not equivalent orders. The same order relation is consistently present in the projected images A', B', C', D'. This is true for almost any position and heading of the eye. We say "almost" because one must exclude the "singular" sensor-landmarks configurations at which A, D, and O (the center of projections) are collinear. There, all landmarks project onto the same image. However, if the landmarks lie on a straight line, the order of their projections is never altered. Collinearity is preserved as G moves around. Conversely, if collinearity is preserved as G's position changes, G's brain can conclude, without need for measures of length or distance, that the landmarks lie on the same straight segment. This is of fundamental importance, because we have now derived the notion of straightness of a line without using any metric concept of length. As we shall see later, the collinearity relation gives us information on the *affine* structure of the external space.

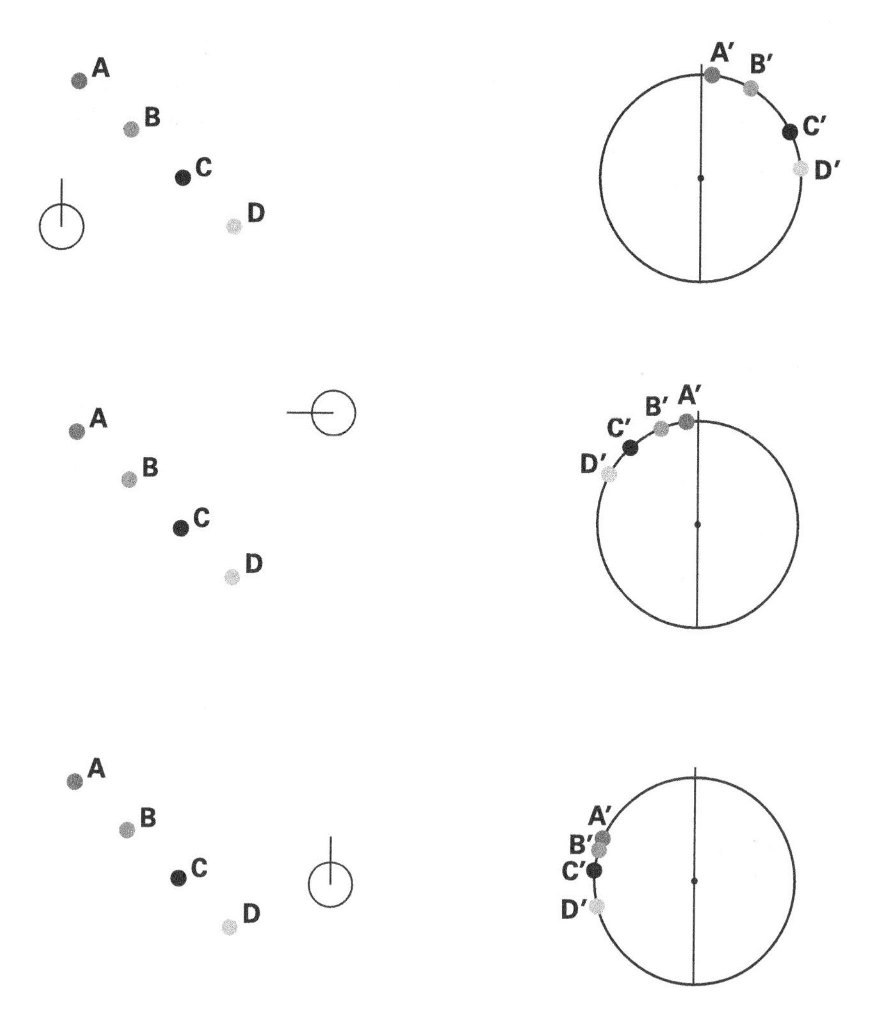

Figure 2.2
Recognizing straightness. If the four points are on a common straight line, the projections preserve the order relation as the observer moves in the environment. There is only a complete reversal, a "mirror symmetry" when the projections cross the midline.

In our example, the eye is a one-dimensional circle, while the external space is two-dimensional. There is an imbalance of dimensions, and this imbalance is reflected by order relations of the projected images, which change with the position of the eye in space. In figure 2.3, the four landmarks are not collinear. We may place them in different orders over different curved lines passing through all the landmarks. As a consequence, the relative order of their images changes for different positions of G relative to the landmarks. This is a cue that G may use to establish that the external world has more than one dimension and that the landmarks are not placed along a straight line. Summing up, it is possible to extract important information about space based only on relations of order between projected points, without recourse to metric operations. However, as we will see next, if one can measure lengths and distances, one can learn more about the structure of external space.

2.4 Distance and Coordinates

Let us look again at G's simplified eye in figure 2.1. It has three particularly important points: the center, O, the "north pole," N, and the "south pole," S. The two poles break the symmetry of the circle. One may say that the poles exist to signify that all directions are not equivalent.

Animals, as well as most human-made vehicles, have a front and a back. Front is sometimes, but not always, defined by the location of the eyes. With some exceptions the orientation of the eyes corresponds to the preferred direction of motion: it is safer to advance where one can see. This forward or frontal direction is what we call "heading." It defines the point N and its opposite, S. Both the anatomical structure and the behavioral preference to move forward contribute to establish a set of particular points on the optical sensor.

Consider two additional stipulations. First, as G moves forward without changing direction, stationary landmarks that project on one side of the $\overline{NS}$ axis will continue to do so. This is simply because a stationary point in a flat environment moves with respect to us parallel to the $\overline{NS}$ axis. If we see a point crossing this axis as we move, we may safely conclude that the corresponding landmark is not at rest.

Second, our movements do not affect the state of the objects around us, unless we come in physical contact with them. Things do not get smaller or bigger because we move. While this is entirely obvious, it has some profound consequences. The perceptual separation between "us" and "environment" is essential for our understanding that the space around us is Euclidean. As G moves, it performs two operations on the positions of the objects relative to itself: translations and rotations. Or, better said, objects that are at rest in the environment rotate and translate relative to G. And G's brain can safely assume that the objects do not change in size or shape. Two objects with the same shape and size are said to be *congruent*, and a

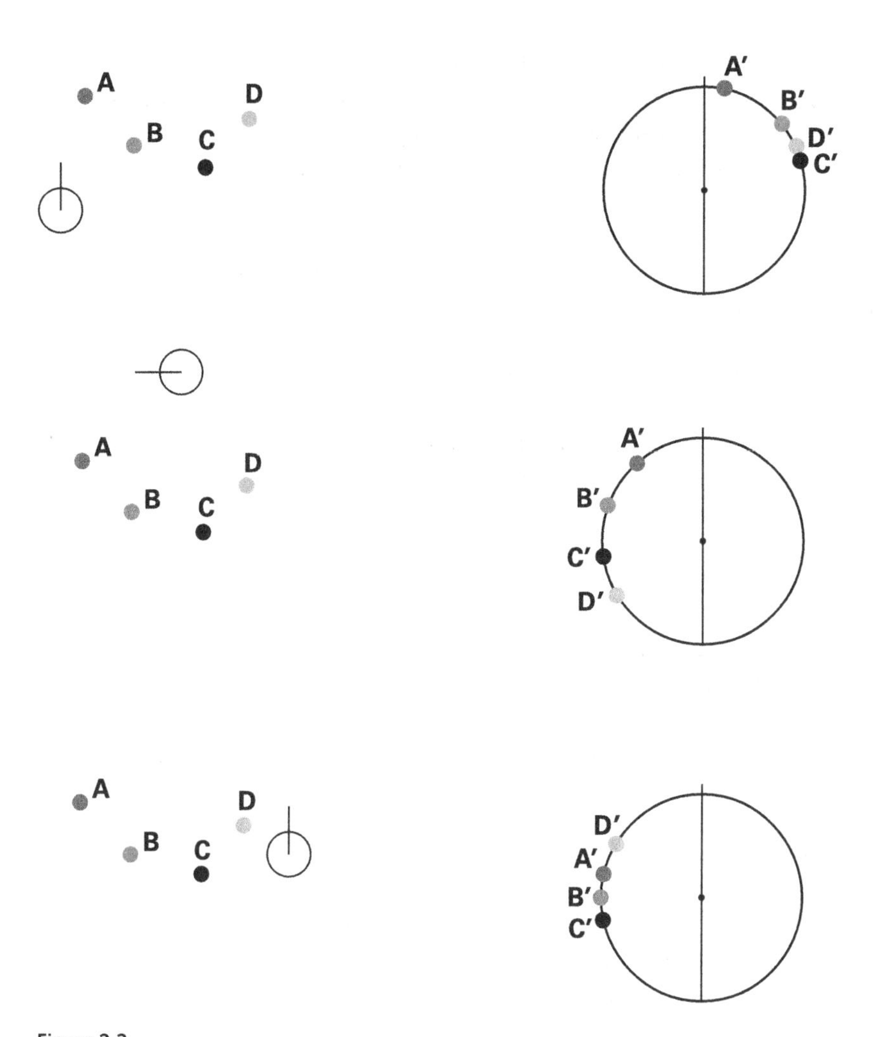

Figure 2.3
Recognizing straightness. The images of the four noncolinear landmarks are ordered in a way that depends on the observer's location and orientation.

transformation that preserves shape and size is called an *isometry*. A related observation is that an object that G is not contacting and is at rest before G starts moving will likely remain at rest. Therefore, in constructing an internal model of the environment, G's brain can take these simple facts (or axioms) into account for extracting spatial information from sensory-motor data.

As G moves forward, the projections of the external landmarks change their position on the eye. G knows that the space around it is more than one-dimensional because it observed that the order of fixed landmark projections over the eye may change as it moves. Now G needs to construct a representation of the landmarks as they are located in the external space. Like the gerbils in Collett's experiments in the previous chapter, G wants to form an extrinsic representation: a representation obtained from its own motion but one that remains invariant as G moves. To this end, G's brain carries out two concurrent operations: (1) keep track of G's location and (2) estimate the distances between G and the landmarks. By combining these two operations, the brain, like a sailor tracking along a coastline, builds and maintains a stable representation of the world.

We begin by establishing a measure of distance that will place the scale of the environment in relation to G's own scale. We associate each landmark projection L' (figure 2.1) with a number expressing the length of the arc $\widehat{NL'}$ in units of radius length. This corresponds to measuring the angle $\widehat{L'ON}$ in radians. Call this angle ξ. We now construct a Cartesian coordinate system Ox^oy^o centered on the origin of the circle (figure 2.4, left). The origin, O, is the center of the eye; the axis y^O points in the north direction; and the axis x^O points to the right (toward the local "east"). The superscripts refer to the particular origin to which the axes are attached. The concept of Cartesian coordinates is a familiar one and does not need to be discussed here in more detail. Its critical importance lies on the possibility to calculate distances between points using the Pythagorean theorem. If two points $P_1 = [x_1,y_1]^T$ and $P_2 = [x_2,y_2]^T$ are given in terms their Cartesian coordinates,[2] their distance, according to Pythagoras, is

$$d(P_1,P_2) = \sqrt{(x_2 - x_1)^2 + (y_2 - y_1)^2}\,. \tag{2.1}$$

Let us begin by assuming that we have an estimate of where we are with respect to a fixed point M in our internal model of the space (figure 2.4, right). We also assume that our internal representation of space is constructed as a two-dimensional Cartesian coordinate system Mx^My^M. We will discuss these assumptions further. Our estimated position in this coordinate system is $\mathbf{r}_0^M = [x_O^M, y_O^M]^T$. Figure 2.4a illustrates the view in the sensor's reference frame. At all times we know the value ξ_L associated with the landmark L. This value is $O \le \xi_L < 2\pi$. The projection of the landmark over the sensor is captured by a function that maps the coordinates of the landmark $[x_L^O, y_L^O]^T$ to the angle/arc-length ξ_L:

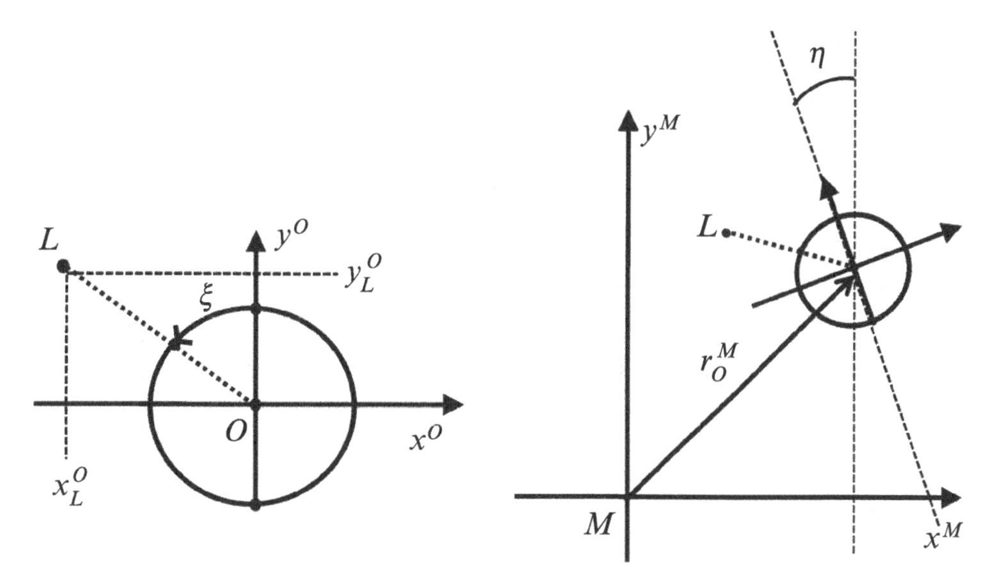

Figure 2.4
Space representation based on metric information. Left: A Cartesian coordinate system, centered on the eye (O) yields ordered pairs of numbers to identify the location of the landmark relative to the observer. Right: A second coordinate frame, centered on a fixed point in the environment, provides an allocentric reference; the observer can recover the location of the landmark in this stationary framework by combining the egocentric representation with a record of motions from M to its current location.

$$\xi_L = \arctan\left(-\frac{x_L^O}{y_L^O}\right) = -\arctan\left(\frac{x_L^O}{y_L^O}\right). \tag{2.2}$$

In more concise and general (but less informative) terms we see that this is a function mapping two spatial coordinates into a single sensor coordinate:

$$\xi = f(x_L^O, y_L^O). \tag{2.3}$$

This function is a nonlinear coordinate transformation, with multiple points mapping to the same projection. Because it maps two variables into one, it does not have a unique inverse. This is another way to state that all points that are collinear with the segment $\overline{OL}$ map to the same sensor coordinate ξ_L. Points on $\overline{OL}$ are equivalent[3] with respect to the coordinate transformation. We obtain a linear transformation by taking the temporal derivative of equation (2.3), that is, by computing how the velocity of a point in space translates into the velocity of its projection on the "retina":

$$\dot{\xi} \equiv \frac{d\xi}{dt} = \frac{\partial \xi}{\partial x_L^0}\dot{x}_L^0 + \frac{\partial \xi}{\partial y_L^0}\dot{y}_L^0 = \begin{bmatrix} \frac{\partial \xi}{\partial x_L^0} & \frac{\partial \xi}{\partial y_L^0} \end{bmatrix} \begin{bmatrix} \dot{x}_L^0 \\ \dot{y}_L^0 \end{bmatrix}. \tag{2.4}$$

The 1×2 matrix

$$J(x_L^0, y_L^0) = \left[\frac{\partial \xi}{\partial x_L^0} \quad \frac{\partial \xi}{\partial y_L^0} \right]$$

is the *Jacobian* of the coordinate transformation, which results in:

$$\dot{\xi} = \left[-\frac{y_L^0}{(x_L^0)^2 + (y_L^0)^2} \quad \frac{x_L^0}{(x_L^0)^2 + (y_L^0)^2} \right] \begin{bmatrix} \dot{x}_L^0 \\ \dot{y}_L^0 \end{bmatrix}. \tag{2.5}$$

Note that this transformation contains nonlinear terms inside the Jacobian. However, unlike equation (2.2), the transformation in equation (2.5) is linear in the velocity coordinates of the landmark, $[\dot{x}_L^O, \dot{y}_L^O]^T$. In general, with a nonlinear coordinate transformation for the representation of a point, one obtains a linear local transformation for the velocity vector representing the motion of the point. The dependence of the Jacobian upon the point at which it is calculated indicates that linearity is achieved locally. As the position of the landmark changes, so does the Jacobian.

2.5 Deriving the Environment from Noise-Free Sensor Data

Now, we wish to use equation (2.5) to obtain the position of the landmark from the observed motion of its projection on the sensor. This task falls in the broader class of "inverse optics" problems. First, we simplify our problem by making some assumptions:

1. That we move along the heading direction, that is, the y-axis of the local frame Ox^oy^o.

2. That the landmark is fixed in the environment. This corresponds to the concept that our own motion does not affect the state of the external world. As a consequence, if we move along the heading direction with a speed v the relative velocity of the landmark in the sensor frame of reference is $-v$.

3. That our dead-reckoning system is accurate. That is we know the position and orientation of the frame Ox^oy^o within the environment frame Mx^My^M.

The first and last hypotheses will later be relaxed to consider translation and rotations and to allow for errors caused by uncertainty about our own state of motion. We take advantage of the first two assumptions to simplify equation (2.5) as

$$\dot{\xi}_L = \left[-\frac{y_L^0}{\rho_L^2} \quad \frac{x_L^0}{\rho_L^2} \right] \begin{bmatrix} 0 \\ -v \end{bmatrix} = -\frac{x_L^0}{\rho_L^2} v. \tag{2.6}$$

Here, we have also taken advantage of the fact that $\rho_L = \sqrt{(x_L^0)^2 + (y_L^0)^2}$ is the distance of the landmark from the center of the sensor along the projecting direction.

We know our speed, v, the projection angle of the landmark, ξ_L, and the rate of change of the projection angle $\dot{\xi}_L$. As an alternative to the information on the temporal derivatives of the landmark and of its projection angle, we may use the changes of these two variables over some small but finite time interval. In this case, however, the greater these changes, the greater the approximation error associated with equation (2.6) expressed in terms of finite differences, $\Delta\xi$ and $\Delta y_L^O = v\Delta t$. We substitute the numerator on the right side of equation (2.6) with its expression in terms of the projection angle, $x_L^O = -\rho_L \sin(\xi_L)$. Then, the unknown distance of the landmark from the eye center, O is

$$\rho_L = \frac{\sin(\xi_L)}{\dot{\xi}_L} v. \tag{2.7}$$

A finite approximation for ρ_L is

$$\rho_L \approx \sin(\xi_L) \cdot \frac{\Delta y_L^0}{\Delta \xi_L}. \tag{2.8}$$

Note that there are two critical situations in which equation (2.7) cannot be used, both related to the vanishing of the image speed, $\dot{\xi}_L$. One is when the landmark is in the heading direction. Then, $\xi_L = \dot{\xi}_L = 0$ and the landmark position can be anywhere along the heading line, either in the N or in the S direction. The other condition ($\xi_L \neq 0, \dot{\xi}_L = 0$) corresponds to the landmark being very far away, ideally at infinity along the ray at ξ_L radians from the heading direction. If either condition occurs, it is impossible to form a model of the landmark location. Otherwise, the local landmark coordinates are

$$\begin{cases} x_L^0 = -\rho_L \sin(\xi_L) \\ y_L^0 = \rho_L \cos(\xi_L) \end{cases}. \tag{2.9}$$

These coordinates are combined with the dead-reckoning information about the position and heading direction of the sensor to form a stable representation of the landmarks in the external space. This representation does not depend upon G's state of motion with respect to the landmarks. The position of the sensor center is a vector $\mathbf{r}_O^M = [x_O^M \quad y_O^M]^T$. The heading direction is the angle η of the oriented $\overrightarrow{ON}$ line with respect to the north direction of the model space. This is also expressed as a unit vector $[-\sin(\eta) \quad \cos(\eta)]^T$. The unit vector describing the sensor x-axis in terms of the model axes is $[\cos(\eta) \quad \sin(\eta)]^T$. Combining this information with the local coordinates of the landmark, we obtain the landmark coordinates in the external space model:

$$\begin{cases} x_L^M = x_O^M + x_L^O \cos(\eta) - y_L^O \sin(\eta) = x_O^M - \rho_L \sin(\xi_L)\cos(\eta) - \rho_L \cos(\xi_L)\sin(\eta) \\ y_L^M = y_O^M + x_L^O \sin(\eta) + y_L^O \cos(\eta) = x_O^M - \rho_L \sin(\xi_L)\sin(\eta) + \rho_L \cos(\xi_L)\cos(\eta) \end{cases}. \tag{2.10}$$

This expression can be written in a more compact form, using a vector/matrix notation:

$$\mathbf{r}_L^M = \mathbf{r}_O^M + R(\eta)\mathbf{r}_L^O, \tag{2.11}$$

where we introduced the rotation matrix

$$R(\eta) = \begin{bmatrix} \cos(\eta) & -\sin(\eta) \\ \sin(\eta) & \cos(\eta) \end{bmatrix}. \tag{2.12}$$

This is a special type of matrix, as will be further discussed, that describes rotations over a plane. The behavior of a rat's place cell is consistent with this operation, since the cell fires when the rat passes at a spatial location that is referred to a fixed frame of reference.

2.6 Rigid Motions and Homogeneous Coordinates

Rigid motions are combination of translations and rotations. These motions are called "rigid" because they do not affect the distances between points in space. In mechanics, a rigid body is a solid object whose points remain always at a fixed distance with respect to each other. Such an object can only undergo translations and rotations, which are therefore called rigid transformations. When we move around a room, everything else remains at rest—assuming that we are not colliding with any object. Therefore, if we look at things from our perspective, we see the stationary environment moving with respect to us as a big "rigid body." But how can we take advantage of this basic element of knowledge to derive our own motion from what we observe?

To approach this issue, it is useful to introduce an algebraic tool that was first conceived by August Ferdinand Moebius, a mathematician known to many for the eponymous Moebius strip, a two-dimensional surface in which up and down cannot be distinguished. Less broadly known is the fact that Moebius introduced homogeneous coordinates to simplify problems of projective geometry. Homogeneous coordinates also provide us with a single framework to describe in matrix form both rotations and translations. It is a nice trick. Consider a point P in a two-dimensional Cartesian space, with coordinates x and y. We can do two types of operations in the Cartesian framework that will change the coordinates. We can apply a transformation such as a stretch, a shear or a rotation, while the origin of the coordinate system remains fixed. These transformations are represented by 2×2 matrices, so that in the new system, the new coordinates of P, $\bar{x}$ and $\bar{y}$, are linear transformations of the old coordinates:

$$\begin{bmatrix} \bar{x} \\ \bar{y} \end{bmatrix} = \begin{bmatrix} m_{1,1} & m_{1,2} \\ m_{2,1} & m_{2,2} \end{bmatrix} \begin{bmatrix} x \\ y \end{bmatrix}. \tag{2.13}$$

Transformations of this kind form an important group, called the general linear group, GL. We obtain a second type of transformation simply by moving, or translating, the origin of the reference frame. If we displace the origin by a vector $-b$ with coordinates $-b_x$ and $-b_y$, then every point in the plane will have new coordinates

$$\begin{bmatrix} \bar{x} \\ \bar{y} \end{bmatrix} = \begin{bmatrix} x \\ y \end{bmatrix} + \begin{bmatrix} b_x \\ b_y \end{bmatrix}. \tag{2.14}$$

Combining a transformation of GL with a translation of the origin and using a more compact notation, one obtains a general affine transformation:

$$\bar{\mathbf{r}} = M\mathbf{r} + b, \tag{2.15}$$

with $\mathbf{r} = [x, y]^T$ and $\bar{\mathbf{r}} = [\bar{x}, \bar{y}]^T$. While this expression looks quite simple, Moebius managed to make it simpler by introducing homogeneous coordinates. In homogeneous coordinates, the general affine transformation is reduced to a single matrix operation.

To obtain this result, we need to change the representation of the points by adding one component to each of them. Here, we will not go into much detail about the significance of this extra component in projective geometry. However, to understand the concept of affine geometry, we need to make a distinction between points and vectors (figure 2.5). A space (or a plane) is a collection of points. A vector is the

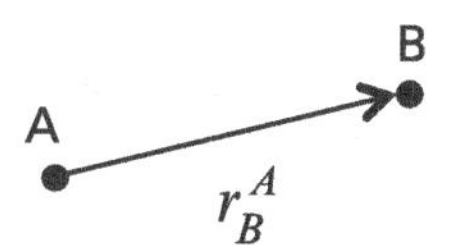

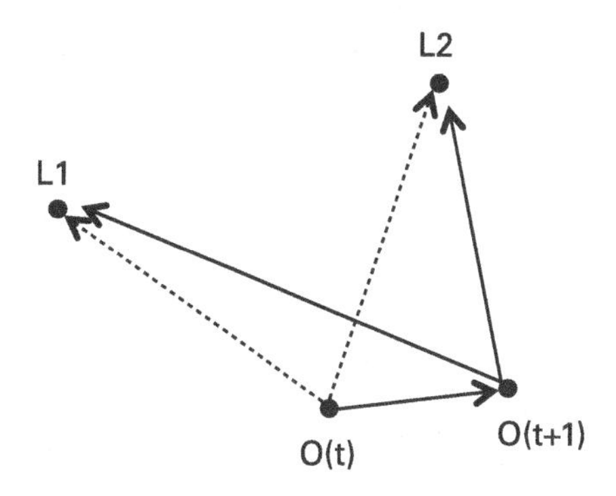

Figure 2.5
Affine space. Top: In affine geometry, a vector is a translation that brings a point into another. Bottom: If the landmarks (*L1* and *L2*) are stationary, the motion of the observer is equal and opposite to the motion of each landmark relative to the observer.

transformation that leads from a point to another. Therefore, a pair of points, A and B, defines the vector $\overrightarrow{AB}$ that transforms A into B. The combined notions of points and vectors constitute what is known as the affine space. Homogeneous coordinates represent points on a plane by three coordinates, which we place into a column vector for performing algebraic operations. The three coordinates are graphically obtained by considering a family of parallel planes intersecting the z-axis at different distances, w, from a center of projection, P. Consider the plane at $w = 1$. Over this plane, a point with Cartesian coordinates (x,y) has homogeneous coordinates $[x, \ y, \ 1]^T$. In projective geometry this point is equivalent to the points on other planes, along the same ray from P. These equivalent points have coordinates $[wx, \ wy, \ w]^T$ where $w \neq 0$ is an arbitrary positive number. Thus, for example, we can represent the point $(3x,3y)$ as $[3x, \ 3y, \ 1]^T$ or, equivalently, as $\left[x, \ y, \ \frac{1}{3}\right]^T$.

Thus, the third component of the homogeneous vector is a scaling factor for the coordinates of the point lying on the plane at $w = 1$. On this plane, the point at infinity along the direction of $[x,y]^T$ has homogeneous coordinates $[x,y,0]^T$. To represent all points in the Euclidean plane at finite distance from the origin, we set $w = 1$. Consider a point P_1 with coordinates $[x_1,y_1,1]^T$ and a point P_2 with coordinates $[x_2,y_2,1]^T$. The vector $\mathbf{d}$ that joins them is

$$\mathbf{d} = \begin{bmatrix} x_2 - x_1 \\ y_2 - y_1 \end{bmatrix}, \tag{2.16}$$

and the distance between the two point is simply the Euclidean norm of $\mathbf{d}$, namely

$$\|\mathbf{d}\| = \sqrt{\mathbf{d}^T \mathbf{d}}. \tag{2.17}$$

Starting from the original representation of a point in two dimension as $\mathbf{r} = [x,y]^T$, we write the representation in homogeneous coordinates as $[x, \ y, \ 1]^T = [\mathbf{r}^T \ 1]^T$. We then derive the general affine transformation in two dimensions by building the matrix

$$H = \begin{bmatrix} m_{1,1} & m_{1,2} & b_x \\ m_{2,1} & m_{2,2} & b_y \\ 0 & 0 & 1 \end{bmatrix} = \begin{bmatrix} M & \mathbf{b} \\ 0 & 1 \end{bmatrix}. \tag{2.18}$$

Applying H to the point in homogeneous coordinates, we obtain:

$$H\begin{bmatrix} \mathbf{r} \\ 1 \end{bmatrix} = \begin{bmatrix} M & \mathbf{b} \\ 0 & 1 \end{bmatrix}\begin{bmatrix} \mathbf{r} \\ 1 \end{bmatrix} = \begin{bmatrix} M\mathbf{r} + \mathbf{b} \\ 1 \end{bmatrix} = \begin{bmatrix} \overline{\mathbf{r}} \\ 1 \end{bmatrix}, \tag{2.19}$$

which is analogous to equation (2.15). Thus, we are now able to express all affine transformations as matrix operations on points in homogeneous coordinates.

Rigid transformations are particular affine transformations that do not affect the distance between points. Consider, again, the points P_1 and P_2 with their distance **d** as in equation (2.17). This distance should not change after a rigid transformation. The coordinates of the two points, after an affine transformation, become

$$\begin{bmatrix} M\mathbf{r}_1 + \mathbf{b} \\ 1 \end{bmatrix} \text{ and } \begin{bmatrix} M\mathbf{r}_2 + \mathbf{b} \\ 1 \end{bmatrix}, \tag{2.20}$$

with $\mathbf{r}_i = [x_i, \ y_i]^T \quad (i = 1,2)$. Therefore the new difference vector is

$$\bar{\mathbf{d}} = M(\mathbf{r}_2 - \mathbf{r}_1) = M\mathbf{d}, \tag{2.21}$$

and the distance is

$$\|\bar{\mathbf{d}}\| = \sqrt{\mathbf{d}^T M^T M\mathbf{d}}. \tag{2.22}$$

The requirement that $\|\mathbf{d}\| = \|\bar{\mathbf{d}}\|$ is evidently satisfied by any translation, since the vector **b** does not appear in equation (2.21). As for the matrix M, the invariance of distances corresponds to requiring that

$$M^T = M^{-1}. \tag{2.23}$$

This is the definition of an orthogonal matrix and is satisfied by rotation matrices, as in equation (2.12). As a result, any rigid motion combines a rotation and a translation and is represented by a single matrix in homogeneous coordinates: the product of a translation matrix and a rotation matrix:

$$\overset{\textit{Rotation}}{\begin{bmatrix} M & 0 \\ 0 & 1 \end{bmatrix}} \overset{\textit{Translation}}{\begin{bmatrix} I & \mathbf{b} \\ 0 & 1 \end{bmatrix}} = \overset{\textit{Rigid Motion}}{\begin{bmatrix} M & \mathbf{b} \\ 0 & 1 \end{bmatrix}}. \tag{2.24}$$

What insight do we derive from this? Consider how the order of two rigid motions affects the final result. Take a step forward and then turn to the left by 90 degrees. Make a note of your position and orientation and start again. Turn left by 90 degrees and then take a step forward. It is evident that we are now in a position that is quite different from the previous one. The two combinations of rotation and step only differ by their order. This effect of the order is typical of matrix multiplications. In general, the product of two matrices is not commutative—that is, $AB \neq BA$ with the exception of some particular cases.

Let us now consider what happens when a generic rotation R and a translation T, both in the plane, are described by the two homogeneous matrices:

$$R = \begin{bmatrix} R_1 & 0 \\ 0 & 1 \end{bmatrix} \tag{2.25}$$

and

$$T = \begin{bmatrix} I & \mathbf{b}_1 \\ 0 & 1 \end{bmatrix}. \tag{2.26}$$

In equation (2.25), R_1 is a 2×2 matrix of the form in equation (2.12), and $\mathbf{b}_1$ is a 2×1 vector. To derive the effect on a vector of a translation followed by a rotation, we write a cascade:

$$RT = \begin{bmatrix} R_1 & \mathbf{0} \\ 0 & 1 \end{bmatrix} \begin{bmatrix} I & \mathbf{b}_1 \\ 0 & 1 \end{bmatrix} = \begin{bmatrix} R_1 & R_1\mathbf{b}_1 \\ 0 & 1 \end{bmatrix}. \tag{2.27}$$

The reverse sequence—rotation followed by translation—is:

$$TR = \begin{bmatrix} I & \mathbf{b} \\ 0 & 1 \end{bmatrix} \begin{bmatrix} R_1 & 0 \\ 0 & 1 \end{bmatrix} = \begin{bmatrix} R_1 & \mathbf{b} \\ 0 & 1 \end{bmatrix}. \tag{2.28}$$

This illustrates that in the step-turn/turn-step example, we end up with the same orientation but in different locations. This lack of commutativity creates an ambiguity that is resolved in a continuous movement where small rotations and small translations along the heading directions are repeated in time. In fact, if we consider very small motions, we see that rotations and translations commute. With a small angle, $\delta\theta$, the homogeneous rotation is approximated by

$$\delta R = \begin{bmatrix} 1 & -\delta\theta & 0 \\ \delta\theta & 1 & 0 \\ 0 & 0 & 1 \end{bmatrix}. \tag{2.29}$$

With a small translation in the heading direction, δy, the homogeneous translation is

$$\delta T = \begin{bmatrix} 1 & 0 & 0 \\ 0 & 1 & \delta y \\ 0 & 0 & 1 \end{bmatrix}. \tag{2.30}$$

Then, combining the two, we obtain

$$\delta R \cdot \delta T = \begin{bmatrix} 1 & -\delta\theta & -\delta\theta \cdot \delta y \\ \delta\theta & 1 & \delta y \\ 0 & 0 & 1 \end{bmatrix} \approx \begin{bmatrix} 1 & -\delta\theta & 0 \\ \delta\theta & 1 & \delta y \\ 0 & 0 & 1 \end{bmatrix} = \delta T \cdot \delta R. \tag{2.31}$$

The approximation corresponds to neglecting second-order terms.

2.7 Updating the Space Model

As we move, we can safely assume that most objects around us remain stationary with respect to each other. Thus they collectively form a frame of reference that we can use to derive and update a model of space. Suppose that our gerbil now takes a small step, δl, in the heading direction and that it also rotates by a small angle $\delta\theta$. How would this added rotation affect G's estimate of the landmark locations? To derive the coordinates of the landmarks in equation (2.9) we assumed a pure translational motion. Now we want to allow for both rotations and translations, under the hypothesis that all landmarks in sight are stationary. Therefore, in G's field of view, the landmarks will all move by the same amount, equal and opposite to G's motion. Using equation (2.31), we derive where G expects to see the landmark at time t:

$$\begin{bmatrix} x_i^O \\ y_i^O \\ 1 \end{bmatrix}(t) = \begin{bmatrix} 1 & \delta\theta & 0 \\ -\delta\theta & 1 & -\delta l \\ 0 & 0 & 1 \end{bmatrix} \begin{bmatrix} x_i^O \\ y_i^O \\ 1 \end{bmatrix}(t-1) = \begin{bmatrix} x_i^O \\ y_i^O \\ 1 \end{bmatrix}(t-1) + \begin{bmatrix} y_i^O(t-1)\cdot\delta\theta \\ -x_i^O(t-1)\cdot\delta\theta - \delta l \\ 1 \end{bmatrix}. \tag{2.32}$$

The change in each landmark's location relative to G is

$$\begin{bmatrix} \delta x_i^O \\ \delta y_i^O \\ 1 \end{bmatrix}(t) = \begin{bmatrix} y_i^O(t-1)\cdot\delta\theta \\ -x_i^O(t-1)\cdot\delta\theta - \delta l \\ 1 \end{bmatrix}. \tag{2.33}$$

We can now abandon the homogeneous coordinate notation, which has fulfilled its role of combining rigid motions. We place the movement commands, δl and $\delta\theta$, in a single command, or "input" array: $\mathbf{u} = [\delta l \quad \delta\theta]^T$. With this, the relative motions of the landmarks become

$$\mathbf{r}_i^O(t) = \mathbf{r}_i^O(t-1) + M(\mathbf{r}_i^O(t-1))\mathbf{u}(t), \tag{2.34}$$

where

$$\mathbf{M} = \begin{bmatrix} 0 & y_i^O(t-1) \\ -1 & -x_i^O(t-1) \end{bmatrix}.$$

Let us go back to the expression in equation (2.5) for the Jacobian of the landmark projections. We use it now to derive the expected change in the projection of landmark i caused by the motion command:

$$\delta\xi_i \approx \begin{bmatrix} -\dfrac{y_i^O}{\rho_i^2} & \dfrac{x_i^O}{\rho_i^2} \end{bmatrix} \begin{bmatrix} y_i^O\delta\theta \\ -x_i^O\delta\theta - \delta l \end{bmatrix} = \frac{-(y_i^O)^2 - (x_i^O)^2}{\rho_i^2}\delta\theta - \frac{x_i^O}{\rho_i^2}\delta l = -\delta\theta + \frac{\sin\xi_i}{\rho_i}\delta l.$$

From this we obtain the new expression for the distance of the landmark:

$$\rho_i \approx \frac{\delta l}{\delta \xi_i + \delta \theta} \sin \xi_i. \tag{2.35}$$

This simply says that in deriving the distance of each landmark one should subtract the projection change $-\delta\theta$ associated to our own rotation. Indeed, our own rotation cannot carry any information about the distance of an object, since the effect is the same for all objects on the same projective line!

Once we have an initial model of the space and the landmarks, we can ask how G can maintain this model by collecting additional information. The problem of deriving a map of space and to localize oneself in this map is an important problem in robotics (Dissanayake et al., 2001; Thrun et al., 2001). Practical applications include the development of autonomous vehicles capable of moving unmanned in a mine, inside a harbor or other dangerous environments to collect and transport items. The environment is populated by various objects and by people moving around. It may be possible, however, to place beacons or other fixtures at a variety of locations. Then the task for the vehicle becomes quite similar to the task faced by the gerbils when looking for seeds in relation to fixed landmarks. Often, robotic engineers have explored these problems from a biomimetic perspective. "Simultaneous localization and map building" or SLAM (Dissanayake et al., 2001) is a term to describe how the problem of navigation is dealt in the mathematical framework of optimal state estimation (we will get to this topic in chapter 4). Here, we merely introduce the general issues encountered in forming and maintaining a map of space.

We build and update G's model of space in the reference frame of the fixed landmarks. This reflects the observation that space-coding cells in the hippocampus and in the entorhinal cortex respond to moving into locations of space that are fixed in some external reference frame. Without getting into the modeling style of artificial neural networks, here we wish only to present some mathematical problems associated with the formation of a spatial map. We begin by establishing an external frame of reference, centered at some point that may either be one of the landmarks or any element of the scene that is stationary with respect to the landmarks. In what follows, we make the assumption that all coordinates are referred to this fixed frame. Thus, the extrinsic description of the space model has a state vector

$$\mathbf{s} = [x_O, \ y_O, \ \eta, \ x_1, \ y_1, \ \cdots, \ x_N, \ y_N]^T = [\mathbf{r}_O^T, \eta, \mathbf{r}_1^T, \cdots, \mathbf{r}_N^T]^T. \tag{2.36}$$

The state vector includes G's position and heading direction together with the position of the N fixed landmarks. In this extrinsic frame, when G makes a movement, $[\delta l, \delta\theta]^T$ its position changes by a translation along the heading direction η:

$$\mathbf{r}_O(t) = \mathbf{r}_O(t-1) + \begin{bmatrix} -\delta l \cdot \sin \eta(t-1) \\ \delta l \cdot \cos \eta(t-1) \end{bmatrix}. \tag{2.37}$$

Then the heading direction is updated:

$$\eta(t) = \eta(t-1) + \delta\theta. \tag{2.38}$$

This can be rewritten in a compact matrix form as

$$\begin{bmatrix} x_O \\ y_O \\ \eta \end{bmatrix}(t) = \begin{bmatrix} x_O \\ y_O \\ \eta \end{bmatrix}(t-1) + \begin{bmatrix} -\sin \eta(t-1) & 0 \\ \cos \eta(t-1) & 0 \\ 0 & 1 \end{bmatrix} \cdot \begin{bmatrix} \delta l \\ \delta\theta \end{bmatrix}. \tag{2.39}$$

By definition, in the extrinsic reference the landmarks do not move. Therefore, the state of the environment is governed by the following equation:

$$\mathbf{s}(t) = \mathbf{s}(t-1) + B(s(t-1))\mathbf{u}, \tag{2.40}$$

where we have introduced the $(N + 3) \times 2$ matrix

$$B = \begin{bmatrix} -\sin \eta(t-1) & 0 \\ \cos \eta(t-1) & 0 \\ 0 & 1 \\ \vdots & \vdots \\ 0 & 0 \end{bmatrix}. \tag{2.41}$$

Equation (2.40) has a deceptive linear appearance. However, it is not a linear equation because the "control matrix" B depends upon one of the state variables, the heading direction. This limits the possibility to apply known linear methods even in this very simple case.

We consider two elements that contribute to the updating of the internal model of the environment. We have described how G expects its position with respect to the fixed landmarks to change in time based on how it thinks it is moving. This is called the "process." The other is a model of the expected sensation caused by G's motion. This is called the "observation." The observation may come in two flavors. One may assume to know how objects generate projected images in the eye and have a model of how such *sensations* are formed. Alternatively, we have a model like the one described here earlier, which generates images of the objects based on sensations from the eye. This is a model of *perception*, and it is the kind of observation model that we consider here. The perception model has the useful geometrical property of generating mathematical objects of the same type as the mathematical objects produced by the process model. Both perception and process models generate hypotheses about the state of navigation. Thus, we can compare their results.

The observation model generates an estimate of the current locations of the landmarks relative to us, in our own frame of reference, $\mathbf{r}_i^O(t)$. We may readily transform these data into an estimate of the positions of the landmarks in the external frame, M:

$$\hat{\mathbf{r}}_i(t) = H(\eta, \xi_i, \delta l(t), \delta\xi_i(t)). \tag{2.42}$$

The hat superscript indicates the data obtained from the observation. To derive this expression more explicitly, we need to know the heading direction and the step δs along this direction. We can apply equation (2.11). Using the homogeneous coordinate notation:

$$H(\eta, \xi_i, \delta l(t), \delta\xi_i(t)) = \begin{bmatrix} R_{-\eta} & -R_{-\eta} r_M^O(t) \\ 0 & 1 \end{bmatrix} \begin{bmatrix} r_i^O(\xi_i, \delta l, \delta\xi_i) \\ 1 \end{bmatrix} =$$

$$= \begin{bmatrix} \cos(\eta) & -\sin(\eta) & -\cos(\eta)x_M^O + \sin(\eta)y_M^O \\ \sin(\eta) & \cos(\eta) & -\sin(\eta)x_M^O - \cos(\eta)y_M^O \\ 0 & 0 & 1 \end{bmatrix} \begin{bmatrix} -\frac{\delta l}{\delta\xi_i}\sin^2(\xi_i) \\ \frac{\delta l}{\delta\xi_i}\sin(\xi_i)\cos(\xi_i) \\ 1 \end{bmatrix} =$$

$$= \begin{bmatrix} -\frac{\delta l}{\delta\xi_i}(\sin^2(\xi_i)\cos(\eta) + \sin(\xi_i)\cos(\xi_i)\sin(\eta)) - \cos(\eta)x_M^O + \sin(\eta)y_M^O \\ \frac{\delta l}{\delta\xi_i}(\sin(\xi_i)\cos(\xi_i)\cos(\eta) - \sin^2(\xi_i)\sin(\eta)) - \sin(\eta)x_M^O - \cos(\eta)y_M^O \\ 1 \end{bmatrix}. \tag{2.43}$$

Note that the displacement term, $r_M^O = [x_M^O, y_M^O]$ is the location of the fixed reference landmark in G's moving frame of reference. Therefore, G's own location is derived by transforming its origin from its own frame (i.e., the point $[0 \quad 0 \quad 1]^T$) to the external frame (figure 2.4):

$$\mathrm{r}_0 = \begin{bmatrix} R_{-\eta} & -R_{-\eta} r_M^O(t) \\ 0 & 1 \end{bmatrix} \cdot \begin{bmatrix} \begin{bmatrix} 0 \\ 0 \end{bmatrix} \\ 1 \end{bmatrix} = \begin{bmatrix} -R_{-\eta} r_M^O(t) \\ 1 \end{bmatrix} = \begin{bmatrix} -\cos(\eta)x_M^O + \sin(\eta)y_M^O \\ -\sin(\eta)x_M^O - \cos(\eta)y_M^O \\ 1 \end{bmatrix}. \tag{2.44}$$

Equation (2.40) and equation (2.42) are two ways for computing the same thing: the structure of the space around G in terms of the landmarks and G's location at different instants of time. The first method is based on predicting how things will look when G moves. The second uses G's observation of the landmarks and of the reference point. Both methods use some form of prior knowledge about the landmarks being stationary and about self-motion. But they do so in different ways. If everything is working then the predictions from the process model and the actual observation must coincide, as shown in figure 2.6. The leftmost panel displays the paths as we are moving. The environment has three fixed landmarks, plus a reference point, indicated by a small square. The reference direction (north) is shown by the arrow in figure 2.6. As we move in this space we form two models, one based on the observation model of equation (2.42) and the other based on the predictive model of equation (2.40). The outcomes of the prediction models are shown on the top panels of figure 2.6. The outcomes of the observation models are shown in the lower panels.

Error Sources

Why can there be a discrepancy between what G expects to observe and what it actually sees? The answer is deceivingly simple: because of noise. But what is noise? This is a more complex question. We can call noise whatever causes unexpected behaviors. If we know exactly the structure of the process, the command that we are issuing and their effect on our movement, and if the images have no blur or unexpected distortion, there would be no question about the fidelity of our internal representation of the environment. Unfortunately, things are not so simple. Any model is likely to have some structural errors. For example, one normally integrates small but finite movements instead of infinitesimal displacements. Figure 2.6 illustrates the effect of increasing amounts of uncertainty on G's position and on the placement of the landmarks within a model of space. The top row demonstrates what would happen if G's model were built based only on what G knows a priori about its movement. Here, G starts from an initial estimate of the landmarks and its own location. It assumes that this initial estimate is correct. Then, each time G takes a step it calculates a new position for itself and for the landmarks, using equation (2.40). However, now there is an unexpected term, $\boldsymbol{\varepsilon}_s(t)$, a random variable representing the uncertainty of the predictions:

$$\mathbf{s}(t) = \mathbf{s}(t-1) + B(\mathbf{s}(t-1))\mathbf{u} + \boldsymbol{\varepsilon}_s(t)\,. \tag{2.45}$$

The random variable may follow some unknown distribution. However, most analyses assume that noise is drawn from a normal distribution with known variance, Q. When the state variable is a vector quantity, the variance is replaced by a covariance matrix of the same dimension. The process noise that was used in the examples of figure 2.6 had two components, one for position uncertainty and one for heading uncertainty. In practical situations, process uncertainty derives not only from a limited knowledge of the actual value of the commands, but also, what is more important, from the unexpected external factors that may affect the outcome of each command. Factors like rough terrain and wind gusts would cause variable degrees

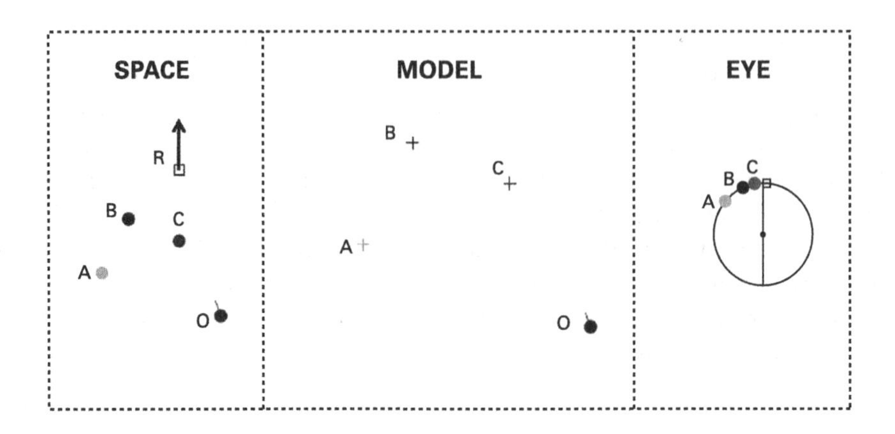

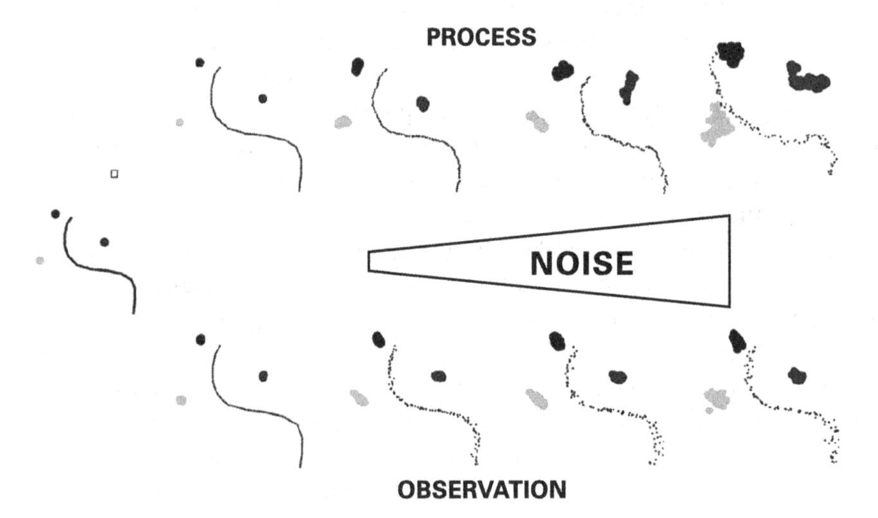

Figure 2.6
A simple navigation. Top left: The external space. The simulated gerbil (G) moves within a planar environment populated by landmarks. These are indicated as points with letters *A*, *B*, *C*. A small square (*R*) and an arrow indicate a reference point in the environment and the actual "north" direction. The point *R* can be thought of as a particular landmark or as point in space with some particular salience. Top middle: Space model obtained from the projected images of the landmarks and of the reference on the eye model (top right). The scales of the model and of the space are deliberately different in this figure. The locations of the landmarks and G are derived from the motions of their projections on the eye after a small translation of the agent in the heading direction. In the model, all locations are referred to the image of the reference, which becomes the origin of the model's coordinates. Bottom: Noise effects as G moves in the environment (bottom left). The four "process" panels represent four internal models of the environment obtained from the iteration of the process model. The models are corrupted by Gaussian noise of increasing amplitude from left to right. The four "observation" panels are model reconstructions based only on the observation of the landmarks. These models are also corrupted by Gaussian noise of increasing amplitude from left to right. The leftmost observation and process models have zero noise. Note that, without noise, the observation model has a minimal amount of error, revealed by the slightly larger images of the landmarks. This is because the observation model has a nonlinear inverse perspective transformation that is affected by the discrete approximation of the positional increments.

of uncertainty on G's predicted position. We are safe to assume that the position of the landmarks is constant. However, knowledge of this position could also be affected by some degree of uncertainty. In generating the trajectories of figure 2.8, the process model assumed that the uncertainty about each landmark position was the same as the uncertainty on G's position. Alternatively, one may assume a lower amount of uncertainty for elements that are known to be stationary. But the really critical issue here is the shape of the expected error distribution. One approach to random variables of uncertain origin is to consider them as normally distributed, with zero mean and variance Q:

$$\varepsilon_s \sim N(0,Q). \tag{2.46}$$

This choice has its main rationale in the central limit theorem of probability theory, which states that the mean of any random variable tends to be normally distributed, as the number of observations grows larger. However, one can be sure that the effect of friction on the movement of a vehicle is always opposite to the direction of motion. Therefore, when looking at this effect as a random variable, it could hardly be described as a noise with zero mean with symmetric distribution. Nevertheless, the standard assumption of equation (2.46) has significant computational advantages, and it is common practice in mathematics to make simplifications of this type. In such cases it remains important to keep in mind what elements of reality are being ignored.

In figure 2.6, the variance Q for the positional noise was varied from 0.25 to 4 units. This being a simulation, the entity of a unit is somewhat arbitrary. To give an idea of the dimensions at play, the distances between pairs of landmarks was about 90 units. The variance for the heading direction varied between 1 and 9 degrees.

Noise is also present in the observation process, starting from the signals originating from sense organs. In our case, G observes the locations of the landmarks and infers its own position by looking at the projections of the landmarks and knowing that the landmarks are fixed in space. In a perception model, one observes variables that are related at all times to one's own state of motion. We have derived a particular expression, equation (2.43), for the observation process that gives the location of the landmarks as a function of G's state of motion. As it was the case for the process model, we now add a random variable ε_r to the deterministic component of the observation model:

$$\hat{\mathbf{r}}_i(t) = H(\eta, \xi_i + \varepsilon_\xi, \delta l(t), \delta\xi_i(t)) + \varepsilon_r(t). \tag{2.47}$$

An additional noise term, ε_ξ appears inside the function H. This represents the uncertainty on the "retinal" signals, ξ, which results into an uncertainty on the reconstructed landmark locations. The observed data about the positions of the landmarks relative to G are then reflected into the uncertainty on G's position. The effects of observation noise are illustrated in the lowest portion of figure 2.6. We generated these examples with the retinal noise varying from 0.05 to 0.2 degrees, the position noise from 0.5 to 1.5 and the heading noise from 0.5 to 1.5 degrees.

2.8 Combining Process and Observation Models

We have considered two models. One generates a prediction about the next state and the other makes an observation of the same state. The two models include some amount of randomness that causes uncertainty on their outcome. Methods of optimal estimation (more details in chapter 4) are based on the idea of combining the outcomes of these two models and a particular way to do so is to require that the combination be convex. A convex combination of two points is a third point that lies between them. The most general form for a convex combination of two points *P1* and *P2* is a point *P3* that lies on the segment joining them:

$$P_3 = \alpha_1 \cdot P_1 + \alpha_2 \cdot P_2 \quad \text{with} \quad \alpha_1 + \alpha_2 = 1. \tag{2.48}$$

Note that this rule applies to points in any number of dimensions. Let us begin by considering a simple one-dimensional case, in which the process model generates the estimate $s(t)$ and the observation generates another estimate $\hat{s}(t)$. Both are one-dimensional real numbers. It seems plausible that the true unknown value may likely fall between these two estimates. However, this is not always the case. So, instead of considering individual trials, we should consider collections of "equivalent" trials. This is easy to do in a simulation, although it is time consuming. All one needs is to repeat each prediction and each observation multiple times and collect some statistics of the outcomes. In doing so, one implicitly assumes that the process under study is *ergodic*. By this, we mean that one can infer the statistical properties of the process from a large number of samples at each point of time. In our case, we repeat each step of the process a number of times and calculate the mean and covariance of the predicted and observed states. Suppose that the observed state has very little variability compared to the predicted state. Because we have assumed that the noise has a Gaussian distribution with zero mean, we could safely conclude that the true state is likely closer to the observed state. Conversely, if the observations are more variable than the predictions, the true state is likely closer to the predicted state. Therefore, variance appears to be a reasonable criterion to establish the position of the final estimate between the observed and the predicted state. One simple way to do so is to give the more weight to the process with smaller variance. That is, let $\beta_O = \frac{1}{\sigma_O^2}$ be the inverse of the variance of the observation $\hat{s}(t)$ and $\beta_P = \frac{1}{\sigma_P^2}$ be the inverse of the variance of the prediction model, $s(t)$. Then, we generate the normalized coefficients

$$\begin{cases} \dfrac{\beta_O}{\beta_O+\beta_P} = \dfrac{\sigma_P^{\,2}}{\sigma_O^{\,2}+\sigma_P^{\,2}} \\ \dfrac{\beta_P}{\beta_O+\beta_P} = \dfrac{\sigma_O^{\,2}}{\sigma_O^{\,2}+\sigma_P^{\,2}} \end{cases},$$

and we derive the state estimate from the convex combination

$$s_E(t) = \frac{\sigma_O^{\,2}}{\sigma_O^{\,2}+\sigma_P^{\,2}}\cdot s(t) + \frac{\sigma_P^{\,2}}{\sigma_O^{\,2}+\sigma_P^{\,2}}\cdot \hat{s}(t). \tag{2.49}$$

A simple algebraic manipulation leads to an expression for the estimated state that is a correction of the predicted state based on the difference between observed and predicted state:

$$\begin{aligned} s_E &= \frac{\sigma_O^{\,2}}{\sigma_O^{\,2}+\sigma_P^{\,2}}\cdot s(t) + \frac{\sigma_P^{\,2}}{\sigma_O^{\,2}+\sigma_P^{\,2}}\cdot s(t) + \frac{\sigma_P^{\,2}}{\sigma_O^{\,2}+\sigma_P^{\,2}}\cdot \hat{s}(t) - \frac{\sigma_P^{\,2}}{\sigma_O^{\,2}+\sigma_P^{\,2}}\cdot s(t) = \\ &= s(t) + K\cdot(\hat{s}(t)-s(t)) \end{aligned}, \tag{2.50}$$

with

$$K = \frac{\sigma_P^{\,2}}{\sigma_O^{\,2}+\sigma_P^{\,2}}. \tag{2.51}$$

We see that the estimated state is obtained from the predicted state with a correction proportional to the difference between predicted and observed states. If the variability of the prediction is much larger than the variability of the observation, then the gain K is close to 1. In the opposite case, K is close to zero.

This approach is illustrated by the examples in figure 2.7. In our case, the state is not a scalar, but the array (2.36) with heterogeneous positional and angular components. To preserve the flavor of a convex combination, we estimate independently each element of the state vector $-\mathbf{r}_O^T, \eta, \mathbf{r}_1^T, \cdots, \mathbf{r}_N^T$ -by taking the trace of the respective covariance matrices for the observed and the predicted values. The more rigorous approach to this type of estimate is presented in chapter 4. Here, we consider two different cases, with different values for the relative variances of the observation and of the prediction processes. The data shown on the top row were obtained with high observation noise and low prediction noise. The trajectories and landmarks derived from the observation and prediction processes are shown together with the trajectory obtained from their combination, using equation (2.49). The graph on the top rightmost panel illustrates the net "space error," that is, the net positional error for the three landmarks and for G's position. As expected, the observation model generates larger errors than the prediction model and the estimated combination has similar performance to the prediction model. The situation in the bottom row is characterized by a similar amount of variance in the prediction and observation

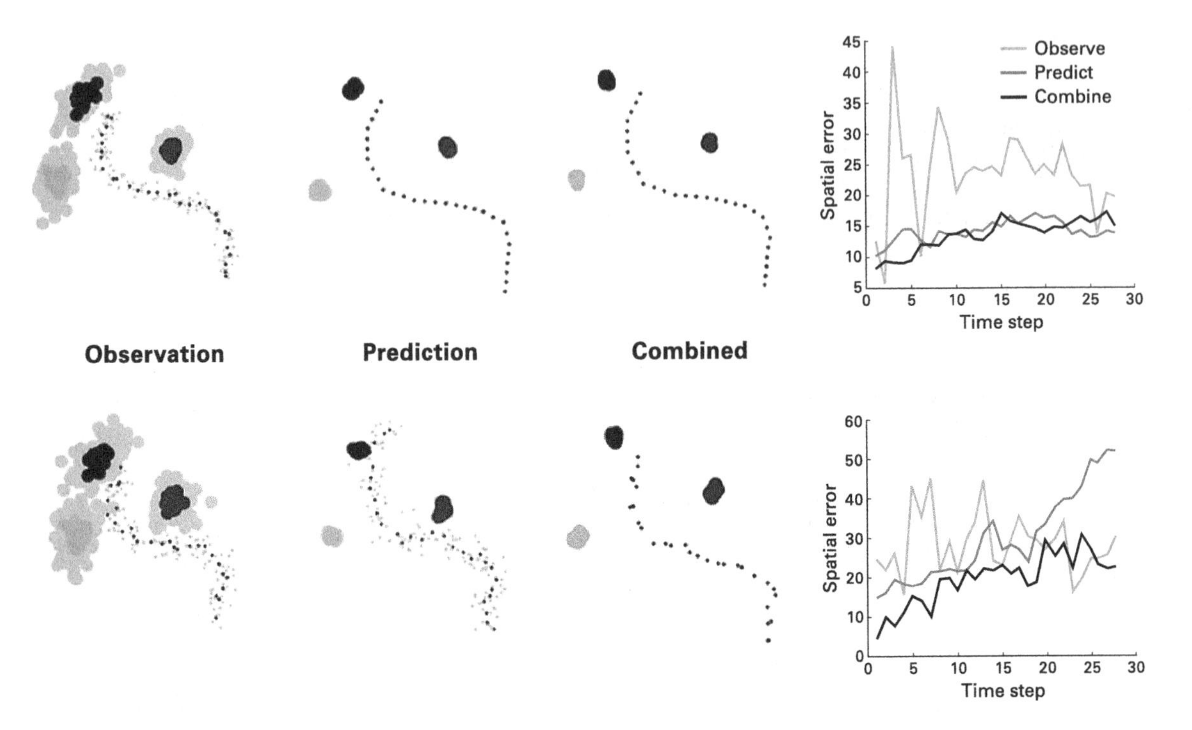

Figure 2.7
Combining predictions and observations. As G moves among landmarks, it forms a map of the landmarks and localizes itself in this map. This figure illustrates the effects of two sources of noise: noise in the internal model of the process and noise in the observation. The top panels correspond to a condition in which the observation noise is large compared to the process noise. The bottom panels describe a situation in which both noise terms are relatively large and similar. The panels on the left show estimated locations of landmarks and G, obtained by the observation system alone. At each time step, the model used five samples of the landmark locations and G's position. These are shown in light gray. The dark colored markers are averages of these individual samples. The second panels from the left show estimated positions obtained from the process model alone. Again, each dark-colored marker is an average over five points. The third panels from the left show the same scenario derived from a convex combination of observed and predicted locations. The combination is based on the relative variance of predicted and observed points, as described in the text. The graphs on the right display the overall reconstruction errors obtained for each of the two noise distributions and for each reconstruction model. Note that most times the combination of observation and prediction models provides a better estimate than either method.

noise. Here, the lower rightmost panel shows that the performance of the combined model is superior to both component models.

As we pointed out earlier, an obvious drawback of this approach stems from the need to calculate means and variances from multiple data, when a process may be allowed to generate only one sample per time interval. In real life, we would not take multiple small steps back and forth along any given trajectory to generate multiple samples of the landmarks and of our own position. Therefore, a great deal of attention has been devoted by signal and control theorists to the problem of estimating the statistics "on the go," one sample at a time. The Kalman filter that will be discussed in chapter 4 is a successful and fundamental algorithm that solves this problem for linear systems with normally distributed zero mean noise. The algorithm of the Kalman filter uses an update expression that is very similar in form and substance to equations (2.50) and (2.51), and its most important and critical part is in the update of the process covariance as the data keep coming in.

2.9 Back to the Gerbils

We have described how a simple combination of geometrical and probabilistic rules is sufficient to reconstruct the spatial distribution of the landmarks around G together with G's own location. Of course, G is only a fictional character based on an oversimplified model of the visuomotor apparatus. Can this model account for real data? Let us recall the experiment by Collett and collaborators that we described in the previous chapter. They placed their Mongolian gerbils inside a circular arena with two distinct landmarks. They hid a seed under the gravel at a fixed location with respect to the landmarks. After some explorations the gerbil found the seed. The gerbil engaged in this search several times, with the seed at the same location. Eventually, the gerbil remembered the location of the seed and went straight to get it. The interesting part of the experiment came after this initial practice. Once the gerbil had learned to find the seed, Collett and colleagues played a revealing trick. They displaced the landmarks and removed the seed. Now the gerbil went in the changed environment looking for food while the scientists recorded where the gerbil would spend time searching. In one case, the distance between the landmarks was doubled. The gerbil responded to that perturbation by searching at two locations, each location corresponding to the learned location of the seed relative to each marker (figure 1.3c). This was seen as evidence that "they treat each landmark independently when planning a path to the goal and formulate a separate trajectory for each landmark." However, this finding appears to be at odds with the result of another experiment, where the landmarks had different appearance and, after training, were rotated by 180 degrees. Then, instead of searching at two locations, the gerbils concentrated their effort at a single location that was also rotated by the same amount (figure 1.3f). As we pointed out in the previous chapter, this result implies that the gerbil's internal model of space was Euclidean. We have included already this assumption in some of the operation of G's rudimentary visual system. Let us now construct a single mathematical function capable of accounting for both of Collet's findings.

G's navigation system has a state that combines G's own location, r_0, and the locations of the markers, $r_1,\ldots,r_N$ in a single vector. In the navigation paradigm that we have described so far, we did not address an important issue: how does G decide where to move? We simply assumed that it moves somewhere. If G were a real gerbil looking for food, we may assume that not knowing the food location, it would engage in some kind of random walk, stopping occasionally to check under the gravel. Things would change when G finds the seed. The seed is a reward that would cause G to store the state vector for future use. As this experience is repeated, this memory is likely to become stronger and more stable, and perhaps also a little blurred, because the state would not be identical from trial to trial. Mathematically, we can represent this memory as a "value" function encoding the probability of finding the reward at one or more locations of the space map. Now, we can formulate a rule to build the value map, given that we found a nutritious seed at a state

$$\hat{\mathbf{s}} = \left[\mathbf{r}_{SEED}^T, \hat{\mathbf{r}}_1^T, \hat{\mathbf{r}}_2^T\right]^T. \tag{2.52}$$

Here, $\mathbf{r}_{SEED}$, $\mathbf{r}_1$, $\mathbf{r}_2$ are the positions of the seed and of the two landmarks. We dropped the heading direction η since it is not relevant to the problem at hand.

We use the three elements of $\hat{\mathbf{s}}$ to build two independent representations of the seed, as shown in figure 2.8. These are:

$$\begin{aligned}
\mathbf{s}_1^0 &= \left[\frac{(\mathbf{r}_0-\mathbf{r}_1)^T(\mathbf{r}_2-\mathbf{r}_1)}{\|\mathbf{r}_2-\mathbf{r}_1\|} \quad \frac{(\mathbf{r}_0-\mathbf{r}_1)\times(\mathbf{r}_2-\mathbf{r}_1)\cdot\hat{\mathbf{k}}}{\|\mathbf{r}_2-\mathbf{r}_1\|}\right]^T \\
\mathbf{s}_2^0 &= \left[\frac{(\mathbf{r}_0-\mathbf{r}_2)^T(\mathbf{r}_1-\mathbf{r}_2)}{\|\mathbf{r}_2-\mathbf{r}_1\|} \quad \frac{(\mathbf{r}_0-\mathbf{r}_2)\times(\mathbf{r}_1-\mathbf{r}_2)\cdot\hat{\mathbf{k}}}{\|\mathbf{r}_2-\mathbf{r}_1\|}\right]^T
\end{aligned} \tag{2.53}$$

The only element here that is in a way extraneous to the state vector (2.52) is the term $\hat{\mathbf{k}}$, which appears on the second term of $\mathbf{s}_0^1$ and $\mathbf{s}_0^2$. This is simply the unit vector perpendicular to the plane in which G moves and pointing upward. The "cross product" indicated by the symbol $\times$ is a standard operator of vector calculus. Given a system of unit axes in three orthogonal directions, $\hat{\mathbf{i}}, \hat{\mathbf{j}}, \hat{\mathbf{k}}$, and two vectors, $\mathbf{v} = v_1\hat{\mathbf{i}} + v_2\hat{\mathbf{j}} + v_3\hat{\mathbf{k}}$ and $\mathbf{w} = v_1\hat{\mathbf{i}} + v_2\hat{\mathbf{j}} + v_3\hat{\mathbf{k}}$, the cross product of v and w is

$$\mathbf{v}\times\mathbf{w} = \begin{bmatrix} \hat{\mathbf{i}} & \hat{\mathbf{j}} & \hat{\mathbf{k}} \\ v_1 & v_2 & v_3 \\ w_1 & w_2 & w_3 \end{bmatrix} = (v_2w_3 - v_3w_2)\hat{\mathbf{i}} + (v_3w_1 - v_1w_3)\hat{\mathbf{j}} + (v_1w_2 - v_2w_1)\hat{\mathbf{k}}. \tag{2.54}$$

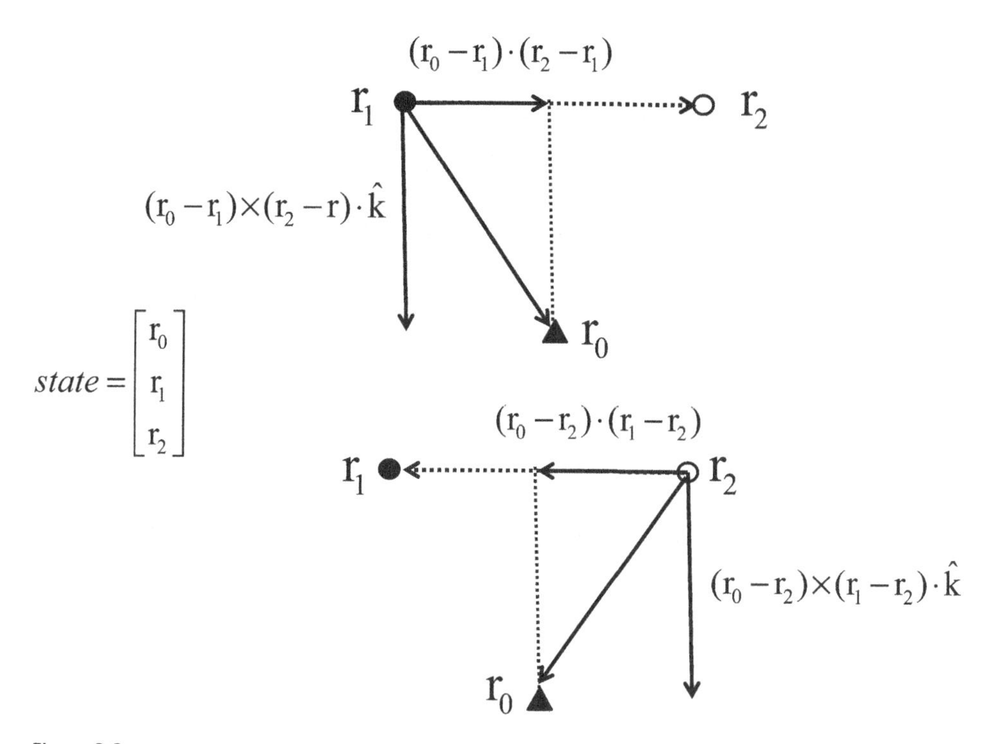

Figure 2.8
Representations of the seed in the reference frames of the landmark. The state vector contains the position of G and of the two landmarks at the time the seed is discovered. Each pair of landmarks defines a reference vector upon which the seed vector is projected by inner product. The outer product operation yields the projection of the seed vector on the orthogonal direction.

We need also to remember that the cross product of two vectors is proportional to the sine of the angle between them:

$$\mathbf{v} \times \mathbf{w} = \|\mathbf{v}\| \times \|\mathbf{w}\| \sin(\widehat{\mathbf{vw}}). \tag{2.55}$$

From both equations (2.54) and (2.55) it follows that the cross product is anticommutative:

$$\mathbf{v} \times \mathbf{w} = -\mathbf{w} \times \mathbf{v}. \tag{2.56}$$

Therefore, the sign of the second components of the "seed vectors" depends on the relative orientation of the landmarks. This is critical for reproducing both the apparently conflicting results of Collett and coworkers, when they increased the distance between the landmarks and when they rotated the landmark pattern.

In our model, when G finds a seed it constructs a value function over the navigation space, expressing the probability to find the hidden reward in relation to each visible landmark. An example of such function with the two-landmark scenario is:

$$V\left(\mathbf{s}\mid \mathbf{s}_1^0,\mathbf{s}_2^0\right)\propto \exp\left(-\frac{\left\|\mathbf{s}_1-\mathbf{s}_1^0\right\|^2}{\sigma^2}\right)+\exp\left(-\frac{\left\|\mathbf{s}_2-\mathbf{s}_2^0\right\|^2}{\sigma^2}\right), \tag{2.57}$$

where $\mathbf{s}_1$ and $\mathbf{s}_2$ are representations of the current position of G with respect to the current landmarks, derived using the same transformation, equation (2.53), that generated the two "seed vectors," $\mathbf{s}_1^0$ and $\mathbf{s}_2^0$. Each landmark, with its neighbors, is an independent reference frame that generates an additive component of the value function. On the next navigation, G will move toward the places that promise the highest reward. In the landmark arrangement of figure 2.9A, the value function has two coincident "hills" (figure 2.9B). Therefore, G will search in the same place where it found the seed in previous trials. If the landmarks are placed at a greater distance, then the two exponentials in the value function will separate—the degree of separation being a function of the uncertainty, σ^2 (figure 2.9C). If instead the landmarks

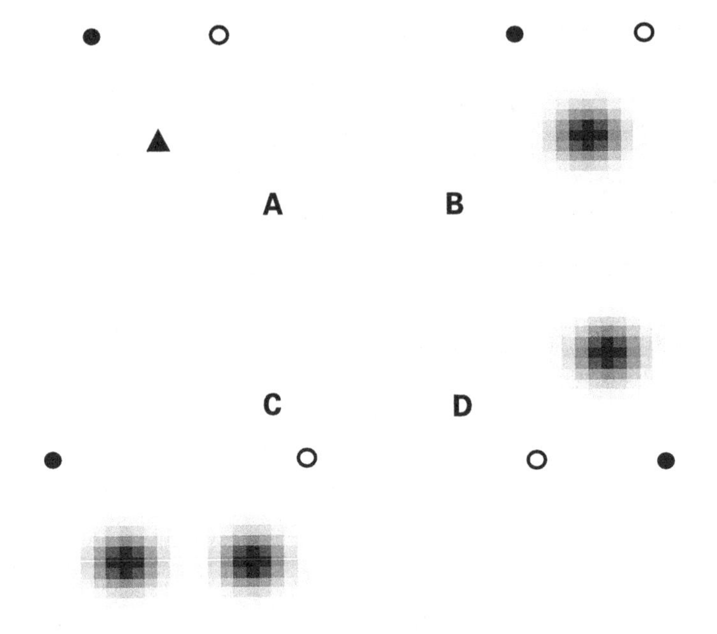

Figure 2.9
Simulation of the experiment by Collett et al. (1986). Compare with figure 2.3. (A) Two landmarks and a hidden seed are arranged in a triangular configuration. (B) As G navigates in the environment it forms a state representation of itself and the landmark, as in figures 2.6 and 2.7. Since the seed is always found at a fixed location with respect to each marker, the value function has a single "hill" centered at the seed's location. (C) If the landmarks are placed at a greater distance, the two representations of the seed separate by an equivalent amount. (D) If the landmark array is rotated by 180 degrees, the value function is also rotated and its true components are still fully overlapped.

are rotated by 180 degrees (or any other angle), the Gaussian contributions to the value function will also rotate accordingly and will map to the same location on the navigation map (figure 2.9D).

The map of space built during navigation by combining the process and the observation models defines a spatial domain that offers a support for storing and retrieving memories of rewarding events, such as the discovery of food, as well as of adverse and dangerous situations. It is therefore not surprising that episodic memory and spatial information processing share common territories in the mammalian brain and that damage to this territory impairs both our ability to remember recent facts and to orient ourselves in space.

Summary

How can we construct a mathematical map of space, starting from sensory and motor information? We consider the simplified model of a gerbil, with a single one-dimensional "eye," moving over a two-dimensional plane. The first problem that we encounter is to extract geometrical information from the projections of the objects on the eye. Important geometrical properties, such as the straightness of a line and the distance between two points on the navigation plane can be reconstructed from the knowledge of our own motion and from a basic assumption that we are looking at objects that are fixed in space. This assumption constrains the relative motions of the objects on the world to be in the class of rigid motions.

Homogeneous coordinates provide us with a compact representation for rigid motions by combining rotations and translations into a single linear operation. With this operation, we may update the state of the navigation environment, which includes the position of the moving gerbil and of the surrounding landmarks. This update constitutes the "process model," yielding a prediction of the future state, given the current state and knowledge of the movement intention.

Sensory information from the visual system also provides an evolving representation of the state of the navigation environment. This is the "observation model." Observation and process model are both affected by uncertainty, caused by different forms of noise. Uncertainties result in variability of the corresponding models. An intuitive way to obtain the best estimate of the state of navigation is by forming a convex combination of the state estimates generated by the observation model and the process model. In this combination, each process contributes in inverse proportion of its own uncertainty.

The state of navigation is effectively a map of the environment in terms of its fixed landmarks and the gerbil's location. When a salient event occurs—such as finding food—this map provides a spatial domain upon which memory can take the form of a reward function. The reward function is a photograph of the location at which the event occurred. It provides a goal for the navigation when the same environment is encountered. From the state of the navigation environment we derive multiple representations of the location of the gerbil with respect to each landmark, at the time the salient event occurred. By a simple additive mechanism it is possible to construct a reward function that reproduces some of the experimental findings described in the previous chapter. This provides a computational rationale for the interaction of spatial and episodic memory.

3 The Space Inside

In the previous two chapters, we considered how the brain forms maps of the environment as the body moves in it. Now and in the following chapters, we will discuss how the brain forms maps of the body in relation to the external world. As we move and act on the environment, our brain is like a crane operator who must act on a myriad of levers for commanding the motions of the limbs. Even the simplest actions, like reaching for a glass of water, require the coordinated control of some thirty-five arm and hand muscles. These muscles are divided in smaller compartments, each compartment operated by a single motor neuron. In the end there are thousands of little levers that our crane operator—our body—must act upon successfully for conducting the simple gestures. The operator receives information about the state of motion of each limb and about the forces applied by the muscles. There are two geometries for this scenario: the geometry of the body (of its motors and sensors), and the geometry of the world outside. We already encountered this duality in the navigation problem, where the geometry of the images projected on the eyes is different from the geometry of the surrounding space. Here, as well as in the navigation, what matters is that the two geometries are mutually connected.

3.1 Geometry vs. Dynamics

Consider the situation depicted in figure 3.1A. It describes an experiment carried out by Pietro Morasso (1981) while he was visiting the laboratory of Emilio Bizzi at MIT. The sketch is the top view of a subject holding the handle of a planar mechanism that was designed by Neville Hogan to study the reaching movements of the hand. We can describe the subject's geometry in two alternative ways. We can focus on his hand, or we can focus on his arm. In the first case, the position of the hand on the plane is determined by the two coordinates (x,y), of its center, taken with respect to a Cartesian system. An alternative way (not the only one) is to look at the two joint angles: the shoulder angle, ϕ, and the elbow angle, θ. These different points of view are related to the different perspective that one may have. Earlier motor physiologists, before Morasso's time, focused on the movements of individual body segments. And, most often, of one body segment at a time, for example the upper arm motions about the elbow. On the other hand, robotics and artificial intelligence placed the focus on the goals of behaviors, for example, moving objects. Morasso came from the school of robotics and artificial intelligence. In his experiments, he paid attention to the motion of the hand.

In the planar configuration of figure 3.1A, the hand coordinates (x,y) and the joint coordinates (ϕ,θ) are connected by simple trigonometry:

$$\begin{cases} x = l_1 \cos(\phi) + l_2 \cos(\phi+\theta) \\ y = l_1 \sin(\phi) + l_2 \sin(\phi+\theta) \end{cases}, \tag{3.1}$$

where l_1 and l_2 are the lengths of the two arm segments. What is the geometrical interpretation of this relation between angular coordinates of the joints and Cartesian coordinates of the hand? First, it is a nonlinear relation. Suppose that we need to move the arm from an initial configuration (ϕ_I,θ_I) to a final configuration (ϕ_F,θ_F) of joint angles. A simple way to do so is by setting

$$\begin{cases} \phi = \phi_I + \Delta\phi \cdot u(t) \\ \theta = \theta_I + \Delta\theta \cdot u(t) \end{cases}. \tag{3.2}$$

Here, $\Delta\phi = \phi_F - \phi_I$, $\Delta\theta = \theta_F - \theta_I$ and $u(t)$ is an arbitrary function of time that varies from 0 to 1 between the start and the end of the movement. In other words, we may drive the two joint synchronously from start to end locations. In this case it is easy to see—by eliminating $u(t)$—that the two angles are linearly related to each other:

$$\phi = m\theta + q \quad \text{with} \quad \begin{cases} m = \dfrac{\Delta\phi}{\Delta\theta} \\ q = \phi_I - \dfrac{\Delta\phi}{\Delta\theta}\theta_I \end{cases}. \tag{3.3}$$

We can apply the same logic to the movements of the hand in Cartesian coordinates and derive a rectilinear motion from (x_I,y_I) to (x_F,y_F).

The two rectilinear motions are not generally compatible. This is illustrated in figure 3.1B, where we see that straight lines in hand coordinates correspond to curved lines in joint coordinates, and vice versa. The diagram in this figure, however, is somewhat misleading. We know that it is appropriate to use Cartesian axes to describe the ordinary space in which the hand moves. We have seen in chapter 2 that Cartesian coordinates capture the metric properties of Euclidean space and, in particular, the independent concepts of length and angle. If we have two points P, Q, their Cartesian coordinates may change, depending on the origin and on the

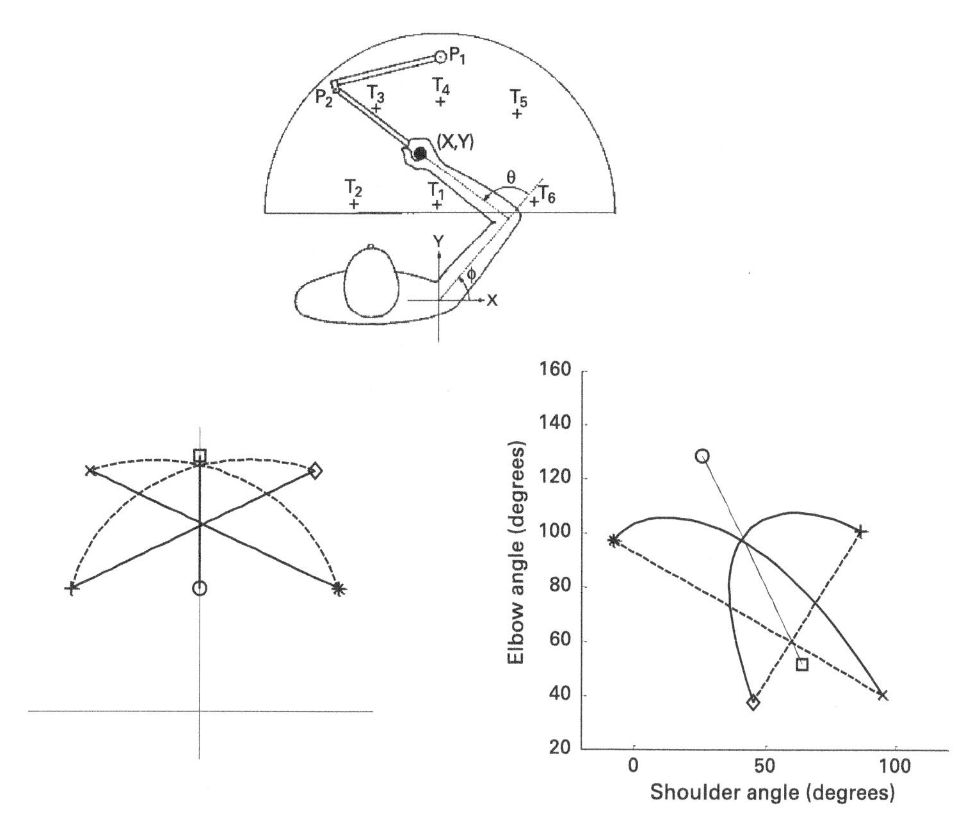

Figure 3.1
Arm kinematics. Top: Sketch of a subject in the experiment by Morasso. The subject holds the handle of a manipulandum that is free to move over the horizontal plane. The subject maintains the arm on the same plane. The manipulandum records two Cartesian coordinates, x and y, that describe the position of the hand with respect to the shoulder. Since the lengths of the subject's upper arm and forearm are known, we can derive the joint angle of the shoulder (Φ) and of the elbow (θ) corresponding to the Cartesian coordinates of the hand. Bottom panels: straight solid lines traced by the subject hand (left) map onto curved solid line in joint angle coordinates (right). Conversely, straight dashed lines in angle coordinates map onto curved dashed line traced by the hand. Note that there is one exception. The line that intersects the subject's shoulder is straight in both coordinate systems. (Modified from Morasso, 1981.)

orientation of the coordinate axes. But the distance computed with the sum-of-squares rule remains invariant.

What can we say about using Cartesian coordinates for the joint angles, ϕ and θ? If angles are represented by real numbers what prevents us from placing these numbers on two lines and call them Cartesian coordinates of an "angle space"? In principle, this could be done (and is routinely done by scientists). However, by doing so, one neglects the critical fact that 0, 2π, 4π, and so on are not really different angles. The numbers are different, but the angles are not. Because of their cyclical character, angular variables are better represented over circles than over lines. Angles have curvature! But then how do we represent two angular variables, as in our case of shoulder and elbow joint coordinates? As shown in figure 3.2, the natural geometrical structure for describing a two-link planar mechanism, like a double pendulum or our simplified arm, is a doughnut-shaped object known as a torus.

A torus, like a sphere is a two-dimensional geometrical object. A fundamental theorem by John Nash (1956) establishes that this type of objects can be placed inside an Euclidean space of higher dimension by an operation called an isometric embedding.[1] By this operation, we can represent the two angles of the arm in figure 3.1A as coordinates over a torus inside a three-dimensional Euclidean space (figure 3.2).

If we take two points, A and B, there is a unique line on the torus that has the smallest length. This is called a *geodesic* line. The linear joint interpolation of equation (3.2) defines such a geodesic line. Since the concept of a manifold is a generalization of the concept of a surface, the Euclidean plane is also a manifold, whose geodesics are straight lines. Summing up, the Cartesian coordinates of the hand (x,y)

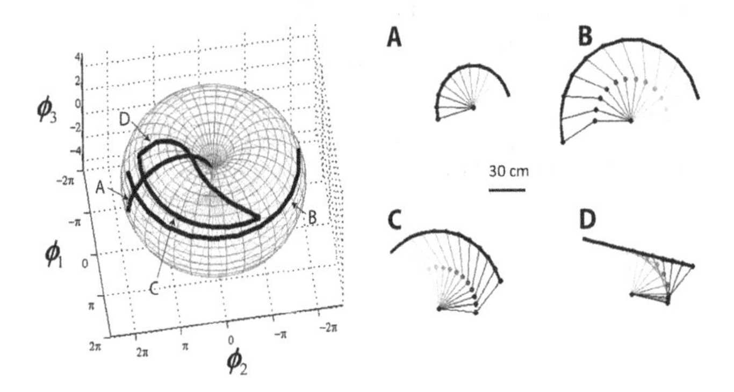

Figure 3.2
Riemannian structure. The configuration space of a two-joint planar mechanism, analogous to the arm in Morasso's experiments, forms a torus (left), a two-dimensional curved manifold embedded in the Euclidean 3D space, reminiscent of the surface of a doughnut. The light gray mesh forms orthogonal geodesics (minimum path length excursions) spanning the torus. The four panels on the right contain trajectories of the endpoint of the arm's endpoint, marked by the thick black lines. The letters of each panel correspond to the trajectories on the torus.

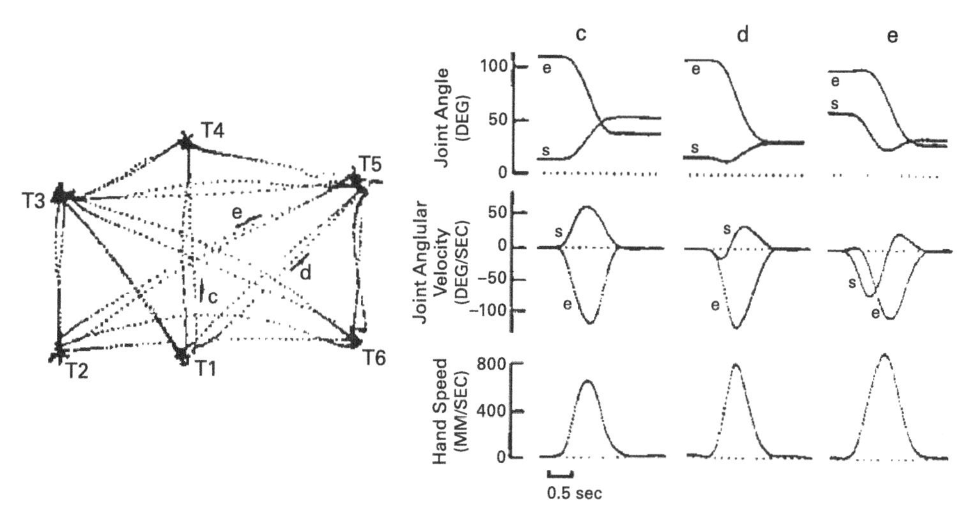

Figure 3.3
Reaching movements. Left: hand trajectories between the targets shown on the top panel of figure 3.1. Joint angles (top, *e*: elbow, *s*: shoulder) and angular velocities (middle) corresponding to the trajectories labeled *c*, *d*, and *e* on the left panel. The traces in the bottom panel are the speed of the hand calculated from the sum of squares of the *x*- and *y*- components of hand velocity. Note how the variety of shapes in joint angle profile corresponds to similar bell-shaped curves for the hand speed. Also note that movements *d* and *e* had reversals of joint motions. Movement *c* is the only one that does not show any reversal. This is consistent with the observation (see figure 3.1) that straight lines of the hand that pass by the shoulder joint map to straight lines in joint coordinates. (From Morasso, 1981.)

and the angular coordinates (ϕ,θ) describe two geometrical domains with different geodesic properties. If we move the hand between two points over a plane, what type of geodesic do we tend to follow? Morasso addressed this question by asking his subject to move his or her hand between targets on the plane. He found that movements tend to occur along straight lines—that is, along Euclidean geodesics (figure 3.3). Tamar Flash and Neville Hogan (1985) interpreted this finding as evidence that the nervous system seeks to optimize the smoothness of hand movements by moving along trajectories that minimize the integrated amplitude of the third time derivative of hand position, or "jerk":

$$\int_0^T \left\{ \left(\frac{d^3x}{dt^3}\right)^2 + \left(\frac{d^3y}{dt^3}\right)^2 \right\} dt. \tag{3.4}$$

The idea that there is a separation between the geometry of movement—the "kinematics"—and its dynamical underpinnings did not go unchallenged and is still an object of controversy. Yoji Uno, Mitsuo Kawato, and Ryoji Suzuki (1989) suggested that, instead of optimizing the smoothness of hand motions, the nervous system was concerned with minimizing variables that are more directly connected with effort. Then, they considered, as cost function, the integral of the net torque changes over each movement:

$$\int_0^T \sum_{i=1}^{n} \left(\frac{d\tau_i}{dt} \right)^2 dt. \tag{3.5}$$

The term τ_i represents the joint torque produced by one of the n arm muscles. One can intuitively see the similarity between the two formulations. Consider a point mass. By Newton's law, the point mass accelerates in proportion to the applied force. If one takes one more time derivative, one obtains a relation between the jerk and the rate of change of the force. This explains why the two optimization criteria yield relatively similar predictions of arm trajectories. There is however an important difference between kinematic and dynamic criteria. The outcome of the dynamical optimization depends on properties such as the inertia of the moving arm. The minimization of the cost integral (3.5) requires knowing the relation between forces and motions, as expressed by the limb's dynamical equations. These equations are quite complex and grow in complexity as more degrees of freedom are introduced. The reader can find explicit forms for two-joint dynamics in standard robotics handbooks. Here, we do not want to get into excessive detail and we limit ourselves the describe dynamics as a second-order nonlinear ordinary differential equation (ODE), whose general form[2] is

$$\mathbf{D}(\mathbf{q}, \dot{\mathbf{q}}, \ddot{\mathbf{q}}) = \mathbf{Q}(t). \tag{3.6}$$

The generalized coordinate $\mathbf{q} = [\phi, \theta]^T$ is the configuration of the arm and the corresponding generalized force is a joint-torque vector $\mathbf{Q} = [\tau_\phi, \tau_\theta]^T$. Thus, with the vector notation of equation (3.6) we are representing a system of two coupled nonlinear ODEs. The optimization by Uno and colleagues requires us to calculate the torque vector based on these ODEs. On the other hand, the kinematic approach of Flash and Hogan is strictly geometrical and does not involve forces (that is, dynamics). Unlike dynamic criteria, the geometrical optimization predicts that neuromuscular activities are coordinated by the brain to generate a rectilinear motion of the hand or of whatever other element plays the role of a controlled "endpoint element," and this result will not be affected by changes in the dynamical structure controlled by the muscles.

Randy Flanagan and Ashwini Rao (1995) performed a simple and elegant variant of Morasso's experiment. They asked a group of volunteers to control a cursor by moving the hand over a plane. However, in this case, the position of the cursor was not an image of the position of a subject's hand but was in direct relation to his or her shoulder and elbow angle. This task was difficult and somewhat confusing for the subjects. However, after some practice, they learned to move the cursor onto the targets. Most remarkably, as they learned to do so, they gradually but consistently and spontaneously learned to produce rectilinear movements of the cursor, at the

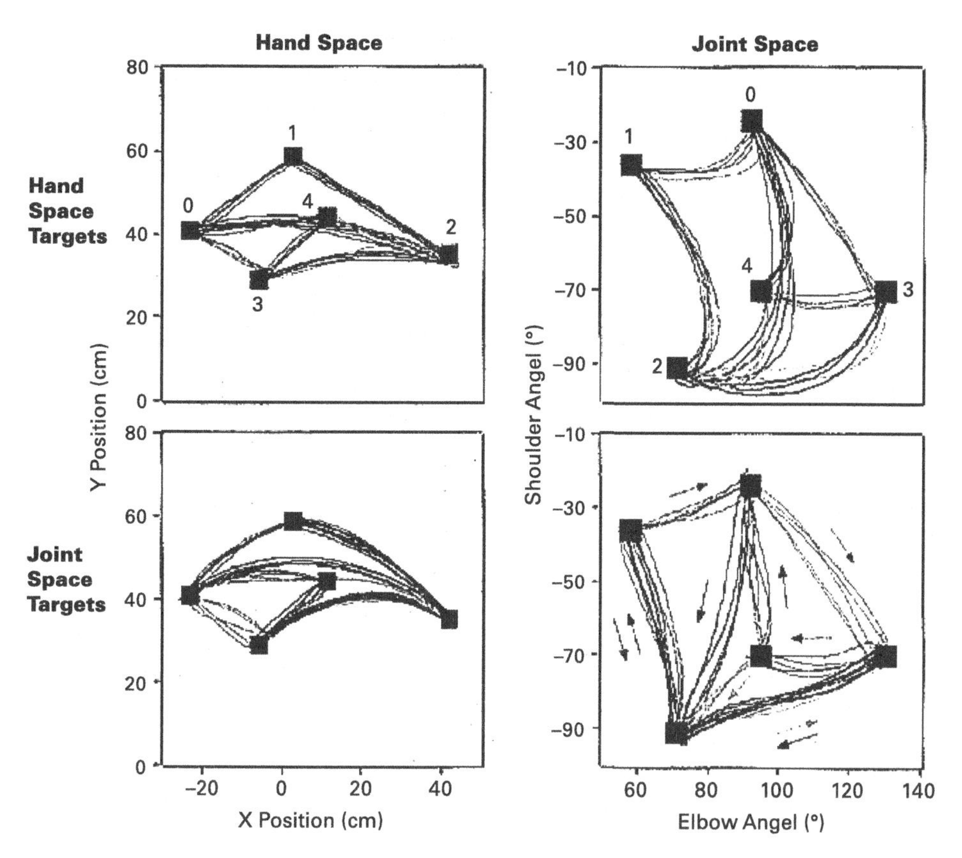

Figure 3.4
Subjects move in straight lines over the visual space. Motion paths represented in hand (left) and joint (right) coordinates for movements between targets presented in either hand coordinates (top) or joint coordinates (bottom). In both cases, subjects learned to organize coordination so as to produce straight-line motions path in the observed space. (From Flanagan and Rao, 1995.)

expenses of more curved motions of the hand (figure 3.4). This result is incompatible with the notion that the nervous system attempts to minimize a dynamic criterion, such as the minimum-torque change of equation (3.5).

3.2 Does the Brain Compute Dynamics Equations?

Until the late 1970s, the study of neurobiology was mostly limited to the control of single muscles or of muscle pairs acting on a single joint. A large volume of work was dedicated to the control of eye movements, where six extrinsic muscles are responsible to move the eye and to control the direction of the gaze. The engineering frameworks for oculomotor control and for the control of limb movements derived from

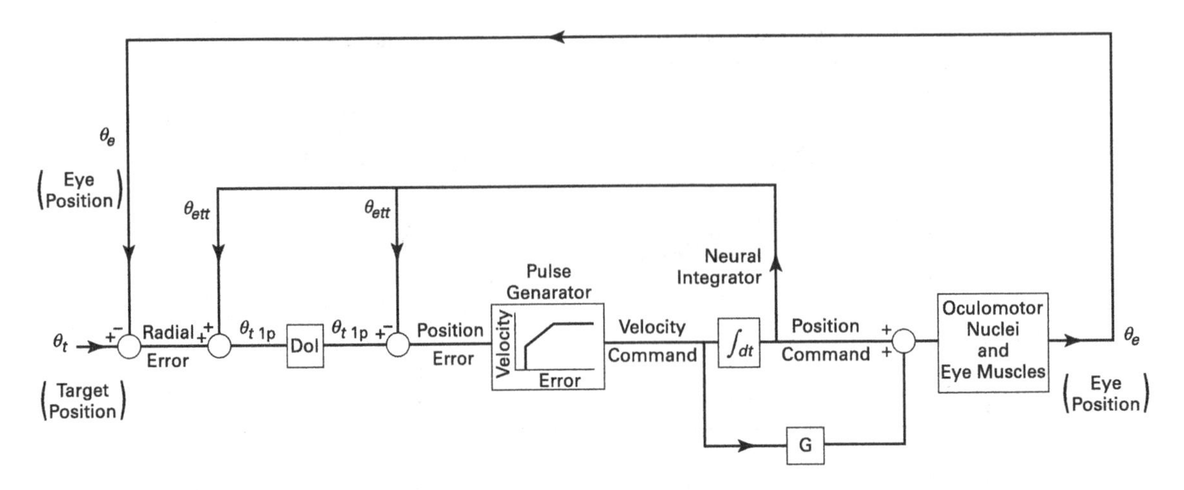

Figure 3.5
Block diagram based on control engineering to describe the operation of the oculomotor system in response to a visual target. (From Zee et al., 1976.)

the theory of linear control systems and of feedback control in particular. Computational models were abundant in "box diagrams" such as the one in figure 3.5. Robotics research and, in particular, the theory of robot manipulators brought to the neurobiology of motor control a new awareness of the dynamical complexity of multiarticular structures such as the human arm. Let us compare the inertial dynamics of the upper arm about the elbow joint and the inertial dynamics of the two-joint arm of figure 3.1. The first are described by the following equation:

$$I\ddot{q} = Q\,. \tag{3.7}$$

This is a straightforward translation of $ma = F$ in angular terms. Here, I is the moment of inertia of the forearm about the axis of rotation of the elbow, $\ddot{q}$ is the angular acceleration about the same axis, and Q is the corresponding joint torque vector. In the two-joint case, we have two angles that we call q_1 and q_2 (instead of ϕ and θ, for notational convenience) and two corresponding torques, Q_1 and Q_2. The inertial dynamics look like this:

$$\begin{aligned}
&\left(I_1 + I_2 + m_2 l_1 l_2 \cos(q_2) + \frac{m_1 l_1^2 + m_2 l_2^2}{4} + m_2 l_1^2\right)\ddot{q}_1 + \left(I_2 + m_2 \frac{l_1 l_2}{2}\cos(q_2) + m_2 \frac{l_2^2}{4}\right)\ddot{q}_2 + \\
&\qquad -\left(m_2 \frac{l_1 l_2}{2}\sin(q_2)\right)\dot{q}_2^2 - (m_2 l_1 l_2 \sin(q_2))\dot{q}_1\dot{q}_2 = Q_1 \\
&\left(I_2 + m_2 \frac{l_1 l_2}{2}\cos(q_2) + m_2 \frac{l_2^2}{4}\right)\ddot{q}_1 + \left(I_2 + m_2 \frac{l_2^2}{4}\right)\ddot{q}_2 + \left(m_2 \frac{l_1 l_2}{2}\sin(q_2)\right)\dot{q}_1^2 = Q_2
\end{aligned} \tag{3.8}$$

The reader who is not acquainted with this equation can find its derivation at www.shadmehrlab.org/book/dynamics.pdf. Here, we wish to focus the attention not on the detailed form of these equations but on a few relevant features. First, note how the transition from equation (3.7) to equation (3.8) has led to a disproportionate increase in size. In the first we have a single multiplication, whereas in the second one we count 15 additions, 42 multiplications, and a couple of trigonometric functions. It is not a pretty sight. If we were to write the detailed equations for the entire arm in 3D, we would fill a few pages of text. To put some logical order in these equations, we may start by noticing that some terms contain accelerations and some contain velocities. The terms that contain acceleration are the first two in each equation. They may be written in vector-matrix form as the left end term of equation (3.7), that is,

$$\begin{bmatrix} I_1 + I_2 + m_2 l_1 l_2 \cos(q_2) + \frac{m_1 l_1^2 + m_2 l_2^2}{4} + m_2 l_1^2 & I_2 + m_2 \frac{l_1 l_2}{2}\cos(q_2) + m_2 \frac{l_2^2}{4} \\ I_2 + m_2 \frac{l_1 l_2}{2}\cos(q_2) + m_2 \frac{l_2^2}{4} & I_2 + m_2 \frac{l_2^2}{4} \end{bmatrix} \begin{bmatrix} \ddot{q}_1 \\ \ddot{q}_2 \end{bmatrix} = \mathbf{I}(\mathbf{q})\ddot{\mathbf{q}}.$$

This is called the *inertial term*. The acceleration appears linearly, as it is typical of all mechanical systems derived from Newton's equation. However, the inertia matrix I depends on arm configuration, unlike the inertial term of equation (3.7), which is constant. In addition to this term, the two-link arm equation also contains terms that depend on the angular velocities. These are the "centripetal" terms, depending on the squares of the velocities and the "Coriolis" term depending on the product of the velocities. We may collect these terms in a single vector $\mathbf{G}(\mathbf{q},\dot{\mathbf{q}})$ and the dynamical equation assumes the much less intimidating form

$$\mathbf{I}(\mathbf{q})\ddot{\mathbf{q}} + \mathbf{G}(\mathbf{q},\dot{\mathbf{q}}) = \mathbf{Q}. \tag{3.9}$$

While this simple algebraic representation provides a valid description for more complex structures, such as a typical robotic manipulator or the whole human arm, it highlights the most relevant difference between the dynamics of a single joint and the dynamics of multiarticular limb and points to some computational challenges. One such challenge stems from the dependence of the torque on one joint upon the state of motion of another. These dynamics cannot be simplified by treating each dimension independently of the others. How important are these cross-dependencies in natural arm movements? Equation (3.8) provides us with a simple and direct way to answer this question. It is sufficient to record the trajectory of a reaching arm movement, derive the corresponding angular motion of the shoulder and elbow joints, take the first and second derivatives, and plug these data into the equations. John Hollerbach and Tamar Flash (1982) performed this experiment, using the same apparatus and task of Morasso. When they calculated the different contributions to the net torques at the shoulder and elbow, they were able to conclude that the

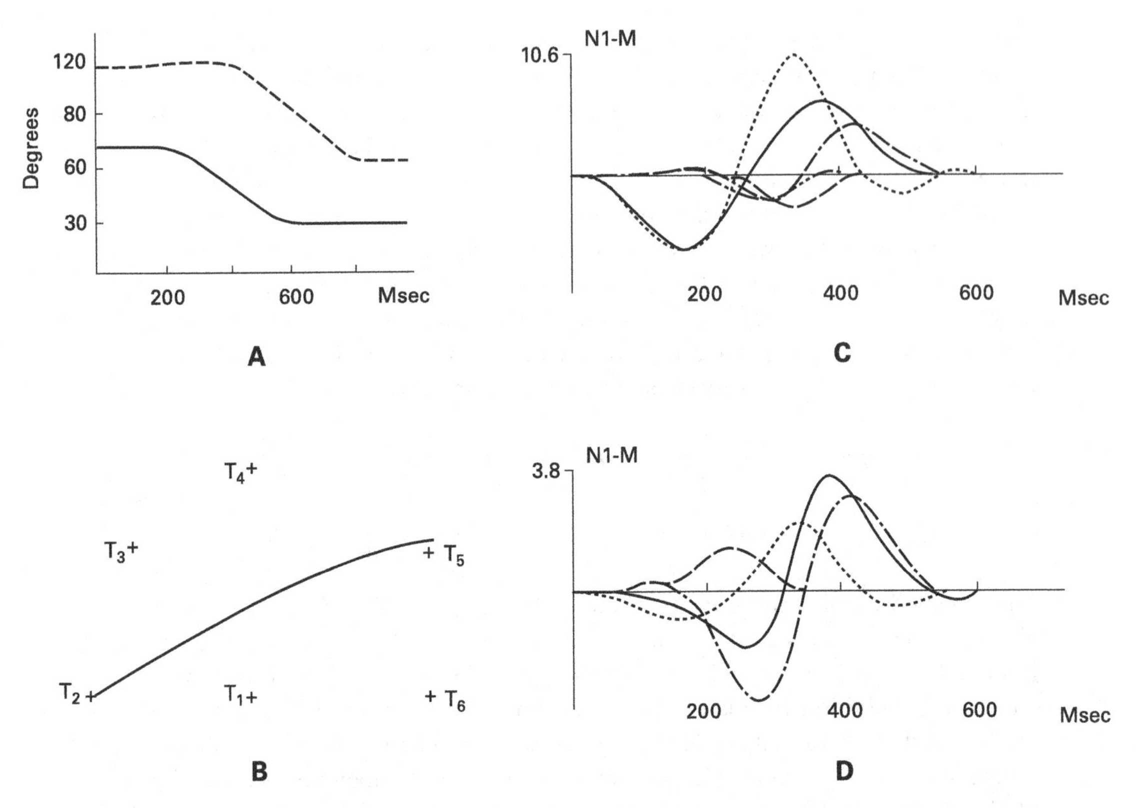

Figure 3.6
The dynamics of a reaching movement. (A) Shoulder and elbow angle traces corresponding to the hand movement in B. Different components of the shoulder (C) and elbow (D) torque were derived by applying the equation of motions to the observed kinematics. Solid lines: net torque at each joint. Dotted line: shoulder inertial torque. Dot-dashed line: elbow inertial torque. Dashes: centripetal torques. Two dots with a dash: Coriolis torque at the shoulder. (From Hollerbach and Flash, 1982.)

interaction components that are summarized by the nonlinear term $\mathbf{G}(\mathbf{q},\dot{\mathbf{q}})$ are quite substantial compared to the pure inertial terms. As an example, consider the trajectory in figure 3.6A and B, which is similar to the movement labeled as "e" in figure 3.3. The movement time is about half a second, corresponding to a relatively rapid but not unusual speed. The different components of the torque profiles at the shoulder and elbow joint are plotted in figure 3.6C and D, respectively. Note the amount of toque that is due to centripetal and Coriolis terms, and also the amount of inertial torque at the shoulder due to the acceleration at the elbow and vice versa. Most important, the same motion of the hand between two different targets would lead to qualitatively different torque profiles. This evidence was sufficient to conclude that arm movements cannot be controlled by some linear feedback mechanism, such as the one depicted in figure 3.5 for the oculomotor system.

3.3 The Engineering Approach

Let us briefly switch our perspective from neuroscience to robotics. Control engineers are primarily concerned with the concept of stability. Loosely speaking, a system is stable if, given some unexpected perturbation it will eventually go back to its unperturbed state. However, "eventually" is not strong enough a requirement for practical applications. A stronger form of stability requires the system to converge exponentially in time to the unperturbed state. One of the objectives, if not the main objective, of feedback control is to insure stability against uncertainties that arise from limited knowledge of the environment and of the controlled "plant." Importantly, robotic engineers have some good prior knowledge of their own devices. So, for example, they would know to good accuracy the parameters in equation (3.9). Given this knowledge, we now want to write the right-hand term, the torque vector, as the driving control input to the arm. Because we are dealing with feedback control, we think of $\mathbf{Q}$ as a function of the state of motion—that is, the vector $[\mathbf{q},\dot{\mathbf{q}}]^T$—and of time. A physical system whose dynamics equations do not depend upon time is said to be *autonomous*. For such system the future is entirely determined by a measure of the state at some initial instant. For example, the equation of a simple oscillator, like a mass attached to a spring, is

$$m\ddot{q} = -kq. \tag{3.10}$$

The solution of this equation is

$$q = A\cos\left(\left(\sqrt{\frac{k}{m}}\right)t + \phi\right). \tag{3.11}$$

The phase and the amplitude are determined by the initial position $q_0 = q(t_0)$ and velocity $v_0 = \dot{q}(t_0)$. Therefore, by setting (or measuring) these initial conditions, we know how the oscillator will behave for all future times. However, if we add an external input, as in

$$m\ddot{q} = -kq + u(t), \tag{3.12}$$

then the trajectory of the system is no longer determined by the initial state alone and depends upon the future values of $u(t)$ as well.[3]

Now, consider the stability problem. Suppose that we have a robotic arm, governed by a feedback controller that attempts to track a desired trajectory, $\hat{\mathbf{q}}(t)$. Then, the combined system+controller equation takes the form

$$\mathbf{I}(\mathbf{q})\ddot{\mathbf{q}} + \mathbf{G}(\mathbf{q},\dot{\mathbf{q}}) = \mathbf{Q}(\mathbf{q},\dot{\mathbf{q}},\hat{\mathbf{q}}(\mathbf{t})). \tag{3.13}$$

Our goal is to design the control function $\mathbf{Q}(\mathbf{q},\dot{\mathbf{q}},\hat{\mathbf{q}}(\mathbf{t}))$ so that the resulting movement is exponentially stable about the desired trajectory. As we stated before, a robotics engineer would have a good model of the controlled mechanical arm. The model would contain the two terms

$$\begin{aligned} \hat{\mathbf{I}}(\mathbf{q}) &\approx \mathbf{I}(\mathbf{q}) \text{ and} \\ \hat{\mathbf{G}}(\mathbf{q},\dot{\mathbf{q}}) &\approx \mathbf{G}(\mathbf{q},\dot{\mathbf{q}}) \end{aligned} \quad (3.14)$$

We can use these terms to design the control function as

$$\mathbf{Q}(\mathbf{q},\dot{\mathbf{q}},\hat{\mathbf{q}}(\mathbf{t})) = \hat{\mathbf{I}}(\mathbf{q})\cdot \mathbf{a}(\mathbf{q},\dot{\mathbf{q}},\mathbf{t}) + \hat{\mathbf{G}}(\mathbf{q},\dot{\mathbf{q}}). \quad (3.15)$$

Substituting this in equation (3.13) with the assumptions of equation (3.14) we get the simple double-integrator system:

$$\ddot{\mathbf{q}} \approx \mathbf{a}(\mathbf{q},\dot{\mathbf{q}},t). \quad (3.16)$$

It is important to observe that the symbol $\approx$ is not the ordinary equals sign, but an approximate equality. We will come back to this in a little while but for the next few lines we will use the standard equality. We begin by giving a particular form to the term $\mathbf{a}(\mathbf{q},\dot{\mathbf{q}},t)$. Starting from the desired trajectory, we calculate the desired velocity, $\dot{\hat{q}}(t)$, and acceleration, $\ddot{\hat{q}}(t)$, by taking the first and second time derivatives of $\hat{q}(t)$. Then, we set

$$\mathbf{a}(\mathbf{q},\dot{\mathbf{q}},t) = \ddot{\hat{\mathbf{q}}}(t) - \mathbf{K_P}(\mathbf{q} - \hat{\mathbf{q}}_d(t)) - \mathbf{K_D}(\dot{\mathbf{q}} - \dot{\hat{\mathbf{q}}}_\mathbf{d}(t)) \quad (3.17)$$

and

$$\eta(\mathbf{q},t) \triangleq \mathbf{q} - \hat{\mathbf{q}}(t). \quad (3.18)$$

The last equation defines the *tracking error* as a function of the current position and time. The two matrices $\mathbf{K}_P$ and $\mathbf{K}_D$ contain gain coefficients that multiply the position and velocity errors.[4] By combining equations (3.18), (3.17), and (3.16) we finally obtain a linear second-order differential equation for the error function:

$$\ddot{\eta} + \mathbf{K_D}\dot{\eta} + \mathbf{K_P}\eta = 0. \quad (3.19)$$

From calculus we know that by setting

$$\begin{aligned} \mathbf{K_P} &= \begin{bmatrix} \omega_1^2 & 0 & \cdots & 0 \\ 0 & \omega_2^2 & \cdots & 0 \\ \cdots & \cdots & \cdots & \cdots \\ 0 & 0 & \cdots & \omega_N^2 \end{bmatrix} = diag(\omega_1^2, \omega_2^2, \ldots, \omega_N^2) \\ \mathbf{K_D} &= 2diag(\omega_1, \omega_1, \ldots, \omega_N) \end{aligned} \quad (3.20)$$

we can rewrite equation (3.19) as N decoupled equations

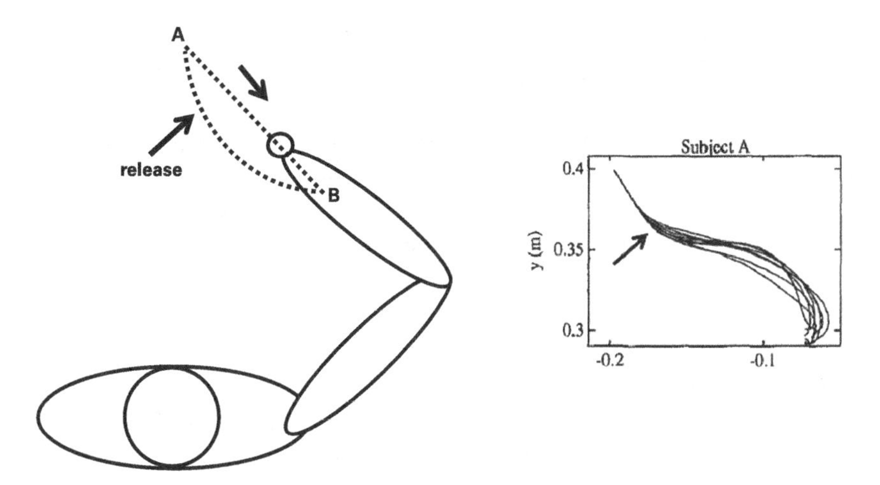

Figure 3.7
Dynamic stability. Right: Subjects practiced moving the hand from A to B while holding the handle of a planar manipulandum. In most trials the manipulandum could be moved freely over the plane and the subjects moved on straight paths (dashed line). In some trials, however, electromagnetic brake blocked the motion of one manipulandum joint at the onset of the reaching movement. In these perturbed trials, the hand was constrained to remain on a circular path (dotted line) until the brake was suddenly released. Right: Constrained and released hand trajectories. The arrow indicates approximately where the release took place. The hand moved almost instantaneously toward the planned straight trajectory from A to B. (From Won and Hogan, 1995.)

$$\ddot{\eta}_i + 2\omega_i \dot{\eta}_i + \omega_i^2 = 0, \tag{3.21}$$

whose general solution is

$$\eta_i = (c_{1,i} + c_{2,i} t) e^{-\omega_i t}. \tag{3.22}$$

Therefore, with the appropriate choice of the two gain matrices, we can insure that the tracking error will go to zero in exponential time. This type of dynamic stability was observed in reaching movements by Justin Won and Neville Hogan (1995). They asked subjects to execute reaching movements of the hand while holding a planar manipulandum (figure 3.7). Most of the times, the manipulandum allowed for free motion of the hand in the plane, and the subjects moved along straight lines from start to end target. In some random trials, one of the joints of the manipulandum was blocked at the onset of the movement by an electromagnetic brake. As a consequence, in these test trials the movement of the hand was initially constrained to remain on a circular arc (dotted line). Shortly after the start, the brake was released. The resulting motion of the hand converged rapidly to the unperturbed path, thus demonstrating a dynamic stability of the planned movement trajectory, similar to the exponential stability discussed earlier.

But now remember that in equations (3.21) and (3.22), instead of = we should have used ≈ as the model of the arm dynamics that can at most approximate the actual value of the parameters. However, for small deviations, the uncertainty of the model can be regarded as an internal perturbation that does not critically compromise stability. The essential point is that if we have an internal representation of the dynamics, a control system can generate behaviors that are stable, in the face of uncertainties concerning the environment, and the controlled structure.

3.4 Does the Brain Represent Force?

The idea of model-based control was born as an engineering idea. What can it tell us about the brain? One of the fathers of computer technology, the mathematician John von Neumann, was driven by the desire to create an artificial brain. And some of the earliest computational neuroscientists, like Warren McCulloch and Walter Pitts, described neurons as logical gates. But it is hard to capture the intricacies of biology within the confines of engineering and mathematics. And the idea that the brain computes something like the dynamics equations for the limb it controls has encountered—and still encounters to this day—some strong resistance. One of the strongest counterarguments to the idea of explicit computations comes from the observation that ordinary physical systems behave in ways that may be described by equations but do not require solving equations with calculus-like rules. If you drop a stone, the stone will follow a simple and regular trajectory that you can derive by integrating the equations of motion in the gravitational field. Yet the stone does not compute integrals! In the same vein, an important concept in science is to seek for principles that can explain observations based on simple rules of causality.

Consider another example. Take the two opposing springs and dampers connected to a mass (figure 3.8A). These are physical elements that generate forces acting on the point mass. The two springs define a static equilibrium point. If an external agent places the point mass away from this equilibrium, the physical system will eventually bring it back there, because the equilibrium is stable. This behavior can be readily implemented by analog or digital elements (figure 3.8B) that produce forces in response to measured position and velocity of the point mass. Thus, we have a physical system and a computational system (which is nothing but another type of physical system!) that do the same job. In 1978, Andres Polit and Emilio Bizzi published a finding that physiologists of that time found hard to digest. They trained monkeys to reach a visual target by a movement of the upper arm about the elbow. A monkey could not see its arm as it was moving under an opaque surface and was rewarded when it pointed correctly to the visual target. Not surprisingly, that monkey learned the task. However, the monkeys were able to perform the task after they were surgically deprived of all sensory information coming from their arm. And they

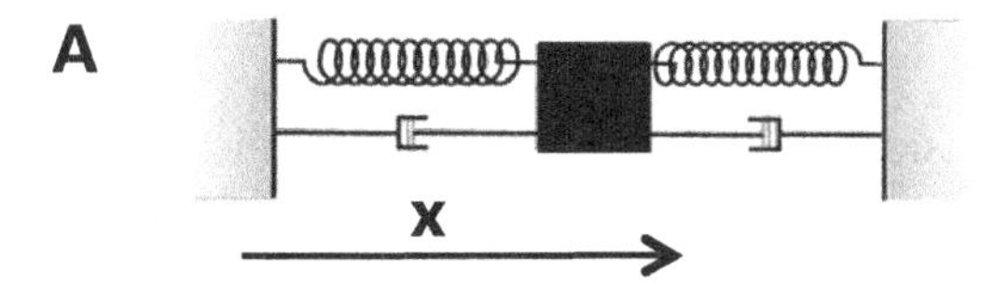

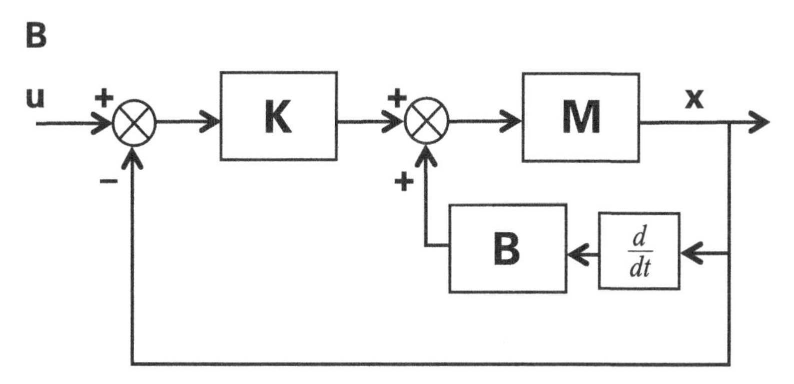

Figure 3.8
Springlike behavior in physical (A) and computational (B) systems. The forces generated by springs on a mass are reproduced by a control systems that multiplies by a gain (K) the difference between the measured position of the mass (M) and a commanded equilibrium position (u). The output of the K is then added to another feedback signal proportional to the measured velocity of the mass and the resulting force is applied to M. The presence of delays along the feedback lines in B may cause unstable oscillations that are not observed in A.

successfully reached the targets even if the arm were unexpectedly displaced by a motor before starting the movement. If one were to think of movement control according to the scheme drawn in figure 3.8B, this result would not make any sense, because now the lines carrying the information about position and velocity were cut! But the scheme of figure 3.8A is perfectly consistent with the finding of Polit and Bizzi: If the muscles act upon the arm, as opposing viscoelastic elements, then all what the nervous system has to do is to set their equilibrium point to the target. No need to monitor the position and state of the arm in order to correct for errors. Instead, in order to move a limb the brain can set the equilibrium point so as to follow a desired movement trajectory.

Anatol Feldman pioneered this idea in 1966. Then, as well as in more recent works, Feldman proposed that we move our bodies by setting the equilibrium postures specified by reflexes and muscle mechanical properties. Does this make dynamic computations unnecessary? Answering this question is not quite as simple as it seems.

One may ask if to move your arm your brain must represent forces, as defined by Newton. Consider the equation that represents the behavior of a spring-mass-damping system, such as the one shown in figure 3.8:

$$M\ddot{x}+B\dot{x}+K(x-u(t))=0. \tag{3.23}$$

We derive this by substituting the expression of the forces generated by the spring and damper in $F=m\ddot{x}$. As a result the equation does not contain force at all. It has only the state (position and velocity) its derivative (acceleration) and an external input, $u(t)$, that sets instant by instant the equilibrium point of the springs. The situation is not different from the one described in the previous paragraph to derive the controller for a robotic arm. We may not need to represent forces to derive a trajectory produced by the input function, but we need to know how the system reacts to applied forces. These are the terms that appear in equation (3.23). This equation allows us to derive the trajectory $x(t)$ that results from the commanded equilibrium-point trajectory $u(t)$. And the same equation allows us to derive the equilibrium-point trajectory from the desired movement. The only condition for this to be possible is $K \neq 0$. Then, given a desired trajectory $\hat{x}(t)$ with first and second time derivatives $\dot{\hat{x}}(t)$ and $\ddot{\hat{x}}(t)$ the commanded equilibrium point-trajectory is

$$u(t)=K^{-1}M\ddot{\hat{x}}(t)+K^{-1}B\dot{\hat{x}}(t)+\hat{x}(t). \tag{3.24}$$

This is not substantially different from what we have done in deriving the engineering controller of equation (3.17). The only new element here is the focus on representing the control signal as a moving equilibrium point that drives the controlled "plant" along the desired trajectory. Intuitively, if K is sufficiently large and if the motion is smooth and slow, the first two terms in equation (3.24) may be small enough to be ignored and then all what one needs to do is to move the equilibrium point along the desired trajectory. This is the attractive promise of equilibrium-point control: there is no need to compute dynamics. The mechanical properties of the body and, in particular, the viscoelastic properties—responsible for the K and B terms—of the neuromuscular system may provide a way to avoid complex computations altogether. Interestingly, this viewpoint emerged vigorously at the same time in which the robotics viewpoint on inverse dynamic computations was becoming influential in the neuroscience of the motor system. But is it true that dynamics computations may be avoided?

Consider the simple movement of the right arm illustrated schematically in figure 3.9. This is a simulation of a simplified model of the arm, based on equation (3.8). The numerical parameters are listed in the box. We rewrite the dynamics equations to include the viscoelastic terms and make it formally similar to equation (3.23):

$$\mathbf{I}(\mathbf{q})\ddot{\mathbf{q}}+\mathbf{G}(\mathbf{q},\dot{\mathbf{q}})+\mathbf{B}\dot{\mathbf{q}}+\mathbf{K}(\mathbf{q}-\mathbf{u}(t))=0. \tag{3.25}$$

An appealing idea is to simply move the equilibrium point along the desired trajectory of the hand. The panels on the left illustrate the outcome of this approach. If the movement is sufficiently slow, for example if one takes two seconds to move the

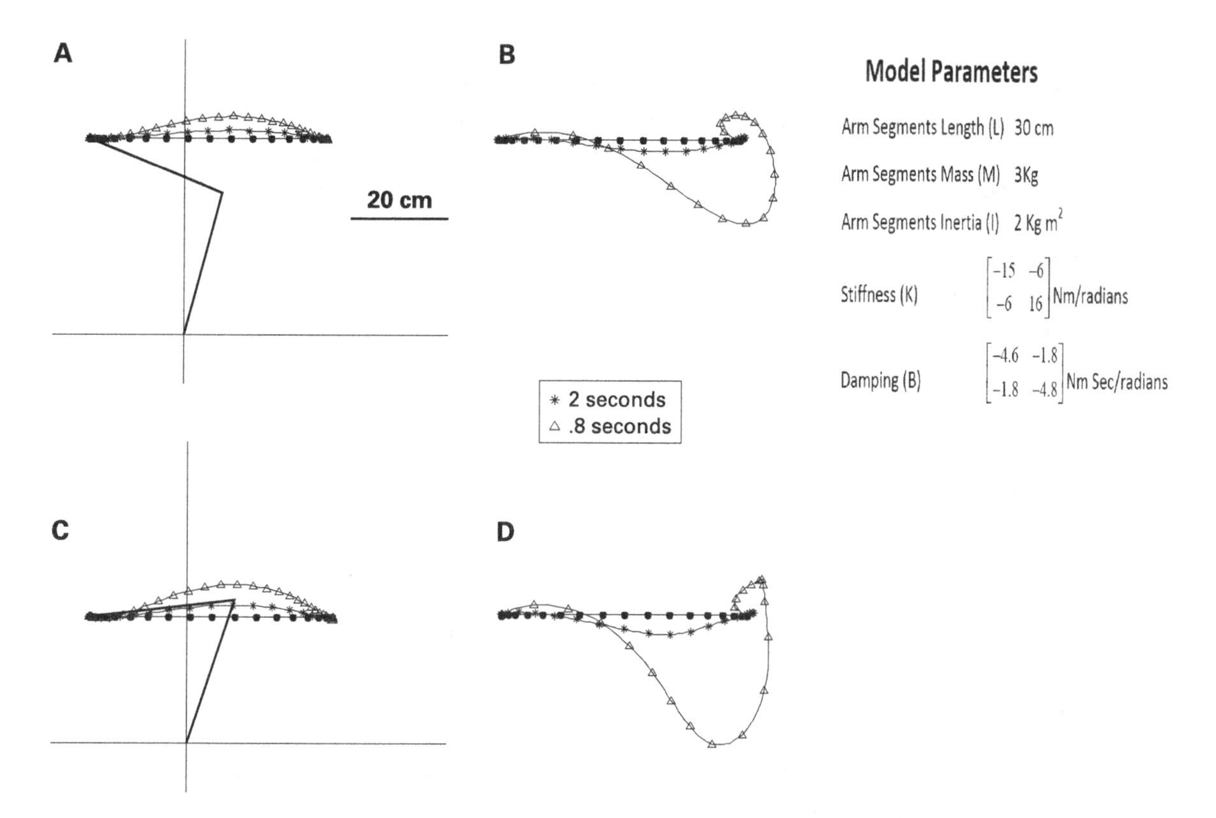

Figure 3.9
Movement simulations. (A, C) Movements of the arm obtained by shifting the static equilibrium point from left to right along a straight line with two different speeds. The equilibrium point motions follow a bell-shaped speed profile with durations of 2 seconds (slow) and .8 seconds (fast). (B, D) Movements of the equilibrium points that would produce a straight line motion of the hand with the slow and fast speed profiles.

equilibrium point about 40 cm across, the hand moves in a nearly rectilinear path. However the deviation from the desired trajectory increases as the planned motion becomes faster (0.8 sec) and when the movement is more proximal to the body (figure 3.9C). Then, we may ask what would it take to change the equilibrium trajectory, $\mathbf{u}(t)$, in such a way to obtain the desired straight motion of the hand. We can certainly move our hand in a nearly straight 40 cm line within less than one second. To find the equilibrium point trajectory that achieves this goal, we begin by transforming the desired rectilinear trajectory of the hand, $\mathbf{r}(t) = [x(t),y(t)]^T$ into the trajectories of the shoulder and elbow joints, $\mathbf{q}(t) = [q_1(t),q_2(t)]^T$ with their first and second time derivatives. Then, we solve equation (3.25) algebraically to obtain

$$\mathbf{u}(t) = \mathbf{K}^{-1}\left(\mathbf{I}(\mathbf{q}(t))\ddot{\mathbf{q}}(t) + \mathbf{G}(\mathbf{q}(t),\dot{\mathbf{q}}(t)) + \mathbf{B}\dot{\mathbf{q}}(t)\right) + \mathbf{q}(t). \quad (3.26)$$

The results of this simulation are shown in the left panels, *B* and *D* for the distal and proximal arm configurations. Only a small correction of the equilibrium point trajectory is sufficient to "straighten" the movement at low speed (2 sec). However, as the movement becomes faster (0.8 sec) the equilibrium point must take a quite distorted path in order to keep the hand on track. It is evident that we could not derive this complex equilibrium trajectory by performing a simple linear manipulation on the errors—shown on the left panels—that we would make by moving the equilibrium on the straight path. Here, to derive the motion of the equilibrium point that produces a desired actual movement of the hand we have explicitly solved an inverse dynamics problem. Our brain is likely using some different method. But, as we are indeed capable to move our limbs with dexterity, this method must effectively solve the same inverse dynamics problem.

3.5 Adapting to Predictable Forces

So far, we have described the basic computational idea of an internal model. Is there a more stringent empirical basis for this idea? Perhaps the first empirical evidence came from an experiment that the authors of this book performed in the early 1990s, while they were in the laboratory of Emilio Bizzi at MIT. We asked volunteers to move their arm while holding the handle of a planar manipulandum (figure 3.10A). They performed reaching movements of the hand, as in Morasso's experiments. The robot, designed by Neville Hogan, was a lightweight two-joint device, equipped with position sensors and with two torque motors. The motors were "backdrivable," meaning that when they were turned off, they generated only minimal resistance to motion. And the manipulator itself was made of light aluminum bars, so that the whole thing did not impede appreciably the free motion of the hand.

Therefore, with the motors turned off, the hand trajectories were rectilinear and smooth (figure 3.10B). After acquiring unperturbed movement data, the motors were programmed to generate a force, $\mathbf{F}$, that depended linearly upon the velocity of the hand, $\dot{\mathbf{r}}$:

$$\mathbf{F} = \mathbf{B} \cdot \dot{\mathbf{r}}. \tag{3.27}$$

This force field was equivalent to a controlled change in the dynamics of the subject's arm. During the initial phase of the experiment, the subject's dynamics were as in equation (3.13). Without making any assumption about the control function, $\mathbf{Q}(\mathbf{q}, \dot{\mathbf{q}}, t)$, that a subject used to perform a particular trajectory $\mathbf{q}_A(t)$, the dynamics equations are reduced to an algebraic identity over this trajectory. If the force field (3.27) is introduced without changing the control function, then the new limb dynamics are

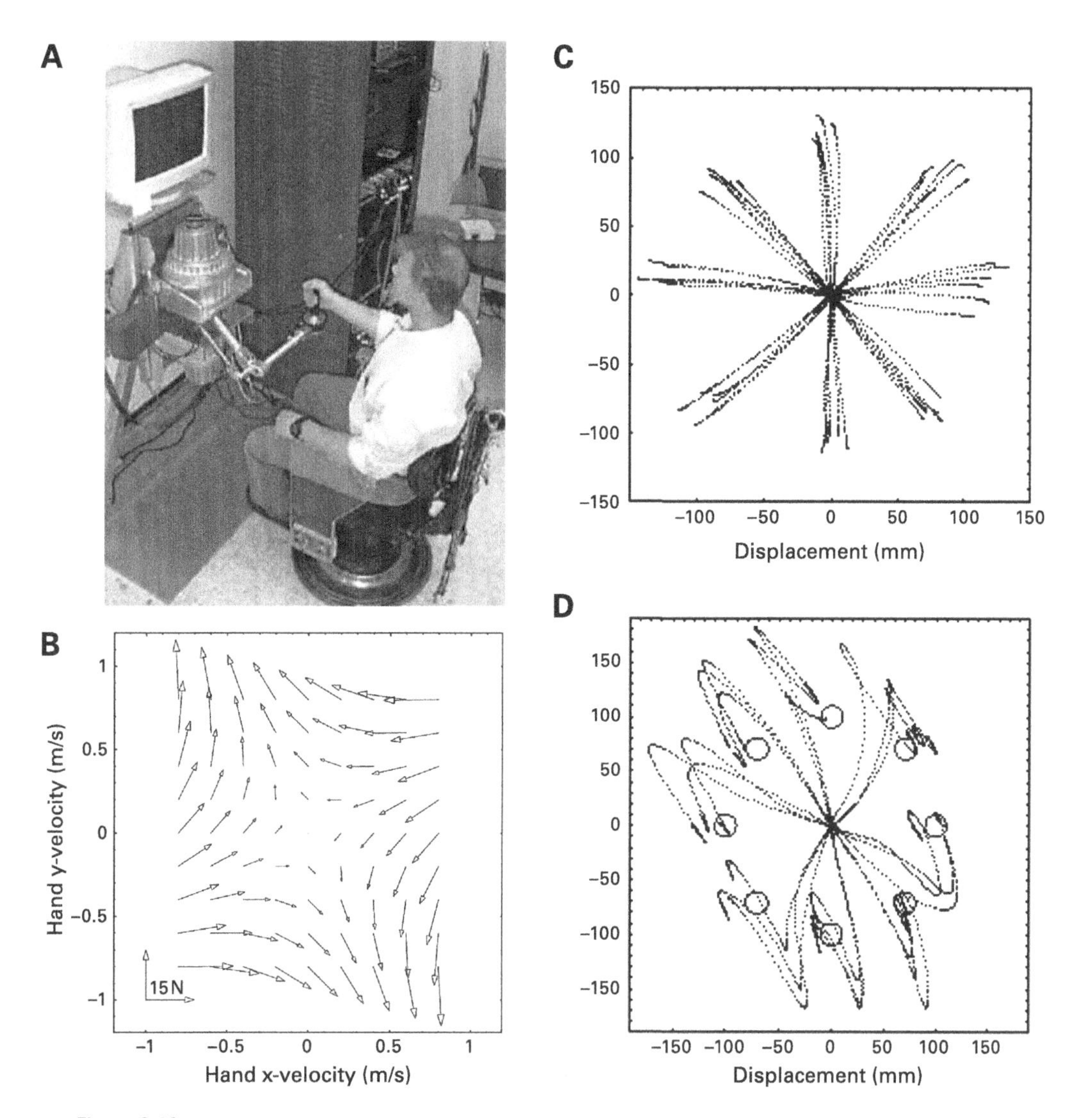

Figure 3.10
Learning to move in a force field. (A) A subject holds the handle of a manipulandum. The computer monitor displays a cursor indicating the position the hand and presents targets to the subjects. Two torque motors are programmed to apply forces that depend upon the velocity of the handle, as shown in (B). (C) When the motors are inactive, subjects perform straight reaching movements to the targets. (D) In random trials the force field shown in (C) is unexpectedly activated. The movement trajectories are strongly perturbed and display a pattern of characteristic hooks. At the end, the movement terminates in the proximity of the targets. (From Shadmehr and Mussa-Ivaldi, 1994.)

$$\mathbf{I}(\mathbf{q})\cdot\ddot{\mathbf{q}}+\mathbf{G}(\mathbf{q},\dot{\mathbf{q}})+\mathbf{J}^{\mathrm{T}}(\mathbf{q})\cdot\mathbf{B}\cdot\mathbf{J}(\mathbf{q})\cdot\dot{\mathbf{q}}=\mathbf{Q}(\mathbf{q},\dot{\mathbf{q}},t), \tag{3.28}$$

where $\mathbf{J}(\mathbf{q})$ is the Jacobian of the kinematic map from joint angles to hand position in Cartesian coordinates. With $\mathbf{r} = [x,y]^T$ and $\mathbf{q} = [q_1,q_2]^T$ the Jacobian is a 2×2 matrix of partial derivatives:

$$\mathbf{J}(\mathbf{q})=\begin{bmatrix}\frac{\partial x}{\partial q_1} & \frac{\partial x}{\partial q_2}\\ \frac{\partial y}{\partial q_1} & \frac{\partial y}{\partial q_2}\end{bmatrix}. \tag{3.29}$$

The Jacobian is a function of the arm configuration (i.e., the joint angles) as the kinematic map is nonlinear. We derived equation (3.28) from the force field equation (3.27) and from the expression of the Jacobian by combining the transformation from joint to hand velocity,

$$\dot{\mathbf{r}}=\mathbf{J}(\mathbf{q})\cdot\dot{\mathbf{q}}, \tag{3.30}$$

with the transformation from hand force to joint torque

$$\boldsymbol{\tau}=\mathbf{J}^{\mathrm{T}}(\mathbf{q})\cdot\mathbf{F}. \tag{3.31}$$

We can therefore regard the field as an unexpected change on the function $\mathbf{G}(\mathbf{q},\dot{\mathbf{q}})$:

$$\mathbf{G}'(\mathbf{q},\dot{\mathbf{q}})=\mathbf{G}(\mathbf{q},\dot{\mathbf{q}})+\mathbf{J}^{\mathrm{T}}(\mathbf{q})\cdot\mathbf{B}\cdot\mathbf{J}(\mathbf{q})\cdot\dot{\mathbf{q}}. \tag{3.32}$$

The solution of equation (3.28) is a new trajectory $\mathbf{q}_\mathbf{B}(t) \neq \mathbf{q}_\mathbf{A}(t)$. Figure 3.10C shows the response of a subject to this sudden dynamic change. It is important to observe that the force field and, accordingly, the change to G vanishes when the hand is at rest. This has two consequences. First, the subjects could not know if the field is on or off before starting to move, and second, the field does not alter the static equilibrium at the end of each movement. Therefore, we were not surprised to see that the effect of the unexpected force field on the trajectories was to cause a hook-shaped deviation, with the hand landing eventually on the planned target. While further studies revealed the presence of online corrections, these are not needed to maintain the final position unchanged by the force field. In the second part of the experiment, subjects were continuously exposed to the force field, with only sporadic "catch trials" in which the field was unexpectedly suppressed.

Figure 3.11 shows the evolution of motor performance across four consecutive blocks of trials. The four trajectories on the left—F1 to F4—were produced by a subject moving in the force field. The trajectories on the right—C1 to C4—were obtained from the same subject in the same four periods, but when the force field was unexpectedly removed. These are the catch trials, where one can observe the aftereffect of learning, that is, how the motor command evolved as the subject

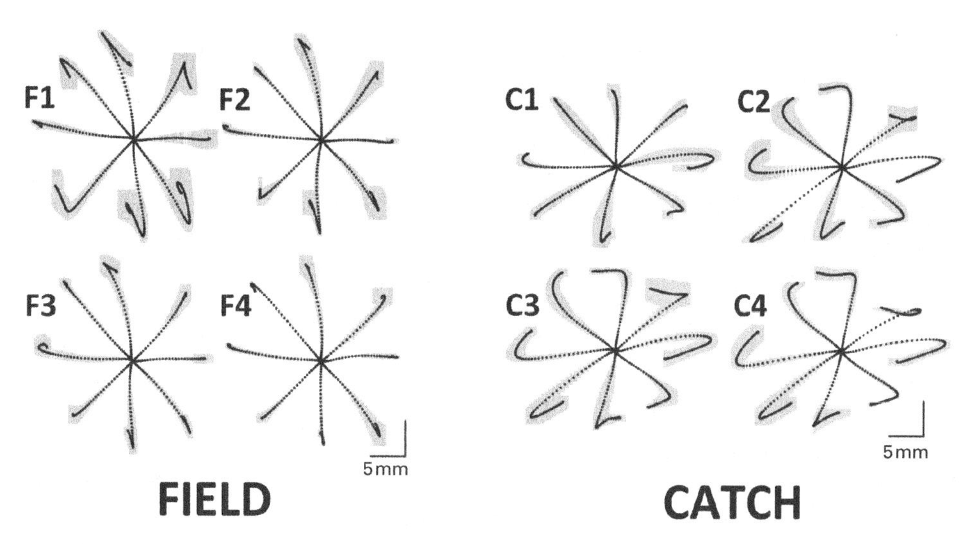

Figure 3.11
Learning the field. Left panels: Average and standard deviation of hand trajectories while a subject was training in the force field of figure 3.10B. Performance plotted during the first, second, third and final 250 targets (F1, F2, F3, and F4). Left panel: Aftereffects were observed when the field was unexpectedly removed during the four training sets. The four panels (C1 to C4) show the average and standard deviation of the hand trajectory while moving in the absence of perturbing forces. As the trajectories recovered the initial rectilinear shape in the field, they developed increasing aftereffects in the catch trials. (From Shadmehr and Mussa-Ivaldi, 1994.)

learned to move inside the force field. If one compares the first set of movements in the field (F1 and figure 3.10D), to the last set of aftereffects (C4) one can immediately observe that the shape of the aftereffect at the end-of learning is qualitatively the mirror image of the initial effect of the exposure to the field. This is easily understood in the dynamical framework. As the subjects learn to move in the field, they compensate for the forces that are generated by it along the desired trajectory. We represent this process by adding a term to the original controller:

$$\mathbf{I}(\mathbf{q})\cdot\ddot{\mathbf{q}}+\mathbf{G}(\mathbf{q},\dot{\mathbf{q}})+\mathbf{J}^{T}(\mathbf{q})\cdot\mathbf{B}\cdot\mathbf{J}(\mathbf{q})\cdot\dot{\mathbf{q}}=\mathbf{Q}(\mathbf{q},\dot{\mathbf{q}},t)+\Delta(\mathbf{q},\dot{\mathbf{q}},t). \tag{3.33}$$

For this equation to admit the original trajectory as a solution it is sufficient that

$$\mathbf{J}^{T}(\mathbf{q})\cdot\mathbf{B}\cdot\mathbf{J}(\mathbf{q})\cdot\dot{\mathbf{q}}=\Delta(\mathbf{q},\dot{\mathbf{q}},t) \tag{3.34}$$

along $\mathbf{q}_A(t)$. This is equivalent to perform a local approximation of the external field. In this case, if the external field is suddenly removed after learning, this approximation—that is, the *internal model* of the field—becomes a perturbation with the opposite sign:

$$\mathbf{I}(\mathbf{q})\ddot{\mathbf{q}} + \mathbf{G}(\mathbf{q},\dot{\mathbf{q}}) - \Delta(\mathbf{q},\dot{\mathbf{q}},t) \approx \mathbf{I}(\mathbf{q})\cdot\ddot{\mathbf{q}} + \mathbf{G}(\mathbf{q},\dot{\mathbf{q}}) - \mathbf{J}^{\mathrm{T}}(\mathbf{q})\cdot\mathbf{B}\cdot\mathbf{J}(\mathbf{q})\cdot\dot{\mathbf{q}} = \mathbf{Q}(\mathbf{q},\dot{\mathbf{q}},t). \quad (3.35)$$

This, however, is not the only way for our motor system to compensate for the external force. We discuss briefly two alternatives: stiffen up, or create a motor tape. The first consists in increasing the rigidity of the moving arm by coactivating opposing muscles. This is what you would do if you want to resist movement when someone shakes your hand. You activate the biceps and the triceps simultaneously. The two muscles oppose each other and the result is that the forearm becomes more rigid. Our brains chose this approach whenever facing an environment that is hard or impossible to predict. For example when holding the rudder of a sailboat and you need to keep the heading as steady as possible. In our force-field experiment, the presence of aftereffects is sufficient to rule out this type of control response: subjects were not merely becoming stiffer but they learned to counteract forces that—by design—were not random but predictable. The second possibility, instead, would be consistent with our observation of aftereffects. Equation (3.34) does not imply that $\mathbf{\Delta}$ must be a representation of the force field on the left-hand side. The applied forces were designed to depend of the state of motion of the arm, $[\mathbf{q}^T, \dot{\mathbf{q}}^T]^T$. However, to recover the initial trajectory it would be sufficient for Δ to be a pure function of time,

$$\Delta(t) = \mathbf{J}^{\mathrm{T}}(\mathbf{q}_{\mathrm{A}}(t))\cdot\mathbf{B}\cdot\mathbf{J}(\mathbf{q}_{\mathrm{A}}(t))\cdot\dot{\mathbf{q}}_{\mathrm{A}}(t). \quad (3.36)$$

This is simply a "motor tape" that plays back the forces encountered along the desired trajectory. The motor tape does not contain any information about the dependence of the external force upon the state variables. A second experiment was needed to address this possibility.

Michael Conditt, Francesca Gandolfo, and Sandro Mussa-Ivaldi (1997) tested the hypothesis of a motor tape with a simple experiment. They trained a group of volunteers to make reaching movements in a force field. The duration of each movement was typically less than one second. Therefore, after learning to reach targets in the field, the motor tapes would have the typical duration of each movement and would be "specialized" to reproduce selectively the force sequences encountered in the training phase. The hypothesis then predicts that subjects would not be able to compensate for forces encountered while performing different movements, like drawing a circle, a triangle or figure-eight pattern. Even if these drawing movements take place in the same region of space and velocity ranges as the reaching movements, the states of motions are traversed in different temporal sequences. The drawing movements last at least twice as long as the reaching movements. Nevertheless, the subjects of this experiment, after training to reach in the force field, were perfectly capable to compensate for the disturbing forces while drawing circles and other figures. A second group of subjects, instead of practicing reaching movements, practiced drawing circles in the force field. The drawing movements of these subjects after training were not distinguishable from the drawing movements of subjects in the first group, who trained with reaching. These findings are sufficient to rule out the idea that subjects learned a motor tape and provide instead additional support to the concept that the brain constructed, through learning a computational representation of the dynamics as a dependence of the actions to be produced by the control system—in this case the function $\Delta(\mathbf{q},\dot{\mathbf{q}})$—upon the state of motion of the limb. While the represented entity does not have to be a force in the Newtonian sense, the resulting behavior is consistent with the representation of movement dynamics as formulated by classical laws.

3.6 Another Type of State-Based Dynamics: Motor Learning

The force-field experiments demonstrate the keen ability of the brain to deal with dynamics and with the concept of "state." In mechanics, the state is a minimal set of variables that is sufficient to predict the future evolution of a system. In a single motor act, such as reaching for an object, the state in question is the mechanical state of the limb. Because the laws of classical mechanics govern the movement, the state is a collection of position and velocity variables. There is another kind of state that is relevant to motor behavior. Learning experiments have demonstrated that it is possible to predict how knowledge and control evolve through time based on past experience and of incoming information. Learning itself can therefore be observed as a dynamical system. This perspective will be developed in the remainder of this book. Here, we briefly introduce some of the key concepts of this approach. Let us start again from Newton's equation for a point mass, $F = m\ddot{x}$. We derive the velocity of the point mass by integrating once this equation:

$$\dot{x}(t) = \dot{x}(t_0) + \int_{t_0}^{t} \frac{F}{m} dt'. \tag{3.37}$$

Instead of looking at the velocity as a continuous function of time, we may take samples at discrete instants of time. Then, provided that these instants are separated by small time intervals, δt, we may use Newton's equation to derive as an approximation the velocity at the next instant $\dot{x}^{(n+1)}$ from the velocity $\dot{x}^{(n)}$ and the external force $F^{(n)}$ at the current instant. This is the *difference equation*:

$$\dot{x}^{(n+1)} = \dot{x}^{(n)} + \frac{F}{m}\delta t. \tag{3.38}$$

If the mechanical work in an infinitesimal displacement is an exact differential, then the force is the gradient of a potential energy function,

$$F(x,t) = -\frac{\partial U(x,t)}{\partial x}. \tag{3.39}$$

We now can rewrite the difference equation for a point mass in a potential field as

$$\dot{x}^{(n+1)} = \dot{x}^{(n)} - \frac{\partial U^{(n)}}{\partial x} \cdot \alpha, \tag{3.40}$$

with

$$\alpha = \frac{\delta t}{m}.$$

Opher Donchin, Joseph Francis, and Reza Shadmehr (2003) proposed to describe motor learning with a differential equation that has a similar form. However, in this case, the state variable in question is not the velocity or the position of the arm, but the internal model of the force field in which the arm is moving. They represented the internal model as a collection of basis functions over the state space of the arm. For example, each basis function could be a Gaussian centered on a preferred velocity. But this is certainly not the only possible choice. Referring to equation (3.34), let us represent the force field model in generalized coordinates as

$$\Delta(\mathbf{q},\dot{\mathbf{q}}) = \mathbf{W} \cdot \mathbf{g}(\mathbf{q},\dot{\mathbf{q}}), \tag{3.41}$$

where $\mathbf{g} = [g_1(\mathbf{q},\dot{\mathbf{q}}), g_2(\mathbf{q},\dot{\mathbf{q}}), \ldots, g_m(\mathbf{q},\dot{\mathbf{q}})]^T$ is a collection of m scalar basis functions and W is a $n \times m$ matrix of coefficients. For a two-joint arm, $n = 2$. But, more generally, n is the dimension of the configuration space or, equivalently, the number of degrees of freedom under consideration. Then, the matrix **W** can also be written as an array of m n-dimensional vectors, one for each basis function

$$\mathbf{W} = [\mathbf{w}_1, \mathbf{w}_2, \ldots, \mathbf{w}_m], \tag{3.42}$$

and the basis functions are scalars that modulate the amplitude of the w vectors over regions of the workspace, in a way similar to the receptive field of a sensory neuron. As we perform a series of reaching movements in the force field, $\mathbf{q}^{(1)}(t), \mathbf{q}^{(2)}(t), \ldots, \mathbf{q}^{(k)}(t), \ldots$ we experience a corresponding sequence of "force trajectories":

$$\mathbf{D}^{(i)}(t) = \mathbf{J}^{\mathrm{T}}(\mathbf{q}^{(i)}(t)) \cdot \mathbf{B} \cdot \mathbf{J}(\mathbf{q}^{(i)}(t)) \cdot \dot{\mathbf{q}}^{(i)}(t). \tag{3.43}$$

Suppose that on the ith movement, we intended to move along the trajectory $q_A(t)$. Then, based on the internal model that we had at that point, we expected to experience a force

$$\Delta^{(i)}(t) = \Delta^{(i)}(\mathbf{q}_{\mathbf{A}}(t), \dot{\mathbf{q}}_{\mathbf{A}}(t)) = \mathbf{W}^{(i)} \cdot \mathbf{g}(\mathbf{q}_{\mathbf{A}}(t), \dot{\mathbf{q}}_{\mathbf{A}}(t)). \tag{3.44}$$

Here, we have that the variable portion of the internal model is the coefficient matrix W. We assume that the basis functions remain fixed through time. This, of course is for conceptual convenience. More complex (and also more realistic) models of learning would allow for variations in all terms of **Δ**. Most importantly, at the end of the ith movement we have a discrepancy—an error—between what we experienced and what we expected to experience. This is a function

$$\mathbf{e}^{(i)}(t) = \mathbf{D}^{(i)}(t) - \Delta^{(i)}(t). \tag{3.45}$$

If we integrate the Euclidean norm of this function along the movement, we obtain a scalar quantity, a real number

$$E^{(i)}(\mathbf{W}^{(i)}) = \int_0^T \mathbf{e}^{(i)T}(t) \cdot \mathbf{e}^{(i)}(t)\, dt. \tag{3.46}$$

This is an overall error that we have experienced across the movement and this error depends on the parameters of the internal models. The purpose of learning is to get better at what we are doing. In this case it means to make the expectation error as small as possible. A straightforward way to do so is by changing the parameters of the internal model so as to minimize the error on the next trial. If we expand the argument of the integral (3.46), we see that it is a quadratic form of the parameters:

$$\mathbf{e}^{(i)T} \cdot \mathbf{e}^{(i)} = \left(\mathbf{D}^{(i)} - \Delta^{(i)}\right)^T \cdot \left(\mathbf{D}^{(i)} - \Delta^{(i)}\right) = \mathbf{D}^{(i)T} \cdot \mathbf{D}^{(i)} - 2 \cdot \mathbf{D}^{(i)T} \cdot \mathbf{W}^{(i)} \cdot \mathbf{g} + \mathbf{g}^T \cdot \mathbf{W}^{(i)T} \cdot \mathbf{W}^{(i)} \cdot \mathbf{g}. \tag{3.47}$$

Therefore, the error function $E^{(i)}(W^{(i)})$ has a global minimum that we reach by changing W along the gradient:

$$\mathbf{W}^{(i+1)} = \mathbf{W}^{(i)} - \frac{\partial \mathbf{E}^{(i)}}{\partial \mathbf{W}^{(i)}} \cdot \alpha. \tag{3.48}$$

This is a difference equation that tells us how the internal model of the field changes in time. We compare it with equation (3.40) that describes the motion of a point-mass in a potential field. The two equations have different order. Newtonian mechanics is of the second order while this simple learning model is of the first. Both equations tell us how a state variable changes in time under the action of an external input. In the case of the mass, the input is the mechanical force, which the gradient of a potential energy function. In the case of learning, the "driving force" is the gradient of the prediction error. Perhaps, the most important common feature of

Newtonian mechanics and the description of learning as a dynamical system is that both allow us to connect theory with experiment. This will be the leading thread of the chapters that follow.

Summary

Space maps are relevant not only to navigation but also to movements in general. Here, we consider the maps that our brains must form to manipulate the environment. The kinematics maps relate the motions of the hand to the motions of the arm and come in two forms: direct kinematics, from arm's joint angles to hand position, and inverse kinematics, from hand position to arm's joint angles. These maps are nonlinear and, as a consequence, a straight line in hand space maps into a curve in arm space and vice versa. This reflects the fact that a space of angular variables is inherently curved, while the Euclidean space where the hand moves is flat. Observation of human movements over a plane demonstrated that the hand tends to follow rectilinear trajectories.

Observations with altered kinematics and with force perturbations demonstrated that the kinematics of arm movement is not a collateral effect of the brain attempting to economize or simplify the generation of muscle forces. Instead, to produce naturally observed movements, the brain must effectively solve a complex problem of dynamics. The solution of this problem does not necessarily involve any explicit representation of Newtonian forces, but it requires the ability to relate the motor commands that reach the muscles to the consequences in terms of changes in state of motion of the arm. This connection is represented in the arm's dynamics equations, which contain terms representing the inertial properties of the arm and terms representing the viscoelastic properties of the muscles. The motor commands need not be a force and may represent instead the static equilibrium point expressed by the neuromuscular system. This is the position at which the arm would eventually settle under the influence of opposing muscle forces. Controlling the motion of this equilibrium point for producing a desired movement of the hand is equivalent to solving an inverse dynamic problem.

The first experimental evidence for internal models of dynamics derived from observing how subjects learned to perform reaching movements in the presence of deterministic force fields. These fields apply to the hand a force that depends upon the hand's state of motion, that is, its position and velocity. The observations were consistent with the subjects gradually developing the ability to predict and cancel the applied forces along the movement trajectories. The sudden removal of the field results in aftereffects of learning, that is, in a perturbation that mirrors the initial effect of the unexpected application of the field. The aftereffects, as well as the pattern of learning generalizations, demonstrate that the subjects learn to predict the functional form of the perturbation, which effectively corresponds to a change in the controlled dynamical system.

We introduce the concept that motor learning is a dynamical process whose state is the associated internal model of the body and its environment. In its simplest form, learning to move in a force field is described by a first-order differential equation whose state variable is the model of the field and whose input is the gradient of the prediction errors. In this respect, movement and learning are two concurrent processes operating over different time scales.

4 Sensorimotor Integration and State Estimation

At a cocktail party, one of the toughest jobs (at least from a motor control standpoint) belongs to the fellow who brings the tray with the drinks. As he holds the tray and you pick up the glass, he needs to compensate for the reduced mass on the tray and not spill the remaining drinks. To convince yourself that that is indeed a tough task, try the following experiment. Put out your hand with the palm side up and place a moderately heavy book on it. Now have a friend stand in front of you and have him or her pick the book up off your hand. You will note that even though you can see your friend reach and pick up the book, you cannot hold your hand perfectly steady—it invariably moves up when the book is removed. Now replace the book on your palm and go ahead and pick it up yourself. When you pick up the book, the hand that used to hold the book remains perfectly still.

This simple experiment suggests that when you send commands to your arm to pick up the book, your brain predicts the exact moment that the mass will be removed off the resting arm and the exact mass of the book, and then it reduces the activity in the muscles that are producing force to hold up the book. In contrast, when someone else picks up the book, you have to rely on your sensory system rather than your predictions. Because of the inherent delay in the sensory system, you are invariably late in reducing the activity of those same muscles. Therefore, the brain appears to predict that lifting the book by one arm has consequences on the other arm, and it compensates before the arrival of sensory feedback. (This is the reason why you should let the waiter pick up the glass and hand it to you.)

The general idea is that our body is a multisegmented structure in which motion of any one segment has consequences on the stability of other segments. To maintain stability, the brain needs to be able to predict how motor commands to one segment affect the states of all other segments. For example, consider an experiment by Paul Cordo and Lewis Nashner (1982) in which people were placed in a standing posture and were asked to pull on a doorknob that was attached to a rigid wall (figure 4.1). Pulling on the knob involves activation of the biceps muscle. However, doing so would not only result on a pulling force on the knob, but also a pulling force on

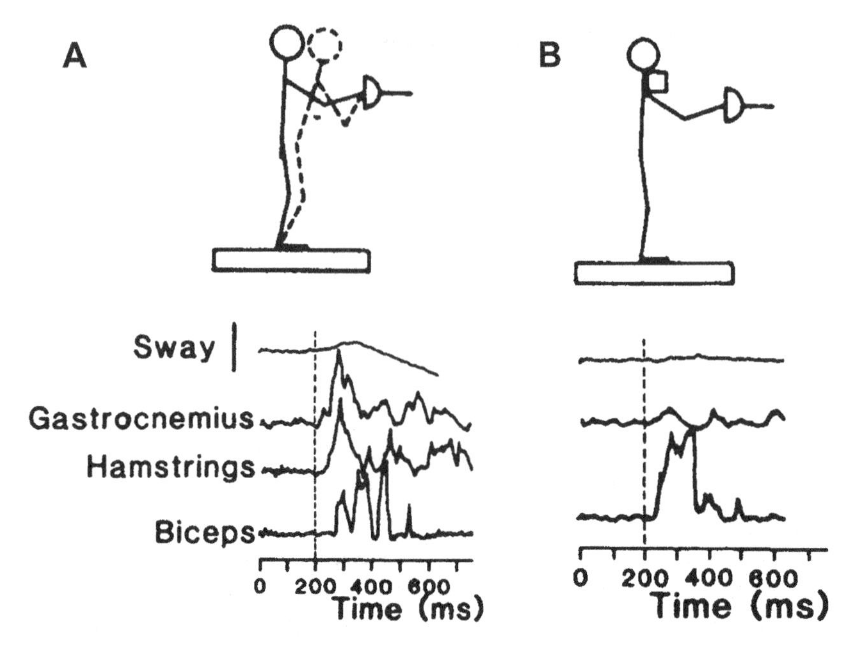

Figure 4.1
Subject was instructed to pull on a knob that was fixed on a rigid wall. (A) EMG recordings from arm and leg muscles. Before biceps is activated, the brain activates the leg muscles to stabilize the lower body and prevent sway due to the anticipated pulling force on the upper body. (B) When a rigid bar is placed on the upper body, the leg muscles are not activated when biceps is activated. (From Cordo and Nashner, 1982.)

your body. If you are not careful, you will end up hitting your forehead on the wall. To maintain stability, you need to activate muscles that resist bending (i.e., flexion) of the ankle and knee joints. Indeed, Cordo and Nashner found that people activated their hamstring (a muscle that produces an extension torque on the ankle) and gastrocnemius (which produces extension torque on the knee) muscles just before they activated their biceps muscle (figure 4.1A). That is, the brain stabilizes the leg just as it pulls on the knob.

This may seem like a "hard-wired" reflex, but it is not. Consider what happens when a rigid bar is placed in front of the chest, preventing the body from swaying forward. In this case, a pull on the doorknob would not cause a sway of the body no matter what you do with your knee and ankle muscles. Indeed, with the rigid bar in place, people no longer activated the hamstring and gastrocnemius muscles as they pulled on the knob (figure 4.1B).

These data suggest that as our brain plans and generates motor commands, it also predicts the sensory consequences and acts on the predicted consequences. When standing upright without a rigid bar to lean against, pulling on the knob will flex the knee and ankles. If we can predict that this is the sensory consequence of pulling on the knob, we can act on it by activating muscles that resist this flexion (the extensors). When we have the rigid bar to lean against, the same pull on the knob will no longer flex the knee and ankles. If we can predict this, we need to do nothing, which seems consistent with the data in figure 4.1.

More direct evidence for the idea that the brain predicts the sensory consequences of motor commands comes from the work of Rene Duhamel, Carol Colby, and Michael Goldberg (1992). They trained monkeys to fixate a light spot and make a saccade to it whenever it jumped from one location to another. They recorded from cells in the posterior parietal cortex (PPC), an area in the brain in which cells have receptive fields that depend on the location of the stimulus with respect to the fixation point. Figure 4.2A illustrates response of a cell that had its receptive field

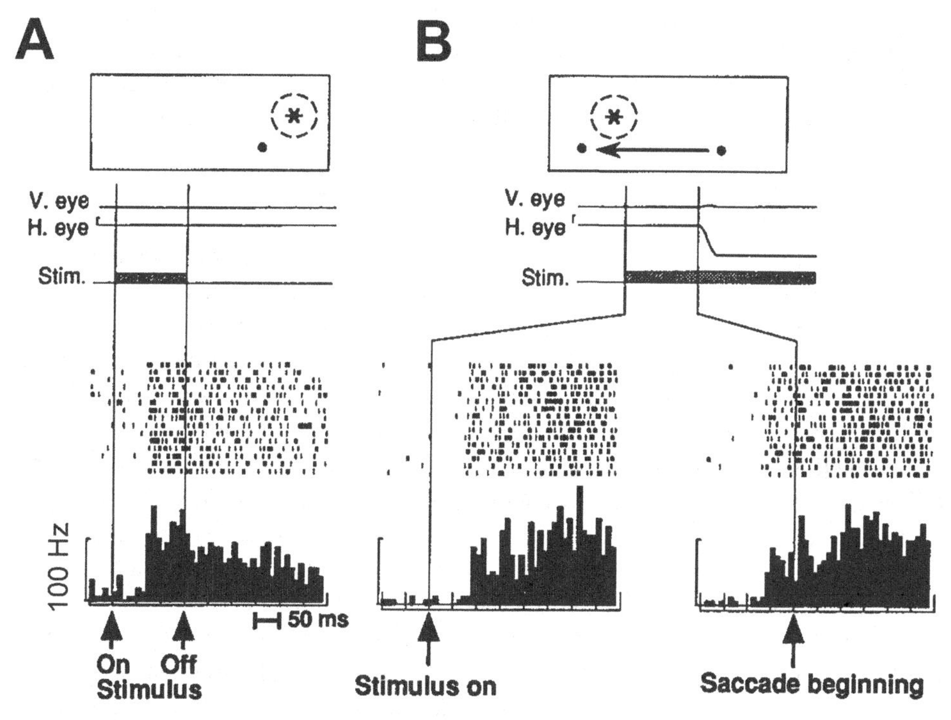

Figure 4.2
Effect of eye movement on the memory of a visual stimulus. In the top panel, the filled circle represents the fixation point, the asterisk indicates the location of the visual stimulus, and the dashed circle indicates the receptive field a cell in the LIP region of the parietal cortex. (A) Discharge to the onset and offset of a visual stimulus in the cell's receptive field. H. eye, horizontal eye position; Stim, stimulus; V. eye, vertical eye position. (B) Discharge during the time period in which a saccade brings the stimulus into the cell's receptive field. The cell's discharge increased before the saccade brought the stimulus into the cell's receptive field. (From Duhamel, Colby, and Goldberg, 1992.)

to the upper right of fixation. When the stimulus was turned on in the receptive field of this cell (circular dashed line), the cell responded after about 60 ms, which is a typical delay period. Next, they had the monkey make a saccade by turning off the fixation point at right and redisplaying it at left (figure 4.2B). They placed the stimulus in the upper right part with respect to the second fixation point. In this way, when the animal was looking at the first fixation point (the one on the right), there was nothing in the cell's receptive field. When the animal looked at the second fixation point (the one on the left), the stimulus fell in the cell's receptive field. However, if the brain predicts the sensory consequences of the motor commands, then one of the consequences of the command to move the eyes from one fixation to another is that after the completion of the movement, the stimulus will fall in the cell's receptive field. Thus, the cell might fire in anticipation of this event, rather than in passive response to the sensory input. Indeed, the cell's response was around the start of the saccade, rather than after its completion (figure 4.2B; note that saccades take about 50 ms to complete). Therefore, some part of the brain predicted that as a consequence of the motor commands that move the eyes, a light stimulus would fall in the receptive field of this cell, and it began to fire in apparent anticipation.

4.1 Why Predict Sensory Consequences of Motor Commands?

Why should the brain predict the sensory consequences of motor commands? In the experiment where you lifted the book off your hand, the clear advantage of making sensory predictions is that the brain does not have to wait for the sensory measurements to know that it needs to shut off the muscles that are holding up the book. The delay in sensory measurements is long enough that it can cause stability problems. Relying on predictions, rather than delayed measurements, allows one to overcome this delay. However, in the case of the saccade shown in figure 4.2, it may be unclear why the brain should predict that the sensory consequences: Why should a visually responsive cell be activated in anticipation of the sensory stimulus, as well as in response to that sensory stimulus? After all, the visual stimulus will appear in the receptive field shortly after the saccade. Why predict this event?

One possibility is that our perception—that is, our ability to estimate the state of our body and the external world—is always a combination of two streams of information: one in which our brain predicts what we should sense, and one in which our sensory system reports what was sensed. The advantage of this is that if our expectations or predictions are unbiased (i.e., their mean is not different from the "true" state), then our perception (and the decisions that are made based on that perception) will be better than if we had to rely on sensory measurements alone. In a sense, our perception will be more accurate (e.g., less variable), if we combine what we predicted with what our sensory system measured. This improved accuracy in perception is a fundamental advantage of making predictions about the sensory consequences of motor commands.

Although this may seem like a fairly new idea, it was first proposed in the eleventh century by the Arab scientist Ibn al-Haytham (known in Europe as Alhazen) in his *Book of Optics*. He was considering the moon illusion, the common belief that the moon looks larger when it is near the horizon. Aristotle and Ptolemy had thought that this was due to a magnification caused by Earth's atmosphere (refraction theory). However, this is not the case. If you were to measure the size of the moon by taking a picture, you would measure a width that is 1.5 percent smaller at the horizon than straight up in the sky. That is, the image that falls on your retina is actually smaller when you are looking at the moon on the horizon (this is because the moon is actually farther away from you by half of the Earth's diameter when it is at the horizon). So despite the fact that the moon at the horizon produces a visual image that is smaller than when it is overhead, we perceive it to be bigger at the horizon. Ibn al-Haytham argued that the moon looked bigger at the horizon because perception was occurring in the brain, and not in the eyes. He recognized that our perception of size depends not only on the size of the image on our retina, but also on our estimate of the object's distance. At the horizon, the brain has cues like trees and buildings to judge distance of objects, whereas in the sky above, these cues are missing. He argued that presence of these cues affected the brain's estimate of distance of the moon, making us perceive it to be much farther away at the horizon than overhead. Perhaps we "see" the moon as being larger at the horizon because we believe it is much farther away at the horizon than overhead.

(The term "belief" is used loosely here. As we will see in the next chapter, different parts of our brain may have differing beliefs about size of a single object. For example, we may verbally indicate that object A is bigger than object B, reflecting the belief of our perceptual system, but when we go to pick up object B, move our fingers apart more for it than when we go to pick up object A, reflecting the belief of our motor system. The root cause of these apparently different beliefs about the property of a single object is poorly understood. It is possible that it has something to do with the fact that each part of the brain may specialize in processing a different part of the sensory data. For example, by focusing on the background of an image, some part of our visual system may get fooled and form an illusion, whereas by focusing on the foreground of an image, another part of our visual system may not form this illusion. So there may not be a single "belief" in our brain about some issue, but instead multiple beliefs. The belief that you express may depend on what part of your brain was queried.)

In 1781, Immanuel Kant, in his theory of *idealism*, claimed that our perceptions are not the result of a physiological process in which, for example, the eyes faithfully transmit visual information to the brain, but rather, our perceptions are a result of a psychological process in which our brain combines what it already thinks, believes, knows, wants to see, with the sensory information to form a perception (Gilbert, 2006). In his *Critique of Pure Reason*, he writes, "The understanding can intuit nothing, the senses can think nothing. Only through their union can knowledge arise."

If we follow this line of reasoning and return to our example of predicting the sensory consequences of the motor commands that move the eyes in a saccade (figure 4.2B), we might guess that during the postsaccadic period, the brain combines what it predicted with what it currently senses. The combination of the two streams of information would allow it to sense the world better than if it only had the sensory information stream. That is, it would be able to 'see' the stimulus better because it had two sources of information about it, rather than just one.

Siavash Vaziri, Jörn Diedrichsen, and Reza Shadmehr (2006) tested this idea. In their control condition (figure 4.3A), subjects fixated a visual target and reached to it. In their static condition, the target appeared in the peripheral vision. The ability to estimate the location of objects that appear in peripheral vision is poor, and so the standard deviation of the reach endpoints was higher for these stimuli in the peripheral vision (figure 4.3B). Therefore, in this experiment, reach variability was a proxy for the goodness with which the brain could estimate the location of the visual target. When the target was in the periphery, subjects could not localize it as well as when they looked straight at it. Now, what if one could predict the location of the target? Based on our theory, the brain should be able to localize it better, and this should result in less variable reaches to that target.

In the remap condition, subjects looked at the target, and then looked away before they reached to the target. When they looked away, the target had disappeared. Presumably, during the saccade the brain predicts the new location of the target with respect to fixation (this is called remapping). So after the saccade completes, it can rely on the predicted target position to reach. The endpoint variance in the remapped condition was much less than in the static condition (figure 4.3B). The crucial test was in the combined condition in which subjects looked away from the target, and in the postsaccadic period the target reappeared for a brief period of time. If the brain combined what it predicted about target position with the actual sensory feedback, then reach endpoint variance in the combined condition should be better than both the remap and the static conditions (figure 4.3B). Indeed, this was the case, consistent with the idea that the brain predicted the sensory consequences of the saccade (remapping of the target) and then combined this prediction with the postsaccadic visual information to estimate the actual position of the target.

Therefore, by predicting the sensory consequences of motor commands, the brain not only can overcome delay in sensory feedback, but perhaps more important, it

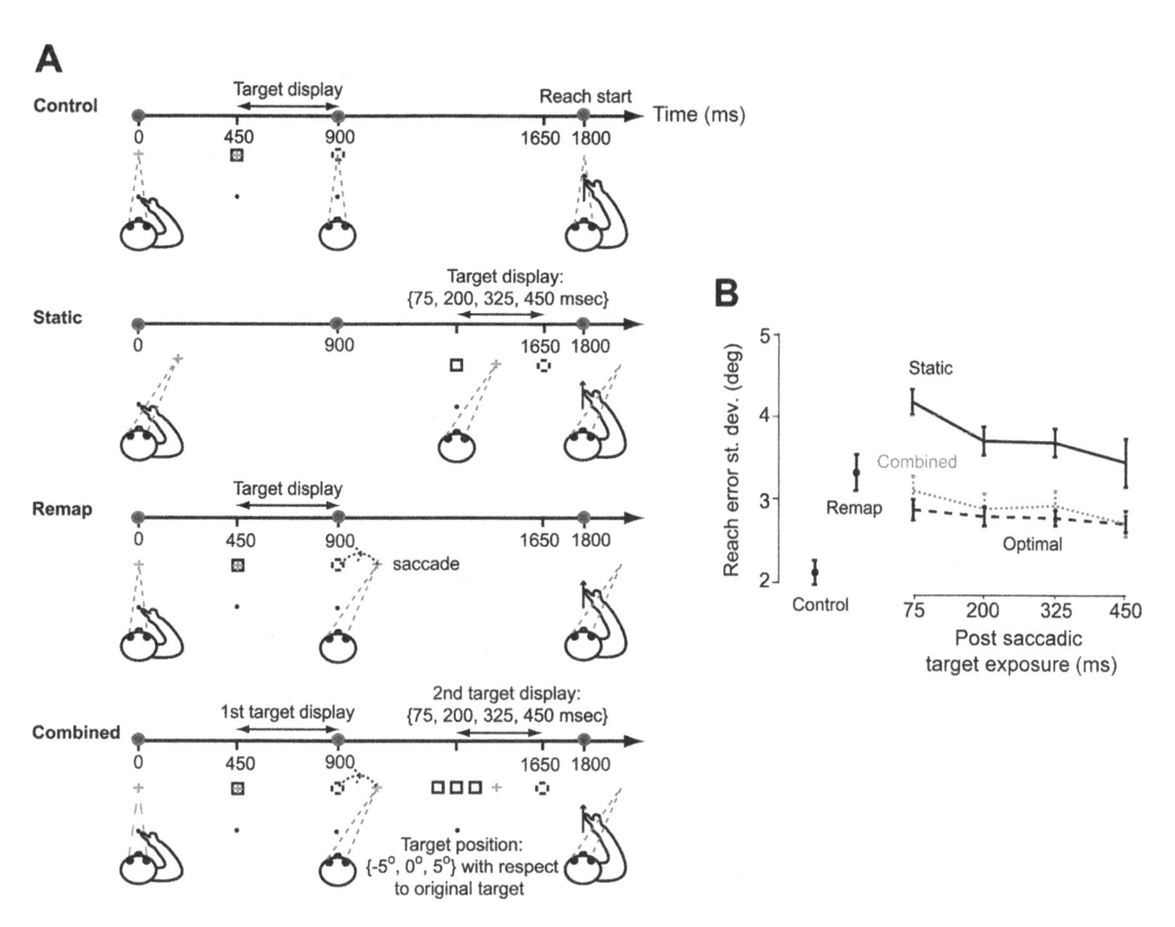

Figure 4.3
Subject reached to the location of a visual stimulus. (A) In the Control condition, subject fixated the target stimulus for a brief period of time, the target disappeared, and after a delay period a reach was made. In the Static condition, subject fixated a secondary stimulus when the target stimulus appeared to one side. In the Remap condition, the subject fixated the target stimulus, then made a saccade to a secondary stimulus (at which point the target stimulus was erased). In the Combined condition, the subject fixated the target stimulus, made a saccade to a secondary stimulus, and then during the delay period was again shown the target stimulus after completion of the saccade. (B) The standard deviation of the reach, that is, a measure of uncertainty about the location of the target stimulus. Optimal refers to a weighted combination of static and remap conditions. (From Vaziri, Diedrichsen, and Shadmehr, 2006.)

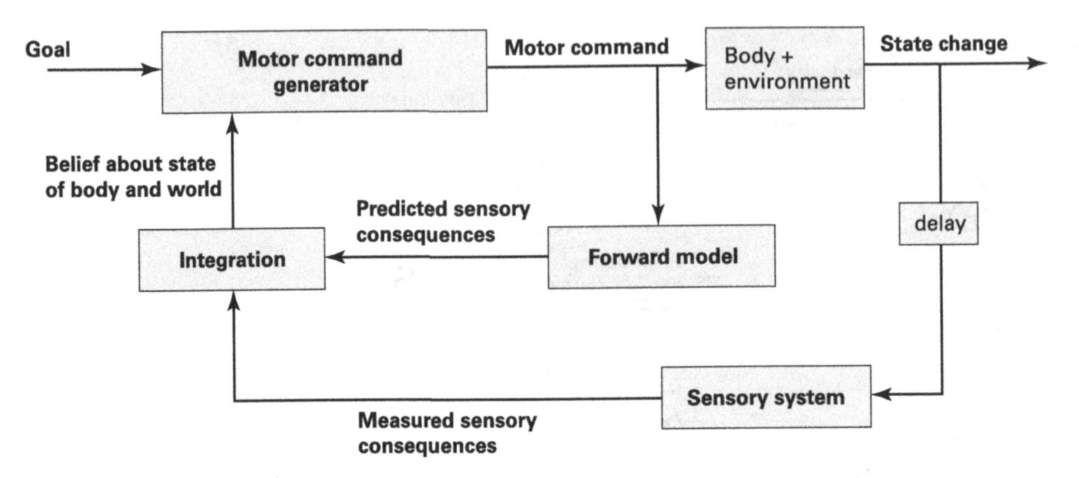

Figure 4.4
Motor commands change the states of our body and the environment around us. These states are transduced by our sensory system and become the measured sensory consequences of our motor commands. As we generate motor commands, the brain also predicts the sensory consequences via an internal model that is called a "forward model." By combining the predicted and measured sensory consequences, we form a belief about the states of our body and the environment. This belief is a reflection of both our predictions and our observations.

can actually sense the world better than is possible form sensory feedback alone. The latter comes about when our brain combines what it has predicted with what it has measured. A diagram that summarizes the idea of predicting the sensory consequences of motor commands is provided in figure 4.4. An internal model that predicts the sensory consequences of motor commands is known as a *forward model.*

4.2 Disorders in Predicting the Sensory Consequences of Motor Commands

The central idea is that our perception is based on a combination of two streams, one that arises from the motor system (predicting the sensory consequences), and the other that arises from the sensory system (measuring the sensory consequences). If our brain could not accurately predict sensory consequences of our motor commands, then we would not be able to sense the world around us in a normal way. An example of this is patient RW, a thirty-five-year-old man who was described by Thomas Haarmeier, Peter Thier, Marc Repnow, and Dirk Petersen (1997). RW suffered a stroke in a region covering parts of the parietal and occipital cortex, centered on an area that contains the vestibular cortex, a location in which cells are sensitive to visual motion. RW complained of vertigo only when his eyes tracked visual objects, but not when his eyes were closed. He explained that when he was watching his son run across a field (a condition in which his eyes smoothly moved to follow his son), he would see the boy running, but he would also perceive the rest of the visual scene (e.g., the trees) smoothly moving in the opposite direction.

Haarmeier and colleagues conjectured that when RW moved his eyes, his brain was unable to predict the sensory consequences of the oculomotor commands. As his eyes moved to follow his son, the trees moved in the opposite direction on his retina. The healthy brain predicts that moving the eyes will have the sensory consequence of shifting the image of the stationary world on the retina. We do not perceive this shifting image as real motion of the world because we predict it to be a consequence of motion of our eyes. In RW, perhaps his vertigo was a symptom of his brain's inability to predict such sensory consequences.

To test this conjecture, Haarmeier and colleagues had RW sit in front of a computer monitor and keep his eyes on a moving cursor (figure 4.5). As the cursor moved smoothly from left to right, random dots were displayed for 300 ms. In some trials the random dots would stay still, and in other trials the dots would move to the right or left with a constant speed. On each trial they asked RW and some healthy volunteers to guess whether the random dots were moving to the left or right. From the response that they recorded the authors estimated the speed of motion of the random dots for which subjects sensed it to be stationary. For healthy volunteers, the speed of subjective stationarity of the random dots was near zero, no matter what the speed of the moving cursor that they were looking at. That is, regardless of whether the eyes moved quickly or slowly, healthy people perceived a stationary collection of dots as stationary (the unfilled circles in figure 4.5C). However, RW saw the collection of dots as being stationary only when the dots moved at the same speed as the eyes (the filled symbols in figure 4.5B). That is, for RW an object was stationary only if during the movement of the eye, its image remained stationary on the retina.

You do not need to have a brain lesion to get a feel for what RW sees when he moves his eyes. Take a camera, aim it at a runner, and try to pan it so that the image of the runner stays at the center of the picture. As you are moving the camera, take a picture. That picture will show a sharply focused runner but a blurry background that appears to be moving in the opposite direction. However, when you are following the runner with your naked eyes, the background appears perfectly still. The reason is because your brain predicts the background image shift that should take place on the retina as you move your eyes and combines it with the actual shift. By combining the observed and predicted images, the parts that agree must have been stationary, and parts that disagree must have moved.

In 1996, Chris Frith put forth the hypothesis that in schizophrenia, the symptoms of delusions—for example, auditory hallucinations in which the patient hears voices or has beliefs about someone else's guiding his or her actions—are potentially related to a disorder in the brain's ability to interpret its own inner voice or covert

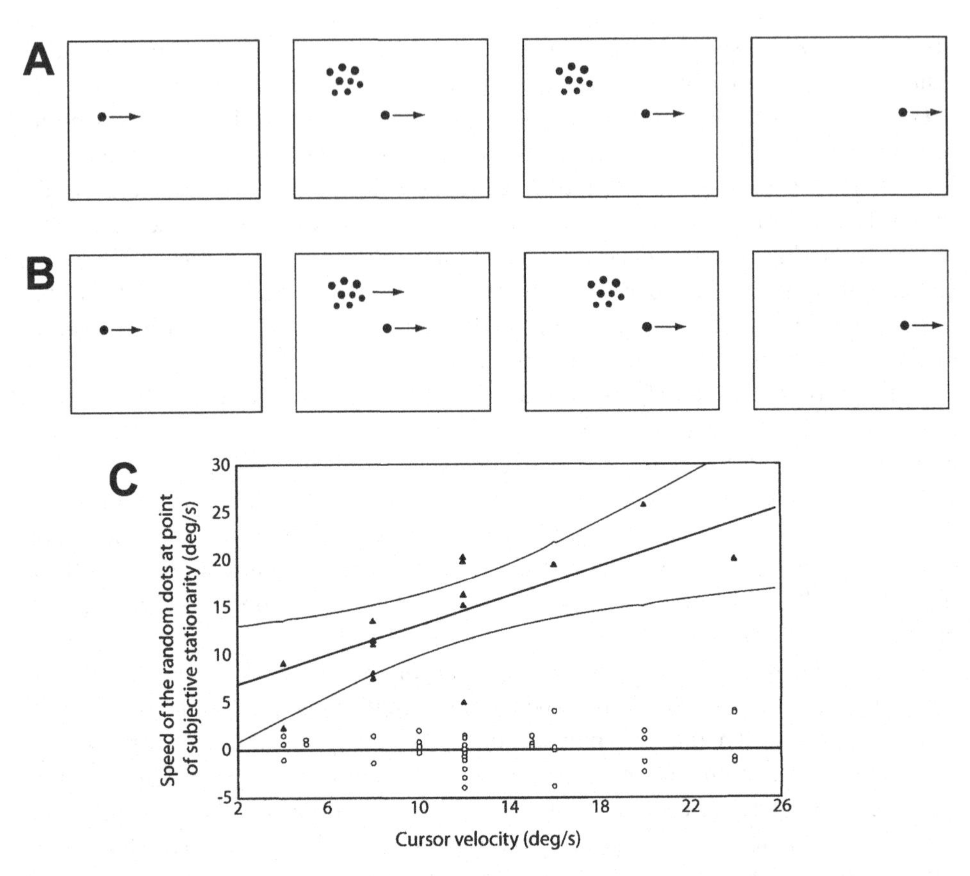

Figure 4.5
Subject looked at a moving cursor while a group of dots appeared on the screen for 300 ms. In some trials the dots would remain still (A) while in other trials they would move together left or right with a constant speed (B). Subject indicated the direction of motion of the dots. From this result, the authors estimated the speed of subjective stationarity, that is, the speed of dots for which the subject perceived them to be stationary. The unfilled circles (C) represent performance of control subjects. Regardless of the speed of the cursor, they perceived the dots to be stationary only if their speed was near zero. The filled triangles represent performance of subject RW. As the speed of the cursor increased, RW perceived the dots to be stationary if their speed was near the speed of the cursor. (From Haarmeier et al., 1997.)

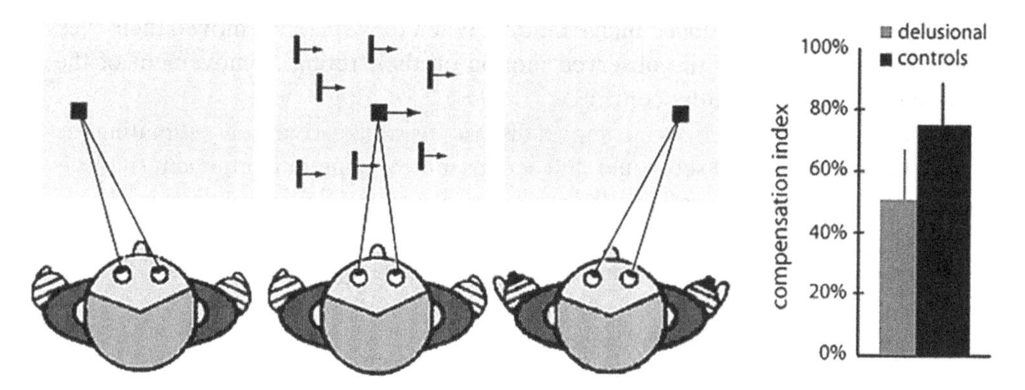

Figure 4.6
Disorders of agency in schizophrenia relate to an inability to compensate for sensory consequences of self-generated motor commands. In a paradigm similar to that shown in figure 4.5, volunteers estimated whether during motion of a cursor the background moved to the right or left. By varying the background speed, at each cursor speed the experimenters estimated the speed of perceptual stationarity, that is, the speed of background motion for which the subject saw the background to be stationary. They then computed a compensation index as the difference between speed of eye movement and speed of background when perceived to be stationary, divided by speed of eye movement. The subset of schizophrenic patients who had delusional symptoms showed a greater deficit than control in their ability to compensate for sensory consequences of self-generated motor commands. (From Lindner et al., 2005.)

action (Frith, 1996). He and his colleagues formalized this idea in terms of a problem in which the brain had an incorrect forward model (Frith, Blakemore, and Wolpert, 2000). This would result in the patients mistaking the sensory consequences of their own actions as consequences of actions of others, that is, a misattribution of the cause of sensory events.

An interesting test of this idea was performed by Axel Lindner and colleagues (2005), who essentially repeated the experiment that they had earlier performed on patient RW on a group of schizophrenic patients. In this experiment, subjects watched a red dot that moved from left to right at a constant velocity (figure 4.6). They then presented a moving background for 200 ms during the sweep of the target and asked the subject to report the direction of the background motion. By varying the background speed, they determined the velocity that produced about equal perception of rightward or leftward motion. At this velocity of perceptual stationarity, they compared the speed of the background image (which ideally should be zero) with the speed of the eye movements (plotted in bar graph in figure 4.6). In this figure, 100 percent implies that the subjects saw the background as stationary when in fact it was stationary. Healthy subjects had a less than 100 percent performance, and this was attributed to the very limited time for which the background was presented. However, for the schizophrenic patients who suffered from delusional symptoms, the performance was significantly worse: They had a harder time compensating for self-induced image motion. When these patients moved their eyes, they attributed more of the observed motion on their retina to movement of the external world than healthy controls.

Of course, schizophrenia is a complex disease for which accurately estimating the sensory consequences of self-generated actions may be a minor component (if this is not the case, then why is patient RW not schizophrenic?). Nevertheless, it is curious that for those patients who suffer from delusional symptoms, there is a tendency to have a motor disorder in perceiving self-generated sensory consequences.

4.3 Combining Predictions with Observations

To combine two streams of information, one needs to apply a weighting to each stream. In principle, the weight should be higher for the more reliable information source. In the experiment shown in figure 4.3, the two sources of information are the remapped target (i.e., predicted target location), and the postsaccadic target (i.e., observed target location). Vaziri et al. (2006) manipulated the reliability of the postsaccadic information by presenting the target for either a short or long period of time. The idea was that the longer the information was available, its reliability would increase, and so the weight that the brain might assign to it should increase. Indeed, with increased postsaccadic target exposure in the combined condition, endpoint variance decreased, suggesting that the brain increased the weighting assigned to the observed sensory information source.

The basic idea that emerges is that our estimate of the state of the world is a combination of two sources of information: what we predicted, and what we observed. Konrad Kording and Daniel Wolpert (2004) varied the reliability of these two hypothetical sources of information and tested the idea that perception was a weighted combination of the two sources. They first trained subjects to reach to a goal location by providing them feedback via a cursor on a screen (the hand was never visible). As the finger moved from the start position, the cursor disappeared. Halfway to the target, the cursor reappeared briefly (figure 4.7A). However, its position was, on average, 1 cm to the right of the actual finger position, but on any given trial the actual displacement was chosen from a Gaussian distribution. The objective was to produce a movement that placed the cursor inside the target.

If you were a subject for this task, you might start by moving the finger straight to the target. You would note that in the middle of the movement the cursor appears about 1 cm to the right, so you would correct by moving the finger a little to the left. After some practice, you would learn that when you produce motor commands that move the hand slightly to the left, you should, on average, see the cursor straight ahead. Because the location of the cursor is probabilistic, the confidence that you have about this predicted sensory consequence of your motor command should be described by the variance of the Gaussian distribution that describes the displacement of the cursor (figure 4.7B, top row). In this way, the experiment controlled the confidence that the brain should have in predicting the sensory consequences of its motor commands.

To control the confidence that the brain should have regarding sensory measurements, they added noise to the display of the cursor: the cursor was displayed as a cloud of dots. This induced uncertainty. On some trials the cursor was shown clearly so the uncertainty regarding its position was low. In other trials the uncertainty was high as the cursor was hidden in a cloud of noise. The idea was that on a given trial, when the subject observes the cursor position midway to the target, they will correct based on two sources of information: what they observed on this trial, and what they predicted regarding where the cursor should be. For example, if on a given trial they see the cursor at 2 cm (figure 4.7B, middle row), they should combine this observation with their prediction (figure 4.7B, top row), to form a belief about cursor position that is somewhere between 2 cm (observed) and 1 cm (predicted). This "belief" would depend on how much weight (or uncertainty) they assign to the observed and predicted sources of information. If the observed cursor is in a noisy cloud, they should rely more on their prediction (σ_∞ line, figure 4.7A). If the observed cursor is clear, they should rely more on their observation (σ_0 line, figure 4.7A). The weighting of the two sources should be inversely related to the variance of the distributions. Indeed, Kording and Wolpert's (2004) experimental data was consistent with this theoretical framework.

In summary, the data suggests that as the brain programs motor commands, it also predicts the sensory consequences. Once the sensory system reports its measurements, the brain combines what it had predicted with the measurements to form a "belief" that represents its estimate of the state of the world. Our actions are not simply based on our current sensory observations. Rather, our actions are based on an integration of the sensory observations with our predictions. In engineering, this is called estimation theory.

4.4 State Estimation: The Problem of Hiking in the Woods

The problem of estimating the state of our body (or state of something else) has two components. The first is associated with learning to accurately predict what the sensory measurements of that state should be—this is our *prior* estimate of state. The second is associated with combining the measured quantities with the predicted one to form a *posterior* estimate of state. The first problem is one of model building, that is, describing an internal model that predicts what our sensors should be telling, called a forward model. The second problem is one of integration, that is, describing how to form an estimate of state based on the two sources of information, the

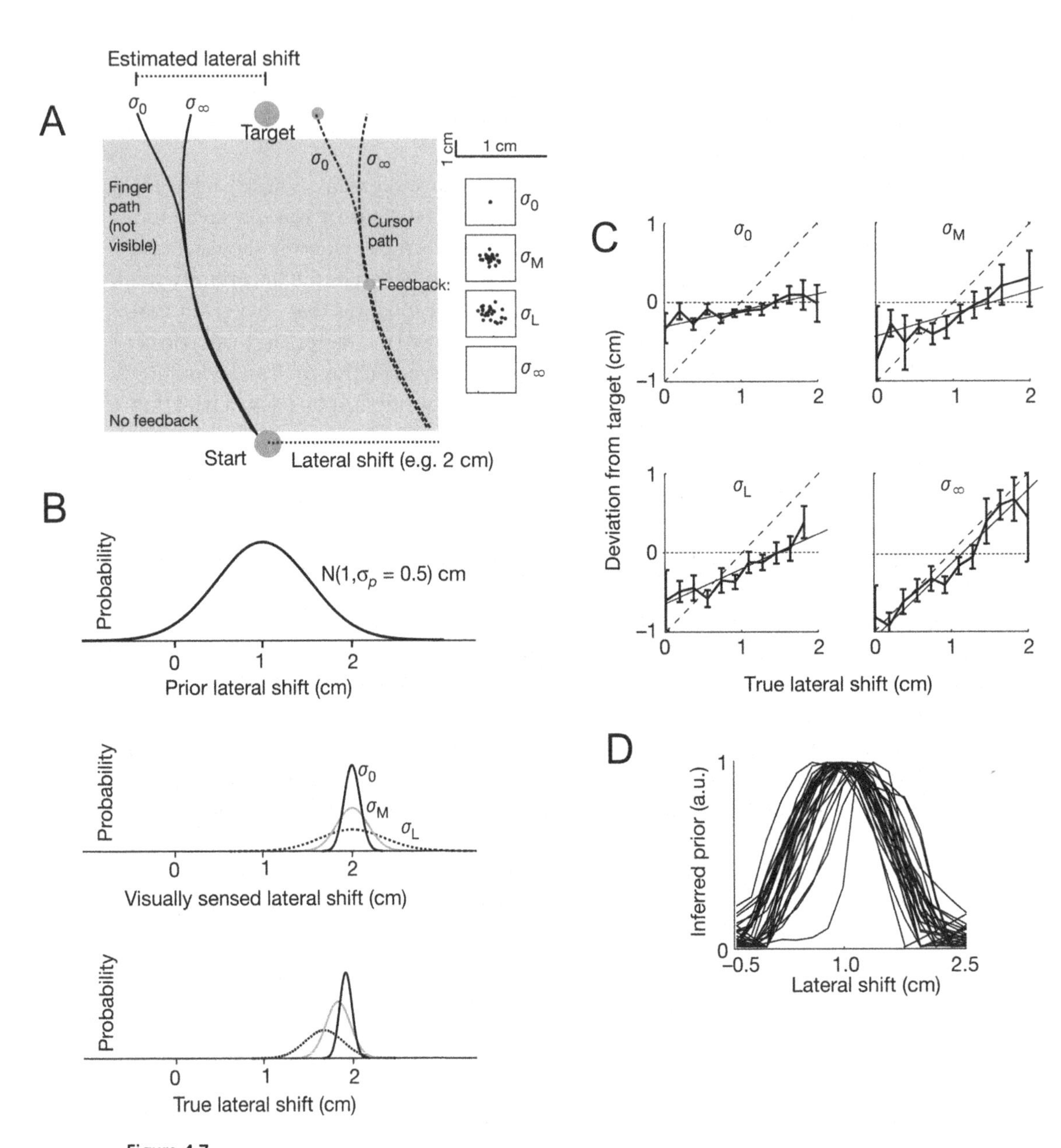

Figure 4.7
Response to a visual stimulus depends on both the noise in the stimulus and the brain's predictions about that stimulus. (A) In a reaching task, feedback about the current position of the finger was provided for only a brief time during the movement. As the finger moved from the starting circle, cursor (indicating current finger position) was extinguished. Halfway to the target, feedback was briefly provided. The position of this feedback was displaced from actual finger position by a random amount, with a mean of 1 cm. The quality of the feedback was controlled: the feedback was either clear (σ_0), or with different

prediction and the observation. Our goal here is to build a mathematical framework in which we can describe this problem. Once the framework is in place, we will use it to try and account for some behaviors in people and other animals.

To illuminate the basic problem that the brain faces, let us consider the following example. Say that you are hiking in the woods and are concerned about getting lost. To help with navigation, you have bought two Global Positioning System (GPS) devices. One of the devices is European-made and uses satellites operated by the European Union. The other is made to work with American satellites. Therefore, you can assume that each uses an independent set of measurements as it provides you with a probability distribution regarding your position (figure 4.8). Device A reports that your position is at coordinates $\mathbf{y}_a = (+1,0)$, with a probability distribution specified by a Gaussian with covariance R_a (a 2×2 matrix). This is the location for which its probability distribution has a peak (i.e., the mean). Device B reports that your position is at coordinate $\mathbf{y}_b = (+4,0)$, with a probability distribution with covariance R_b. Interestingly, your most likely location is not somewhere between the two centers of probability distributions. Rather, given the uncertainties associated with each device, as depicted by the distributions shown in figure 4.8, your most likely location is probably at coordinates (+2.5, –1.5).

To see why this is the case, suppose that the state that we wish to estimate (our position) is described by a 2×1 vector $\mathbf{x}$, and the reading from each device is described by 2×1 vectors $\mathbf{y}_a$ and $\mathbf{y}_b$. To estimate our position, we need to put forth a hypothesis regarding how the devices' readings are related to our position. Suppose that our measurement, denoted by 4×1 vector $\mathbf{y} = [\mathbf{y}_a \quad \mathbf{y}_b]$, is related to a hidden state (our position) $\mathbf{x}$ by the following equation:

$$\mathbf{y} = C\mathbf{x} + \boldsymbol{\varepsilon}. \tag{4.1}$$

This is our internal model. It describes our belief about how the data that we observe is related to the hidden state that we wish to estimate. In our internal model, we believe that the devices are unbiased. Therefore, we set

degrees of blur (σ_M and σ_L), or withheld (σ_∞). The paths illustrate typical trajectories for a displacement of 2 cm. (B) Top subplot: For 1000 trials, subjects trained with the illustrated distribution of lateral shifts, that is, a Gaussian with a mean of 1 cm. This constitutes the prior probability of the displacement. Middle subplot: A diagram of various probability distributions associated with the current measurement. This distribution is shown for the clear and the two blurred feedback conditions for a trial in which the true shift was 2 cm. Bottom subplot: The estimate of displacement for an optimal observer that combines the prior with the evidence. (C) The lateral deviation of the cursor at the end of the reach as a function of the imposed lateral shift for a typical subject. The horizontal line at 0 would indicate full compensation to the observed error. The dash line would indicate complete denial of the observed error. The solid line is the Bayesian model with the level of uncertainty fitted to the data. For example, when the feedback was clear, i.e., σ_0, the subject compensated almost fully, nearly hitting the target in all conditions. When the feedback was uncertain, i.e., σ_L, the subject missed the target in most conditions. (D) The inferred priors for each subject and condition. The true distribution is shown in (B). (From Kording and Wolpert, 2004.)

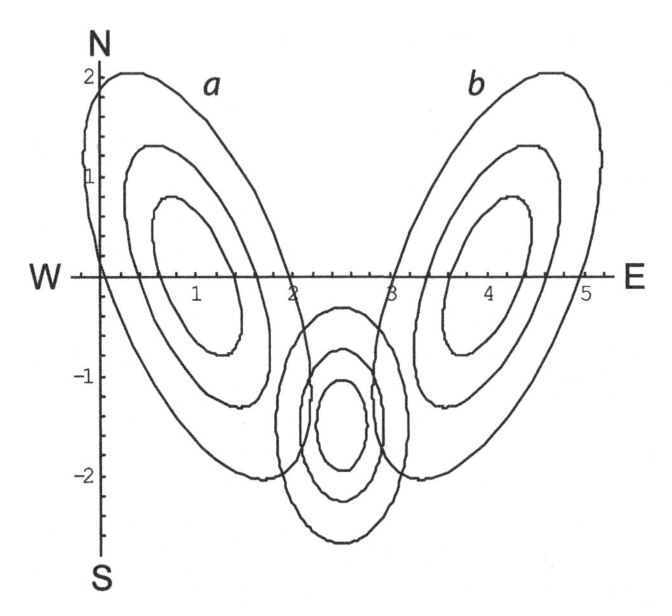

Figure 4.8
Device A and device B provide independent estimates of a hidden variable (position on a map). Each device has a Gaussian noise property with a covariance $R_a = [1,-1;-1,3]$ and $R_b = [1,1;1,3]$. The ellipses describe the region centered on the mean of the distribution that contains 10 percent, 25 percent, and 50 percent of the data under the distribution. The maximum likelihood estimate of the hidden variable is marked by the distribution at the center.

$$C = \begin{bmatrix} 1 & 0 \\ 0 & 1 \\ 1 & 0 \\ 0 & 1 \end{bmatrix} = \begin{bmatrix} I_{2\text{x}2} \\ I_{2\text{x}2} \end{bmatrix}, \tag{4.2}$$

where I is a 2×2 identity matrix. We also believe that the noises inherent in the measurements of the devices are independent, zero mean, and distributed as a Gaussian:

$$\begin{aligned} &\boldsymbol{\varepsilon} \sim N(0,R) \\ &R = \begin{bmatrix} R_a & 0 \\ 0 & R_b \end{bmatrix}, \end{aligned} \tag{4.3}$$

where R_a and R_b are 2×2 symmetric matrices. The expected value of $\mathbf{y}$, written as $E(\mathbf{y})$, is $E(\mathbf{y}) = C\mathbf{x}$. The variance of $\mathbf{y}$, written as $\text{var}(\mathbf{y})$, is $\text{var}(\mathbf{y}) = C\text{var}(\mathbf{x})C^T + R$. The probability distribution of $\mathbf{y}$ is specified by a Gaussian:

$$\mathbf{y} \sim N\left(C\mathbf{x}, C\,\text{var}(\mathbf{x})C^T + R\right). \tag{4.4}$$

This implies that if we "knew" our position **x** with certainty, i.e., if var(**x**) = 0 (or alternatively, we stayed still and kept on taking measurements), the measurements from the devices would have the following distribution:

$$p(\mathbf{y}|\mathbf{x}) = N(C\mathbf{x}, R). \tag{4.5}$$

This equation is called a *likelihood*. It describes the probability distribution of an observation given that the thing that we want to estimate (the hidden state **x**) is at a particular value. To find our most likely position, we find the value for **x** that maximizes this likelihood:

$$p(\mathbf{y}|\mathbf{x}) = \frac{1}{\sqrt{(2\pi)^4 |R|}} \exp\left[-\frac{1}{2}(\mathbf{y}-C\mathbf{x})^T R^{-1}(\mathbf{y}-C\mathbf{x})\right]. \tag{4.6}$$

It is convenient to take the log of the above expression, arriving at

$$\ln p(\mathbf{y}|\mathbf{x}) = -2\ln(2\pi) - \frac{1}{2}\ln|R| - \frac{1}{2}(\mathbf{y}-C\mathbf{x})^T R^{-1}(\mathbf{y}-C\mathbf{x}). \tag{4.7}$$

To find the location **x** for which this quantity is maximum, we find the location at which its derivative is zero:

$$\frac{d}{d\mathbf{x}}\ln p(\mathbf{y}|\mathbf{x}) = \left(C^T R^{-1}\mathbf{y} - C^T R^{-1} C\mathbf{x}\right)$$

$$\hat{\mathbf{x}} = \left(C^T R^{-1} C\right)^{-1} C^T R^{-1}\mathbf{y}. \tag{4.8}$$

If we note that $R^{-1} = \begin{bmatrix} R_a^{-1} & 0 \\ 0 & R_b^{-1} \end{bmatrix}$ and $C^T R^{-1} = \begin{bmatrix} R_a^{-1} & R_b^{-1} \end{bmatrix}$, then we can rewrite equation (4.8) as:

$$\hat{\mathbf{x}} = \left(R_a^{-1} + R_b^{-1}\right)^{-1}\left(R_a^{-1}\mathbf{y}_a + R_b^{-1}\mathbf{y}_b\right) \tag{4.9}$$

Equation (4.9) describes our maximum likelihood estimate of **x**. Now if we simply stay still and keep taking measurements, our readings **y** will keep changing. These changes are due to the inherent noise in the devices, and will produce changes in our estimate $\hat{\mathbf{x}}$. Therefore, $\hat{\mathbf{x}}$ is a random variable with a distribution. Its expected value is specified by equation (4.8), and its variance is:

$$\mathrm{var}(\hat{\mathbf{x}}) = \left(C^T R^{-1} C\right)^{-1} C^T R^{-1}\,\mathrm{var}(\mathbf{y}) R^{-T} C \left(C^T R^{-1} C\right)^{-T}. \tag{4.10}$$

Assuming that we stay still and do not move around, then var(**y**) = R. Because R is symmetric (as are all variance-covariance matrices), the above equation simplifies to:

$$\mathrm{var}(\hat{\mathbf{x}}) = \left(C^T R^{-1} C\right)^{-1}. \tag{4.11}$$

Note that $C^T R^{-1} = \begin{bmatrix} R_a^{-1} & R_b^{-1} \end{bmatrix}$ simplifies the preceding equation to:

$$\text{var}(\hat{\mathbf{x}}) = \left(R_a^{-1} + R_b^{-1}\right)^{-1}. \tag{4.12}$$

The result in equation (4.9) indicates that our most likely location is one that weighs the reading from each device by the inverse of the device's probability covariance. In other words, we should discount the reading from each device according to the inverse of each device's uncertainty. Using equation (4.9) and equation (4.12), we have drawn the mean and variance of our maximum likelihood estimate in figure 4.8 (the distribution in the middle of the figure). It is quite unlikely that you are somewhere between the centers of left and right distributions, because $p(\mathbf{y}|\mathbf{x})$ is quite low there. The most likely location, it turns out, is a bit south of the center of each distribution, at (+2.5, −1.5). Another important point to note is that the estimate of your position has a tighter distribution than either of the two sensors, that is, it has a "smaller" variance. Therefore, when you combined these two pieces of information, your result was an estimate that had a lower uncertainty than either of the two initial measures. It is better to have two GPS readings than one.

4.5 Optimal Integration of Sensory Information by the Brain

It turns out that this framework is quite relevant to how the brain perceives the environment and processes sensory information. After all, we have multiple sensors. For example, when we examine an object, we do so with both our hands and our eyes. Marc Ernst and Marty Banks (2002) were first to demonstrate that when our brain makes a decision about a physical property of an object, it does so by combining various pieces of sensory information about that object in a way that is consistent with maximum likelihood state estimation.

Ernst and Banks began by considering a hypothetical situation in which one has to estimate the height of an object. Suppose that you use your index and thumb to hold an object. Your somatosensory/haptic system reports its height. Let us represent this information as a random variable y_h (a scalar quantity) that has as a distribution described by a Gaussian with variance σ_h^2. Similarly, your visual system provides you with information y_v, which has a variance σ_v^2. Our internal model is:

$$\mathbf{y} = \mathbf{c}x + \boldsymbol{\varepsilon}, \tag{4.13}$$

where x is the true height of the object, $\mathbf{y} = [y_h \quad y_v]^T$, $\mathbf{c} = [1 \quad 1]^T$ (which implies that your sensors are unbiased), $\boldsymbol{\varepsilon} \sim N(0,R)$, and $R = [\sigma_h^2, 0; 0, \sigma_v^2]$. According to equation (4.9), your maximum likelihood estimate of the height of the object has the following distribution:

$$
\begin{aligned}
E(\hat{x}) &= \frac{1/\sigma_h^2}{1/\sigma_h^2 + 1/\sigma_v^2} y_h + \frac{1/\sigma_v^2}{1/\sigma_h^2 + 1/\sigma_v^2} y_v \\
\operatorname{var}(\hat{x}) &= \frac{1}{1/\sigma_h^2 + 1/\sigma_v^2}
\end{aligned}
\tag{4.14}
$$

If the noise in the two sensors is equal, then the weights that you apply to the sensors (the coefficients in front of y_h and y_v in equation 4.14) are equal as well. This case is illustrated in the left column of figure 4.9. On the other hand, if the noise in y_h is larger than y_v, your uncertainty for the haptic measure is greater and so you apply a smaller weight to its reading (illustrated in the right column of figure 4.9). If one were to ask you to report the height of the object, of

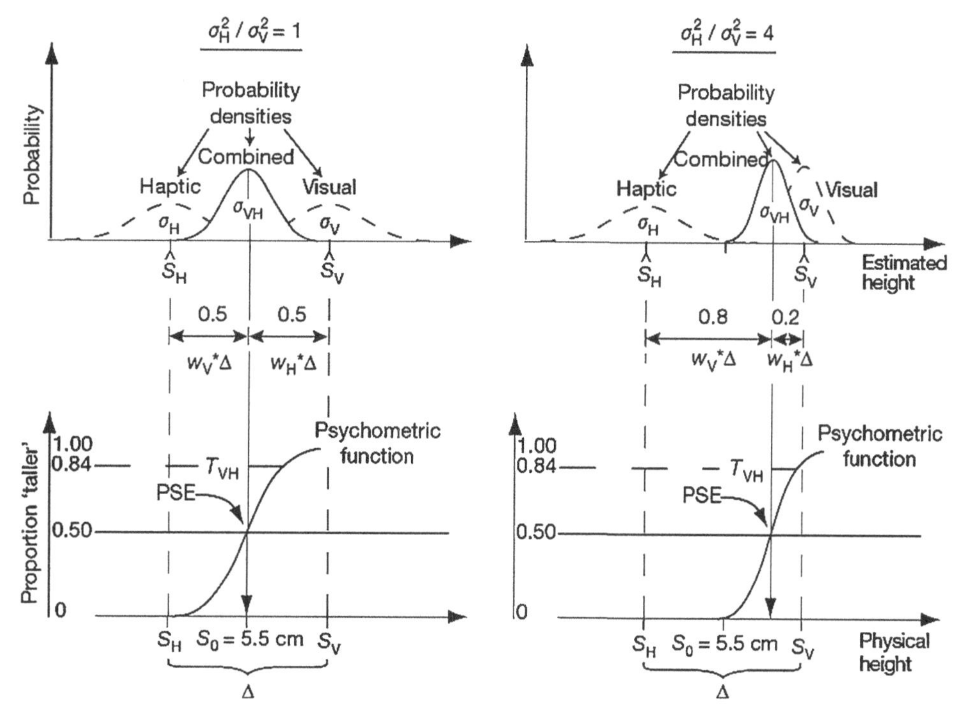

Figure 4.9
Maximum likelihood integration of two sensory modalities. Visually and haptically specified heights of an object differ by Δ. On the left columns, the visual and haptic variances are equal. The mean of the combined probability density is equal to the mean of the visual and haptic densities. The variance of the combined density is half of the visual (or haptic) density. If the judgment of relative height is based on the combined density, the psychometric function is the cumulative Gaussian (bottom left) with the point of subjective equality (PSE) equal to the average of the visual and haptic heights. In the right column of figures, the noise in the haptic sensor is four times the visual noise. The psychometric function is shifted so that the PSE is closer to the visual height. (From Ernst and Banks, 2002.)

course you would not report your belief as a probability distribution. To estimate this distribution, Ernst and Banks acquired a psychometric function, shown in the lower part of the graph in figure 4.9. To acquire this function, they provided their subjects a standard object of height 5.5 cm. They then presented a second object of variable length and asked whether it was taller than the first object. If the subject represented the height of the standard object with a maximum likelihood estimate with a distribution of equation (4.14), then the probability of classifying the second object as being taller is simply the cumulative probability distribution of $\hat{x}$. This is called a psychometric function, shown in the lower row of figure 4.9. The point of subject equality (PSE) is the height at which the probability function is at 0.5. Note that this point shift toward the estimate from vision when one is more certain about the visual measurement (right column of figure 4.9).

The task at hand is to test whether the brain combines haptic and sensory information in a way that is consistent with maximum likelihood. To test the theory, the first step is to estimate the noise in the haptic and visual sensors. To do so, Ernst and Banks considered a situation in which both the standard and the second stimulus were of the same modality. For example, a robot presented a virtual object (standard, of length μ_1) and then a second object (of length μ_1 + Δ). (People held the handle of the robot, and when they moved it, they felt a stiff surface of a given length. So there was no real object there, just a sensation associated with running your hand along the surface of an object.) The subject responded by choosing one that had a longer height. To make this decision, suppose that the subject's haptic sense represented the objects as y_1 and y_2, where $y_1 \sim N(\mu_1, \sigma_h^2)$ and $y_2 \sim N(\mu_1 + \Delta, \sigma_h^2)$. The estimate of the difference between the two objects is $\hat{\Delta} = y_2 - y_1$, and this random variable is a Gaussian with the following distribution: $\hat{\Delta} \sim N(\Delta, 2\sigma_h^2)$. The probability of picking the second object as being taller is:

$$\Pr(y_2 > y_1) = \Pr(\hat{\Delta} > 0) = \int_0^{\infty} N(x; \Delta, 2\sigma_h^2)\, dx. \tag{4.15}$$

The term $N(x; \Delta, 2\sigma_h^2)$ represents a normal distribution of random variable x with mean Δ and variance $2\sigma_h^2$. To compute this integral, we use the "error function":

$$\mathrm{erf}(x) = \frac{2}{\pi}\int_0^{\infty} e^{-x^2} dx$$

$$\int_{-\infty}^{x} N(t; \mu, \sigma^2)\, dt = \frac{1}{2}\left(1 + \mathrm{erf}\left(\frac{x-\mu}{\sigma\sqrt{2}}\right)\right)$$

And so we have:

$$\Pr\left(\hat{\Delta}>0\right)=1-\frac{1}{2}\left(1+\operatorname{erf}\left(\frac{-\Delta}{2\sigma_h}\right)\right). \tag{4.16}$$

For example, suppose that the second object is 3 mm longer than the standard object, i.e., $\Delta = 3$. From equation (4.16), we would predict that if $\sigma_h^2 = 1$ (resulting in the two distributions plotted in figure 4.10A), then the subject should pick object 2 as being larger with 98 percent probability. In contrast, if $\sigma_h^2 = 2$, then the subject should pick object 2 as being larger with 69 percent probability, and so on. The condition for which $\Delta = 3$ and $\sigma_h^2 = 1$ is plotted in figure 4.10A. The resulting distribution for $\hat{\Delta}$ is plotted in figure 4.10B. The probability of equation (4.16) is plotted as a function of Δ, i.e., the difference in the lengths of the two objects, for various

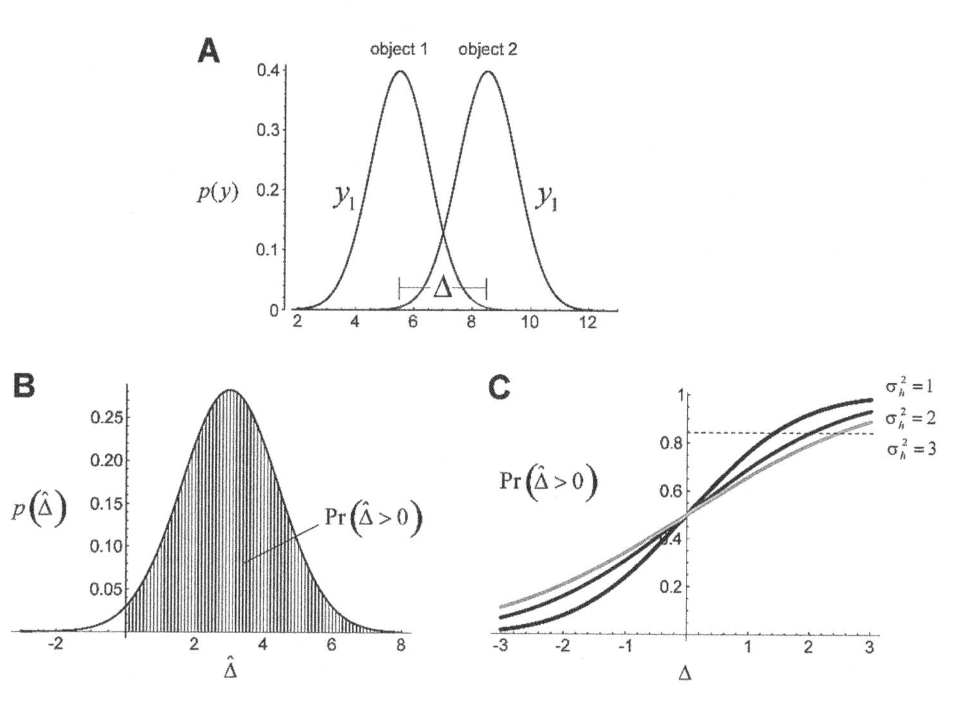

Figure 4.10
Procedure for determining the noise in the haptic sensory modality. (A) Two objects that differ in size by amount $\Delta = 3$ are presented and a subject holds each object to estimate its size. The estimate of size for each object, $y_1 \sim N(\mu_1,\sigma_h^2)$ and $y_2 \sim N(\mu_1+\Delta,\sigma_h^2)$ are shown (here, we assumed $\sigma_h^2 = 1$). (B) The estimate of the difference between the two objects is $\hat{\Delta} = y_2 - y_1$ and this random variable is a Gaussian with the following distribution: $\hat{\Delta} \sim N(\Delta, 2\sigma_h^2)$. The probability of picking the second object as being larger, $\Pr(\hat{\Delta}>0)$, is the integral of $p(\hat{\Delta})$ from zero to infinity. (C) The probability $\Pr(\hat{\Delta}>0)$ rises faster as a function of Δ when sensory noise σ_h^2 is small. When $\Delta = \sqrt{2}\sigma_h$, $\Pr(\hat{\Delta}>0) = 0.84$, as indicated by the dashed line.

σ_h^2 in figure 4.10C. As the noise in the haptic sensor increases, the subject has more difficulty dissociating the two objects at a given length difference.

In equation (4.16), notice that if we set $\Delta = \sqrt{2}\sigma_h$, then the probability of picking object two as being larger is always 84 percent. So Ernst and Banks presented second objects of various Δ and found the one for which the subject was 84 percent correct (call it Δ^*). From the results in figure 4.11a, we see that this object was about 5 mm longer than the standard, i.e., $\Delta^* \approx 5$. Therefore, the noise in the haptic sensory pathway must be:

$$\sigma_h^2 = \frac{1}{2}(\Delta^*)^2.$$

A similar procedure was performed to estimate the noise in the visual pathway. Two objects were presented visually, and, as figure 4.11a suggests (0 percent noise curve), 84 percent probability of correct choice was for $\Delta^* \approx 2.5$. Therefore, $\sigma_h^2 \approx 4\sigma_v^2$. In the final step of the experiment, Ernst and Banks used these noise parameters to predict how subjects would estimate the height of an object when the visual and haptic senses were both present (as in figure 4.9). For example, subjects were presented with a standard object for which the haptic information indicated a height of μ_1 and visual information indicated a height of $\mu_1 + \Delta$. Equation (4.14) predicted that subjects would assign a weight of around 0.8 to the visual information and around 0.2 to the haptic information. To estimate these weights, they presented a second object (for which the haptic and visual information agreed) and ask which one was taller. The probability of the second object being taller is plotted in figure 4.11b, and the weight assigned to each sensory modality is plotted in figure 4.11c. The observed weights (figure 4.11c) agree quite well with the predicted values. Note that for the 0 percent visual noise condition, the curve is shifted toward the mean of the visual information. One can also predict the variance of the estimate (equation 4.14), which specifies the rate of rise in the psychometric curve. A proxy for this rate is a "discrimination threshold," defined as the difference between the point of subject equality and the height of the second stimulus when it is judged to be taller than the standard stimulus 84 percent of the time. As the variance of the estimate increases, the rate of rise decreases, and the discrimination threshold increases. Figure 4.11d shows that the predicted value nicely matches the observed value. They next repeated these steps for conditions in which noise was added to the visual display. This reduced the weighting of the visual information, making the subjects rely more on the haptic information. The predicted values continued to match the measured quantities.

Therefore, the results showed that the brain combined visual and haptic information in a way that was similar to a maximum-likelihood estimator: A weight was assigned to each sensory modality and this weight was a function of the uncertainty in that modality.

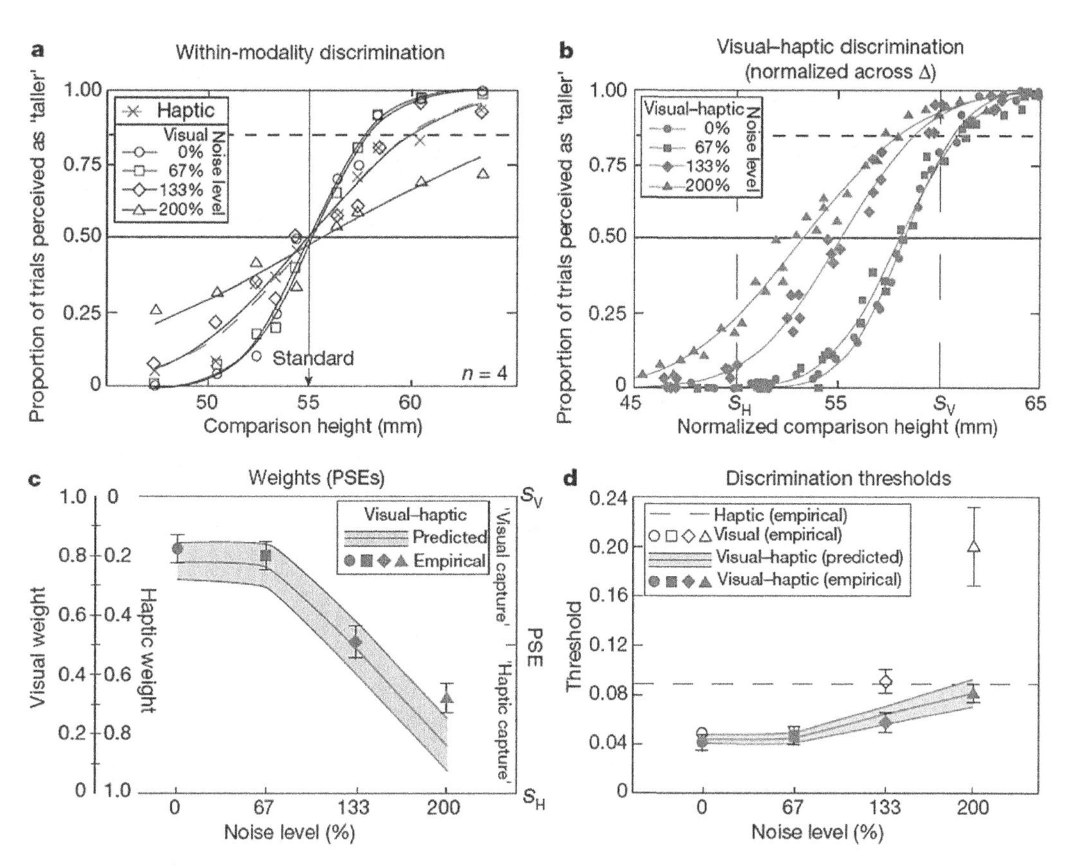

Figure 4.11
Experimental data from an experiment on haptic and visual sensory integration. (a) Within-modality experimental results. Proportion of trials in which the second stimulus was perceived as taller than the standard stimulus is plotted against the height of the second stimulus. Four noise levels were considered for the visual condition. (b) Visual-haptic discrimination. (c) Predicted and experimental weights and PSEs. The shaded area represented predicted weights from within-modality discrimination. The height of the shaded area represents predicted errors given the standard errors of within modality discrimination. (d) Combined and within modality discrimination thresholds. Thresholds are from the psychometric function in (a) and (b). The dashed line represents haptic-alone threshold. (From Ernst and Banks, 2002.)

4.6 Uncertainty

How does uncertainty come about? It has something to do with the precision with which one knows some bit of information. In particular, it seems to have something to do with the history of how that information was acquired. In our hiking-in-the-woods example, the uncertainty of each GPS device had a particular shape because there were correlations along the two dimensions in which each device was making measurements. Perhaps these correlations came about because of the position of the satellites, giving rise to the shape of the probability distribution. Let us explore this idea a bit further in a scenario in which we control how we acquire information about a quantity that we wish to estimate.

In figure 4.12 we have drawn points from three normal distributions. In each distribution, the vector $\mathbf{x} = [x_1 \quad x_2]^T$ is drawn from a Gaussian: $\mathbf{x} \sim N(\boldsymbol{\mu}, R)$, with mean zero $\boldsymbol{\mu} = [0 \quad 0]^T$ and covariance R. This covariance is defined as:

$$R = E\left[(\mathbf{x}-\boldsymbol{\mu})(\mathbf{x}-\boldsymbol{\mu})^T\right]$$
$$= \begin{bmatrix} \mathrm{var}[x_1] & \mathrm{cov}[x_1, x_2] \\ \mathrm{cov}[x_2, x_1] & \mathrm{var}[x_2] \end{bmatrix}.$$

In figure 4.12, the distribution on the left has a covariance matrix in which x_1 and x_2 are negatively correlated (the off-diagonal elements of the covariance matrix are negative). This roughly corresponds to our GPS device A in figure 4.8. We can now infer that the measurements that device A was taking along the east/west (x_1) and north/south (x_2) dimensions were showing negative correlations: as

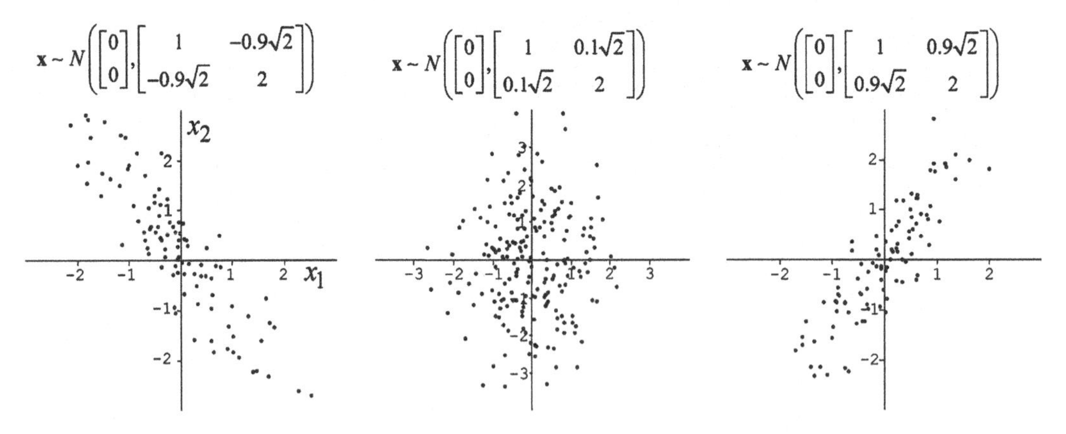

Figure 4.12
Samples from three Gaussian distributions.

the measurement along one dimension increased, the measurement along the other dimension decreased. So when device A reported its measurement, it was uncertain along the northwest-southeast dimension, but very certain along the southwest-northeast dimension.

In figure 4.12, the distribution at right has a covariance in which x_1 and x_2 are positively correlated. This roughly corresponds to our device B, indicating that for this device, as the measurement along one dimension increased, so did the measurement along the other dimension. The overlap between the leftmost distribution and rightmost distribution (assuming that the centers are not at the same location) would be at a maximum (i.e., the most number of points are likely to overlap) somewhere along their diagonals.

Our simple example illustrates the idea that in order for any system to combine two pieces of information about a single quantity, it needs to have a measure of uncertainty about each quantity. That is, it needs to "know" the two pieces of information in terms of their probability distributions. If the distribution is Gaussian, all we need to know are the mean and covariance.

Now to illustrate how these correlations—these uncertainties—arise during the process of acquiring information, let us consider the classic problem in linear regression. On each trial i, you measure a scalar quantity $y^{(i)}$. However, your measurements are noisy. The "true" value on that trial is $y^{*(i)}$, and your measure is corrupted by a Gaussian noise that has zero mean and σ^2 variance:

$$y^{(i)} = y^{*(i)} + \varepsilon \quad \varepsilon \sim N\left(0, \sigma^2\right). \tag{4.17}$$

This means that the probability to find a value of $y^{(i)}$ within a given interval of the true value is

$$\Pr[y^{*(i)} + \varepsilon_1 \le y^{(i)} \le y^{*(i)} + \varepsilon_2] = \frac{1}{\sigma\sqrt{2\pi}} \int_{\varepsilon_1}^{\varepsilon_2} e^{-\frac{\varepsilon^2}{\sigma^2}} d\varepsilon. \tag{4.18}$$

We assume that the term $y^{*(i)}$ is linearly related to a known quantity expressed as a N-dimensional vector, $\mathbf{x}^{(i)}$:

$$y^{*(i)} = \mathbf{w}^T \mathbf{x}^{(i)}. \tag{4.19}$$

In the language of systems theory, $\mathbf{x}^{(i)}$ is an input vector and $y^{*(i)}$ is the corresponding output generated by the linear transformation. In this case the unknown multiplicative term $\mathbf{w}^T$ is a $1 \times N$ matrix—that is, a row vector, mapping an N-dimensional vector into a single number. According to this simple model, noise is only affecting the observation of the output variable $y^{(i)}$. More complex situations may involve noise in the input vector $\mathbf{x}^{(i)}$ and in the vector $\mathbf{w}$. Our objective is to estimate the vector $\mathbf{w}$ from a data set that includes pairs of input/output measures:

$D = (\{\mathbf{x}^{(1)}, y^{(1)}\}, \{\mathbf{x}^{(2)}, y^{(2)}\}, \cdots, \{\mathbf{x}^{(n)}, y^{(n)}\})$. Given our model in equations (4.17) and (4.19), we can write the probability distribution of observing $y^{(i)}$, given input $\mathbf{x}^{(i)}$ and parameters $\mathbf{w}$ and σ^2:

$$p(y^{(i)} | \mathbf{x}^{(i)}) = N(\mathbf{w}^T \mathbf{x}^{(i)}, \sigma^2). \tag{4.20}$$

That is, we would expect $y^{(i)}$ to be normally distributed with mean $\mathbf{w}^T\mathbf{x}^{(i)}$ and variance σ^2. The probability density function for observing the specific data that we were given is simply the joint probability density of all data points:

$$\begin{aligned} p(y^{(1)}, \cdots, y^{(n)} | \mathbf{x}^{(1)}, \cdots, \mathbf{x}^{(n)}, \mathbf{w}, \sigma) &= \prod_{i=1}^{n} p(y^{(i)} | \mathbf{x}^{(i)}, \mathbf{w}, \sigma) \\ &= \prod_{i=1}^{n} \frac{1}{(2\pi\sigma^2)^{1/2}} \exp\left(-\frac{1}{2\sigma^2} (y^{(i)} - \mathbf{w}^T \mathbf{x}^{(i)})^2 \right). \\ &= \frac{1}{(2\pi\sigma^2)^{n/2}} \exp\left(-\frac{1}{2\sigma^2} \sum_{i=1}^{n} (y^{(i)} - \mathbf{w}^T \mathbf{x}^{(i)})^2 \right) \end{aligned} \tag{4.21}$$

The quantity in equation (4.21) is a likelihood. It describes how likely it is that given our model and its parameters, we would observe the specific collection of data $\{y^{(1)}, \ldots, y^{(n)}\}$. Let us refer to our likelihood as $L(\mathbf{w}, \sigma)$. The parameters that we would like to estimate are $\mathbf{w}$ and σ. If we have no other information about the parameters that we wish to estimate (i.e., no priors), then the best are those that maximize the likelihood. We start by observing that the exponential function is monotonic in its argument. Therefore, the exponential is at a maximum when its argument is also at a maximum. Because in the normal distribution of equation (4.20) the argument is always negative, the maximum of the likelihood function is attained when the argument of the exponential reaches a minimum. To put into a formula, we take the natural logarithm of the function that we wish to maximize:

$$\begin{aligned} \log L(\mathbf{w}, \sigma) &= \log \prod_{i=1}^{n} p(y^{(i)} | \mathbf{x}^{(i)}, \mathbf{w}, \sigma) \\ &= -\frac{1}{2\sigma^2} \sum_{i=1}^{n} (y^{(i)} - \mathbf{w}^T \mathbf{x}^{(i)})^2 - \frac{n}{2} \log(2\pi\sigma^2) \end{aligned} \tag{4.22}$$

and look for its maximum. To further simplify things, let us write the first sum in the second line of the above equation in matrix form. Suppose we use vector $\mathbf{y}$ and matrix X to refer to collection of $y^{(i)}$ and $\mathbf{x}^{(i)}$ (the vector $\mathbf{x}^{(i)}$ becomes a row in the matrix X), and then rewrite equation (4.22). This requires a little manipulation, but the final result is worth it. We begin by collecting all the $\mathbf{x}^{(i)}$ vectors into a matrix, one vector per row:

$$X = \begin{bmatrix} x_1^{(1)} & x_2^{(1)} & \cdots & x_N^{(1)} \\ x_1^{(2)} & x_2^{(2)} & \cdots & x_N^{(2)} \\ \cdots & \cdots & \cdots & \cdots \\ x_1^{(n)} & x_2^{(n)} & \cdots & x_N^{(n)} \end{bmatrix} \tag{4.23}$$

and all the output data $y^{(i)}$, in a column vector

$$\mathbf{y} = \begin{bmatrix} y^{(1)} \\ y^{(2)} \\ \cdots \\ y^{(n)} \end{bmatrix}. \tag{4.24}$$

Then, the vector

$$\begin{aligned} \mathbf{y} - X\mathbf{w} &= \begin{bmatrix} y^{(1)} \\ y^{(2)} \\ \cdots \\ y^{(n)} \end{bmatrix} - \begin{bmatrix} x_1^{(1)} & x_2^{(1)} & \cdots & x_N^{(1)} \\ x_1^{(2)} & x_2^{(2)} & \cdots & x_N^{(2)} \\ \cdots & \cdots & \cdots & \cdots \\ x_1^{(n)} & x_2^{(n)} & \cdots & x_N^{(n)} \end{bmatrix} \cdot \begin{bmatrix} w_1 \\ w_2 \\ \cdots \\ w_N \end{bmatrix} \\ &= \begin{bmatrix} y^{(1)} - \sum_{k=1}^{N} w_k \, x_k^{(1)} \\ y^{(2)} - \sum_{k=1}^{N} w_k \, x_k^{(2)} \\ \cdots \\ y^{(n)} - \sum_{k=1}^{n} w_k \, x_N^{(n)} \end{bmatrix} = \begin{bmatrix} y^{(1)} - w^T \cdot \mathbf{x}^{(1)} \\ y^{(2)} - w^T \cdot \mathbf{x}^{(2)} \\ \cdots \\ y^{(n)} - w^T \cdot \mathbf{x}^{(n)} \end{bmatrix} \end{aligned} \tag{4.25}$$

has a Euclidean norm

$$\|\mathbf{y} - X \cdot \mathbf{w}\|^2 = (\mathbf{y} - X \cdot \mathbf{w})^T (\mathbf{y} - X \cdot \mathbf{w}) = \sum_{i=1}^{n} \left(y^{(i)} - w^T \mathbf{x}^{(i)}\right)^2. \tag{4.26}$$

Note that this provides a way to rewrite the sum on the left side of equation (4.22). With this substitution, we obtain:

$$\begin{aligned} \log L(\mathbf{w}, \sigma) &= -\frac{1}{2\sigma^2} (\mathbf{y} - X\mathbf{w})^T (\mathbf{y} - X\mathbf{w}) - \sum_{i=1}^{n} \log\left(2\pi\sigma^2\right)^{1/2} \\ &= -\frac{1}{2\sigma^2} \left(\mathbf{y}^T \mathbf{y} - \mathbf{y}^T X\mathbf{w} - \mathbf{w}^T X^T \mathbf{y} + \mathbf{w}^T X^T X\mathbf{w}\right) \; . \\ &\quad -\sum_{i=1}^{n} \left(\log(2\pi)^{1/2} + \log\sigma\right) \end{aligned} \tag{4.27}$$

To find $\mathbf{w}$ at which this function is maximum, we find its derivative and set it equal to zero:

$$\begin{aligned}\frac{d\log L(\mathbf{w},\sigma)}{d\mathbf{w}} &= -\frac{1}{2\sigma^2}\left(-2X^T\mathbf{y}+2X^TX\mathbf{w}\right)=0\\ X^TX\mathbf{w} &= X^T\mathbf{y}\\ \hat{\mathbf{w}} &= \left(X^TX\right)^{-1}X^T\mathbf{y}\end{aligned} \quad (4.28)$$

We observe that this solution for $\mathbf{w}$ corresponds to a maximum of the log likelihood, because the second derivative (or Hessian) of log L:

$$\frac{d^2\log L(\mathbf{w},\sigma)}{d\mathbf{w}^2} = -\frac{1}{\sigma^2}X^TX \quad (4.29)$$

is negative definite, if X is full row rank (i.e., if $rank(X) = N$). This result is interpreted intuitively by observing that the matrix $(X^TX)^{-1}X^T$ is the left inverse of the input data matrix, X. Therefore the maximum likelihood estimation of $\mathbf{w}$ is analogous to a ratio of the output to the input. We derive the same solution by looking for the parameters that minimize the square difference from the observed outputs and the outputs that are obtained by assuming the model of equation (4.19). Thus, we have found a correspondence between the maximum likelihood estimation and the least-squares estimation, in this linear case with Gaussian noise.

Equation (4.28) represents our maximum likelihood estimate of $\mathbf{w}$. Similarly, to find the maximum likelihood estimate of noise parameter σ we take the derivative of equation (4.27) with respect to σ and set it equal to zero.

$$\begin{aligned}\frac{dl}{d\sigma} &= \frac{1}{\sigma^3}(\mathbf{y}-X\mathbf{w})^T(\mathbf{y}-X\mathbf{w})-\sum_{i=1}^{n}\frac{1}{\sigma}=0\\ &= \frac{1}{\sigma^2}(\mathbf{y}-X\mathbf{w})^T(\mathbf{y}-X\mathbf{w})-n=0\\ \hat{\sigma}^2 &= \frac{1}{n}(\mathbf{y}-X\mathbf{w})^T(\mathbf{y}-X\mathbf{w})\end{aligned} \quad (4.30)$$

However, it is important to note that because $\mathbf{y}$ is a random variable whose value is corrupted by random noise, our estimates $\hat{\mathbf{w}}$ and $\hat{\sigma}$ are also random variables. To see why this is true, consider how we got our data set D: for some known set of inputs $\mathbf{x}^{(1)},\cdots,\mathbf{x}^{(n)}$, we made a set of measurements $y^{(1)},\cdots,y^{(n)}$ that were generated by the "true" parameter vector $\mathbf{w}$ but were corrupted with random noise $\varepsilon^{(1)},\cdots,\varepsilon^{(n)}$. If someone gave the same inputs again to the system, it would not generate exactly the same measurements as before. That is, given a set of inputs $\mathbf{x}^{(1)},\cdots,\mathbf{x}^{(n)}$ in data set $D^{(1)}$, we can estimate an optimal $\hat{\mathbf{w}}$ as in equation (4.28). But if the same inputs are used for generating a second data set $D^{(2)}$, our estimate $\hat{\mathbf{w}}$ would generally be different from $D^{(1)}$. Indeed if we do this over and over, we would see that $\hat{\mathbf{w}}$ has a distribution that depends upon the statistics of the noise on the data. We can compute this distribution as follows:

$$\begin{aligned}\hat{\mathbf{w}} &= \left(X^T X\right)^{-1} X^T \mathbf{y} \\ &= \left(X^T X\right)^{-1} X^T \mathbf{y}^* + \left(X^T X\right)^{-1} X^T \boldsymbol{\varepsilon}. \\ &= \mathbf{w} + \left(X^T X\right)^{-1} X^T \boldsymbol{\varepsilon}\end{aligned} \tag{4.31}$$

In equation (4.31), the vector y^* is the vector of "true" outputs $\left[y^{*(1)}, \ y^{*(2)}, \ \ldots, \ y^{*(n)}\right]^T$ and the term $\boldsymbol{\varepsilon}$ is the corresponding vector random variable whose elements are the noises ε in equation (4.17). From equation (4.31) we can compute the probability distribution of $\hat{\mathbf{w}}$:

$$\hat{\mathbf{w}} \sim N\left(\mathbf{w}, \operatorname{var}\left(\left(X^T X\right)^{-1} X^T \boldsymbol{\varepsilon}\right)\right). \tag{4.32}$$

The term *var* inside the parenthesis is the variance-covariance matrix of the distribution. We compute it as follows:

$$\begin{aligned}\operatorname{var}\left(\left(X^T X\right)^{-1} X^T \boldsymbol{\varepsilon}\right) &= \left(X^T X\right)^{-1} X^T \operatorname{var}(\boldsymbol{\varepsilon}) X \left(X^T X\right)^{-T} \\ &= \left(X^T X\right)^{-1} X^T I \sigma^2 X \left(X^T X\right)^{-T} \\ &= \sigma^2 \left(X^T X\right)^{-1}\end{aligned} \tag{4.33}$$

So, the probability distribution of our estimate $\hat{\mathbf{w}}$ has a mean equal to the "true" value $\mathbf{w}$, but a covariance that depends both on the measurement noise σ^2 and the specific inputs $\mathbf{x}^{(i)}$ (recall that the rows of matrix X consists of $\mathbf{x}^{(i)}$). Intuitively, the variance of the vector $\mathbf{w}$ is simply the ratio of the output variance to the input covariance. Because we have assumed that the input variable is noiseless, then a large excursion in the input vector value—corresponding to higher covariance values, would lead to a diminished uncertainty on the estimate $\hat{\mathbf{w}}$. Larger variability of the input corresponds to a broader exploration of the domain over which the transformation from $\mathbf{x}$ to $\mathbf{y}$ is defined. Basically, this means that when we are given input data that ranges over a large region, we are more certain about our estimate $\hat{\mathbf{w}}$. Thus, it is not surprising that the estimate of $\mathbf{w}$ becomes more accurate. In contrast, large amount of output variability can only reduce our confidence on the estimate of $\mathbf{w}$.

Now, returning to the question at the beginning of this section: where does uncertainty in our estimate $\hat{\mathbf{w}}$ come from? From equation (4.33) we note that some of the uncertainty comes from the fact that our measurements $y^{(i)}$ were affected by noise (with variance σ^2). If this noise was large, we would be more uncertain about our estimate. But this point is trivial. More interestingly, the uncertainty regarding $\hat{\mathbf{w}}$ also depends on the history of inputs $\mathbf{x}^{(i)}$, which are the

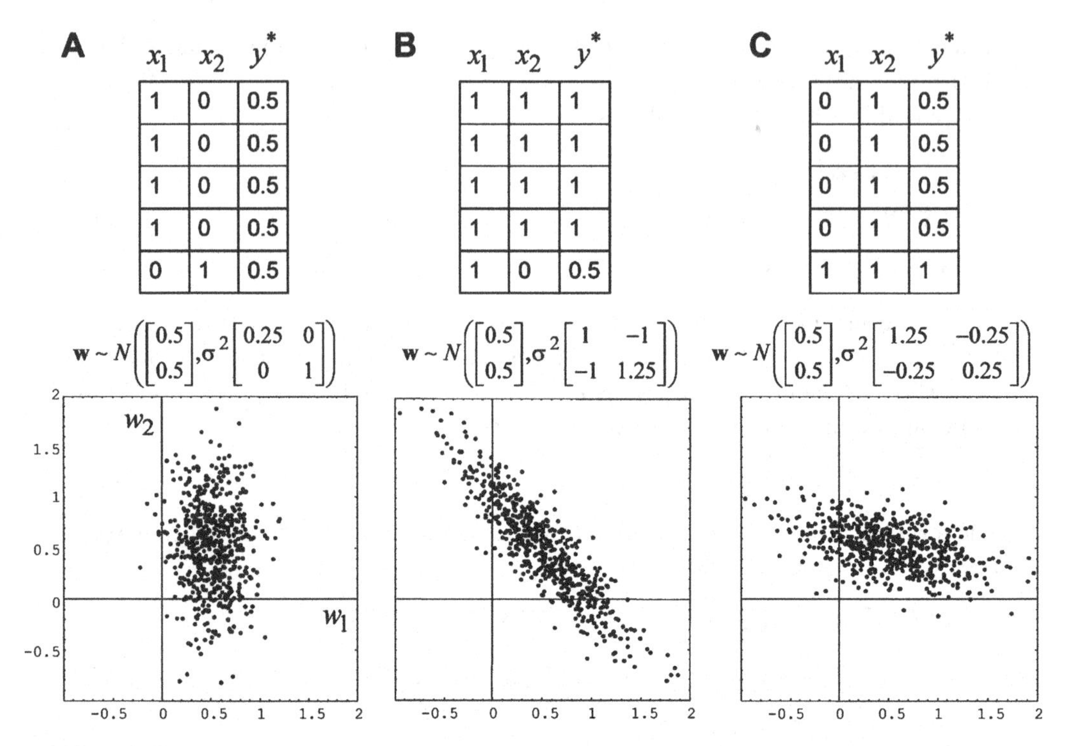

A

x_1	x_2	y^*
1	0	0.5
1	0	0.5
1	0	0.5
1	0	0.5
0	1	0.5

B

x_1	x_2	y^*
1	1	1
1	1	1
1	1	1
1	1	1
1	0	0.5

C

x_1	x_2	y^*
0	1	0.5
0	1	0.5
0	1	0.5
0	1	0.5
1	1	1

Figure 4.13
Distribution of parameter **w** in a linear regression problem. In all three cases the mean value of the distribution is [0.5, 0.5], but the covariance of the distribution depends on the specific data that were used to form the estimate.

elements of the matrix X. This is a very important idea and one that is worth thinking about.

To see how the history of inputs $\mathbf{x}^{(i)}$ affects the probability distribution of $\hat{\mathbf{w}}$, let us try some example data sets. Suppose that we are given five data points, as shown in the table at the top part of figure 4.13A. Each row is a data point, where $\mathbf{x}^{(i)} = \left[x_1^{(i)} \quad x_2^{(i)}\right]^T$ and $y^{(i)} = y^{*(i)} + \varepsilon$ (the noise is not explicitly specified in the table, but it has a normal distribution). What we need to estimate is $\mathbf{w} = [w_1 \quad w_2]^T$ in our model $y^{(i)} = \mathbf{w}^T\mathbf{x}^{(i)} + \varepsilon$. We notice that in our data, x_1 is "on" most of the time, x_2 is on once, and x_1 and x_2 are never on together. So when we go and estimate w_1 and w_2, we should be able to estimate w_1 with a lot of confidence, but perhaps not so for w_2. If we generate lots of data sets from this table (that is, we keep $\mathbf{x}^{(i)}$ as specified in this table and generate $y^{(i)} = y^{*(i)} + \varepsilon$ by adding random noise), and find $\hat{\mathbf{w}}$ for each data set, we will end up with a distribution shown in figure 4.13A. The distribution demonstrates that we can be pretty certain of $\hat{w}_1$ but we will be fairly uncertain for $\hat{w}_2$. This is simply a reflection of the amount of information we received for x_1

and x_2 (we got more information, in some vague sense, about x_1 than x_2). Furthermore, the distribution has zero covariance between $\hat{w}_1$ and $\hat{w}_2$, which is a reflection of the fact that x_1 and x_2 were never on together. In the case of the data set in figure 4.13B, x_1 and x_2 are mostly on together. The resulting $\hat{\mathbf{w}}$ will have the same mean as in the data in figure 4.13A (that is, the distribution is centered on +0.5, +0.5). However, now all that we can say with certainty is that as $\hat{w}_1$ increases, $\hat{w}_2$ should decrease. This is reflected in the negative covariance of the distribution.

Therefore, in linear regression the probability distribution associated with the parameter that we are trying to estimate is implicitly a reflection of the statistics of inputs, that is, the data that we were given to form our estimate. In a sense, the covariance of the probability distribution keeps a record of this data. In the hiking problem, the uncertainty of each device was due to the history of its measurements. Similarly in the regression problem, the uncertainty of our estimate is due to the history of our measurements. In the next chapter we will consider the problem of learning, and we will see that when animals learn, they too appear to keep a measure of uncertainty about their estimates, and this uncertainty is a reflection of the history of the stimuli that they observed.

4.7 State Estimation and the Kalman Filter

In the hiking problem, we used the uncertainties of each device to combine the two measurements. The problem is basically the same when you want to combine a predicted value for a quantity with the measured value for that same quantity. For example, the evidence in figure 4.7 suggests that when the brain estimates the position of the hand-held cursor, it does so by optimally combining the value that it predicted with the value that it measured in such a way as to render minimal the variance of this estimate. The result is a "belief" or estimate that is a weighted combination of the predicted and measured values.

Unfortunately, our framework is still quite weak as it suffers from two flaws. First, we cannot incorporate our prior knowledge into the problem of estimation. For example, if we have hiked in the region before, we have some idea of where we might be. It is not clear how to combine this information with our readings from the GPS devices. Second, when we make a movement, predictions and measurements are not quantities that occur once; they are continuous streams of information. We need to continuously combine our predictions with observations, form beliefs, and then update our predictions for the future.

To give you a simple example of this, consider the problem of lifting one of those insulated coffee cups that they sell for taking your coffee with you for a drive. These cups have a lid that makes it impossible to see how much liquid is inside. Suppose that at a coffee shop, the person behind the counter just placed such a cup on the counter, so you believe it to be full, but it is in fact empty. When you go to lift it, your hand will jerk upward. Your prediction about the cup's weight was larger than the amount reported by the proprioception from your arm. Your belief about the actual weight will shift from your prediction toward the observation. As the hand keeps moving upward, this belief will converge to the mean value reported by the sensory apparatus. The rate of this convergence will have a lot to do with how strongly the brain believed in its own predictions, that is, the uncertainty of the prediction. Said in simple terms, if you have a belief that is very certain, it will take a lot of evidence (i.e., data) to change your mind.

To modify our framework so we can approach this problem, it is useful to consider again the problem of regression, but now in a trial-by-trial basis rather than in a batch form in which all the data were given to us at once. Suppose that on trial i, we are given input $\mathbf{x}^{(i)}$ and we use our current estimate $\hat{\mathbf{w}}^{(i)}$ to predict that the output should be $\hat{y}^{(i)} = \mathbf{x}^{(i)T}\hat{\mathbf{w}}^{(i)}$. However, we observe the quantity $y^{(i)}$ instead. So we have a difference between what we predicted and what we observed. Note that earlier we wrote it differently as $\hat{y}^{(i)} = \hat{\mathbf{w}}^{(i)T}\mathbf{x}^{(i)}$. If two vectors have only real components, then the scalar product is symmetric, since $\hat{\mathbf{w}}^{(i)T}\mathbf{x}^{(i)} = \mathbf{x}^{(i)T}\hat{\mathbf{w}}^{(i)}$. The first form, $\hat{y}^{(i)} = \mathbf{x}^{(i)T}\hat{\mathbf{w}}^{(i)}$, will turn out to be convenient in the following derivation. We need to combine the two pieces of information in order to update our estimate:

$$\hat{\mathbf{w}}^{(i+1)} = \hat{\mathbf{w}}^{(i)} + \mathbf{k}^{(i)}\left(y^{(i)} - \mathbf{x}^{(i)T}\hat{\mathbf{w}}^{(i)}\right). \tag{4.34}$$

It seems rational that the term $\mathbf{k}^{(i)}$ (our "sensitivity to prediction error") should somehow reflect our uncertainty about our estimate. This term is a column vector with as many entries as the parameter vector $\hat{\mathbf{w}}^{(i)}$. The quantity in parenthesis in equation (4.34) is a single number, the prediction error. If $\mathbf{x}^{(i)}$ was oriented along a dimension for which $\hat{\mathbf{w}}^{(i)}$ was uncertain then we should learn a lot from the prediction error. If, on the other hand, $\mathbf{x}^{(i)}$ was oriented along a dimension for which our estimate of $\hat{\mathbf{w}}^{(i)}$ was quite certain, then perhaps our prediction error would simply be due to measurement noise, and we should basically ignore it. Of course, this is a simplification. A large error in a dimension in which we have great certainty would actually be likely to induce a large change in the estimate. However, this would be a rather "catastrophic" or discontinuous change, whereas here we are considering only smooth gradual changes. The term $\mathbf{k}^{(i)}$ in equation (4.34) is called a Kalman gain. It was named after Rudolph Kalman, an electrical engineer and mathematician working at the Research Institute for Advanced Study in Baltimore. He discovered a principled way to set this sensitivity (Kalman, 1960), reflecting both the uncertainty in the prediction and the uncertainty in the measurement.

Let us change slightly the notation and refer to our prior knowledge about parameter $\mathbf{w}$ on trial n as $\hat{\mathbf{w}}^{(n|n-1)}$. That is, our estimate of $\mathbf{w}$ on trial n, given the past $n-1$ trials, is $\hat{\mathbf{w}}^{(n|n-1)}$. After we make an observation (i.e., we measure $y^{(n)}$), we form a new or "posterior" estimate. We can rewrite equation (4.34) using this new terminology:

$$\hat{\mathbf{w}}^{(n|n)} = \hat{\mathbf{w}}^{(n|n-1)} + \mathbf{k}^{(n)}\left(y^{(n)} - \mathbf{x}^{(n)T}\hat{\mathbf{w}}^{(n|n-1)}\right). \tag{4.35}$$

We now express the value of the output $y^{(n)}$ in terms of the true value of $\mathbf{w}$ and its associated uncertainty. This is done by inserting equations (4.17) and (4.19) into (4.35):

$$\begin{aligned}\hat{\mathbf{w}}^{(n|n)} &= \hat{\mathbf{w}}^{(n|n-1)} + \mathbf{k}^{(n)}\left(\mathbf{x}^{(n)T}\mathbf{w} + \varepsilon^{(n)} - \mathbf{x}^{(n)T}\hat{\mathbf{w}}^{(n|n-1)}\right) \\ &= \left(I - \mathbf{k}^{(n)}\mathbf{x}^{(n)T}\right)\hat{\mathbf{w}}^{(n|n-1)} + \mathbf{k}^{(n)}\varepsilon^{(n)} + \mathbf{k}^{(n)}\mathbf{x}^{(n)T}\mathbf{w}\end{aligned}. \tag{4.36}$$

Let us define $P^{(n|n-1)}$ to be the variance (i.e., uncertainty) of our prior estimate $\hat{\mathbf{w}}^{(n|n-1)}$:

$$P^{(n|n-1)} \equiv \text{var}\left(\hat{\mathbf{w}}^{(n|n-1)}\right). \tag{4.37}$$

Similarly, we define $P^{(n|n)}$ to be the variance of our posterior estimate $\hat{\mathbf{w}}^{(n|n)}$:

$$P^{(n|n)} \equiv \text{var}\left(\hat{\mathbf{w}}^{(n|n)}\right). \tag{4.38}$$

From equation (4.36), we can write our posterior variance as a function of the prior variance:

$$\begin{aligned}P^{(n|n)} &= \left(I - \mathbf{k}^{(n)}\mathbf{x}^{(n)T}\right)P^{(n|n-1)}\left(I - \mathbf{k}^{(n)}\mathbf{x}^{(n)T}\right)^T + \mathbf{k}^{(n)}\,\text{var}\left(\varepsilon^{(n)}\right)\mathbf{k}^{(n)T} \\ &= \left(I - \mathbf{k}^{(n)}\mathbf{x}^{(n)T}\right)P^{(n|n-1)}\left(I - \mathbf{k}^{(n)}\mathbf{x}^{(n)T}\right)^T + \sigma^2\mathbf{k}^{(n)}\mathbf{k}^{(n)T}\end{aligned}. \tag{4.39}$$

(The last term in equation (4.36) depends on $\mathbf{w}$, which is not a random variable and has zero variance.) Now our problem is to set the term $\mathbf{k}^{(n)}$ in such a way so that our posterior estimate $\hat{\mathbf{w}}^{(n|n)}$ is as certain as possible, that is, of minimum variance. Therefore, our problem is to find $\mathbf{k}^{(n)}$ so to minimize $P^{(n|n)}$. However, $P^{(n|n)}$ is a matrix, so we need to clarify what we mean by "minimizing" it. This requires us to first define the size or "norm" of a matrix. The norm of a matrix must be a positive number whose value is zero only when the matrix is zero. There are several definitions that fit this criterion. One, for example, is the largest singular value. However, it would be difficult to use it for our purpose. Another kind of "norm" is simply the trace of a matrix. This is clearly not a valid norm for all matrices, because a nonzero matrix can have a zero trace. However, the variance matrix can only have positive or zero terms along the main diagonal. And it is easy to verify that if all the terms along the diagonal are zero, that is, if the trace is zero, the whole variance matrix must also be zero (because the diagonal elements are σ_1^2, σ_2^2, etc., and the off-diagonal terms are $\rho_{1,2}\sigma_1\sigma_2$, $\rho_{2,1}\sigma_1\sigma_2$ etc., where $\rho_{1,2}$ is the correlation between $\hat{w}_1$ and $\hat{w}_2$). Therefore the trace is a very simple way to measure the size of a variance matrix. If we minimize the trace of the matrix $P^{(n|n)}$, we minimize the sum of the diagonal elements. The diagonal elements of $P^{(n|n)}$ are variances of the individual elements of the vector $\hat{\mathbf{w}}$.

By minimizing the diagonal elements, it may seem that one is ignoring the covariance terms. But this is not really the case, because the covariance terms are proportional to the square root of the individual variances. Kalman's approach was to set the term **k** so to minimize the trace of the posterior uncertainty. Multiplying out the terms in equation (4.39) and then finding their trace results in the following:

$$\begin{aligned} tr\left[P^{(n|n)}\right] = tr\left[P^{(n|n-1)}\right] - tr\left[P^{(n|n-1)}\mathbf{x}^{(n)}\mathbf{k}^{(n)T}\right] - tr\left[\mathbf{k}^{(n)}\mathbf{x}^{(n)T}P^{(n|n-1)}\right] \\ + tr\left[\mathbf{k}^{(n)}\left(\mathbf{x}^{(n)T}P^{(n|n-1)}\mathbf{x}^{(n)} + \sigma^2\right)\mathbf{k}^{(n)T}\right] \end{aligned} . \tag{4.40}$$

The trace is a linear operator with some nice properties that we can use to simplify equation (4.40). For example, $tr(A) = tr(A^T)$. Therefore:

$$\begin{aligned} tr\left[P^{(n|n)}\right] = tr\left[P^{(n|n-1)}\right] - 2tr\left[\mathbf{k}^{(n)}\mathbf{x}^{(n)T}P^{(n|n-1)}\right] \\ + tr\left[\mathbf{k}^{(n)}\left(\mathbf{x}^{(n)T}P^{(n|n-1)}\mathbf{x}^{(n)} + \sigma^2\right)\mathbf{k}^{(n)T}\right] \end{aligned} . \tag{4.41}$$

Noting that the term $\mathbf{x}^{(n)T}P^{(n|n-1)}\mathbf{x}^{(n)}+\sigma^2$ is a scalar quantity and that $tr[aA] = a\ tr[A]$, we have:

$$\begin{aligned} tr\left[P^{(n|n)}\right] = tr\left[P^{(n|n-1)}\right] - 2tr\left[\mathbf{k}^{(n)}\mathbf{x}^{(n)T}P^{(n|n-1)}\right] \\ + \left(\mathbf{x}^{(n)T}P^{(n|n-1)}\mathbf{x}^{(n)} + \sigma^2\right) tr\left[\mathbf{k}^{(n)}\mathbf{k}^{(n)T}\right] \end{aligned} . \tag{4.42}$$

One last important step can be taken by observing that the trace of the external product of two vectors (a column vector multiplying a row vector) is simply the dot product of the same vectors. Noting that the second and third terms of equation (4.42) contain traces of such external products, we can further simplify the expression:

$$\begin{aligned} tr\left[P^{(n|n)}\right] = tr\left[P^{(n|n-1)}\right] - 2\mathbf{k}^{(n)T}P^{(n|n-1)}\mathbf{x}^{(n)} \\ + \left(\mathbf{x}^{(n)T}P^{(n|n-1)}\mathbf{x}^{(n)} + \sigma^2\right)\mathbf{k}^{(n)T}\mathbf{k}^{(n)} \end{aligned} . \tag{4.43}$$

To minimize this expression, we find its derivative with respect to **k**. We have:

$$\frac{d}{d\mathbf{k}^{(n)}} tr\left[P^{(n|n)}\right] = -2P^{(n|n-1)}\mathbf{x}^{(n)} + \left(\mathbf{x}^{(n)T}P^{(n|n-1)}\mathbf{x}^{(n)} + \sigma^2\right)\left(2\mathbf{k}^{(n)}\right). \tag{4.44}$$

We set equation (4.44) equal to zero and solve for **k**, resulting in:

$$\mathbf{k}^{(n)} = \frac{P^{(n|n-1)}}{\left(\mathbf{x}^{(n)T}P^{(n|n-1)}\mathbf{x}^{(n)} + \sigma^2\right)}\mathbf{x}^{(n)}. \tag{4.45}$$

Because the second derivative of equation (4.43) is positive definite:

$$\frac{d^2}{d\mathbf{k}\mathbf{k}^T} tr\left[P^{(n|n)}\right] = 2\left(\mathbf{x}^{(n)T}P^{(n|n-1)}\mathbf{x}^{(n)} + \sigma^2\right), \tag{4.46}$$

the expression in equation (4.45) corresponds to a minimum.

Note that the measurement uncertainty as expressed by the variance of the output is:

$$\mathrm{var}\left(y^{(n)}\right) = \mathrm{var}\left(\mathbf{w}^{(n|n-1)T}\mathbf{x}^{(n)} + \varepsilon\right) = \mathbf{x}^{(n)T}P^{(n|n-1)}\mathbf{x}^{(n)} + \sigma^2. \tag{4.47}$$

Therefore, the optimum sensitivity to prediction error, or Kalman gain, is a ratio between our prior uncertainty on the parameter that we are trying to estimate (numerator of equation 4.45), and our measurement uncertainty (the denominator). If we are uncertain about what we know (the numerator is "large" relative to the denominator), we should learn a lot from the prediction error. If we are uncertain about our measurement (the denominator is large relative to the numerator), we should ignore the prediction error. (You can perhaps already guess that the Kalman gain formula in equation (4.35) is a very useful model of biological learning, which is something that we will explore in detail in the next chapter.)

Our final step is to formulate the posterior uncertainty, i.e., the variance of $\mathbf{w}^{(n|n)}$. We insert equation (4.45) into equation (4.39), and after a bit of simplification[1] we arrive at the posterior uncertainty:

$$P^{(n|n)} = \left(I - \mathbf{k}^{(n)}\mathbf{x}^{(n)T}\right)P^{(n|n-1)}. \tag{4.48}$$

Since the term $\mathbf{k}^{(n)}\mathbf{x}^{(n)T}$ is positive definite, equation (4.48) implies that typically, our uncertainty declines as we make more observations.

In figure 4.14 we have summarized the problem that we just solved. We begin with the assumption that there was a hidden variable $\mathbf{w}$ that linearly interacts with a known quantity $\mathbf{x}^{(n)}$ on each trial to produce the measurement $y^{(n)}$. To estimate $\mathbf{w}$, we have a prior estimate $\hat{\mathbf{w}}^{(1|0)}$ and prior uncertainty $P^{(1|0)}$. We compute $\mathbf{k}^{(1)}$ using equation (4.45) and the posterior uncertainty using equation (4.48) and then update our estimate using the prediction error (equation 4.35). Because we assume that the variable $\mathbf{w}$ did not change from trial to trial, we compute the prior estimate for the next trial by simply setting it to the equal to the posterior estimate of the last trial:

$$\begin{aligned} \hat{\mathbf{w}}^{(n+1|n)} &= \hat{\mathbf{w}}^{(n|n)} \\ P^{(n+1|n)} &= P^{(n|n)} \end{aligned}. \tag{4.49}$$

What if we believe that the variable $\mathbf{w}$ will not stay constant from trial to trial? For example, in figure 4.15 we have assumed that $\mathbf{w}$ changes from trial to trial. This change is simply governed by a matrix A (assumed to be known), and random noise ε_w. Now our *generative model* (that is, the model that we assume is responsible for generating our data) becomes:

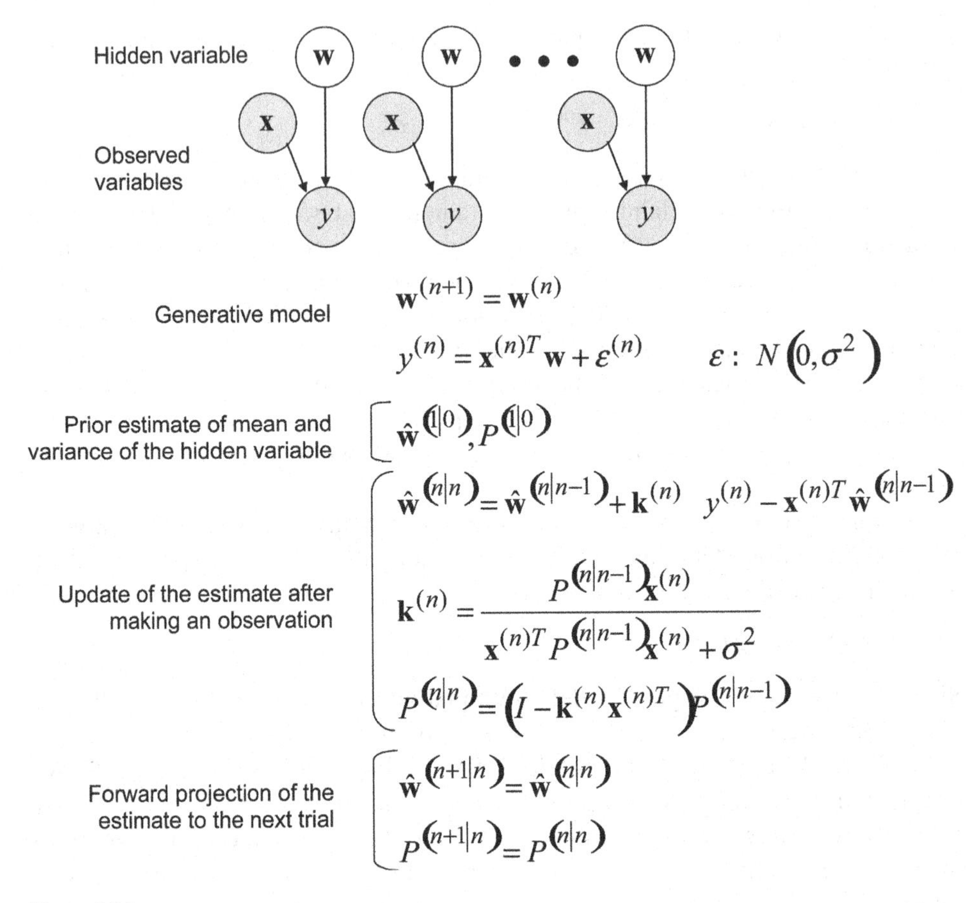

Figure 4.14
Estimation of a hidden variable via Kalman filter. In this example, the hidden state **w** is assumed to be invariant. The open circles indicate hidden variables. The gray circles indicate observed variables.

$$\begin{aligned} \mathbf{w}^{(n+1)} &= A\mathbf{w}^{(n)} + \boldsymbol{\varepsilon}_w^{(n)} & \boldsymbol{\varepsilon}_w &\sim N(0,Q) \\ y^{(n)} &= \mathbf{x}^{(n)T}\mathbf{w}^{(n)} + \varepsilon_y^{(n)} & \varepsilon_y &\sim N(0,\sigma^2) \end{aligned} . \tag{4.50}$$

In equation (4.50), the noise $\boldsymbol{\varepsilon}_w$ in the state update equation reflects our uncertainty about how **w** might change from trial to trial. The effect of this assumption is to alter how we project our estimates $\hat{\mathbf{w}}$ from the posterior estimate of one trial to the prior estimate of the next trial:

$$\begin{aligned} \hat{\mathbf{w}}^{(n+1|n)} &= A\hat{\mathbf{w}}^{(n|n)} \\ P^{(n+1|n)} &= AP^{(n|n)}A^T + Q \end{aligned} . \tag{4.51}$$

The algorithm for solving this version of our problem is summarized in figure 4.15.

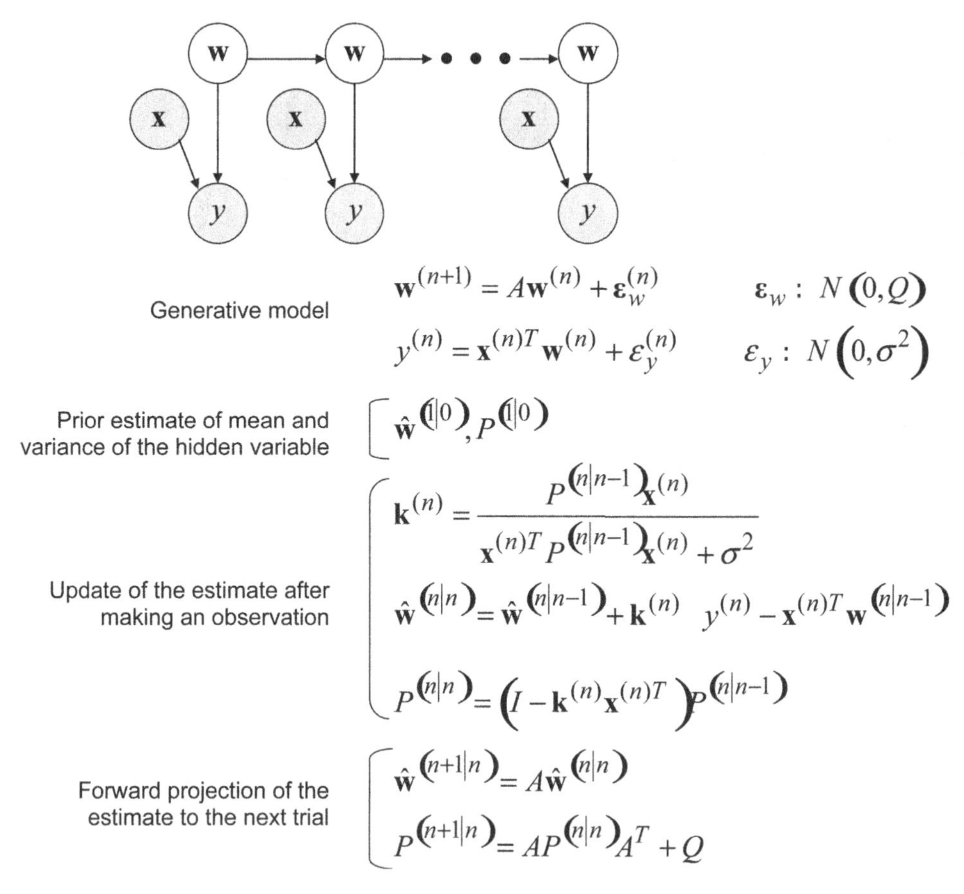

Figure 4.15
Estimation of a hidden variable via Kalman filter. In this example, the hidden state $\mathbf{w}$ is assumed to change from trial to trial.

4.8 Combining Predictions with Delayed Measurements

We now have an algorithm to continuously combine our prior predictions with observations, form posterior beliefs, and then form predictions for the future observations. We have the tools in place to test a simple prediction of our framework: when the brain estimates a value but this value differs from a measured value, the rate at which belief is corrected and converges onto the measured value will depend on the uncertainty of the predictions—the higher the certainty of prediction, the slower the convergence. In other words, the more confident you are about your predictions, the more evidence you will need to change your mind.

Jun Izawa and Shadmehr (Izawa and Shadmehr, 2008) tested this idea by having people move a handheld LED to the center of a target (figure 4.16A). People could

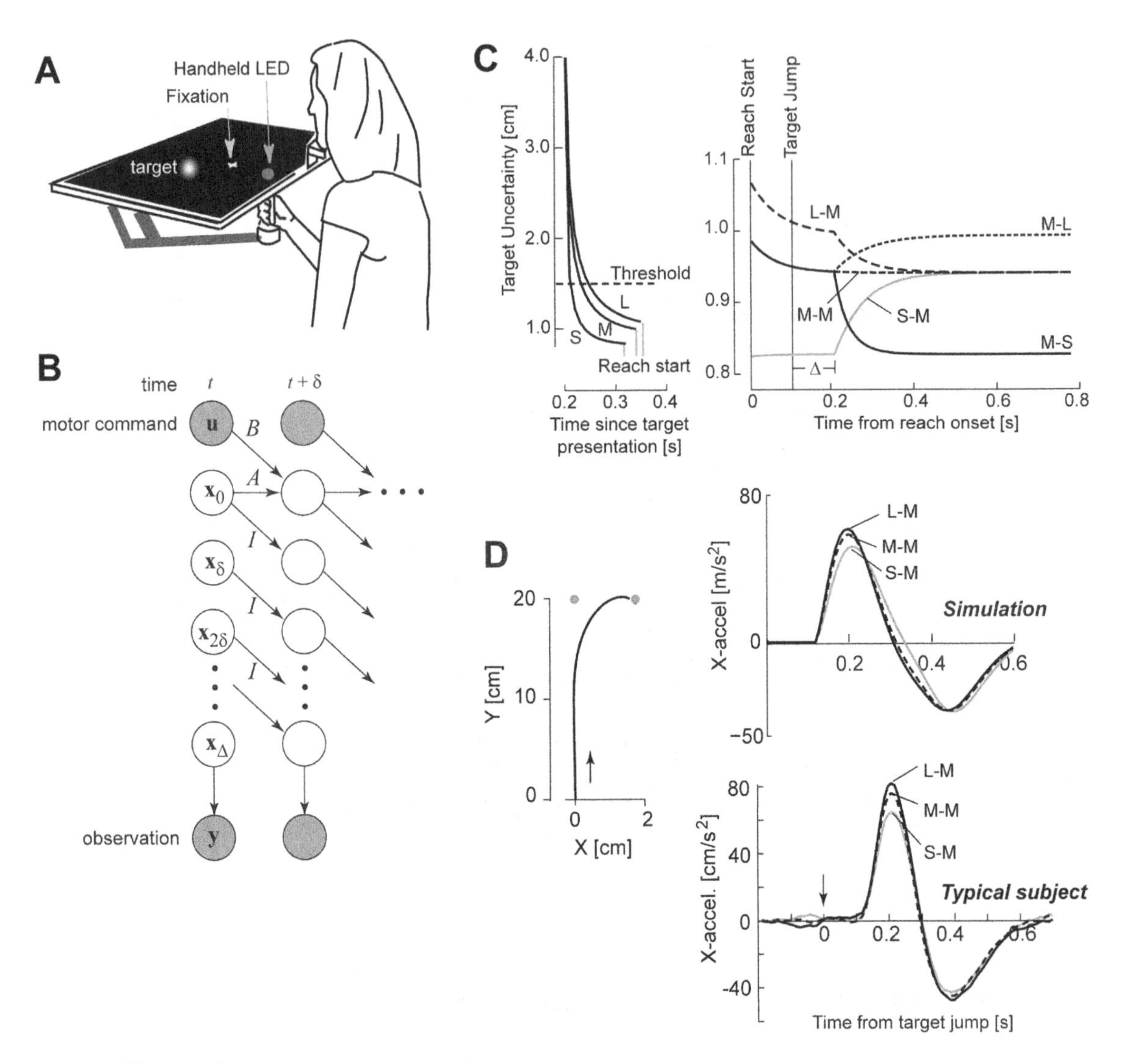

Figure 4.16
The rate at which belief converges onto the measured value depends on the uncertainty of the prior belief. (A) Subjects reached to a bloblike target, with visual characteristics that were described by a Gaussian of small, medium, or large variance. The objective was to place the handheld LED in the center of the target. (B) A generative model describing the relationship between input (motor commands), state (position of the target and hand), and observation (delayed measurement of state). (C) Model predictions. Uncertainty of the state of the target as a function of time. Movement starts earlier for a small variance target. When the target jumps, the prior uncertainty is specified by the first target, and uncertainty converges to the variance of the second target. (D) Example data. When the target jumped, the hand path was corrected to the new target. However, for a given second target (medium uncertainty here), the rate of correction was largest when the first target had the largest uncertainty. (From Izawa and Shadmehr, 2008.)

view the handheld LED at all times, but the position of the target was made uncertain by displaying it as a "blob" (a display in which pixel intensity was described by a Gaussian). This noise affected the certainty with which the subject could estimate the target position. For example, in trials in which the target's position was certain, reaction time (time to start the reach) was shorter and endpoint variance was small. In a fraction of the trials, as the movement was unfolding the target changed position by a small amount. When it changed position, the blob's characteristics also changed (e.g., the distribution that describes the blob became tighter, that is, easier to estimate its center). According to the framework presented earlier, the brain's belief about target position could not change instantaneously to the new position reported by the sensory system. Rather, the rate at which the belief would converge to this new position should depend on the uncertainty of the prior belief, controlled by the characteristics of the first target. For example, if the first target had low uncertainty but the second target has medium uncertainty, the rate of change in hand position should be slow. However, if the first target had high uncertainty, the rate of change in hand position toward the same second target should be fast.

One can describe the estimation problem with the generative model shown in figure 4.16B. The states that we wish to estimate include position of the target, and position and velocity of the handheld LED, all represented by vector $\mathbf{x}$. Our observation $\mathbf{y}$ is a noisy, delayed version of these states. With our motor commands $\mathbf{u}$ to the arm we manipulate the position of the handheld LED. The state update equation becomes:

$$\mathbf{x}^{(k+1)} = A\mathbf{x}^{(k)} + B\mathbf{u}^{(k)} + \varepsilon^{(k)}. \tag{4.52}$$

To represent the fact that our observation is a delayed version of the states, a simple technique is to extend the state vector by its copies. Suppose that an event at time step k was sensed by the first stage of sensory processing and then transmitted to the next stage at time $k + \delta$, where in this case $\delta = 10$ ms. This information continues to propagate and becomes "observable" with a delay of Δ. Thus, we can represent the delay by first extending the state vector $\mathbf{x}_e^{(k)} = \left[\mathbf{x}_0^{(k)}, \mathbf{x}_\delta^{(k)}, \mathbf{x}_{2\delta}^{(k)}, \cdots, \mathbf{x}_\Delta^{(k)}\right]^T$ and then allowing the system to observe only the most delayed state $\mathbf{x}_\Delta^{(k)}$. The sparse matrix C takes care of this last step:

$$\begin{aligned} \mathbf{x}_e^{(k+1)} &= A_e\mathbf{x}_e^{(k)} + B\mathbf{u}^{(k)} + \varepsilon_e^{(k)} \\ \mathbf{y}^{(k)} &= C\mathbf{x}_e^{(k)} + \omega^{(k)} \end{aligned} . \tag{4.53}$$

The basic prediction of this model is that when our sensory system reports that some Δ time ago the target moved to a new position, our estimate of that position does not change immediately to this new position. Rather, our belief converges to the observed position with a rate that depends on our uncertainty about our current estimate. A simulation that shows this idea is provided in figure 4.16C. When the first target is displayed, uncertainty about its position rapidly declines. The rate of decline is fastest for the target with the smallest variance (blob variance is S, M, or L). If we set an arbitrary threshold on our uncertainty, we can see that crossing this threshold is earlier for the small variance target, potentially explaining the faster reaction time. As the movement unfolds, the target jumps to a new location. If the second target has a larger variance, uncertainty will increase after a delay period Δ (for example, condition S-M or M-L). Let us compare the S-M condition with the L-M condition. Reaction to the target jump is most influenced by the prior uncertainty at the time of the target jump. The S-M condition has a low uncertainty when one senses the target jump, and therefore a small Kalman gain. The L-M condition has a high uncertainty and a larger Kalman gain. The large Kalman gain will produce a rapid reaction to the information regarding the changed target position. This is shown in the simulated hand acceleration traces in figure 4.16D. Indeed, when the first target position had a small uncertainty and the second target had medium uncertainty (S-M condition, figure 4.16D typical subject), people corrected the handheld LED's trajectory gradually toward the second target. However, if the first target's uncertainty was large, reaction to that same medium uncertainty second target (L-M condition, figure 4.16D) was quite strong: people corrected the hand path vigorously.

4.9 Hiking in the Woods in an Estimation Framework

Let us now return to the problem of combining multiple sources of information about a hidden variable—the hiking-in-the-woods problem—and recast it in the Kalman framework. Our objective is to show that the Kalman gain is precisely the weights in equation (4.9) that we earlier assigned to the two devices. That is, when our only source of information is from our observation, and we have no prior beliefs, then the Kalman gain is the maximum likelihood estimate.

A graphical representation of the generative model for the hiking in the woods problem is shown in figure 4.17. The hidden variable $\mathbf{x}$ is observed via two independent measurements $\mathbf{y}_a$ and $\mathbf{y}_b$. Our generative model takes the form:

$$\begin{aligned} \mathbf{x}^{(n+1)} &= A\mathbf{x}^{(n)} + \boldsymbol{\varepsilon}_x^{(n)} \qquad \boldsymbol{\varepsilon}_x \sim N(\mathbf{0}, Q) \\ \mathbf{y}^{(n)} &= C\mathbf{x}^{(n)} + \boldsymbol{\varepsilon}_y^{(n)} \qquad \boldsymbol{\varepsilon}_y \sim N(\mathbf{0}, R) \end{aligned} . \tag{4.54}$$

We begin with a prior estimate $\mathbf{x}^{(1|0)}$ and our uncertainty $P^{(1|0)}$. The general form of the Kalman gain and posterior uncertainty is:

$$\begin{aligned} \mathbf{k}^{(n)} &= P^{(n|n-1)}C^T \left(CP^{(n|n-1)}C^T + R\right)^{-1} \\ P^{(n|n)} &= \left(I - \mathbf{k}^{(n)}C\right)P^{(n|n-1)} \end{aligned} . \tag{4.55}$$

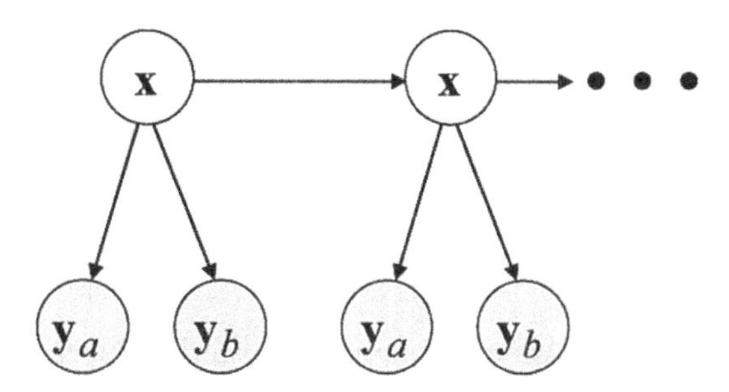

Figure 4.17
A graphical representation of the generative model for the hiking-in-the-woods problem. In this problem, we have two sensors $\mathbf{y}_a$ and $\mathbf{y}_b$ that provide us with independent measures of a hidden state $\mathbf{x}$.

Suppose that we have no idea where we are, that is, $P^{(1|0)} = \infty$. In this case, we cannot compute the Kalman gain from equation (4.55). However, we can proceed by first computing the posterior estimate (the second line in equation 4.55) in terms of the prior estimate, and then rewrite the Kalman gain (the first line in equation 4.55) in terms of the posterior rather than prior uncertainty. We begin by expressing the posterior uncertainty in terms of the prior:

$$\begin{aligned} P^{(n|n)} &= P^{(n|n-1)} - \mathbf{k}^{(n)} C P^{(n|n-1)} \\ &= P^{(n|n-1)} - P^{(n|n-1)} C^T \left(C P^{(n|n-1)} C^T + R \right)^{-1} C P^{(n|n-1)} \end{aligned} \tag{4.56}$$

We can simplify the second line of equation (4.56) by using the matrix inversion lemma. This lemma states that:

$$\left(Z - X Y^{-1} X^T \right)^{-1} = Z^{-1} + Z^{-1} X \left(Y - X^T Z^{-1} X \right)^{-1} X^T Z^{-1}. \tag{4.57}$$

Let us set $-Z^{-1} = P^{(n|n-1)}$, $X = C^T$, and $Y = R$. We can rewrite the second line in equation (4.56) as:

$$P^{(n|n)} = -Z^{-1} - Z^{-1} X \left(Y - X^T Z^{-1} X \right)^{-1} X^T Z^{-1}. \tag{4.58}$$

Therefore we have:

$$-P^{(n|n)} = \left(Z - X Y^{-1} X^T \right)^{-1} = \left(-\left(P^{(n|n-1)} \right)^{-1} - C^T R^{-1} C \right)^{-1},$$

which we can simplify to:

$$\left(P^{(n|n)} \right)^{-1} = \left(P^{(n|n-1)} \right)^{-1} + C^T R^{-1} C. \tag{4.59}$$

Equation (4.59) explains that if our prior uncertainty is infinite, then the inverse of our posterior uncertainty is simply $C^T R^{-1} C$. Now, let us express the Kalman gain in terms of the posterior uncertainty. We begin by multiplying both sides of the first line in equation 4.55 by the term in the parenthesis:

$$\mathbf{k}^{(n)}\left(CP^{(n|n-1)}C^T+R\right)=P^{(n|n-1)}C^T. \tag{4.60}$$

We next multiply both sides of the above expression by R^{-1} and then after a little rearrangement arrive at:

$$\begin{aligned}\mathbf{k}^{(n)}&=P^{(n|n-1)}C^TR^{-1}-\mathbf{k}^{(n)}CP^{(n|n-1)}C^TR^{-1}\\&=\left(P^{(n|n-1)}-\mathbf{k}^{(n)}CP^{(n|n-1)}\right)C^TR^{-1}\end{aligned}. \tag{4.61}$$

The term in the parenthesis in the above expression is simply the posterior uncertainty, allowing us to express the Kalman gain for the system of equation (4.54) in this way:

$$\mathbf{k}^{(n)}=P^{(n|n)}C^TR^{-1}. \tag{4.62}$$

If we insert our prior $P^{(1|0)}=\infty$ into equation (4.59), our posterior becomes

$$P^{(1|1)}=\left(C^TR^{-1}C\right)^{-1}. \tag{4.63}$$

The Kalman gain becomes:

$$\mathbf{k}^{(1)}=\left(C^TR^{-1}C\right)^{-1}C^TR^{-1}. \tag{4.64}$$

Our estimate of the hidden variable (our position) is:

$$\mathbf{x}^{(1|1)}=\mathbf{x}^{(1|0)}+\mathbf{k}^{(1)}\left(\mathbf{y}^{(1)}-C\mathbf{x}^{(1|0)}\right). \tag{4.65}$$

Assuming that our prior $\mathbf{x}^{(1|0)}$ was zero, the above expression reduces to:

$$\mathbf{x}^{(1|1)}=\left(C^TR^{-1}C\right)^{-1}C^TR^{-1}\mathbf{y}^{(1)}. \tag{4.66}$$

This expression is our maximum likelihood estimate in equation (4.8). Furthermore, the variance of our estimate, as expressed in equation (4.63), is the variance of our maximum likelihood estimate in equation (4.11). Therefore, if we are naïve in the sense that we have no prior knowledge about the state that we wish to estimate, then a weighted combination of the two sources of information is the optimal solution. On the other hand, if we also have a prior—for example, we have hiked this path before and have some idea of where we might be—then the Kalman framework gives us the tools to weigh in this additional piece of information.

From a practical point of view, a useful result that we derived in this section is with regard to the prior uncertainty. Often when we try to estimate something we are at a loss as to what the prior uncertainty should be. If we are completely naïve, then equation (4.63) tells us what the uncertainty will be after the first data point is observed. A common technique is to set the prior uncertainty to this value.

4.10 Signal-Dependent Noise

Thus far we have considered processes that have additive Gaussian noise. With this kind of noise, the variance of the variable that we are observing is independent of its mean. If you think about it, this is a bit odd: the signal varies by some amount and the variance is the same whether the signal has a small amplitude or a large amplitude. Biological systems seem to have a different kind of noise: variance varies with the size of the signal. For example, Richard Schmidt and colleagues (Schmidt et al., 1979) asked volunteers to make rapid reaching movements to a target and measured the variability of the endpoint (figure 4.18). They found that for a given target distance, the smaller durations (i.e., faster movements) were associated with larger endpoint variance. As movement duration decreased, the force required to make that movement would of course increase. Richard Schmidt hypothesized that the noise in a muscle was likely dependent on the force developed by the muscle (Schmidt, 1991). He wrote: "Movement's inaccuracy increases as movement time decreases, primarily because of the increased noise involved in the stronger muscle contractions." That is, noise associated with the motor commands was likely an increasing function of the magnitude of those commands.

We can examine the noise properties of muscles in an experiment in which volunteers produced a force by pushing with their thumb on a transducer while viewing the resulting force on a video monitor. Kelvin Jones, Antonia Hamilton, and Daniel Wolpert (2002) measured this force and focused on the standard deviation of this

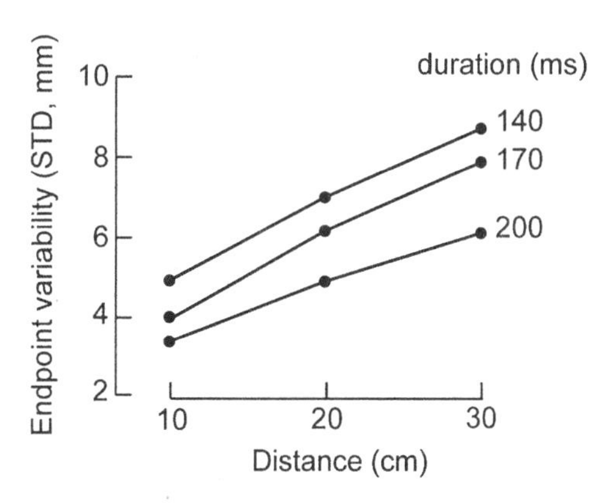

Figure 4.18
Endpoint variability of rapid aiming movements (shown as standard deviation of the endpoint position) as a function of movement duration and amplitude. Reduced duration increases endpoint variability. (Data from Schmidt et al., 1979.)

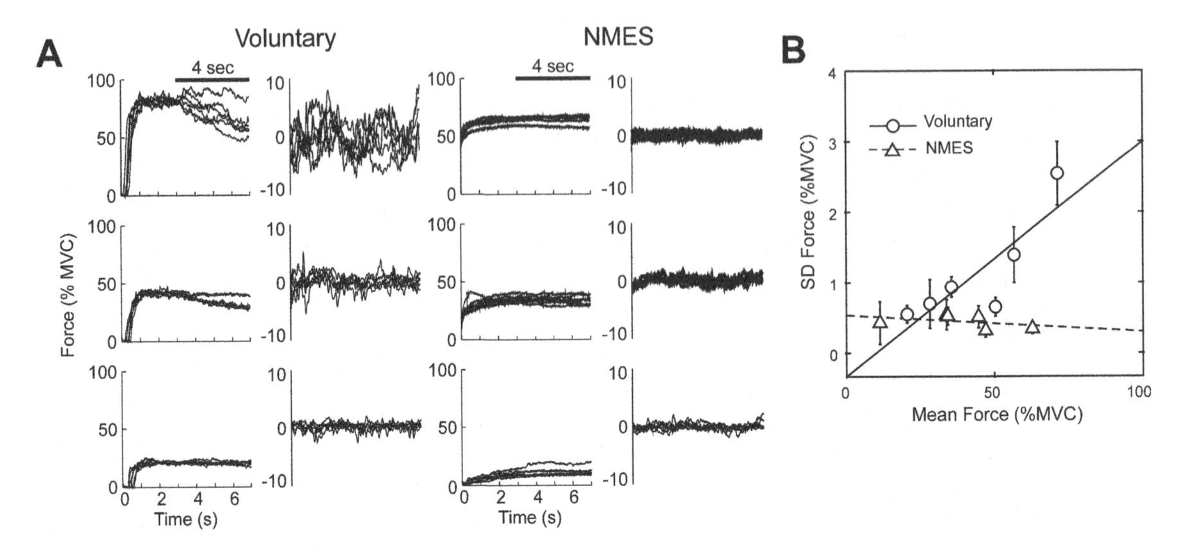

Figure 4.19
The standard deviation of noise grows with mean force in an isometric task. Participants produced a given force with their thumb flexors. In one condition (labeled "voluntary"), the participants generated the force, whereas in another condition (labeled "NMES") the experimenters stimulated the muscles artificially to produce force. To guide force production, the participants viewed a cursor that displayed thumb force, but the experimenters analyzed the data during a 4-s period in which this feedback had disappeared. (A) Force produced by a typical participant. The period without visual feedback is marked by the horizontal bar in the first and third columns (top right) and is expanded in the second and fourth columns. (B) When participants generated force, noise (measured as the standard deviation) increased linearly with force magnitude (with a slope of ~1). NMES, neuromuscular electrical stimulation; MVC, maximum voluntary contraction. (From Jones, Hamilton, and Wolpert, 2002.)

variable during a 4 sec period in which the visual feedback was eliminated (shown in the second column of figure 4.19A). They found that the standard deviation of the force grew roughly linearly as a function of mean force (figure 4.19B). The change in the standard deviation of force may have been because of the transduction of neural signal into force in the muscle, or from the neural signal itself. To resolve this question, Jones and colleagues electrically stimulated the thumb flexor muscle (shown in the third and fourth column of figure 4.19A). They did not observe an increase in variance with increased stimulation strength. This result suggested that the noise originated from the neural motor commands and was not due to the muscle and how it produced force.

The term *signal dependent noise* refers to a process in which the standard deviation of the noise depends on the mean of the signal. For example, the force produced by muscle i may be related to its input u_i via a noise process that has the following form:

$$f_i = u_i(1 + c_i\phi_i) \qquad \phi_i \sim N(0,1). \tag{4.67}$$

The term c_i indicates the rate at which the noise grows with the signal. That is, the variance of force increases as a function of the signal u_i:

$$\mathrm{var}[f_i] = c_i^2 u_i^2. \tag{4.68}$$

And so the standard deviation of force increases linearly with u_i, with slope c_i.

How would we estimate the state of a system if it suffers from signal dependent noise? The Kalman framework that we have been using will need to be modified. Suppose that we have a system in which we produce motor commands **u** (e.g., force in the muscles), and this affects the state of our system **x** (e.g., position, velocity, of our body), resulting in sensory feedback **y** (e.g., proprioception, vision). Suppose that the system is of the form:

$$\begin{aligned} \mathbf{x}^{(k+1)} &= A\mathbf{x}^{(k)} + B\left(\mathbf{u}^{(k)} + \boldsymbol{\varepsilon}_u^{(k)}\right) + \boldsymbol{\varepsilon}_x^{(k)} \\ \mathbf{y}^{(k)} &= H\left(\mathbf{x}^{(k)} + \boldsymbol{\varepsilon}_s^{(k)}\right) + \boldsymbol{\varepsilon}_y^{(k)} \end{aligned}, \tag{4.69}$$

where $\boldsymbol{\varepsilon}_x$ and $\boldsymbol{\varepsilon}_y$ are zero mean Gaussian noise:

$$\begin{aligned} \boldsymbol{\varepsilon}_x &\sim N(\mathbf{0}, Q_x) \\ \boldsymbol{\varepsilon}_y &\sim N(\mathbf{0}, Q_y) \end{aligned} \tag{4.70}$$

and $\boldsymbol{\varepsilon}_u$ and $\boldsymbol{\varepsilon}_s$ are zero mean signal dependent noise terms, meaning that noise depends on the motor commands **u**, and state **x**, respectively:

$$\begin{aligned} \boldsymbol{\varepsilon}_u^{(k)} &\equiv \begin{bmatrix} c_1 u_1^{(k)} \phi_1^{(k)} \\ c_2 u_2^{(k)} \phi_2^{(k)} \\ \vdots \\ c_n u_n^{(k)} \phi_n^{(k)} \end{bmatrix} & \boldsymbol{\varepsilon}_s^{(k)} &\equiv \begin{bmatrix} d_1 x_1^{(k)} \mu_1^{(k)} \\ d_2 x_2^{(k)} \mu_2^{(k)} \\ \vdots \\ d_m x_m^{(k)} \mu_m^{(k)} \end{bmatrix}. \\ \phi &\sim N(0,1) & \mu &\sim N(0,1) \end{aligned} \tag{4.71}$$

The signal dependent motor noise $\boldsymbol{\varepsilon}_u$ affects the state **x** and the signal dependent sensory noise $\boldsymbol{\varepsilon}_s$ affects the observation **y**. It is useful to express the signal dependent noise terms as a linear function of **u** and **x**. To do so, we can define:

$$\begin{aligned} C_1 &\equiv \begin{bmatrix} c_1 & 0 & 0 \\ 0 & 0 & 0 \\ 0 & 0 & \ddots \end{bmatrix} & C_2 &\equiv \begin{bmatrix} 0 & 0 & 0 \\ 0 & c_2 & 0 \\ 0 & 0 & \ddots \end{bmatrix}. \\ D_1 &\equiv \begin{bmatrix} d_1 & 0 & 0 \\ 0 & 0 & 0 \\ 0 & 0 & \ddots \end{bmatrix} & D_2 &\equiv \begin{bmatrix} 0 & 0 & 0 \\ 0 & d_2 & 0 \\ 0 & 0 & \ddots \end{bmatrix} \end{aligned} \tag{4.72}$$

So we have:

$$\begin{aligned}\boldsymbol{\varepsilon}_u^{(k)} &= \sum_{i=1}^{m} C_i \mathbf{u} \phi_i^{(k)} \\ \boldsymbol{\varepsilon}_s^{(k)} &= \sum_{i=1}^{n} D_i \mathbf{x} \mu_i^{(k)}\end{aligned}. \tag{4.73}$$

In equation (4.73), m is the size of the vector $\mathbf{u}$ and n is the size of the vector $\mathbf{x}$. And so we can rewrite the system equations as:

$$\begin{aligned}\mathbf{x}^{(k+1)} &= A\mathbf{x}^{(k)} + B\mathbf{u}^{(k)} + \boldsymbol{\varepsilon}_x^{(k)} + B\sum_i C_i \mathbf{u}^{(k)} \phi_i^{(k)} \\ \mathbf{y}^{(k)} &= H\mathbf{x}^{(k)} + \boldsymbol{\varepsilon}_y^{(k)} + H\sum_i D_i \mathbf{x}^{(k)} \mu_i^{(k)}\end{aligned}. \tag{4.74}$$

On trial k, we have a prior belief $\hat{\mathbf{x}}^{(k|k-1)}$, with uncertainty $P^{(k|k-1)}$, and we make an observation $\mathbf{y}^{(k)}$. To update our belief, we have:

$$\hat{\mathbf{x}}^{(k|k)} = \hat{\mathbf{x}}^{(k|k-1)} + K^{(k)}\left(\mathbf{y}^{(k)} - H\hat{\mathbf{x}}^{(k|k-1)}\right). \tag{4.75}$$

How should we set $K^{(k)}$? As before, we will set it in such a way as to minimize the trace of the posterior uncertainty $P^{(k|k)}$. However, when we do this, we will see that the Kalman gain will now depend on both the state $\mathbf{x}$ and the motor commands $\mathbf{u}$.

Rewriting equation (4.75), we have:

$$\begin{aligned}\hat{\mathbf{x}}^{(k|k)} &= \hat{\mathbf{x}}^{(k|k-1)} + K^{(k)}\left(H\mathbf{x}^{(k)} + \boldsymbol{\varepsilon}_y^{(k)} + H\sum_i D_i \mathbf{x}^{(k)} \mu_i^{(k)} - H\hat{\mathbf{x}}^{(k|k-1)}\right) \\ &= \left(I - K^{(k)}H\right)\hat{\mathbf{x}}^{(k|k-1)} + K^{(k)}\left(H\mathbf{x}^{(k)} + \boldsymbol{\varepsilon}_y^{(k)} + H\sum_i D_i \mathbf{x}^{(k)} \mu_i^{(k)}\right)\end{aligned}. \tag{4.76}$$

The variance of our posterior estimate is:

$$\begin{aligned}P^{(k|k)} &= \left(I - K^{(k)}H\right)P^{(k|k-1)}\left(I - K^{(k)}H\right)^T + K^{(k)}Q_y K^{(k)T} \\ &\quad + \sum_i KHD_i \mathbf{x}^{(k)} \mathbf{x}^{(k)T} D_i^T H^T K^T\end{aligned}. \tag{4.77}$$

Equation (4.77) can be written as:

$$\begin{aligned}P^{(k|k)} &= P^{k|k-1} - 2P^{k|k-1}H^T K^{(k)T} \\ &\quad + K^{(k)}\left(HP^{k|k-1}H^T + Q_y + \sum_i HD_i \mathbf{x}^{(k)} \mathbf{x}^{(k)T} D_i^T H^T\right)K^{(k)T}\end{aligned}. \tag{4.78}$$

The derivative of the trace of the preceding equation with respect to $K^{(k)}$ is:

$$\begin{aligned}\frac{d}{dK^{(k)}} tr\left[P^{(k|k)}\right] &= -2P^{(k|k-1)}H^T \\ &\quad + 2K^{(k)}\left(HP^{(k|k-1)}H^T + Q_y + \sum_i HD_i \mathbf{x}^{(k)} \mathbf{x}^{(k)T} D_i^T H^T\right)\end{aligned}. \tag{4.79}$$

Setting this expression to zero and solving for $K^{(k)}$, we have:

$$K^{(k)} = P^{(k|k-1)}H^T\left(HP^{(k|k-1)}H^T + Q_y + \sum_i HD_i \mathbf{x}^{(k)} \mathbf{x}^{(k)T} D_i^T H^T\right)^{-1}. \tag{4.80}$$

Notice that because of signal dependent noise, the Kalman gain is a function of the state $\mathbf{x}$. (In practice, we replace the term $\mathbf{x}^{(k)}\mathbf{x}^{(k)T}$ in equation (4.80) with our estimate $E[\mathbf{x}^{(k)}\mathbf{x}^{(k)T}]$). In fact, when we compute the uncertainties, we see that the Kalman gain is also a function of the motor commands. Substitute the expression in equation (4.80) in equation (4.78), and we have:

$$P^{k|k} = P^{k|k-1}\left(I - H^T K^{(k)T}\right). \tag{4.81}$$

The prior uncertainty in step $k + 1$ becomes:

$$P^{(k+1|k)} = AP^{(k|k)}A^T + Q_x + \sum_i BC_i\mathbf{u}^{(k)}\mathbf{u}^{(k)T}C_i^T B^T. \tag{4.82}$$

Therefore, the state uncertainty increases with the size of the motor commands, and the Kalman gain decreases with the size of the state.

An implication of our derivation is that if we are pushing a large mass (producing relatively large motor commands), then we will have a larger uncertainty regarding the consequences of these commands (as compared to pushing a small mass with a smaller amount of force). As a result, when we are producing large forces we should rely more on the sensory system and our observations and less on our predictions.

Summary

As our brain plans and generates motor commands, it also predicts the sensory consequences and acts on the predicted consequences. One clear advantage of making sensory predictions is that the brain does not have to wait for the sensory measurements to know that it needs to intervene. The delay in sensory measurements is long enough that it can cause stability problems. Relying on predictions, rather than delayed measurements, allows one to overcome this delay. A second advantage of making sensory predictions is that the brain can combine its predictions with the sensory measurements (when they arrive). The combination of the two streams of information would allow the brain to sense the world better than if it only had the sensory information stream. That is, we are able to "see" the stimulus better because we have two sources of information about it, rather than just one. These two streams also allow the brain to separate the sensory data that is a result of self-generated motion from data that is due to external events. If the process of state estimation is damaged due to disease, then the result may be symptoms of delusions, that is, an inability to assign agency.

To combine two streams of information, one needs to apply a weight to each stream. In principle, the weight should be higher for the more reliable information source. State estimation theory provides a framework in which to describe this problem. In this framework, the objective is to estimate the state of our body or the world around us from our sensory measurements. This problem has two components. The first is associated with learning to accurately predict the future, that is, what the sensory measurements should be—this is our *prior* estimate of state. The second is associated with optimally combining the measured quantities with the predicted one to form a *posterior* estimate of state. Kalman first solved this problem by finding a posterior estimate that for a linear system with Gaussian noise produced a posterior estimate that had the minimum variance.

5 Bayesian Estimation and Inference

We started the last chapter by approaching the problem of state estimation from a maximum likelihood perspective. That is, for some hidden states $\mathbf{x}$ and observations $\mathbf{y}$, we formulated a generative model that described a probability distribution $p(\mathbf{y}|\mathbf{x})$, and then we found an estimate $\hat{\mathbf{x}}$ that maximized the probability of our observations. Formally, this is stated as

$$\hat{x} = \arg\max_{\mathbf{x}} (p(y \mid \mathbf{x})). \tag{5.1}$$

The problem with our approach was that we could not incorporate a prior belief[1] into this formulation, and this was a serious flaw because everything that we perceive is likely an integration of a prior belief with observations. To remedy the situation, we considered the Kalman framework. In this framework, we found the mixing gain $\mathbf{k}^{(n)}$, which allowed us to integrate our prior belief $\hat{\mathbf{x}}^{(n|n-1)}$ with our observation $\mathbf{y}^{(n)}$ to form a posterior belief $\hat{\mathbf{x}}^{(n|n)}$. The gain that we found was one that minimized the trace of the variance of the posterior uncertainty $P^{(n|n)}$. This uncertainty represented the variance of the Gaussian distribution $p(\mathbf{x}^{(n)}|\mathbf{y}^{(n)})$. Finally, we found a relationship between the Kalman gain and maximum likelihood: We found that if we are naïve and have no prior beliefs about the hidden states, then the Kalman gain is in fact the mixing gain that we derived in the maximum likelihood approach.

Our approach thus far is a bit curious because what we are really after is the posterior probability distribution associated with our hidden states $p(\mathbf{x}^{(n)}|\mathbf{y}^{(n)})$. That is, we have some prior belief about the state $p(\mathbf{x}^{(n)})$, we make an observation $\mathbf{y}^{(n)}$, and now we want to update our belief based on what we observed. We have not shown what this posterior probability distribution looks like. For example, if we apply the Kaman gain to the generative model that had Gaussian noise and form a posterior belief $\hat{\mathbf{x}}^{(n|n)}$, is this the expected value of $\mathbf{x}$ in the distribution $p(\mathbf{x}^{(n)}|\mathbf{y}^{(n)})$? Is the uncertainty of our posterior belief $P^{(n|n)}$ the variance in the distribution $p(\mathbf{x}^{(n)}|\mathbf{y}^{(n)})$? Here, we will formulate the posterior and we will see that indeed $\hat{\mathbf{x}}^{(n|n)}$ and $P^{(n|n)}$ are the mean and variance of it. The approach is called *Bayesian state estimation*.

5.1 Bayesian State Estimation

Bayesian estimation has its roots on a very simple and fundamental theorem, first discovered by the English mathematician Thomas Bayes shortly before his death in 1761. In modern terms things go as follows. Consider two random variables, x and y. Suppose that x can take one of N_X values and y can take one of N_Y values. The joint probability of observing $x = x_i$ and $y = y_j$ is $\Pr(x = x_i, y = y_j)$. If x and y are statistically independent, this joint probability is just the product $\Pr(x = x_i)\ \Pr(y = y_j)$. If the x and y are not independent, then one has to multiply the probability that $x = x_i$ given that $y = y_j$ by the probability that $y = y_j$. Of course, this is the most general thing, since the conditional probability $\Pr(x = x_i|y = y_j)$ coincides with $\Pr(x = x_i)$ if the two variables are independent. Bayes's theorem is a direct consequence of the intuitive fact that the joint probability is commutative:

$$\Pr(x = x_i, y = y_j) = \Pr(y = y_j, x = x_i).$$

Then, expanding each side with the corresponding expression on conditional probability, one obtains

$$\Pr(x = x_i \mid y = y_j)\Pr(y = y_j) = \Pr(y = y_j \mid x = x_i)\Pr(x = x_i).$$

Bayes's theorem is then obtained by rearranging the terms as

$$\Pr(x = x_i \mid y = y_j) = \Pr(y = y_j \mid x = x_i)\frac{\Pr(x = x_i)}{\Pr(y = y_i)}.$$

It is important to note that this simple algebra applies not only to probability values, but also to probability distributions. So, if x and y are continuous random variables, then Bayes's theorem allows us to derive the relation between the respective distributions as

$$p(x \mid y) = p(y \mid x)\frac{p(x)}{p(y)}.$$

If the variable x represents a "model" variable—for example, the state of a dynamical system—and y represents an observed variable—for example, the output of a sensor—then

(1) $p(y|x)$ is the likelihood of an observation given that the underlying model is true.

(2) $p(x|y)$ is the probability distribution of the model given the observations. This gives the probability that the model is correct "after the fact" that we have collected an observation. Therefore it is called the *posterior* distribution.

(3) $p(x)$ is the probability distribution of the model independent of any observation, or the *prior* of x.

(4) The prior probability to make an observation, $p(y)$, or *marginal* probability, is generally derived as a normalization factor, to insure that all distributions integrate to 1.

In the following discussion, we take advantage not of Bayes's theorem in its standard form, but of the underlying rule expressing the joint probability from the product of the posterior and the marginal distributions, that is, from

$$p(x,y) = p(x \mid y)p(y).$$

To formulate the posterior distribution, we start with the prior and the likelihood. Say that our prior estimate of the hidden state is normally distributed with mean $\hat{\mathbf{x}}^{(n|n-1)}$ and variance $P^{(n|n-1)}$:

$$p(\mathbf{x}) = N\left(\hat{\mathbf{x}}^{(n|n-1)}, P^{(n|n-1)}\right). \tag{5.2}$$

Further assume that our measurements are related to the hidden states via the following relationship:

$$\begin{aligned} \mathbf{y} &= C\mathbf{x} + \boldsymbol{\varepsilon}_y \\ \boldsymbol{\varepsilon}_y &\sim N(\mathbf{0}, R) \end{aligned}. \tag{5.3}$$

Therefore, the expected value and variance of observation $\mathbf{y}$ are:

$$\begin{aligned} E(\mathbf{y}) &= C\hat{\mathbf{x}}^{(n|n-1)} \\ \mathrm{var}(\mathbf{y}) &= C\,\mathrm{var}(\mathbf{x})C^T + 2\,\mathrm{cov}(C\mathbf{x}, \boldsymbol{\varepsilon}) + \mathrm{var}(\boldsymbol{\varepsilon}). \\ &= CP^{(n|n-1)}C^T + R \end{aligned}$$

We have the distribution $p(\mathbf{y})$:

$$p(\mathbf{y}) = N\left(C\hat{\mathbf{x}}^{(n|n-1)}, CP^{(n|n-1)}C^T + R\right). \tag{5.4}$$

Our next step is to compute the joint probability distribution $p(\mathbf{x},\mathbf{y})$. To form this distribution, we need to know the covariance between $\mathbf{x}$ and $\mathbf{y}$, which is computed:

$$\begin{aligned} \mathrm{cov}(\mathbf{y},\mathbf{x}) &= \mathrm{cov}(C\mathbf{x} + \boldsymbol{\varepsilon}, \mathbf{x}) = E\left[(C\mathbf{x} + \boldsymbol{\varepsilon} - CE(\mathbf{x}))(\mathbf{x} - E(\mathbf{x}))^T\right] \\ &= E\left[C\mathbf{x}\mathbf{x}^T - C\mathbf{x}E(\mathbf{x})^T - CE(\mathbf{x})\mathbf{x}^T + CE(\mathbf{x})E(\mathbf{x})^T + \boldsymbol{\varepsilon}\mathbf{x}^T - \boldsymbol{\varepsilon}E(\mathbf{x})^T\right]. \\ &= CE\left[\mathbf{x}\mathbf{x}^T\right] - CE(\mathbf{x})E(\mathbf{x})^T \\ &= C\,\mathrm{var}(\mathbf{x}) = CP^{(n|n-1)} \end{aligned}$$

The joint probability distribution becomes:

$$p(\mathbf{x},\mathbf{y}) = p\left(\begin{bmatrix}\mathbf{x}\\ \mathbf{y}\end{bmatrix}\right) = N\left(\begin{bmatrix}\mathbf{x}^{(n|n-1)}\\ C\mathbf{x}^{(n|n-1)}\end{bmatrix},\begin{bmatrix}P^{(n|n-1)} & P^{(n|n-1)}C^T\\ CP^{(n|n-1)} & CP^{(n|n-1)}C^T+R\end{bmatrix}\right). \tag{5.5}$$

The Gaussian distribution in equation (5.5) is equal to the product of the posterior probability $p(\mathbf{x}|\mathbf{y})$ and $p(\mathbf{y})$, that is:

$$p(\mathbf{x},\mathbf{y}) = p(\mathbf{x}|\mathbf{y})\,p(\mathbf{y}). \tag{5.6}$$

The item that we are looking for is the posterior probability $p(\mathbf{x}|\mathbf{y})$. We have the joint probability $p(\mathbf{x},\mathbf{y})$ (in equation 5.5), and we also have $p(\mathbf{y})$ in equation (5.4). If we could factor equation (5.5) so that it becomes a multiplication of two normal distributions, one of which is $p(\mathbf{y})$, then we will have the posterior probability that we are looking for.

The general problem that we are trying to solve is to factor a normal distribution $p(\mathbf{x},\mathbf{y})$. In this distribution, $\mathbf{x}$ is a $p \times 1$ vector and $\mathbf{y}$ is a $q \times 1$ vector, and the distribution has the following general form:

$$p(\mathbf{x},\mathbf{y}) = N\left(\begin{bmatrix}\boldsymbol{\mu}_x\\ \boldsymbol{\mu}_y\end{bmatrix},\begin{bmatrix}\Sigma_{11} & \Sigma_{12}\\ \Sigma_{21} & \Sigma_{22}\end{bmatrix}\right), \tag{5.7}$$

in which our *marginal distributions* are

$$\begin{aligned}\mathbf{x} &\sim N(\boldsymbol{\mu}_x,\Sigma_{11})\\ \mathbf{y} &\sim N(\boldsymbol{\mu}_y,\Sigma_{22})\end{aligned} \tag{5.8}$$

and $\mathrm{cov}(\mathbf{x}, \mathbf{y}) = \Sigma_{12} = \Sigma_{21}$. We hope to find a *conditional distribution* $p(\mathbf{x}|\mathbf{y})$.

Our first step is to block-diagonalize the variance-covariance matrix in equation (5.5). To see how to do this, assume we have a matrix M composed of following blocks:

$$M = \begin{bmatrix}E & F\\ G & H\end{bmatrix}. \tag{5.9}$$

If we right and left multiply M by the following two matrices, each of which has an identity determinant, we will end up with a diagonalized version of M:

$$\begin{aligned}\begin{bmatrix}I & -FH^{-1}\\ 0 & I\end{bmatrix} M \begin{bmatrix}I & 0\\ -H^{-1}G & I\end{bmatrix} &= \begin{bmatrix}E-FH^{-1}G & F-FH^{-1}H\\ G & H\end{bmatrix}\begin{bmatrix}I & 0\\ -H^{-1}G & I\end{bmatrix}\\ &= \begin{bmatrix}E-FH^{-1}G & 0\\ 0 & H\end{bmatrix}\end{aligned}. \tag{5.10}$$

The term M/H is called the Schur complement of M and is defined as:

$$M/H \equiv E - FH^{-1}G. \tag{5.11}$$

If we now take the determinant of the matrix in equation (5.10), we have:

$$\begin{aligned}\det(M) &= \det\left(E - FH^{-1}G\right)\det(H)\\ &= \det(M/H)\det(H)\end{aligned}. \tag{5.12}$$

This equality relies on the fact that the determinant of a block-triangular matrix is the product of the determinants of the diagonal blocks.

Our second step is to compute the inverse of matrix M. We will do this by taking advantage of the diagonalization that we did in equation (5.10). Suppose that we call X the matrix that we left-multiplied as M in equation (5.10), with the right multiple as Z. Equation (5.10) is simply:

$$\begin{aligned}XMZ &= W\\ Z^{-1}M^{-1}X^{-1} &= W^{-1}\\ M^{-1} &= ZW^{-1}X\end{aligned}. \tag{5.13}$$

That is, the inverse of matrix M is:

$$M^{-1} = \begin{bmatrix} I & 0 \\ -H^{-1}G & I \end{bmatrix}\begin{bmatrix} (M/H)^{-1} & 0 \\ 0 & H^{-1} \end{bmatrix}\begin{bmatrix} I & -FH^{-1} \\ 0 & I \end{bmatrix}. \tag{5.14}$$

In the expression that describes a normal distribution, we have a determinant of the covariance matrix, and we have an inverse of the covariance matrix. We will use the results in equation (5.12) to factor the determinant term, and the result in equation (5.14) to factor the inverse term.

The distribution that we wish to factor has the following form:

$$p(\mathbf{x},\mathbf{y}) = (2\pi)^{-(p+q)/2}|\Sigma|^{-1/2}\exp\left\{-\frac{1}{2}\left(\begin{bmatrix}\mathbf{x}\\ \mathbf{y}\end{bmatrix}-\begin{bmatrix}\boldsymbol{\mu}_x\\ \boldsymbol{\mu}_y\end{bmatrix}\right)^T \Sigma^{-1}\left(\begin{bmatrix}\mathbf{x}\\ \mathbf{y}\end{bmatrix}-\begin{bmatrix}\boldsymbol{\mu}_x\\ \boldsymbol{\mu}_y\end{bmatrix}\right)\right\}, \tag{5.15}$$

where Σ is the variance-covariance matrix of the preceding distribution, that is,

$$\Sigma \equiv \begin{bmatrix} \Sigma_{11} & \Sigma_{12} \\ \Sigma_{21} & \Sigma_{22} \end{bmatrix}.$$

Using equation (5.12), the determinant and the constants in the joint probability distribution can be factored as:

$$(2\pi)^{-(p+q)/2}\left|\begin{bmatrix} \Sigma_{11} & \Sigma_{12} \\ \Sigma_{21} & \Sigma_{22} \end{bmatrix}\right|^{-1/2} = (2\pi)^{-p/2}|\Sigma/\Sigma_{22}|^{-1/2}(2\pi)^{-q/2}|\Sigma_{22}|^{-1/2}. \tag{5.16}$$

In equation (5.16), the term Σ/Σ_{22} is the Schur complement of the matrix Σ. Using equation (5.14), the exponential term in the joint probability distribution can be factored as:

$$\exp\left\{-\frac{1}{2}\begin{bmatrix}\mathbf{x}-\boldsymbol{\mu}_x\\ \mathbf{y}-\boldsymbol{\mu}_y\end{bmatrix}^T\begin{bmatrix}I & 0\\ -\Sigma_{22}^{-1}\Sigma_{21} & I\end{bmatrix}\begin{bmatrix}(\Sigma/\Sigma_{22})^{-1} & 0\\ 0 & \Sigma_{22}^{-1}\end{bmatrix}\begin{bmatrix}I & -\Sigma_{12}\Sigma_{22}^{-1}\\ 0 & I\end{bmatrix}\begin{bmatrix}\mathbf{x}-\boldsymbol{\mu}_x\\ \mathbf{y}-\boldsymbol{\mu}_y\end{bmatrix}\right\}=$$
$$\exp\left\{-\frac{1}{2}\left(\mathbf{x}-\left(\boldsymbol{\mu}_x+\Sigma_{12}\Sigma_{22}^{-1}(\mathbf{y}-\boldsymbol{\mu}_y)\right)\right)^T(\Sigma/\Sigma_{22})^{-1}\left(\mathbf{x}-\left(\boldsymbol{\mu}_x+\Sigma_{12}\Sigma_{22}^{-1}(\mathbf{y}-\boldsymbol{\mu}_y)\right)\right)\right\} \quad . \tag{5.17}$$
$$\exp\left\{-\frac{1}{2}(\mathbf{y}-\boldsymbol{\mu}_y)^T\Sigma_{22}^{-1}(\mathbf{y}-\boldsymbol{\mu}_y)\right\}$$

Therefore, we factored the joint probability distribution into two terms:

$$\begin{aligned}p(\mathbf{x},\mathbf{y}) &= N\left(\begin{bmatrix}\boldsymbol{\mu}_x\\ \boldsymbol{\mu}_y\end{bmatrix},\begin{bmatrix}\Sigma_{11} & \Sigma_{12}\\ \Sigma_{21} & \Sigma_{22}\end{bmatrix}\right)\\ &= N\left(\boldsymbol{\mu}_x+\Sigma_{12}\Sigma_{22}^{-1}(\mathbf{y}-\boldsymbol{\mu}_y),\Sigma/\Sigma_{22}\right)N(\boldsymbol{\mu}_y,\Sigma_{22})\\ &= N\left(\boldsymbol{\mu}_x+\Sigma_{12}\Sigma_{22}^{-1}(\mathbf{y}-\boldsymbol{\mu}_y),\Sigma_{11}-\Sigma_{12}\Sigma_{22}^{-1}\Sigma_{21}\right)N(\boldsymbol{\mu}_y,\Sigma_{22})\\ &= p(\mathbf{x}|\mathbf{y})p(\mathbf{y})\end{aligned} \quad . \tag{5.18}$$

The posterior probability that we were looking for is the first of the two normal distributions in equation (5.18):

$$p(\mathbf{x}|\mathbf{y}) = N\left(\boldsymbol{\mu}_x+\Sigma_{12}\Sigma_{22}^{-1}(\mathbf{y}-\boldsymbol{\mu}_y),\Sigma_{11}-\Sigma_{12}\Sigma_{22}^{-1}\Sigma_{21}\right). \tag{5.19}$$

In the distribution of equation (5.19), we have the following terms:

$$\begin{aligned}\boldsymbol{\mu}_x &= \hat{\mathbf{x}}^{(n|n-1)}\\ \boldsymbol{\mu}_y &= C\hat{\mathbf{x}}^{(n|n-1)}\\ \Sigma_{11} &= P^{(n|n-1)}\\ \Sigma_{12} = \Sigma_{21} &= P^{(n|n-1)}C^T\\ \Sigma_{22} &= CP^{(n|n-1)}C^T+R\end{aligned} \quad .$$

Rewriting the distribution in equation (5.19) with the preceding terms, we have our usual Kalman filter estimate of the posterior:

$$\begin{aligned}E\left[\mathbf{x}^{(n)}|\mathbf{y}^{(n)}\right] &= \hat{\mathbf{x}}^{(n|n-1)}+P^{(n|n-1)}C^T\left(CP^{(n|n-1)}C^T+R\right)^{-1}\left(\mathbf{y}^{(n)}-C\hat{\mathbf{x}}^{(n|n-1)}\right)\\ &= \hat{\mathbf{x}}^{(n|n-1)}+K^{(n)}\left(\mathbf{y}^{(n)}-C\hat{\mathbf{x}}^{(n|n-1)}\right)\\ \mathrm{var}\left[\mathbf{x}^{(n)}|\mathbf{y}^{(n)}\right] &= P^{(n|n-1)}-P^{(n|n-1)}C^T\left(CP^{(n|n-1)}C^T+R\right)^{-1}CP^{(n|n-1)}\\ &= \left(I-K^{(n)}C\right)P^{(n|n-1)}\end{aligned} \quad . \tag{5.20}$$

So indeed we see that the Kalman gain in equation (4.55) is the term $\Sigma_{12}\Sigma_{22}^{-1}$ in equation (5.19), the posterior belief $\hat{\mathbf{x}}^{(n|n)}$ is the expected value of $p(\mathbf{x}|\mathbf{y}^{(n)})$, and the uncertainty of our posterior belief $P^{(n|n)}$ is the variance of $p(\mathbf{x}|\mathbf{y}^{(n)})$. By using Kalman's approach, we are computing the mean and variance of the posterior probability of the hidden state that we are trying to estimate.

5.2 Causal Inference

In the hiking problem that we described in the last chapter, we had two GPS devices that measured our position. We combined the reading from the two devices to form an estimate of our location. For example, in figure 4.8, the estimate of our location ended up being somewhere in between the two readings. This approach makes sense if our two readings are close to each other (that is, if the two GPS devices are providing us with estimates that pretty much agree with each other). However, we can hardly be expected to combine the two readings if one of them is telling us that we are on the north bank of the river and the other is telling us that we are on the south bank. We know that we are not somewhere in the river! In this case, the idea of combining the two readings makes little sense.

Consider another example. Say you are outside and see lightning, and soon afterward you hear thunder. It seems reasonable that your two sensors (vision and audition) were driven by a single cause: lightning occurring at a specific location. However, if the two sensory events are separated by a long time interval or the sound appears to come from a different direction than the light, then you would be less likely to believe that there was a single cause for the two observations. In principle, when the various sensors in our body report an event, the probability that there was a single source responsible for them should depend on the temporal and spatial consistency of the readings from the various sensors. This probability of a single source—that is, the probability of a single cause—should then play a significant role in whether our brain will combine the readings from the sensors or leave them apart.

Wallace and colleagues (2004) examined this question by placing people in a room where LEDs and small speakers were placed around a semicircle (figure 5.1A). A volunteer sitting in the center of the semicircle held a pointer in hand. The experiment began by the volunteer fixating a location (fixation LED, figure 5.1A). An auditory stimulus was presented from one of the speakers, and then one of the LEDs was turned on 200, 500, or 800 ms later. The volunteer estimated the location of the sound by pointing (pointer, figure 5.1A). Then he pressed a switch with his foot if he thought that the light and the sound came from the same location. The results of the experiment are plotted in figure 5.1B and C. The combination of the audio and visual stimuli in a single percept—perception of unity—was strongest when the two events occurred in close temporal and spatial proximity. When the volunteers perceived a common source, it is important to note, their localization of

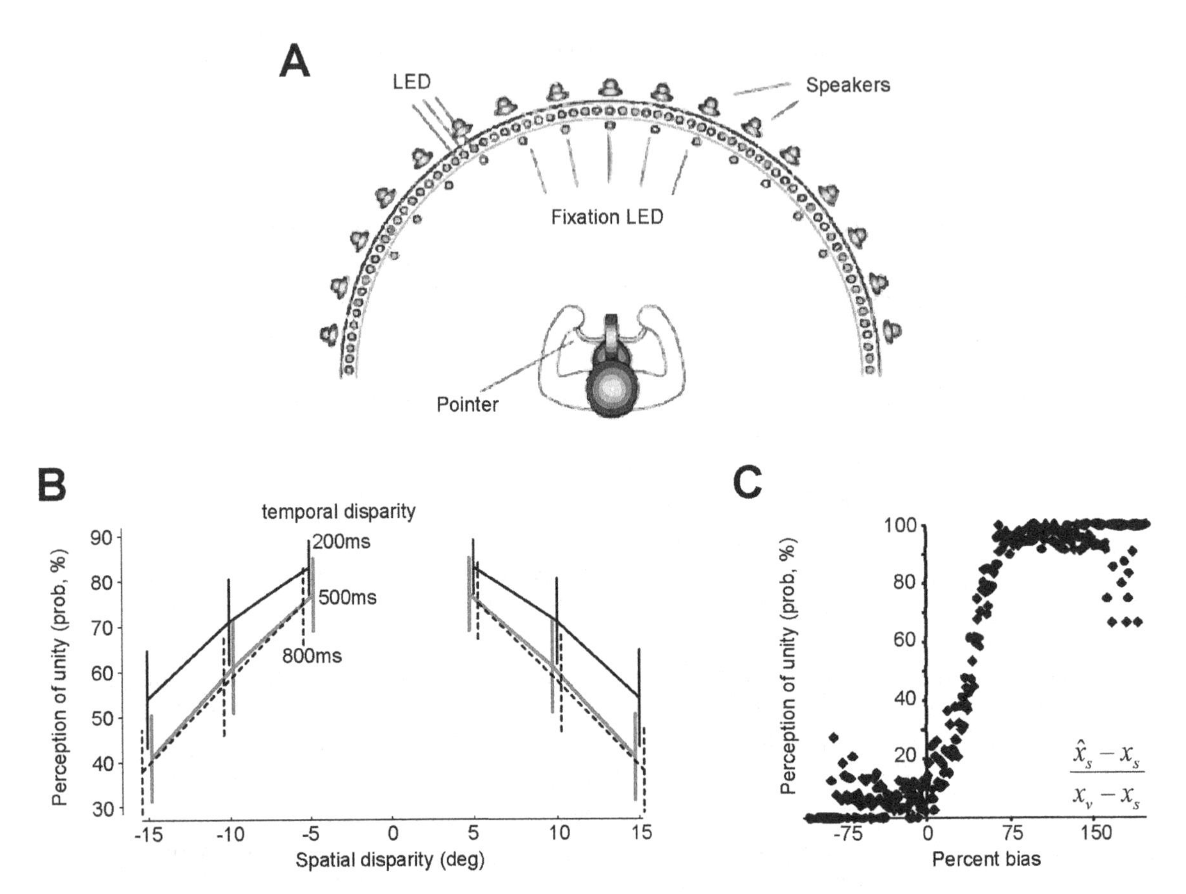

Figure 5.1
People combine visual and auditory information if they believe that the two sensors were driven by a common spatial source. (A) Volunteers were placed in the center of a semicircle and heard an auditory stimulus followed by a light from one of the LEDs. The onset of the two cues was separated in time by 200–800 ms. They then pointed to the location of the sound and pressed a switch if they thought that the light and sound came from the same location. (B) Probability of perceiving a common source as a function of the temporal and spatial disparity between the sound and visual cues. (C) As the probability of a common source increased, the perceived location of sound $\hat{x}_s$ was more strongly biased by the location of light x_v. (From Wallace et al., 2004.)

the sound was highly affected by the location of the light. That is, if x_s represents the location of the sound and x_v represents the location of the LED, the estimate of the location of the sound $\hat{x}_s$ (i.e., where the subject pointed) was biased by x_v when the volunteer thought that there was a common source (figure 5.1C). This bias fell to near zero when the volunteer perceived light and sound to originate from different sources.

The experiment in figure 5.1 suggests that when our various sensory organs produce reports that are temporally and spatially in agreement, we tend to believe that there was a single source that was responsible for both observations (figure 5.2A). In this case, we combine the readings from the sensors to estimate the state of

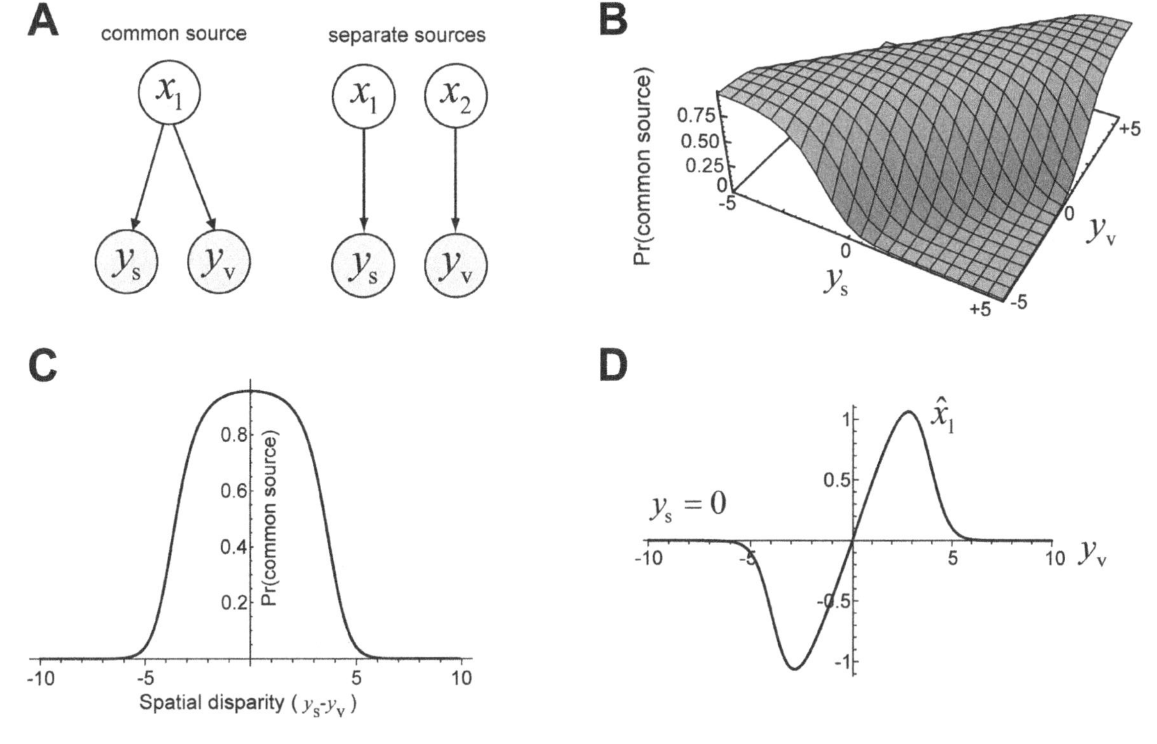

Figure 5.2
Estimating the state when there are two potential generative models. (A) The visual and sound sensor may be driven by a common source, or by two different sources. (B) The probability of a common source, given the sensory measurements. This plot is equation (5.23). When the two measurements y_s and y_v are close to each other, the probability of a common source is nearly one. When they are far from each other, the probability is close to zero. (C) As the spatial disparity between the two measurements increases, it is less likely that one is observing the consequences of a common source. (D) The estimate of the location of the sound $\hat{x}_1$ when the sound is heard at position zero $y_s = 0$ but the light is observed at various displacements y_v. When y_v and y_s are near each other, estimated location of the sound is affected by the observed location of the light.

the source. On the other hand, if our sensory measurements are temporally or spatially inconsistent, then we view the events as having disparate sources (figure 5.2A), and we do not combine the sources. Therefore, the nature of our belief as to whether there was a common source or not is not black or white. Rather, there is some probability that there was a common source. In that case, this probability should have a lot to do with how we combine the information from the various sensors.

The idea is that in principle, there are many generative models that could explain our observations. For example, we could have a model that says that the two observations come from the same source. We could also have another model that says that the two observations are independent. The one that we pick, or rather the probabilities that we assign to the various potential generative models, will determine how we will form our belief.

Konrad Kording and colleagues (Kording et al., 2007) suggested a simple way to frame this problem. Suppose that the binary random variable z specifies whether there is a single source ($z = 1$), or whether there are two distinct sources that drive our sensors ($z = 0$). If $\Pr(z = 1|\mathbf{y}) = 1$, then our visual and sound measurements are reflecting a common source:

$$\begin{aligned}\mathbf{y} &= \begin{bmatrix} y_s \\ y_v \end{bmatrix} = \begin{bmatrix} 1 & 0 \\ 1 & 0 \end{bmatrix}\begin{bmatrix} x_s \\ x_v \end{bmatrix} + \boldsymbol{\varepsilon} \qquad \boldsymbol{\varepsilon} \sim N(0, R) \\ &= C_1\mathbf{x} + \boldsymbol{\varepsilon}\end{aligned}. \tag{5.21}$$

On the other hand, if $\Pr(z = 1|\mathbf{y}) = 0$, then our measurements are reflecting different sources:

$$\begin{aligned}\mathbf{y} &= \begin{bmatrix} y_s \\ y_v \end{bmatrix} = \begin{bmatrix} 1 & 0 \\ 0 & 1 \end{bmatrix}\begin{bmatrix} x_s \\ x_v \end{bmatrix} + \boldsymbol{\varepsilon} \qquad \boldsymbol{\varepsilon} \sim N(0, R) \\ &= C_2\mathbf{x} + \boldsymbol{\varepsilon}\end{aligned}. \tag{5.22}$$

Starting with a prior belief about the location of the two stimuli $\mathbf{x} \sim N(\boldsymbol{\mu}, P)$, and a prior belief about the probability of a common source $\Pr(z = 1)$, we can compute the posterior probability of a common source after making a measurement $\mathbf{y}$:

$$p(z=1|\mathbf{y}) = \frac{p(\mathbf{y}|z=1)\Pr(z=1)}{p(\mathbf{y}|z=0)\Pr(z=0) + p(\mathbf{y}|z=1)\Pr(z=1)}. \tag{5.23}$$

The probability distribution of our measurements given that there is a common source can be computed from equation (5.21) as follows:

$$p(\mathbf{y}|z=1) = N\left(C_1\boldsymbol{\mu}, C_1PC_1^T + R\right). \tag{5.24}$$

Similarly, from equation (5.22) we have:

$$p(\mathbf{y}|z=0) = N\left(C_2\boldsymbol{\mu}, C_2PC_2^T + R\right). \tag{5.25}$$

In figure 5.2B we plotted $p(z = 1|\mathbf{y})$ for various values of our two sensory measurements (in computing equation 5.23, we assumed that $\Pr(z = 1) = 0.5$, that is, a common source was just as likely as two independent sources). When the two measurements y_s and y_v are close to each other, the probability of a common source is nearly one. When they are far from each other, the probability is close to zero. That is, as the spatial disparity between the two measurements increases, it is less likely that we are observing the consequences of a common source (figure 5.2C).

Let us now return to the experiment in figure 5.1A in which we hear a sound and see a light and need to know where the sound came from. Equation (5.23), as plotted in figure 5.2C, gives us a probability of a common source as a function of spatial disparity between the two sensory measurements. The question is how to use the probability of a common source to compute the location of the auditory stimulus. Let us suppose that $\Pr(z = 1|\mathbf{y}) = 1$. In that case we would use the single source model (equation 5.21) to find the posterior probability $p(\mathbf{x}|\mathbf{y})$. Our best estimate for $\mathbf{x}$ is the expected value of this posterior distribution, which is the usual Kalman estimate derived from the model in equation (5.21), that is, one in which we combine the two measurements to estimate the location of the sound. On the other hand, if $\Pr(z = 1|\mathbf{y}) = 0$, then equation (5.22) is the model that we should use, and the Kalman gain here would treat the two measurements as independent and not mix them to estimate the location of the sound. If the probability of a common source is somewhere between 0 and 1, then a rational thing to do would be to use this probability to weigh each of our two estimates for the location of the sound. To explain this in detail, let us begin by computing $p(\mathbf{x},\mathbf{y})$. If there is a common source, then from equation (5.21) we have:

$$p(\mathbf{x},\mathbf{y}) = N\left(\begin{bmatrix} \boldsymbol{\mu} \\ C_1\boldsymbol{\mu} \end{bmatrix}, \begin{bmatrix} P & PC_1^T \\ C_1P & C_1PC_1^T + R \end{bmatrix}\right). \tag{5.26}$$

This implies that if there was a common source, our best estimate is the usual mixing of the two sources. From equation (5.20) we have:

$$E[\mathbf{x}|\mathbf{y}, z = 1] = \boldsymbol{\mu} + PC_1^T\left(C_1PC_1^T + R\right)^{-1}(\mathbf{y} - C_1\boldsymbol{\mu}). \tag{5.27}$$

On the other hand, if there is not a common source, our best estimate is:

$$E[\mathbf{x}|\mathbf{y}, z = 0] = \boldsymbol{\mu} + PC_2^T\left(C_2PC_2^T + R\right)^{-1}(\mathbf{y} - C_2\boldsymbol{\mu}). \tag{5.28}$$

In general, then, our estimate of $\mathbf{x}$ should be a mixture of these two estimates, weighted by the probability of a common source:

$$\begin{aligned} &\Pr(z = 1|\mathbf{y}) = a \\ &\quad E[\mathbf{x}|\mathbf{y}] = aE[\mathbf{x}|\mathbf{y}, z = 1] + (1 - a)E[\mathbf{x}|\mathbf{y}, z = 0] \end{aligned}. \tag{5.29}$$

We have plotted an example of this estimate of $\hat{x}_s$ (location of the speaker) in figure 5.2D. We assumed that the actual location of the sound (i.e., the speaker) was always at zero ($y_s = 0$), while the LED (y_v) was located at various displacements. We see that when the two sources are near each other—that is, when y_v values are small—the estimate of the location of sound $\hat{x}_s$ is highly influenced by the location of the LED. However, as the LED moves farther away, the estimate of the speaker location returns to zero. When there is a large discrepancy between the two measurements y_v and y_s, there is little chance that they are coming from a single source, and so the system does not combine the two measures.

We can now see that in many of the previous experiments in which people were presented multiple cues and they performed "optimally" by combining the cues (some of these experiments were reviewed in the previous chapter, as, for example, in figure 4.7 and figure 4.9), they were doing so because the disparity between the cues was small. If the disparity is large, it is illogical to combine the two cues. The single-source models in those experiments are special cases of the more general causal inference model (figure 5.2) (Kording et al., 2007).

In summary, if we believe that our sensors are reporting the consequences of a common event, then our brain combines information from our various sensors. This belief regarding a common source is itself driven by the disparity between the measurements (i.e., their temporal and spatial agreement).

5.3 The Influence of Priors

When at the coffee shop the attendant hands you a cup full of tea, your brain needs to guess how heavy the drink is. This guess needs to be accurate, otherwise you would have trouble grasping the cup (activating your finger muscles in a way that they does not let it slip out of your hand) and holding it steady (activating your arm muscles so the cup does not rise up in the air or fall down). The only cue that you have is the size information provided by vision. Fortunately, the other piece of information is the prior experience that you have had with cups of tea. The fact that most people have little trouble holding cups that are handed to them in coffee shops suggests that they are making accurate estimates of weight. How are they making these guesses?

A useful place to start is by stating in principle how people should make guesses, whether it be regarding weight of a cup of tea or something else. Consider the cup-of-tea problem. Suppose we label the weight of the cup of tea as x. We have some prior belief about the distribution of these weights, $p(x)$, that is, how much cups of tea weigh in general. We have some prior belief about the relationship between visual property (size) of a teacup s and the weight of tea that it can hold, $p(s|x)$. And we have some prior belief about the distribution of tea cup sizes $p(s)$. Then our guess about weight of this particular cup of tea should be based on the posterior distribution:

$$p(x|s) = \frac{p(s|x)p(x)}{p(s)}. \tag{5.30}$$

In equation (5.30), what we are doing is transforming a prior belief $p(x)$ about the state of something (the weight of a cup) into a posterior belief, after we made an observation or measurement (in this case, s, the size of the cup that we saw). Let us do a thought experiment to consider how prior beliefs should affect people's guesses about weights of cups. In America, people are familiar with the rather large cups that are used for serving soft drinks and other refreshments. For example, in some convenience stores there are drinks called "super big gulps," and they hold something like 1.2 liters of soda. (Single-serving bottles in which fruit juice is sold in America tend to labeled as family-size in Europe.) The prior distribution $p(x)$ for someone who frequents such places in America would be skewed toward large weights. In contrast, someone from another country in which big gulps are not available would have $p(x)$ skewed toward smaller masses. If we now take these two people to a lab and present them with a regular-sized soft drink cup, upon visual inspection the subject for whom $p(x)$ is skewed toward heavier weights (the American fellow) should estimate the weight of the cup to be heavier than the subject for whom $p(x)$ is skewed toward lighter weights (the European chap). The belief about the larger weight should be reflected in the larger force that the American fellow uses to pick up the cup. That is, the prior belief should affect the prediction.

Indeed, prior beliefs do play a very strong role in how people interact with objects in everyday scenarios. For example, consider the task of using your fingers to pick up a small object as compared to picking up a slightly larger object. To pick up an object, you will need to apply a grip force (so the object does not slip out of your fingers) and a load force (so you can lift the object), as shown in figure 5.3A. Suppose you walk into a lab and are given a device like that shown in figure 5.3A. This device is attached to either a small box or a large box. You should apply larger grip and load forces to the larger object. This is indeed what Andrew Gordon, Roland Johansson, and colleagues (1991) observed when they presented volunteers with three boxes that *weighed exactly the same* but were of different sizes. People applied a larger grip force and a larger load force to lift the larger box (figure 5.3B). The result was the familiar scenario in which you go to pick up a bottle that you think is full, but is actually empty: The hand accelerates up faster than you intended.

Another way to explore prior beliefs about physics of objects is with regard to how objects move in a gravitational field: Objects fall with a constant acceleration of $g = 9.8$ m/s^2. For example, when a ball is released from rest and falls toward your hand, your prediction regarding when it will reach you will determine when you will

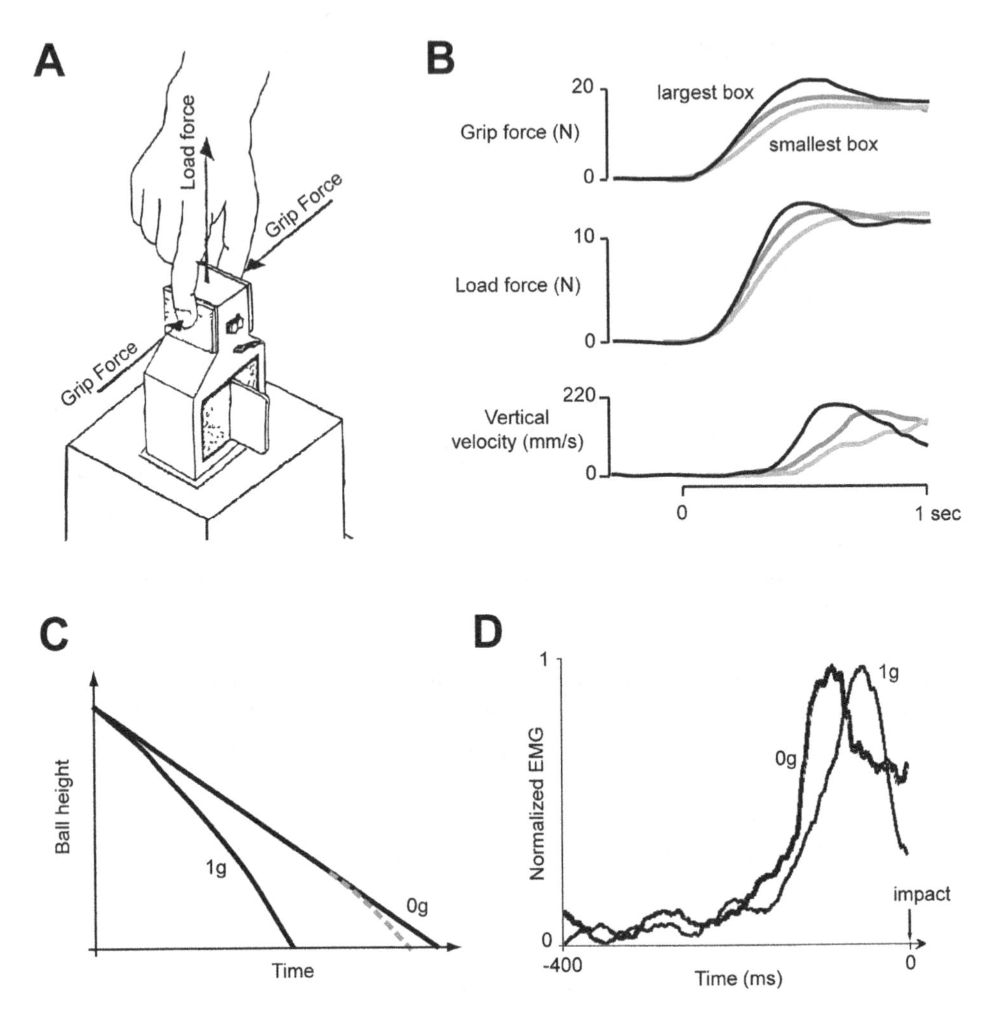

Figure 5.3
The effect of prior beliefs during interactions with everyday objects. (A) Volunteers were asked to use their fingers to lift up a small, medium, or a large box. The instrumented device measured grip and load forces. The three boxes were the same weight. (B) People tended to produce the smaller grip and load forces for the smallest box, resulting in large lift velocities for the largest box (from Gordon et al., 1991). (C) The ball starts with a nonzero velocity from a given height and falls in 0g or 1g gravity. In the 0g scenario (i.e., in space), the recently arrived astronaut will use a 1g internal model to predict the ball's trajectory, expecting it to arrive earlier (dashed line). (D) EMG activity from arm muscles of an astronaut in 0g and 1g. In 0g, the arm muscles activate sooner, suggesting that the astronaut expected the ball to arrive sooner than in reality (from McIntyre et al., 2001).

open your hand. Almost all of us spend our entire lives here on Earth, so presumably our brain has formed an internal model of falling objects. Let us briefly sketch this internal model. Suppose the state (position and velocity) of the ball is labeled as $\mathbf{x}(t)$, our measurement of that state is labeled $\mathbf{y}(t)$, and our goal is to predict the future state $\mathbf{x}(t+\Delta)$. In a 1g environment, the state of the ball can be modeled as:

$$\begin{bmatrix} x(t+\Delta) \\ \dot{x}(t+\Delta) \end{bmatrix} = \begin{bmatrix} 1 & \Delta \\ 0 & 1 \end{bmatrix} \begin{bmatrix} x(t) \\ \dot{x}(t) \end{bmatrix} + \begin{bmatrix} 0 \\ g\Delta \end{bmatrix} + \boldsymbol{\varepsilon}_x. \tag{5.31}$$

Written in a more compact way, we have:

$$\begin{aligned} \mathbf{x}(t+\Delta) &= A\mathbf{x}(t) + \mathbf{b} + \boldsymbol{\varepsilon}_x \qquad & \boldsymbol{\varepsilon}_x &\sim N(0,Q) \\ \mathbf{y}(t) &= \mathbf{x}(t) + \boldsymbol{\varepsilon}_y & \boldsymbol{\varepsilon}_y &\sim N(0,R) \end{aligned}. \tag{5.32}$$

If we assume that the noises $\boldsymbol{\varepsilon}_x$ and $\boldsymbol{\varepsilon}_y$ are Gaussian, then we can use the Kalman framework to estimate the state of the ball. We start with a prior belief, described by mean $\hat{\mathbf{x}}(t)$ and variance $P(t)$. We use this prior belief to predict the expected value of what our sensors should be measuring:

$$\hat{\mathbf{y}}(t) = \hat{\mathbf{x}}(t). \tag{5.33}$$

The difference between what we expected and what we measured allows us to update our estimate of the current state and predict the future. The expected values of our predictions are:

$$\begin{aligned} \hat{\mathbf{x}}(t|t) &= \hat{\mathbf{x}}(t) + K(\mathbf{y}(t) - \hat{\mathbf{y}}(t)) \\ \hat{\mathbf{x}}(t+\Delta|t) &= A\hat{\mathbf{x}}(t|t) + \mathbf{b} \end{aligned}. \tag{5.34}$$

In equation (5.34), K is the Kalman gain, which depends on our prior uncertainty (variance):

$$K = P(t)(P(t)+R)^{-1}. \tag{5.35}$$

The variance of our posterior probability distribution is:

$$\begin{aligned} P(t|t) &= (I-K)P(t) \\ P(t+\Delta|t) &= AP(t|t)A^T + Q \end{aligned}. \tag{5.36}$$

Now if you are an astronaut in an orbiting ship, this 1g internal model should bias your predictions about falling objects in space. In particular, you should predict that the falling ball will reach you earlier than in reality. Joe McIntyre, Mirka Zago, Alan Berthoz, and Francesco Lacquaniti (2001) tested this idea by having astronauts catch balls in both 0g and 1g. An example of a ball's trajectory in 1g and 0g is shown in figure 5.3C. Suppose that the ball starts with a nonzero initial velocity. In 1g, the ball accelerates. However, in 0g the ball velocity remains constant (the black lines in figure 5.3C). If one uses a 1g internal model to predict the state of the ball in 0g, one would predict that the ball will reach the hand sooner than in reality. That is, the prior belief about the behavior of the ball will produce an earlier preparation for ball hitting the hand. McIntyre and colleagues quantified this reaction to the ball motion by recording EMG activity from hand and elbow muscles. When the data were aligned to the moment of ball impact on the hand, they saw that the astronauts in 0g prepared the hand much sooner than in 1g, suggesting that they expected the ball to hit their hand earlier. Interestingly, this pattern continued for the fifteen days that the astronauts were in space—that is, the fifteen days of being in space was not enough to significantly alter the 1g internal model. Although different from the optimal response in 0g, the anticipatory behavior was not considered by the astronauts' brains as an error requiring correction.

(The ability of a good baseball pitcher to strike out a batter relies to a great extent on the prior belief that batters have regarding gravity and how it should affect the ball during its flight. In a fastball, the ball has backspin, giving it lift so that it falls slower than expected, whereas in a curve ball, the ball has topspin, giving it downward force so that it falls faster than expected. The force caused by rotation of the ball is called Magnus force, and Isaac Newton himself studied it on a tennis ball.)

These two examples of picking up objects and predicting state of falling balls demonstrate that our brain relies on prior experience to make predictions. You do not need to read this book to know this point, which is obvious. The more useful question is whether the process of prediction resembles Bayesian state estimation, as in equations (5.30) and (5.34). To explore this question, an interesting experiment was performed by Harm Slijper, Janneke Richter, Eelco Over, Jeroen Smeets, and Maarten Frens (2009). Their idea was that the prior statistics of everyday movements might affect how people move a computer mouse to a given stimulus. That is, the prior statistics should bias the response to the stimulus. They installed "spyware" software on the computers of a group of consenting volunteers and recorded their mouse movements on random days over a fifty-day period. The distribution of movement amplitudes is shown in figure 5.4A. The majority of movements were 3 mm or less. The distribution of movement directions, represented as an angle θ_e (angle of a line connecting the start to the endpoint), is plotted in figure 5.4B. The endpoint directions θ_e were clustered along the primary axes—that is, most of the movements had endpoints that were up/down or left/right with respect to start position. Slijper and colleagues noticed that while some of the movements were straight, many of the movements had an initial angle θ_i that was somewhat different than θ_e. This is illustrated by the cartoon drawing in figure 5.4D. They wondered whether this difference $\theta_i - \theta_e$ was due to the distribution of movements that people made. Their idea was that the movements that were straight were the ones that

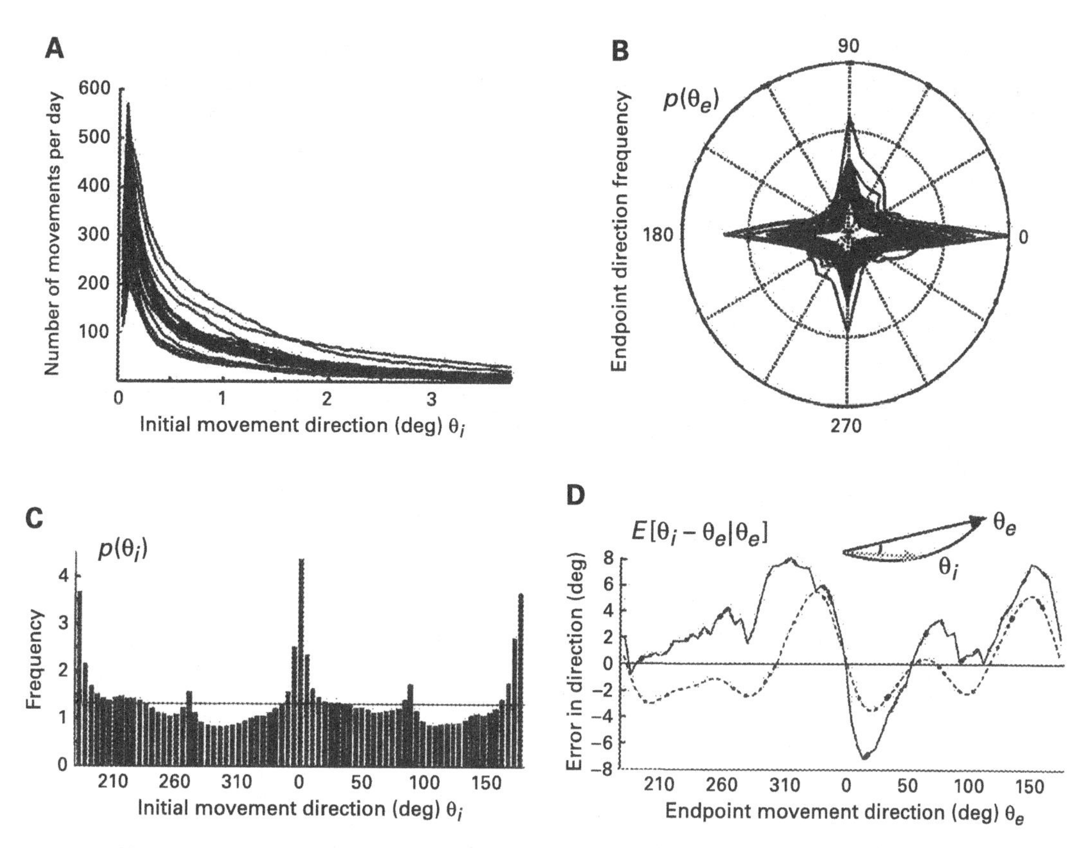

Figure 5.4
The prior statistics of everyday movements affect how people move a computer mouse to a given stimulus. (A) The distribution of movement amplitudes as measured over a multiday period. (B) The distribution of movement directions, represented as an angle θ_e. This is the angle of a line connecting the start to the endpoint. (C) While some of the movements were straight, many of the movements had an initial angle θ_i that was somewhat different than θ_e. The prior probability distribution of θ_i is shown. (D) The expected value of equation (5.37), $E[\theta_i - \theta_e|\theta_e]$, is plotted as the dashed line. The measured value $\theta_i - \theta_e$ is plotted as a solid line. (From Slijper et al., 2009.)

people tended to repeat a lot, whereas movements that were not straight were "attracted" toward the nearby movement directions that were repeated a lot.

When a movement is not straight, there is a difference between θ_i and θ_e. To explain why there was a difference between θ_i and θ_e for some movements but not others, Slijper and colleagues posed the problem in the following way. Suppose that given a desired endpoint at direction θ_e, the probability of moving the mouse in an initial direction θ_i is specified by $p(\theta_i|\theta_e)$. This distribution can be written as:

$$p(\theta_i|\theta_e) = \frac{p(\theta_e|\theta_i)\,p(\theta_i)}{p(\theta_e)}. \tag{5.37}$$

The prior probability of moving in an initial direction is specified by $p(\theta_i)$ and is plotted in figure 5.4C. The most frequent initial movement directions are along the left/right axis. That is, the prior has a large peak at 0 degrees and 180 degrees. Next, they assumed that the likelihood $p(\theta_e|\theta_i)$ was simply a normal with mean at θ_i and variance of a few degrees. Now suppose that we consider making a movement to an endpoint θ_e at 10 degrees. The prior distribution $p(\theta_i)$ has a very large peak at $\theta_i = 0$. Even though the likelihood $p(\theta_e|\theta_i)$ has its peak at $\theta_e = \theta_i$, making it so that the maximum likelihood estimate is simply $\theta_e = \theta_i$ (that is, the movement should be straight to the target), the strong prior at $\theta_i = 0$ will bias the posterior probability. Intuitively, we can see that the initial angle θ_i would be biased toward the large peak at 0 degrees in the prior distribution $p(\theta_i)$. That is, $p(\theta_i|\theta_e)$ would have its expected value somewhere between 0 degrees and 10 degrees. This makes the error in movement direction $\theta_i - \theta_e$ negative. The expected value of equation (5.37), $E[\theta_i - \theta_e|\theta_e]$ is plotted as the dashed line in figure 5.4D. This theoretical prediction matched reasonably well with the actual error $\theta_i - \theta_e$, plotted as the solid line in figure 5.4D.

It is instructive to approach this problem a bit more rigorously because by doing so we can consider probability distribution of a random variable defined on the circle. The circular normal distribution is:

$$p(\theta) = \frac{1}{2\pi I_0(\kappa)} \exp(\kappa \cos(\theta - \mu)). \tag{5.38}$$

In equation (5.38), the function $I_0(\kappa)$ is a normalization constant, called the modified Bessel function of order zero. The mean of the distribution in equation (5.38) is at μ and the variance increases with decreasing κ, as shown with an example in figure 5.5A. We can approximate the prior probability $p(\theta_i)$, that is, the data in figure 5.4C, as a sum of four circular normal distributions with means at μ_1, μ_2, μ_3, and μ_4, with $\mu_i = (i-1)\frac{\pi}{2}$, and a uniform distribution, normalized for the sum to have an integral of one:

$$p(\theta_i) = \frac{1}{m}\left[b + \sum_{j=1}^{4} \frac{a_j}{2\pi I_0(\kappa)} \exp(\kappa \cos(\theta_i - \mu_j)) \right]. \tag{5.39}$$

Our model of the prior $p(\theta_i)$ is plotted in figure 5.5B. Let us consider what happens when we intend to move to endpoint $\theta_e = 0.1$ radians (about 6 degrees). What will be the initial movement direction θ_i? The prior $p(\theta_i)$ has a large peak at $\theta_i = 0$. The posterior distribution $p(\theta_i|\theta_e = 0.1)$ is approximately a Gaussian, with its peak at around 3 degrees, which is the expected value of the posterior. The initial movement direction is skewed toward the prior.

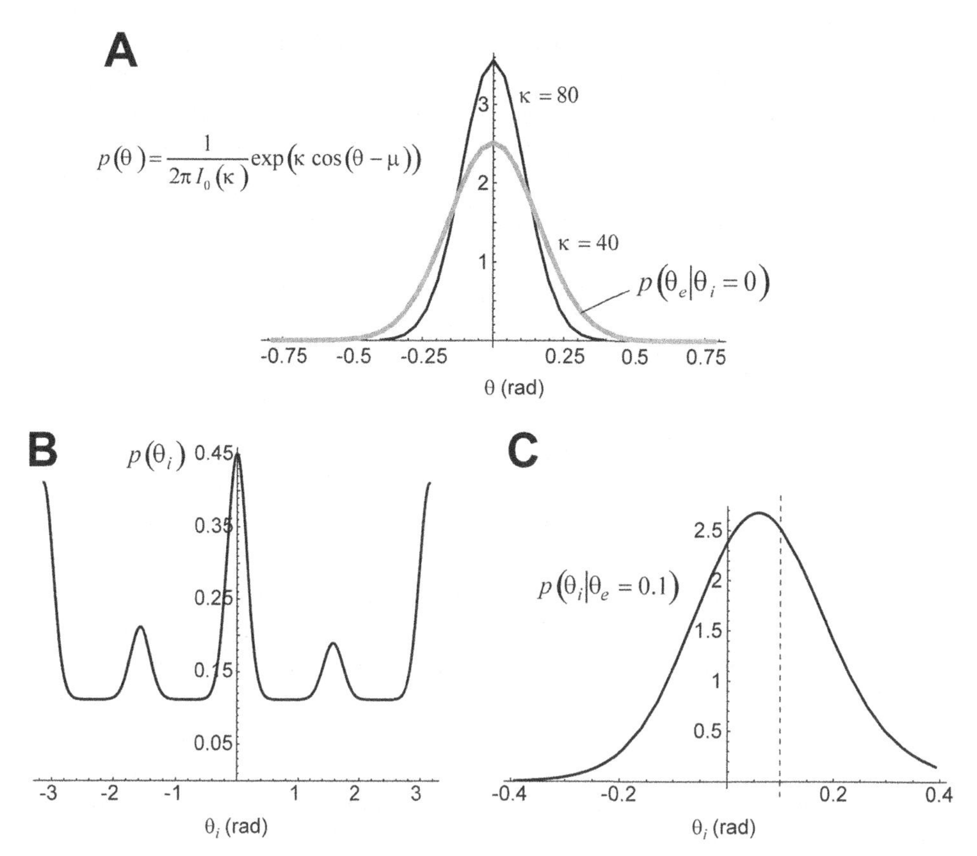

Figure 5.5
Modeling the data in figure 5.4. (A) A normal distribution for a random variable defined on a circle. The mean of the distribution in equation (5.38) is at μ, and the variance increases with decreasing κ. For the example in figure 5.4, we assume that $p(\theta_e|\theta_i)$ is a normal with mean at θ_i and variance of a few degrees. (B) The data in figure 5.4C—that is, the prior probability $p(\theta_i)$—approximated as the sum of four normal distributions and a uniform distribution, as in equation 5.39. (C) $p(\theta_i|\theta_e)$. When the target is at 0.1 radians (about 6 degrees), as indicated by the dashed line, the initial movement direction θ_i is likely to be toward 3 degrees, which is the expected value of the posterior.

In summary, our prior experience with everyday objects such as teacups, balls, and computer devices produce internal models of physics that appear to strongly affect how we interact with these objects. These internal models act as priors that bias our ability to use current observations. If the internal models are correct (as in the distribution of weight of teacups), they aid in control because they allow us to correctly estimate property of the current object. However, if the internal models are incorrect (as in the 1g physics being applied by the astronaut to a 0g environment), then we make mistakes in our estimations. This implies that priors have to continuously change with experience. That is, our internal models need to continuously learn from observations. We will pick up this topic in the next chapter when we consider the problem of adaptation.

5.4 The Influence of Priors on Cognitive Guesses

If we extend our interest a bit outside of motor control, we find evidence that in situations in which people make a "cognitive" guess about an ordinary thing, their guess appears to be consistent with a Bayesian framework. Let us consider the problem of guessing the lifespan x of a person (that is, how many years someone will live), given that they are now t years old. Thomas Griffiths and Joshua Tenenbaum (2006) asked a large group of undergraduates to guess the lifespan of someone who is now eighteen, thirty-nine, sixty-one, eighty-three, or ninety-six years old (each student made a guess only about a single current age t). Their results are shown in figure 5.6A. The students guessed that the eighteen- and thirty-nine-year-olds would probably live to be around seventy-five, which is about the mean lifespan of a male in the United States. The sixty-one-year-old will probably live a little bit longer, around seventy-six years, but the eighty-three-year-old will likely live to around the age of ninety, and the ninety-six-year-old will likely live to around the age of one hundred. The interesting thing about these guesses is the shape of the function that specifies the guess about lifespan as a function of current age (the line in figure 5.6A): The line starts out flat and then rises with a slope that is always less than one. Importantly, according to the students who took this survey, the ninety-six-year-old has less time to live than the eighty-three-year-old.

This pattern of guessing of course makes a lot of sense. However, Griffiths and Tenenbaum showed that it is exactly how you should guess if you have prior beliefs about life spans that resemble reality (as shown in figure 5.6B). Let us go through the mathematics and build a model of this guessing process. Suppose that we model the relative frequency of life spans using a Gaussian function:

$$p(x) = \frac{1}{\sqrt{2\pi}\sigma}\exp\left(-\frac{1}{2\sigma^2}(x-\mu)^2\right) \qquad \mu = 75, \ \ \sigma = 15. \tag{5.40}$$

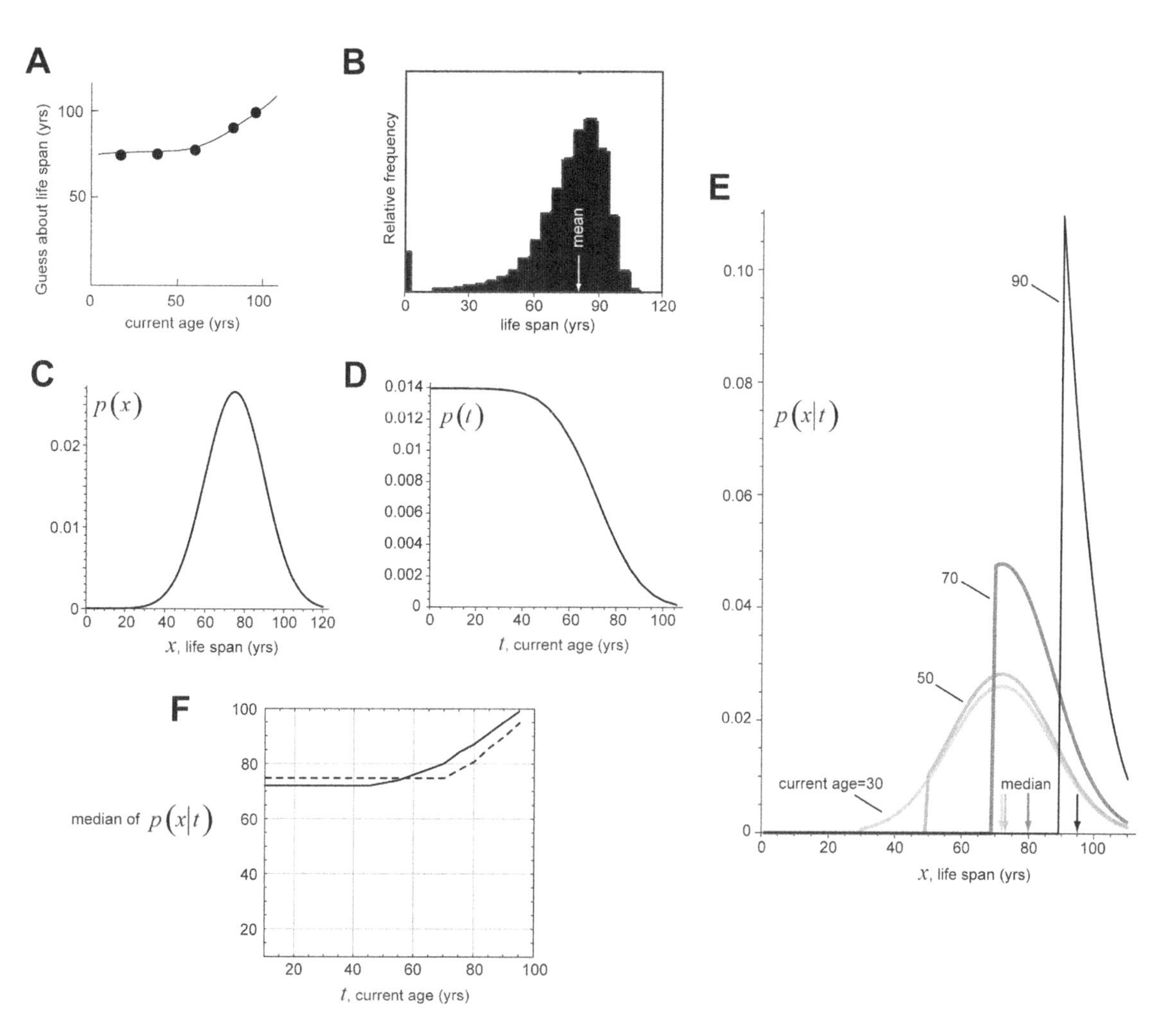

Figure 5.6
The problem of guessing the lifespan x of a person (i.e., how many years someone will live), given that they are now t years old. (A) Guesses from a group of students about lifespan, given current age. (B) The distribution of lifespan for a person born in America, i.e., $p(x)$. (C) A simplified model of lifespan distribution, equation (5.40). (D) Age distribution of people alive today, as in equation (5.42). (E) The posterior probability $p(x|t = 30)$, $p(x|t = 50)$, etc. The median for each distribution is marked by an arrow. (F) The median for the posterior distribution $p(x|t)$ as a function of current age t is plotted with the solid line. This indicates the best guess regarding lifespan for a prior probability $p(x)$ as shown in (C). If the prior probability $p(x)$ is narrower than that shown in (C) but has the same mean, the posterior (dashed line) is altered.

This distribution is plotted in figure 5.6C. (In using this estimate, we are ignoring the fact that the actual distribution of life spans cannot be Gaussian, since the variable x can only be positive and cannot be greater than some finite maximal value. Furthermore, the actual distribution includes a large number of infants who died near birth, which means that the actual distribution [figure 5.6B] is skewed toward dying young.) Further, suppose that we model the conditional probability of someone currently being t years old, given that his or her lifespan is x years old:

$$p(t|x) = \frac{1}{x} \text{ if } x \geq t,\ 0 \text{ otherwise}. \tag{5.41}$$

Equation (5.41) is our likelihood. Here, it implies that if the lifespan is seventy years, then the likelihood of currently being at any age less than or equal to seventy is simply 1/70, but of course 0 for values larger than 70. Finally, we model the probability of someone being currently at any age t as:

$$\begin{aligned} p(t) &= \int_0^\infty p(t|x)p(x)\,dx \\ &= \int_t^\infty p(t|x)p(x)\,dx \end{aligned} \tag{5.42}$$

Equation (5.42) is called the marginal probability. It describes the age distribution of people who are alive today, as shown in figure 5.6D. The integral's boundaries in equation (5.42) are at t and ∞ (rather than 0 and ∞) because of the constraint on x in equation (5.41). The posterior probability becomes:

$$\text{if } x \geq t \quad p(x|t) = \frac{\frac{1}{x}\frac{1}{\sqrt{2\pi}\sigma}\exp\left(-\frac{1}{2\sigma^2}(x-\mu)^2\right)}{\int_t^\infty p(t|x)p(x)\,dx}. \tag{5.43}$$

If $x < t$, then of course $p(x|t) = 0$. Now suppose that we know that a person is thirty years old. The posterior probability $p(x|t = 30)$ is plotted in figure 5.6E. What should be our guess about this person's lifespan? Say that we guess $\hat{x} = 35$. The probability of this person living thirty-five years or less is:

$$\Pr(x \leq 35|t = 30) = \int_{30}^{35} p(x|t = 30)\,dx. \tag{5.44}$$

The probability of this person living more than thirty-five years is:

$$\Pr(x > 35|t = 30) = \int_{36}^{\infty} p(x|t = 30)dx. \tag{5.45}$$

By looking at the area under the curve in figure 5.6E for $p(x|t = 30)$, it should be clear that the probability of living thirty-five years or longer is a lot higher than the probability of living thirty-five years or less. Therefore, $\hat{x} = 35$ is a bad guess. The best guess $\hat{x}$ that we could make is one that makes the probability of living less than $\hat{x}$ equal to the probability of living longer than $\hat{x}$. The $\hat{x}$ that we are looking for is the median of the distribution $p(x|t = 30)$. If the median value is labeled as m, then

$$\Pr(x \le m|t) = \Pr(x \ge m|t) = \int_{-\infty}^{m} p(x|t)dx = \frac{1}{2}. \tag{5.46}$$

Because $p(x|t = 30)$ is quite similar to a Gaussian, peak of the density as well as its mean and median all correspond to the same value, which is a bit less than 75 (i.e., the mean of our prior). However, if the current age is fifty, the posterior probability no longer resembles a Gaussian. Rather, $p(x|t = 50)$ is a truncated Gaussian with a median that is no longer at the peak of the density. Regardless, the median of $p(x|t = 50)$ is still quite close to 75, and this is still our best guess. However, if our subject is seventy years old, then the median (around 80) is now quite far from the peak of the posterior distribution. This captures the intuition that if someone has already lived to be seventy, then he is more likely to live beyond the average life span than not. The median for the posterior distribution as a function of current age t is plotted with the solid line in figure 5.6F, illustrating a form similar to the guesses that people had made (figure 5.6A).

The exact form of the function in figure 5.6F is strongly dependent on the shape of the prior $p(x)$. For example, suppose that instead of the Gaussian prior with a broad standard deviation that we assumed in figure 5.6C we had chosen a narrow Gaussian with the same mean but standard deviation $\sigma = 5$ years (rather than 15). Based on this prior, the best guess for lifespan of someone who is currently eighty years old is around eighty-one, which is inconsistent with the guesses that people made.

This kind of data is of course not conclusive evidence that people are Bayesian estimators, because it reflects the "wisdom of the masses" rather than individuals (because the data in figure 5.6A were averaged guesses from a group of people) (Mozer, Pashler, and Homaei, 2008). Yet, it is consistent with the assumption that when people make guesses about every day questions, they are doing so by relying on a prior internal model (figure 5.6C) that resembles reality (figure 5.6B), and by performing a computation that is consistent with a Bayesian process.

This conclusion appears to be challenged by a simple game that the reader can try with a group of friends. The game is known as the Monty Hall problem. One of its variants is as follows. You are in a room with three closed doors, labeled d1, d2, and d3. Behind one of the doors is a million-dollar prize. You have to guess which door it is, and you do this in two steps. First, you form an initial hypothesis, for example, d1. There is an oracle in the room, who knows where the prize is. You tell the oracle your initial choice and ask her to reveal one of the losing doors among those that you did not select. In this case, it could either be d2 or d3. Suppose that the oracle declares that d3 does not have the coveted prize. At this point you must make your final choice. There are two possibilities: either you stick with the original choice (d1) or you switch to the only remaining alternative (d2). These are the choices, and the question is: "Which is the best strategy? Sticking with the initial choice, switching, or, it really cannot be decided because they are equally likely to win or lose?" Almost invariably, the most popular answer is the last. And the argument appears to be compelling. You have two options, a winning option and a losing option. So, each must have a 50–50 chance to win. The correct answer: Sticking with d1 has a 1/3 probability to win, while switching has a 2/3 probability to win. So, switching is the right thing to do. We leave the calculation of these probabilities, using Bayes's rule, as an exercise. Instead we ask: does the failure to answer correctly reveal that the human mind is non-Bayesian?

To see how this is not the case, you may try a simple variant of the Monty Hall game. The original version has only $N = 3$ doors. Let us try again with a bigger number, say $N = 1000$. The problem is as follows. There are 1,000 doors, labeled $d1,d2,\ldots,d1000$. Behind one of these lies a million-dollar prize. Step 1: Choose a door (for example, d1). The game has 999 losing doors, and now you ask the oracle to eliminate 998 of these. So, now you are left with two doors, d1 and—say—d376. What would you do? Facing with this variant, most people have no trouble recognizing that switching is the best strategy. This is simply because in this case, the prior is much stronger. In the classic Monty Hall problem, the prior probability of being correct was 1/3. Now it is 1/1000. When you had to make your initial choice, your expectation of being defeated is much stronger, as when you play the national lottery. It seems reasonable to assume that this feeling has a stronger and more evident persistence when you are faced with the final choice, compared to a situation where the difference between ignoring and considering the prior is $\frac{1}{2}-\frac{1}{3}$. So, in a way the outcome of the test, in its original and modified form provides more support to the mind's ability to deal with Bayes's rule. However, the "strength" of the prior plays a major role in the final decision making.

5.5 Behaviors That Are Not Bayesian: The Rational and the Irrational

While there are many behaviors in which it appears that the brain acts as a Bayesian state estimator, integrating prior beliefs with observations to estimate the state of the body or of the world, there are some curious behaviors that do not fit the theory. In these behaviors, the behaviors are perplexing, and seemingly illogical. It is interesting to ask why people behave in this way. Let us consider some of the more prominent examples.

Suppose that you were presented with two objects. They have the same shape and color, for example two yellow cubes, made of the same material. But the one is small and the other large. Despite their different sizes, the objects weigh the same (the larger cube is hollow and some of the core material was removed). You pick up one of the objects, place it back down, and then pick up the other object. It is very likely that you will feel that the small object weighed more than the large object. This is called the *size-weight illusion,* something that was first reported over a century ago (Charpentier, 1891). Your belief that the small object weighed more is in fact the opposite of what a Bayesian state estimator should do. To see why, let us consider this problem in detail.

When we first see an object, the visual input y_v allows the brain to make an estimate of its weight w (figure 5.7A). We think this is true because the grip and load forces that people use to pick up objects are affected by what the object looks like (figure 5.3A): If the object looks large, we expect it to be heavier than an object that looks small. Generally, as volume of an object increases, so does its weight. The slope of this relationship depends on the materials in the object. For example, if the object is made of aluminum, the weight-volume relationship grows faster than if it is made of balsa wood (figure 5.7B). The visual property indicates the class of weight-volume relationships that we should expect, which is the basis with which we form prior beliefs about the weights of objects. For example, if we think the object is aluminum, then the relationship between the hidden state w and the volume y_v that we see is:

$$y_v = c_{al} w + \varepsilon_v. \tag{5.47}$$

Whereas if it looks like the object is made of balsa wood,

$$y_v = c_b w + \varepsilon_v \qquad c_b > c_{al}. \tag{5.48}$$

Before we see the object, we have some prior belief $\hat{w}^{(1|0)}$ (the distribution of weights of objects in general) and $\hat{c}^{(1|0)}$ (the distribution of weight-volume slopes in general). When we see the object (gather information about its volume y_v), we form a posterior estimate $\hat{c}^{(1|1)}$ and $\hat{w}^{(1|1)}$. That is, given its visual properties, we form a belief about what it is made of and how much it weighs. Propagating this forward in time, we

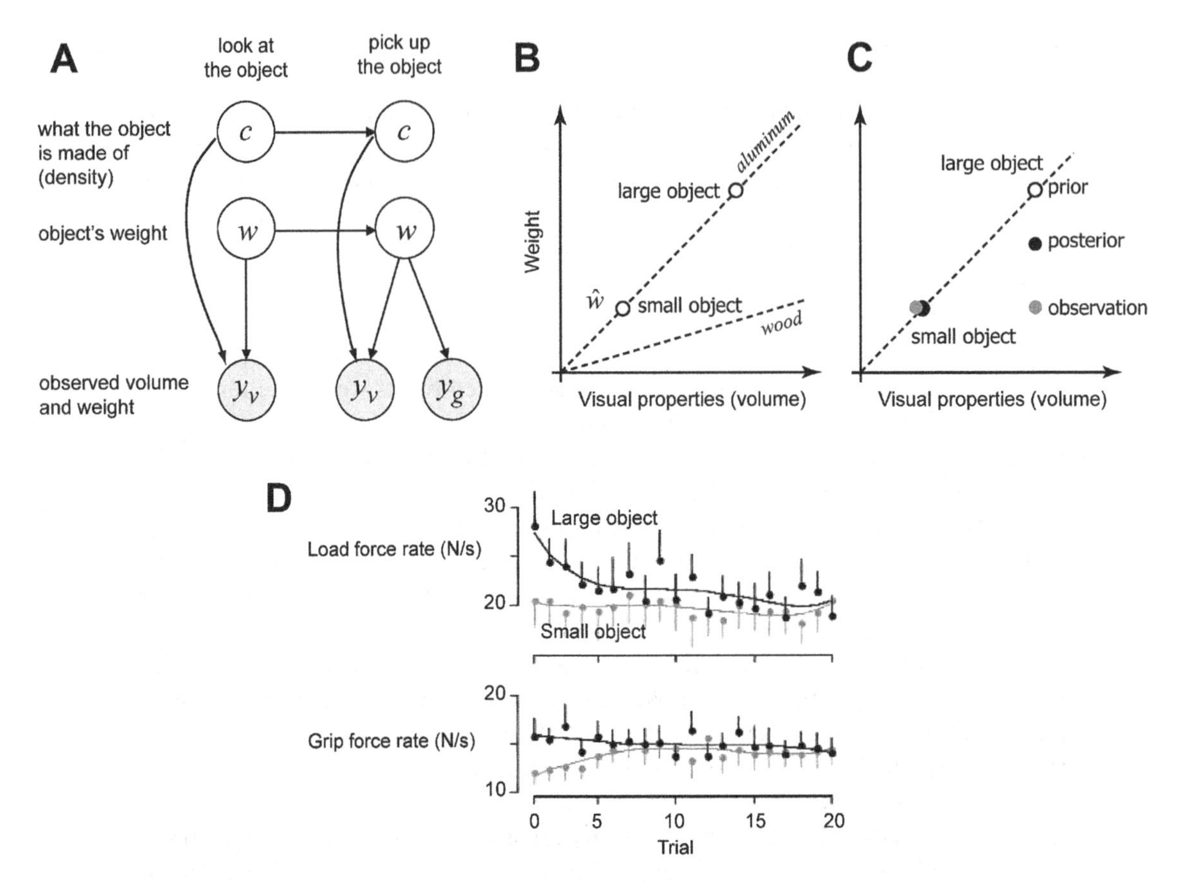

Figure 5.7
The size-weight illusion. (A) A generative model to estimate the weight of an object. y_v refers to observations from the visual sensors and y_g refers to observations from the Golgi tendon (force) sensors. (B) The prior belief regarding the relationship between weight, volume, and material that the object is made of. The term c describes the slope of the weight-volume relationship. When we see two objects that look like are made of aluminum, we expect the larger one to weigh more. (C) Suppose that the larger object is actually made of wood and weighs the same as the smaller object. When we pick up the larger object, the measured weight is smaller than we expected. A Bayesian estimator would believe that the larger object weighs somewhere in between what it predicted (heavy weight), and what is observed (light weight). In all cases, this posterior estimate of weight should be larger than for the smaller object. (D) While people "feel" that the small object weighs more than the large object, the motor system in fact uses a greater amount of load force rate and grip force rate to pick up the larger object. After about 8–10 trials, the force rates for the larger object converge to the small object, consistent with the fact that the two objects weigh the same. Despite this, people still "feel" that the smaller object weighs more, and this feeling lasts for up to hundreds of trials. (From Flanagan and Beltzner, 2000.)

have the prior estimate $\hat{w}^{(2|1)}$ that predicts the force that we should feel via our arm's Golgi tendon organs y_g as we pick up the object:

$$y_g = dw + \varepsilon_g. \tag{5.49}$$

Now suppose that we see two objects, one small and one large. They both look like they are made of aluminum, but actually the larger one is made of balsa wood, and the two are exactly the same weight. When we pick up the large object, we expect it to be heavy, and there will be a prediction error, that is, a difference between the expected force $\hat{y}_g$ and the observed force y_g, as $\hat{y}_g > y_g$. Because y_g is less than what we expected, the posterior $\hat{w}^{(2|2)}$ should be smaller than the prior $\hat{w}^{(2|1)}$. This is shown in figure 5.7C by the gray (observation) and black (posterior) circles. Now suppose that we are given a small object that is made of aluminum and weighs the same as the large object that we just picked up. When we pick up the small object, we will have little or no prediction errors, resulting in a weight estimate that is close to our prior (shown by the black circle in figure 5.7C). If we now estimate which object weighs more, we should clearly guess that the larger object weighs more. But this is not what happens. In fact, people consistently report that the smaller object weighs more. This judgment is irrational from a Bayesian estimation perspective, but it is in fact the way our brain works. What is going on?

Look at the data in figure 5.3B. Clearly, people are gripping the larger object with greater force and are applying a larger load force to pick it up. On the first try, the motor system certainly seems to believe that the larger object weighs more. If after this first try the brain really believes that the smaller object weighs more (for this is what subjects tell you after they pick up the objects), then on the second try the grip forces should be higher for the smaller object. But this is not what happens! Figure 5.7D shows the force data from twenty consecutive attempts to pick up small and large objects that weigh the same. Clearly, the smaller object is never experiencing the larger load force. In fact, by the eighth or tenth trial the forces are about the same, which makes sense, since the two objects really do weigh the same. The remarkable fact is that the illusion that the smaller object weighs more (as verbalized by subjects) persists even after the motor system continues to demonstrate that it believes that the two objects weigh the same, an observation that was nicely quantified by Randy Flanagan and Michael Beltzner (2000).

In summary, these results show that our motor system (as assayed by the forces that we produce with our hands) never believes that the small object weighs more than the larger one. However, apparently our motor system is not responsible for what we verbalize, because we consistently say that "the smaller object feels heavier." It appears that the brain does not have a single estimate of an object's weight. There appear to be two such estimates: one that is used by our "declarative" system to state (in words) how much it thinks the object weighs, and one that is used by our "motor" system to state (in actions) how much it thinks the object weighs. In these experiments, the rational part of the brain is the one that programs the motor commands, while the seemingly irrational one is the one that verbalizes what it thinks the objects weigh.

More recent experiments have shed some light on the mechanism that our brain uses to verbalize opinions about objects and their weights. Apparently, this declarative system also relies on a volume-weight prior belief, but this belief changes much more slowly than the one used by the motor system (it still remains unclear how the prior is integrated with observations). To explore this issue, Randy Flanagan, Jennifer Bittner, and Roland Johansson (2008) trained people on a set of objects that had an unusual property: the larger the volume, the smaller the weight (figure 5.8A). People lifted these objects hundreds of times a day for up to twelve days. At the end of the first day, the experimenters gave the subjects a small and a large object and asked them to indicate their relative weight. The two objects weighed the same, but the subjects reported that the smaller object weighed more (figure 5.8B). This is the usual size-weight illusion that we have seen before. However, after a few more days of lifting objects, the illusion subsided, and by the eleventh day it had reversed direction, so that they now perceived the larger object to weigh slightly more.

During this period of training, the prior that the motor system used for estimating weight from volume changed. On the first trial of the first day, people expected the small object to be light, and so they produced a small load force (early trial, figure 5.8C). By the eighth trial, the load force had substantially increased (late trial, figure 5.8C). On the second day, the motor system remembered this unusual, inverted relationship between volume and weight: From the very first trial, the motor system produced a larger force for the small object than the larger object (figure 5.8D). Therefore, a few trials of training were sufficient to teach the motor system that this class of objects had an unusual property that increased volume produced reduced weight. The declarative system too relied on a prior model, one in which weight increased with volume. However, this model appeared to change much more slowly, since it took eleven days before the illusion reversed. Therefore, if the perceptual system is not acting irrationally, it follows that the experiential structure that form the basis for its internal model is different from the experiential structure associated with motor learning.

5.6 Multiple Prior Beliefs

It is curious indeed that one should believe that a small object feels heavier than a large object and yet consistently act as if believing that the weight of the small object is the same as the large object. Perhaps in our brain there are distinct and sometimes

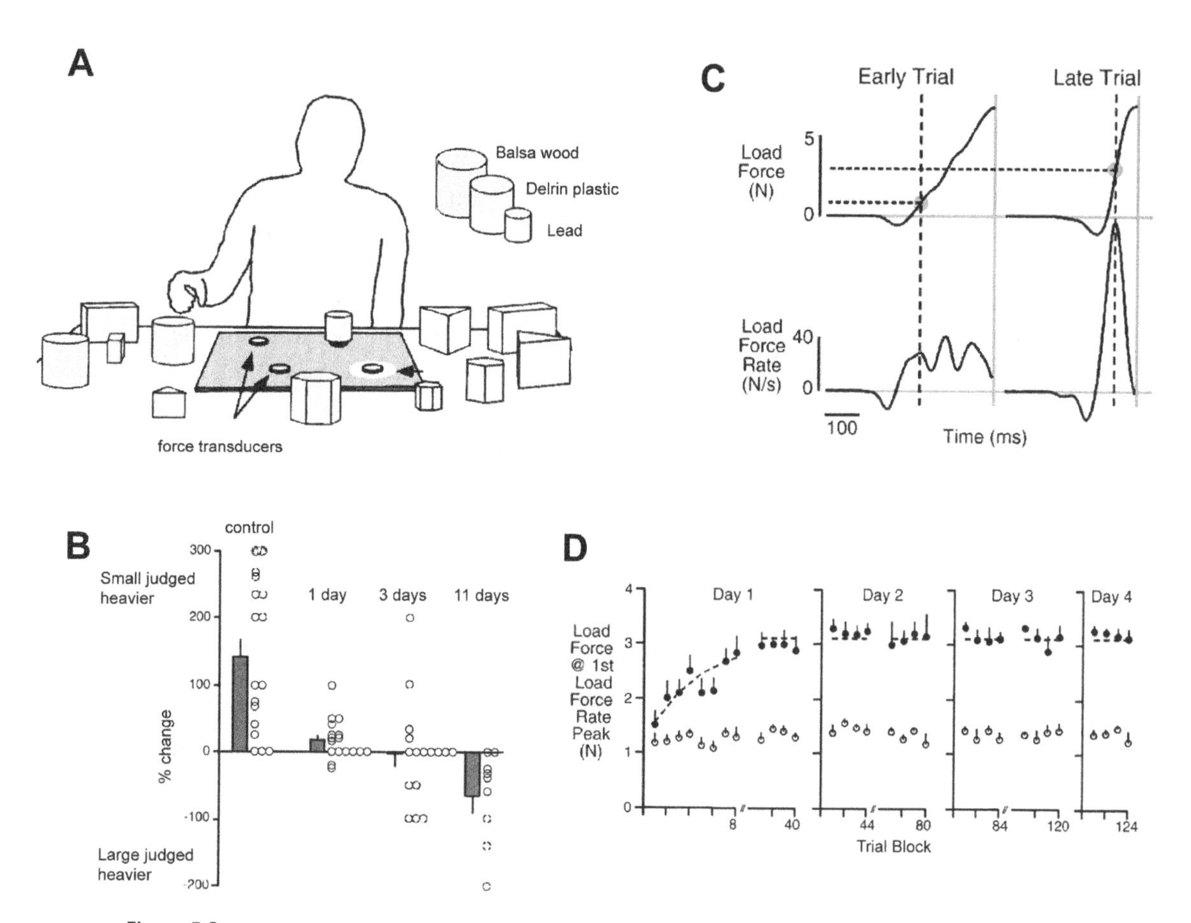

Figure 5.8
People were trained to pick up a set of objects that had an unusual property: the larger the volume, the smaller the weight. (A) Experimental setup. People lifted and placed these objects hundreds of times a day for twelve or more days. (B) At the end of the first day, the experimenters gave the subjects a small and a large object and asked them to indicate their relative weight. The two objects weighed the same, but the subjects reported that the smaller object weighed more. By the eleventh day, the illusion had reversed direction, so that they now perceived the larger object to weigh slightly more. (C) On the first trial of the first day, people expected the small object to be light, and so they produced a small load force (early trial). By the eighth trial, the load force had substantially increased. The black dashed vertical lines mark the time of initial peak in load-force rate, and the horizontal dashed lines mark the load force at the time of initial peak in load-force rate. The gray vertical lines mark the time of liftoff. (D) Load force at the time of the initial peak in load-force rate for the small (filled circle, heavy object) and the medium (open circle, light object) objects. Each point represents the average across participants for five consecutive trials. The motor system learned that the smaller object weighed more, and remembered this from day to day. (From Flanagan et al., 2008.)

conflicting beliefs about the properties of single objects. Perhaps depending on how our brain is queried, we express one belief or the other. An elegant experiment by Tzvi Ganel, Michal Tanzer, and Melvyn Goodale (2008) lends support to this counterintuitive conjecture.

In the experiment (figure 5.9A), two lines of unequal size were presented on a screen. The background was manipulated to give cues suggesting object 1 to be closer than object 2. As a result, most people would estimate object 2 to be taller than object 1. In fact, object 2 was about 5 percent shorter than object 1, as shown in the figure without the illusory background. On each trial, people were instructed to pick up the taller or the shorter object. When they were instructed to pick up the taller object, in about 90 percent of the trials they picked the shorter object, and similarly in about 90 percent of the trials people picked the taller object when they were instructed to pick the shorter one. That is, the background clearly produced a strong illusion.

As the subjects reached toward their selected object, the authors recorded the distance between the fingers. By choosing which object to pick, the subjects expressed their belief about which object was taller. By moving their fingers apart during the reach, they expressed their belief about the height of the object. Interestingly, they found that the distance between the fingers was not affected by the illusion: The aperture was small when picking up the short object, despite the fact that subjects were picking up that object because they thought it was the taller of the two objects (figure 5.9B). Similarly, the aperture was large when picking up the tall object, despite believing that it is the shorter object. Control experiments in which the visual feedback from the object and hand were removed confirmed this result. The motor commands that controlled the fingers in the task of picking up the object were not fooled by the visual cues that caused the illusion.

The same people were then asked to use their fingers to show their estimate of the size of the objects. With the illusory background in place, people were asked to estimate size of the shorter object. To convey their decision regarding which object they believed to be shorter, they moved their hand 5 cm to the right of the object that they chose and then split their fingers apart to show their estimate of its size. As before, people chose the tall object when they were instructed to estimate size of the shorter object. However, now they had their fingers apart by a smaller amount than when they were asked to estimate the size of the taller object. That is, in all cases, their perception of which object was smaller was affected by the illusory background. However, when they were asked to pick up the object, they moved their fingers apart in a way that suggested they were not fooled by the visual cues. In contrast, when they were asked to move their fingers apart so to estimate the size of the object, they were fooled. Finally, when the illusory background was removed and the two objects were displayed on a normal background (middle plot,

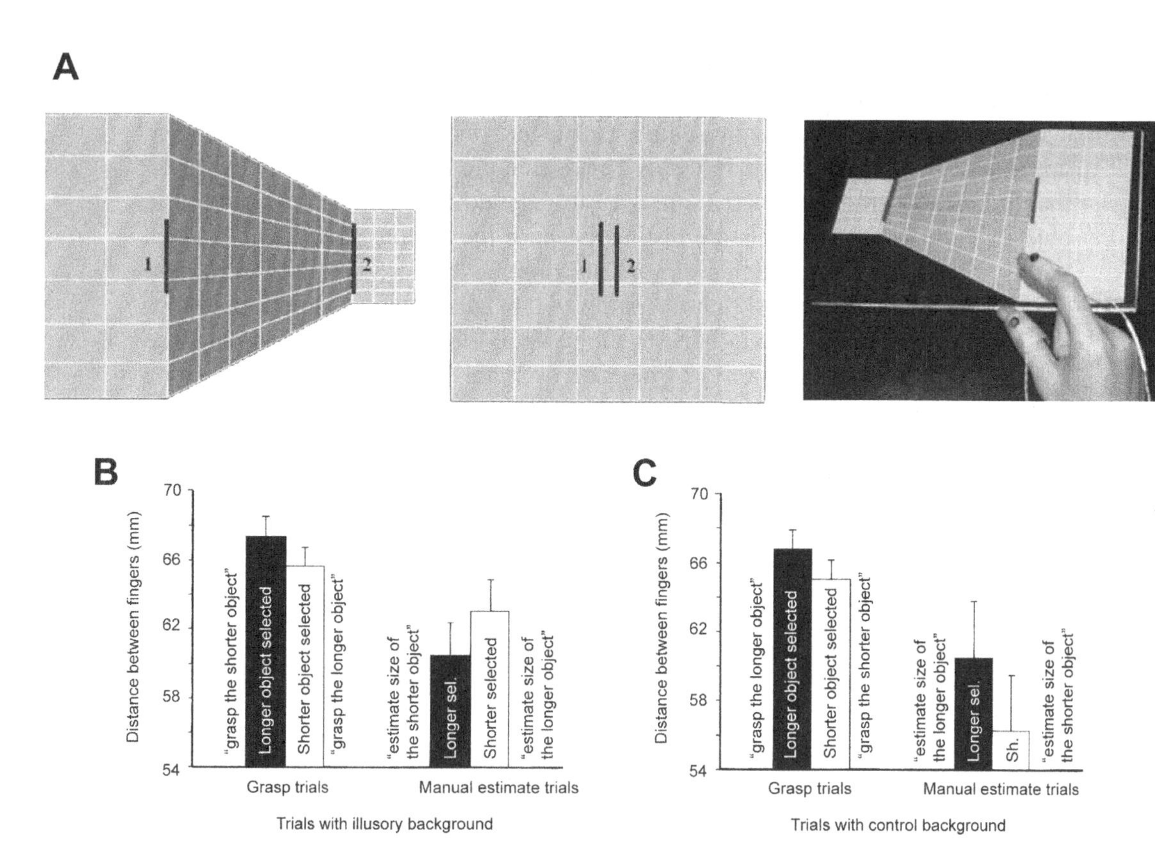

Figure 5.9
Visual illusions affect perception but not action. (A) The background was manipulated so that line 2 appears to be longer than line 1. In fact, line 1 is about 5 percent longer than line 2. People were instructed to reach and pick up the shorter or the longer line. (B) In about 90 percent of the trials, people reached for the shorter line when instructed to pick up the longer line, and vice versa. However, grasp size during the act of picking up the object was not fooled by the visual illusion (bars on the left). In contrast, when subjects were asked to estimate the size of the objects, they were fooled by the visual illusion (bars on the right). (C) Control experiments without an illusory background. (From Ganel et al., 2008.)

figure 5.9A), the grip sizes in the grasp trials and in estimation trials accurately reflected the relative object sizes.

One way to make sense of this data is to imagine that when we respond to "pick up the object," our actions are based on beliefs that are formed in parts of our brain that are distinct from beliefs that are used to respond to "show me the size of the object." Perhaps these beliefs are distinct because the various brain regions focus on distinct parts of the available sensory information. Melvyn Goodale and David Milner (1992) have proposed that the pathway that carries visual information from the visual areas in the occipital lobe to the parietal lobe (dorsal pathway) and the pathway that carries information from the occipital lobe to the temporal lobe (ventral pathway) build fundamentally distinct estimates of the object properties in the visual scene. In a sense, the actions that we perform based on the belief of the ventral pathway can be different than actions that we perform based on belief of the dorsal pathway. They have argued that when we pick up the object, we are relying on internal models in the dorsal pathway. This pathway is less affected by the background. When we use our hand to show an estimate of the size of the object, we are relying on internal models in the ventral pathway. This pathway is more affected by the background.

Summary

In Bayesian estimation, the objective is to transform a prior belief about a hidden state by taking into account an observation, forming a posterior belief. The Kalman gain is a weighting of the difference between the predictions and observations, which when added to a prior, transforms it to the expected value of the posterior. Here, we linked Bayesian estimation with the Kalman gain by showing that the posterior belief $\hat{\mathbf{x}}^{(n|n)}$ is the expected value of $p(\mathbf{x}|\mathbf{y}^{(n)})$, and the uncertainty of our posterior belief $P^{(n|n)}$ is the variance of $p(\mathbf{x}|\mathbf{y}^{(n)})$.

When our various sensory organs produce information that are temporally and spatially in agreement, we tend to believe that there was a single source that was responsible for our observations. In this case, we combine the readings from the sensors to estimate the state of the source. On the other hand, if our sensory measurements are temporally or spatially inconsistent, then we view the events as having disparate sources (figure 5.2A), and we do not combine the sources. The probability of a common source depends on the temporal and spatial alignment of our various sensory measurements, and this probability describes how we will combine our various observations.

Prior beliefs do play a very strong role in how people interact with objects in everyday scenarios. We expect larger things to weigh more than smaller things. We expect objects to fall at an acceleration of 1g. When objects behave differently than we expected, we combine our observations with our prior beliefs in a manner that resembles Bayesian integration. Our guesses about everyday things such as how long someone is likely to live is also consistent with a Bayesian process that depends on a prior belief.

The motor system appears to be rational in the sense that it estimates properties of objects by combining prior beliefs with measurements to form posterior beliefs in a Bayesian way. However, as the size-weight illusion demonstrates, our verbal estimate of an object's relative weight (i.e., whether it is heavier or lighter than another object) is not the same as the motor system's estimate. The verbal estimate appears to rely on a separate internal model, one that changes much more slowly than the internal model that the motor system relies upon. It is possible that our brain has multiple internal models that describe properties of a single object, and depending on how we are asked to interact with that object, we may rely on one or the other model. The distinct pathways that carry visual information in the parietal and temporal lobes may be a factor in these distinct internal models. This may explain the fact that visual cues that produce perceptual illusions about an object's properties often do not affect the motor system's abilities to interact with that object.

6 Learning to Make Accurate Predictions

In the previous couple of chapters we suggested that as the brain sends a motor command to the muscles, it also predicts the sensory consequences. When the sensory system reports on the consequences, the brain combines the observations with its own predictions to form an estimate of the state of the body and the environment. However, combining predictions with observations makes sense only if the predictions are unbiased estimates of the observations. If trial after trial there are persistent differences between predictions and observations—that is, the brain's predictions are consistently biased—then there is something wrong in these predictions.

Occasionally, our predictions are quite wrong. For example, one of us has a couple of teenagers at home who occasionally raid the refrigerator, drinking the milk carton almost dry. On more than one occasion, picking up the carton has resulted in his hand jerking upward. The brain predicted the carton to be heavier than reality, and so the motor commands produced an unexpected motion, resulting in a difference between the expected sensory feedback and the observed quantities. Combining a biased prediction with an unbiased sensory feedback results in a biased estimate of state, something that we would generally want to avoid. Therefore, it seems rational that the brain should somehow change its predictions so that eventually the predictions agree with what the sensory system reports. How should it do this? That is, how should the brain go about improving its predictions?

The problem of forming unbiased predictions of sensory observations is one of the fundamental problems of learning (the other is learning how to produce useful motor commands, something that we will tackle in a few chapters). For example, when you have formed an accurate representation—that is, an *internal model*—of your arm, you can apply motor commands to it and (on average) accurately predict the sensory consequences of those commands in terms of what you will see and what you will feel. An internal model is simply a map that mimics—in its relation between input and output signals—certain aspects of reality in which the brain is immersed. For example, if you hold a golf club in your hand, learning an accurate

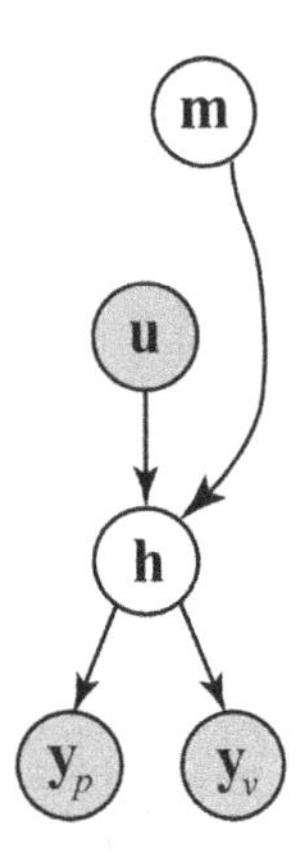

Figure 6.1
A graphical model that associates motor commands, hand position, and mass of an object, with sensory observations via vision and proprioception.

internal model of your arm and the club allows you to apply motor commands and produce a motion of the club, and the ball, that agrees with your predictions. Professional athletes are particularly good at making predictions about the sensory consequences of their motor commands. In tennis, a good player will know whether the ball will land in or out at the moment he or she hits the ball. However, note that having an accurate internal model does not mean that you can hit the ball well; it means only that given some motor commands that you produced, you can predict the sensory consequences.

Let us consider the problem of picking up a carton of milk. Our internal model is a *generative model*, that is, a model that is capable of generating data (things that we can observe, or measure) according to a set of rules, as shown in figure 6.1. According to the model in figure 6.1, when we send motor command u to our arm, it produces a change in the state of our hand h. This state depends on our motor command, as well as the mass that the hand may be holding, m. The consequence of our motor command is observed via our visual and proprioceptive sensors, y_v and y_p. What we have expressed in figure 6.1 is an internal model that describes how motor commands produce sensory consequences. In this internal model there are two hidden states: position of our hand and mass of the object. Using our observations we can estimate these hidden states. We begin with some prior belief about the position of our hand and mass of the object. We have an objective—for example, bringing the state of the hand to some position. Based on our prior belief regarding state of our hand and the mass of the object, we generate motor command u. For example, if our objective is to lift the object, a prior belief that m is large would dictate that u should be large as well. Using our internal model, we predict the sensory consequences $\hat{y}_v$ and $\hat{y}_p$. If the mass is less than what we had expected, there will be a difference between the predicted and actually observed sensory consequences. Using this prediction error, we update our belief about the state of our hand and the mass of the object, forming a posterior belief. This posterior becomes the prior for our next attempt to lift the carton. As a result, the motor commands that we will generate to pick up the carton for a second time will be different than the first time (because our prior associated with m will have changed). The trial-to-trial change in the motor commands will reflect our learning.

The idea is that the brain is constantly trying to predict things that it can observe. To make accurate predictions, it builds an internal model of the process that generates the data that it is observing. We can represent this generative model as a collection of observable quantities as well as a group of hidden states that cannot be observed. The problem of learning can be viewed as the problem of estimating these hidden states. The driving force for learning is the difference between the predicted sensory measurements and the actually measured quantities, that is, the sensory prediction error.

We saw in the last chapter that there are Bayesian algorithms that can help us solve the estimation problem. Therefore, if we frame the problem of learning in terms of state estimation, then we can apply these same algorithms. Here, we are interested in asking whether people and other animals learn internal models in ways that resemble optimal estimation. To help answer our question, it will be useful to consider other algorithms that can solve our learning problem, in particular those that are not Bayesian, so we will have a point of comparison.

To start our discussion, we will begin with some very simple learning problems. These problems come from experiments in which animals learn to predict the relationship between stimuli and reward or punishment. The experiments are called *classical conditioning*. We will see that even in this very simple paradigm, animal behavior exhibits surprisingly complex characteristics. The estimation framework that we developed in the last two chapters does a fairly good job of explaining the behaviors. We will then apply the same framework to experiments in which people learn to control novel tools or adapt to novel distortions on their sensory feedback.

6.1 Examples from Animal Learning

Let us start with a simple example from the animal learning literature called the *blocking* experiment, introduced by Leon Kamin (1968). In this experiment, the animal (a rat) was trained to press a lever to receive food. After completion of this baseline training, a light was turned on for 3 minutes, and during the last half-second of that period, a mild electric shock was given to the foot. In response to the shock, the animal reduced the lever-press activity. As the light-shock pairing was presented again, the animal learned to associate the light with the shock and in anticipation stopped pressing the lever (freezing behavior). After completion of this training, a second stimulus (a tone) was presented with the light. That is, the light and a tone were presented together, followed by the shock. Pairing of light with shock and pairing of light + tone with shock constituted set 1 and set 2 of the training. The experiment concluded with a test period, during which only the tone was presented. The crucial result was that during this test period, the animal did not reduce its lever pressing activity, despite the fact that in set 2 the tone was present when it was being shocked. Somehow, the animal did not fear the tone.

In control animals that had only experienced a tone paired with shock, in the test period there was reduced lever pressing in response to the tone. Therefore, the animal could easily learn to associate the tone with the shock. However, in set 1 the light was paired with shock. This prior pairing somehow blocked the ability of the brain to associate at least some of the shock to the tone in the subsequent light + tone training. The effect is called *Kamin blocking*.

Our objective here is to describe this learning, and other examples like it, in an estimation framework, and then test the idea that the animals are implicitly doing optimal estimation: computing uncertainties, forming a learning rate (i.e., the sensitivity to error or the Kalman gain) that varies according to this uncertainty, and then applying this learning rate to their prediction errors. In comparing the experimental data with the Kalman algorithm, it is useful to have an alternate hypothesis regarding how the learning might take place. Let us sketch one such alternate hypothesis, called the *LMS* (least mean squared) *algorithm*.

6.2 The LMS Algorithm

Suppose that we have a very simple internal model that relates inputs (stimuli: light and tone) on trial n, written as $\mathbf{x}^{(n)}$ with an output (e.g., shock) $y^{(n)}$. That is, the two input quantities are x_1 and x_2 (representing light and tone). When light is on, $x_1 = 1$, and when it is off, $x_1 = 0$. Similarly, when tone is on, $x_2 = 1$. There is one output quantity y^*, representing shock. When we make an estimate $\hat{y}^{(n)} = \mathbf{x}^{(n)T}\hat{\mathbf{w}}^{(n)}$ and observe a prediction error $y^{(n)} - \hat{y}^{(n)}$, we face a credit assignment problem: How should we assign responsibility for the error on our estimate $\hat{\mathbf{w}}^{(n)}$? That is, which element of $\hat{\mathbf{w}}^{(n)}$ should get blamed most for the error, and which should get blamed the least? To answer this question, let us consider a simple learning algorithm called LMS. In our problem, we have $\mathbf{x}$ and $\hat{\mathbf{w}}$ as 2×1 vectors, as shown in figure 6.2. When we project vector $\hat{\mathbf{w}}$ onto $\mathbf{x}$, that is, the dot product of the two vectors $\left(\mathbf{x}^T \cdot \hat{\mathbf{w}}\right)$, we get a scalar quantity p such that $p = \left\|\hat{\mathbf{w}}^{(n)}\right\|\cos\alpha$, where α is the angle between the vectors $\hat{\mathbf{w}}^{(n)}$ and $\mathbf{x}^{(n)}$, and the double vertical lines refer to the magnitude of the vector. The quantity $\hat{y}^{(n)}$ is related to p:

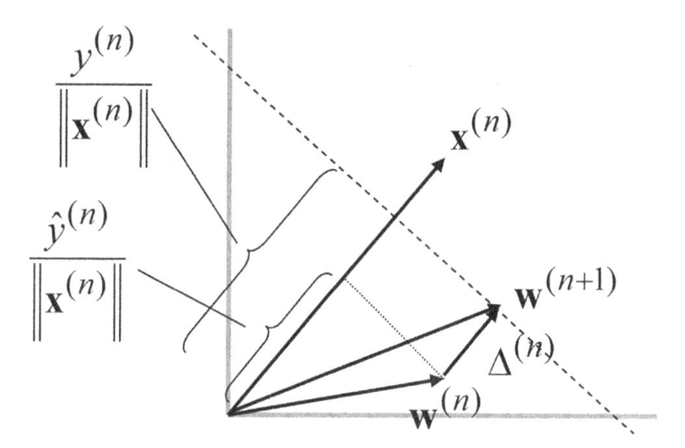

Figure 6.2
A graphical interpretation of the LMS algorithm.

$$\begin{aligned}\cos\alpha &= \frac{\hat{\mathbf{w}}^{(n)T}\mathbf{x}^{(n)}}{\|\hat{\mathbf{w}}^{(n)}\|\|\mathbf{x}^{(n)}\|} \\ p &= \frac{\hat{y}^{(n)}}{\|\mathbf{x}^{(n)}\|}\end{aligned}. \tag{6.1}$$

Equation (6.1) implies that we want to change $\hat{\mathbf{w}}^{(n)}$ by amount $\Delta^{(n)}$ so that when we project $\hat{\mathbf{w}}^{(n)} + \Delta^{(n)}$ onto $\mathbf{x}^{(n)}$, we get a length specified by the dashed line in figure 6.2, that is, $y^{(n)}\|\mathbf{x}^{(n)}\|^{-1}$. Note that there is no unique vector $\Delta^{(n)}$ that will do this for us. As long as we choose $\Delta^{(n)}$ so that $\hat{\mathbf{w}}^{(n)} + \Delta^{(n)}$ touches any point on the dashed line, we have one solution. A reasonable approach would be to add to $\hat{\mathbf{w}}^{(n)}$ a vector parallel to $\mathbf{x}^{(n)}$, with a magnitude that is proportional to $y^{(n)} - \hat{y}^{(n)}$. By making $\Delta^{(n)}$ parallel to $\mathbf{x}^{(n)}$, we are adding the smallest possible vector length to $\hat{\mathbf{w}}^{(n)}$:

$$\begin{aligned}\Delta^{(n)} &= \left(\frac{y^{(n)}}{\|\mathbf{x}^{(n)}\|} - \frac{\hat{y}^{(n)}}{\|\mathbf{x}^{(n)}\|}\right)\frac{\mathbf{x}^{(n)}}{\|\mathbf{x}^{(n)}\|} \\ &= \frac{1}{\|\mathbf{x}^{(n)}\|^2}\left(y^{(n)} - \hat{y}^{(n)}\right)\mathbf{x}^{(n)}\end{aligned}. \tag{6.2}$$

The quantity $\mathbf{x}^{(n)}\|\mathbf{x}^{(n)}\|^{-1}$ is a unit vector along the direction of $\mathbf{x}^{(n)}$. If we change $\hat{\mathbf{w}}^{(n)}$ by the vector specified in equation (6.2), we will completely compensate for the error. In practice, the step size is usually a fraction of this quantity, where $0 < \eta < 1$:

$$\hat{\mathbf{w}}^{(n+1)} = \hat{\mathbf{w}}^{(n)} + \frac{\eta}{\|\mathbf{x}^{(n)}\|^2}\left(y^{(n)} - \hat{y}^{(n)}\right)\mathbf{x}^{(n)}. \tag{6.3}$$

The algorithm in equation (6.3) is the estimation method presented by Widrow and Hoff (1960). It is widely known as the Delta rule or LMS (least mean squared) rule in the engineering literature and the Rescorla-Wagner rule in the psychological literature (Rescorla and Wagner, 1972). A less graphical but more immediate way to derive the LMS rule is by taking the gradient of the squared error with respect to the parameter vector and then "descending" by a small increment in that direction. The gradient of the squared error, $\varepsilon^{(n)2} = \left(y^{(n)} - \hat{y}^{(n)}(\mathbf{w})\right)^2$ is derived by the chain rule:

$$-\nabla_{\mathbf{w}}\varepsilon^{(n)2} = -\frac{\partial \varepsilon^{(n)2}}{\partial \mathbf{w}} = 2\varepsilon^{(n)}\frac{\partial \hat{y}^{(n)}}{\partial \mathbf{w}} = 2(y^{(n)} - \hat{y}^{(n)})\mathbf{x}^{(n)}. \tag{6.4}$$

The quantity of the left-hand side—modulo the step size—is the step of the LMS rule in equation (6.3). How does LMS compare to the Kalman approach?

6.3 Learning as State Estimation

To solve the same problem using state estimation, the basic idea is to describe our observations y as a function of some hidden states $\mathbf{w}$, that is, we start with a generative model. The generative model that describes our observations can be as follows:

$$\begin{aligned} \mathbf{w}^{(n+1)} &= \mathbf{w}^{(n)} + \boldsymbol{\varepsilon}_w^{(n)} & \boldsymbol{\varepsilon}_w &\sim N(0, Q) \\ y^{(n)} &= \mathbf{x}^{(n)T}\mathbf{w}^{(n)} + \varepsilon_y^{(n)} & \varepsilon_y &\sim N(0, \sigma^2) \end{aligned} . \tag{6.5}$$

The model is shown graphically in figure 6.3, in which the shaded circles are the observed variables. Our objective is to find some estimate of the hidden states $\hat{\mathbf{w}}$ that can predict our observation y. On trial n we start with a prior estimate $\hat{\mathbf{w}}^{(n|n-1)}$ and uncertainty $p^{(n|n-1)}$. We make a prediction $\hat{y}^{(n)} = \mathbf{x}^{(n)T}\hat{\mathbf{w}}^{(n|n-1)}$, and then learn from the sensory prediction error $y^{(n)} - \hat{y}^{(n)}$:

$$\hat{\mathbf{w}}^{(n|n)} = \hat{\mathbf{w}}^{(n|n-1)} + \mathbf{k}^{(n)}\left(y^{(n)} - \hat{y}^{(n)}\right). \tag{6.6}$$

The key comparison is between the terms that multiply the prediction error $y^{(n)} - \hat{y}^{(n)}$ in equation (6.3) and the Kalman gain in equation (6.6). In LMS, the terms that multiply the prediction error depend only on the current input $\mathbf{x}^{(n)}$. In the Kalman gain, we have:

$$\mathbf{k}^{(n)} = \frac{P^{(n|n-1)}\mathbf{x}^{(n)}}{\left(\mathbf{x}^{(n)T}P^{(n|n-1)}\mathbf{x}^{(n)} + \sigma^2\right)}. \tag{6.7}$$

In effect, in the Kalman gain we weigh the prediction error based on the ratio of the uncertainty in the state that we wish to predict $P^{(n|n-1)}$, and our uncertainty about our measurements (the denominator in equation 6.7). An important fact is that the

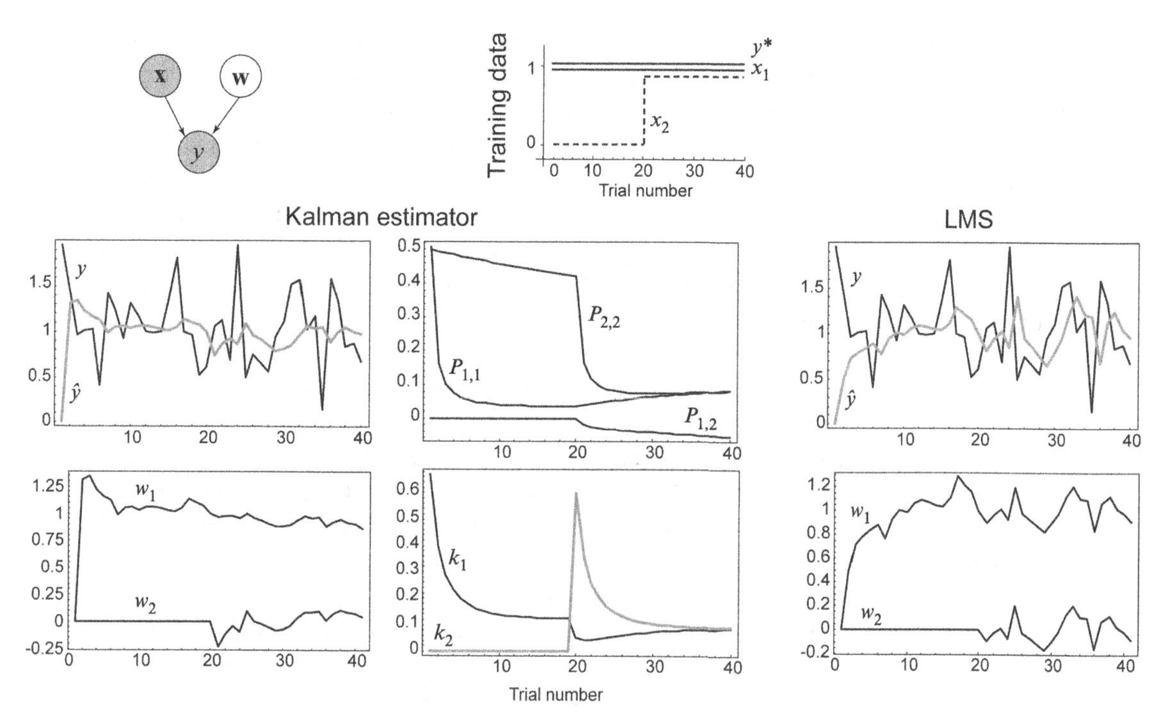

Figure 6.3
A comparison of the Kalman framework and LMS for the Kamin blocking experiment. Training data (top plot): x_1 and x_2 represent light and tone, and y^* represents shock. In the first twenty trials, x_1 is paired with shock. During the last twenty trials, both x_1 and x_2 are paired with shock. Kalman estimator: y and $\hat{y}$ are observed and output quantities, $\hat{w}_1$ and $\hat{w}_2$ are weights associated with x_1 and x_2. The Kalman gain and uncertainty are also plotted. During the last twenty trials, $\hat{w}_2$ does not change significantly because there are little or no prediction errors. LMS: the LMS algorithm also produces little or no changes in $\hat{w}_2$.

state uncertainty matrix P keeps a history of the variance and covariance in the prior inputs. This uncertainty affects how we will respond to the prediction errors. In contrast, in LMS the history of the prior inputs does not play a role in the problem of learning form error. Furthermore, while LMS leaves the step size somewhat unspecified—by allowing for an arbitrary step parameter η—Kalman offers a rigorous way to determine the step size, as well as its direction, based on the input's statistics.

Now let us simulate the Kamin blocking experiment and compare performance of the two learning algorithms. In the first set of training, suppose that we have twenty trials in which $x_1 = 1$ (light is on), $x_2 = 0$ (tone is off), and $y^* = 1$ (shock is present). In the second set of training (another twenty trials), x_1 and y^* remain unchanged but $x_2 = 1$. Our objective is to estimate w_1 and w_2, the weights associated with x_1 and x_2, and then predict behavior in the test period during which $x_1 = 0$ and

$x_2 = 1$ (light is off, tone is on, not shown in figure 6.2). Figure 6.3 shows the simulation results for the Kalman algorithm and LMS. We begin with $\hat{w}_1$ and $\hat{w}_2$ set to zero, that is, with no association between inputs and shock. In the first set, when $x_1 = 1$, both algorithms learn to set $\hat{w}_1 = 1$. That is, the light fully predicts the shock. In set 2, when light and tone are both on, there is no prediction error to learn from as the light fully predicts the shock, and therefore the weights remain unchanged. So by the end of set 2, the weight associated with tone $\hat{w}_2$ is unchanged from before the experiment began; it is still around zero. In the subsequent test period when tone appears alone, its weight is near zero, and the animal does not "freeze." In this case, the two algorithms behave similarly.

A number of observations in this simulation are worth highlighting because, as we will see shortly, they play a crucial role in the success of the Kalman algorithm and the failure of LMS in the next experiment. In the Kalman portion of figure 6.3, the terms k_1 and k_2 are the elements of the Kalman gain $\mathbf{k}$. They specify how much of the prediction error is assigned to $\hat{w}_1$ and $\hat{w}_2$. The terms $P_{1,1}$, $P_{1,2}$, and $P_{2,2}$ are elements of matrix P, describing the variance and covariance of $\hat{w}_1$ and $\hat{w}_2$. In set 1(the first twenty trials), when $x_1 = 1$ but $x_2 = 0$, uncertainty about w_2 is large because its value does not affect the output, as reflected in the large value of $P_{2,2}$. In set 2, when $x_1 = 1$ and $x_2 = 1$, the value of $P_{1,2}$ becomes negative because of the covariance of $\hat{w}_1$ and $\hat{w}_2$ (if $\hat{w}_1$ increases, then $\hat{w}_2$ must decrease). That is, when the inputs are "on" together, they produce a negative covariance on the weights associated with them: To maintain a correct output, if a weight increases, the other must decrease. We saw this phenomenon earlier in figure 4.13B. The elements of matrix P keep a record of the history of the prior inputs, whereas no such record is kept in the LMS algorithm. Let us now consider an experiment in which this record plays an important role.

The next experiment is called *backward blocking*. In this experiment, during set 1 the animal is trained with both stimuli, that is, $x_1 = 1$ and $x_2 = 1$. During set 2, only one of the stimuli remains present, i.e., $x_1 = 1$ but $x_2 = 0$. The Kalman and LMS algorithms now make very different predictions (figure 6.4). The Kalman algorithm, like LMS, predicts that during set 1, the system will learn to assign some weight to both stimuli, that is, $\hat{w}_1 = 0.5$ and $\hat{w}_2 = 0.5$. In set 2 when only x_1 is present, there is a prediction error. Both LMS and Kalman increase the weight of $\hat{w}_1$ to one. However, because x_2 is not present, LMS does not alter $\hat{w}_2$. The Kalman algorithm, however, reduces $\hat{w}_2$ to zero. Therefore, LMS predicts that at the end of set 2, the animal will show as much freezing in response to x_2 (tone) as it did at the end of set 1. In contrast, Kalman predicts that at the end of set 2, in response to x_2 the animal will show significantly less freezing. Kalman makes the unusual prediction that if light predicts the shock, then tone must be ineffective, despite the fact that earlier the animal was shocked in the presence of both light and tone.

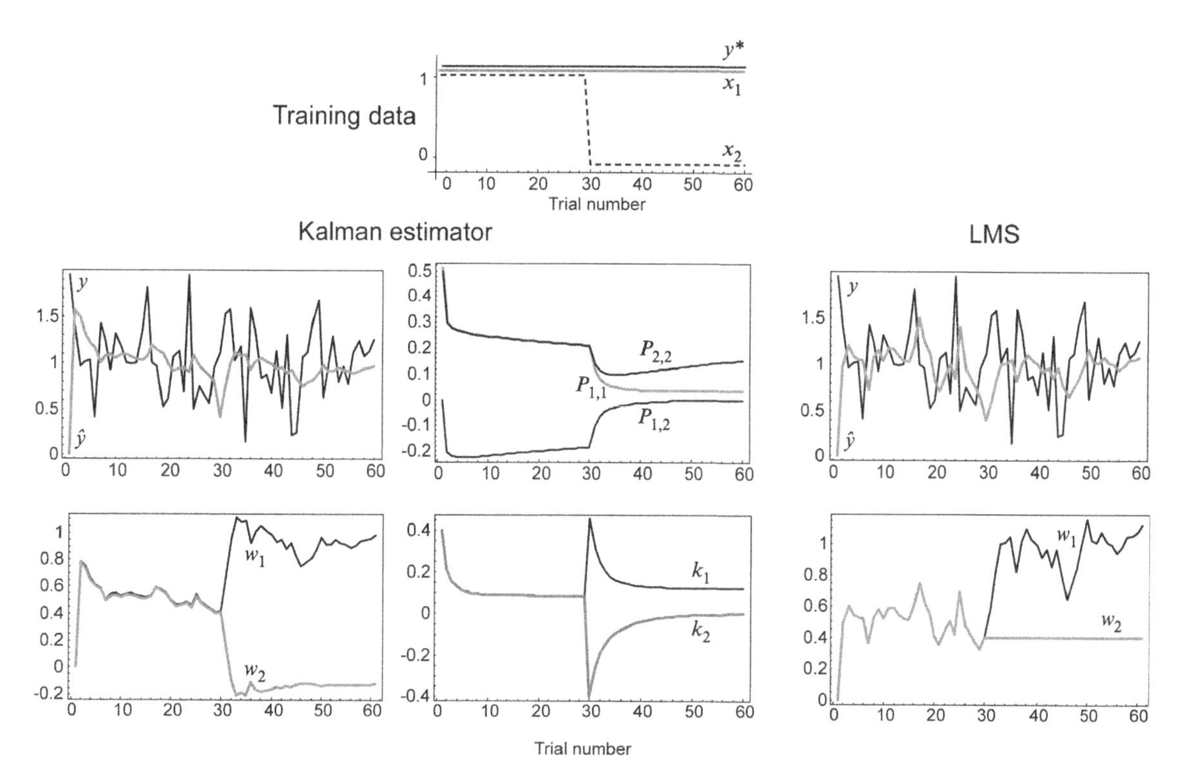

Figure 6.4
A comparison of the Kalman framework and LMS for the backward blocking experiment. During the first twenty trials of training, both x_1 and x_2 are paired with shock y^*. During the last twenty trials, only x_1 is paired with shock. In the Kalman framework, the covariance component of the uncertainty P_{12} becomes negative in the first twenty trials. This then affects the Kalman gain k_2 in the subsequent trials, causing "unlearning" in $\hat{w}_2$ despite the fact that during the last twenty trials x_2 is not present. LMS, on the other hand, does not alter the weight associated with x_2 during the last twenty trials.

The reason for this is that in set 1, the Kalman algorithm sets the covariance of $\hat{w}_1$ and $\hat{w}_2$ to be negative (the term P_{12}). This means that in set 2, when $\hat{w}_1$ is increased (the weight of light), absent of other evidence $\hat{w}_2$ (weight of tone) must decrease. This is what covariance implies. In effect, the prior history causes the Kalman algorithm to alter the weight of $\hat{w}_2$ in set 2, even though in that set x_2 is not present to directly cause learning from the prediction error.

Perhaps we should note that the solution given by the Kalman algorithm is not better than the solution of the LMS algorithm in an absolute sense. It is plausible to encounter situations in which either one is more effective than the other. However, the Kalman algorithm is consistent with Bayesian inference of the posterior probability to observe a shock given the history of experienced stimuli, in the context of our generative model. More interestingly, it nicely agrees with

Group	Treatment: Phase 1	Phase 2	Phase 3	Tests
Backward blocking	AX -> B	A -> B	B -> US	X? A?
Control	AX -> B	A -> C	B -> US	X? A?

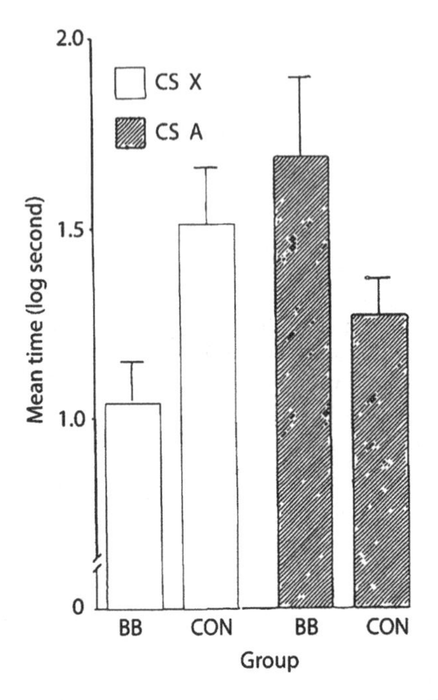

Figure 6.5
The "backward blocking" experiment in rats. The stimuli A, B, C, and X represent different kinds of sounds (A = buzzer, B = tone, C = white noise, X = click). In phase 1, A and X are simultaneously presented and paired with B. In phase 2, A is paired with B. In phase 3, B is paired with shock. The prediction of the Kalman framework is that as the animal learns that A fully predicts B, the prior association between X and B should be weakened. As a result, the animal should show reduced freezing in response to X than in control animals that had not learned A-B association (control group). In the experimental results, the bars depict mean freezing time in log seconds. The solid bars are the freezing time in the presence of stimulus X, and the striped bars are freezing time in presence of stimulus A. Longer times suggest a stronger association. Error bars: SEM. CS, conditioned stimulus; US, foot shock; BB, the backward blocking group; CON, the control group. (From Miller and Matute, 1996.)

behavior of rats. Ralph Miller and Helena Matute (Miller and Matute, 1996) performed the backward blocking experiment in a group of rats (figure 6.5). They had three stimuli: A, B, and X (three different kinds of auditory stimuli). In phase 1, they paired A and X with B, so that AX→B. In phase 2, they paired A with B: A→B. In phase 3, they paired B with shock. In the test period, they presented X or A alone and tested the time period in which the animal stopped moving (freezing period). They found that the animals froze much longer in response to stimulus A than stimulus X in the test period. If we label stimulus A as x_1 and stimulus X as x_2, then their results are consistent with the Kalman algorithm and inconsistent with LMS.

In our example, the sensitivity to prediction error—that is, the learning rate—therefore depended on the history of the past observations. In the Kalman framework, this history defined an uncertainty that dictated the sensitivity to prediction error.

6.4 Prediction Errors Drive Adaptation of Internal Models

In the preceding examples, the tasks were forms of classical conditioning. The animals did not have to do anything to get reward or avoid punishment; the best they could do was predict the consequences of an externally controlled set of stimuli. A more interesting scenario is one in which the brain needs to figure out the consequences of a self-generated action and produce actions that maximize some measure of performance. An example of this is tasks in which people learn to control movements of their bodies.

During development, bones grow and muscle mass increases, changing the relationship between motor commands (e.g., torques) and motion of the limb (position and velocity). In addition to such gradual variations, the arm's dynamics change over a shorter timescale when we grasp objects and perform manipulation. It follows that in order to maintain a desired level of performance, our brain needs to be "robust" to these changes. This robustness may be achieved through an updating, or adaptation, of an internal model that predicts the sensory consequences of motor commands. Indeed, people appear to excel in the ability to adapt rapidly to the variable dynamics of the arm as the hand interacts with the environment: we easily switch from moving a small pen to moving a long stick to moving a baseball bat. A task in which people use their hands to interact with a novel environment might be a good candidate to study how the brain learns internal models.

There are two well-studied versions of the adaptation paradigm. In one version, called *visuomotor adaptation*, the investigator introduces a perturbation that distorts the visual consequences of the motor commands but leaves the proprioceptive consequences unchanged. This is typically done by wearing prism goggles or having people move a cursor on the screen in which the relationship between cursor position and hand position is manipulated (figure 6.6A). In another version of the adaptation paradigm, called *force field adaptation*, the investigator introduces a perturbation that alters both the visual and proprioceptive consequences of motor commands. This is typically done by having the volunteer hold the handle of an actuated manipulandum (a robotic arm) that produces forces on the hand (figure 6.6B). This type of adaptation can also be done by having people reach in a rotating room (the rotation imposes novel forces on the hand), or even in microgravity in which the usual forces are removed. In both the visuomotor and force field experiments, learning depends on sensory prediction errors.

The oldest record of visuomotor adaptation experiment that we know of is an 1867 report by Hermann von Helmholtz. In that work, he asked subjects to point

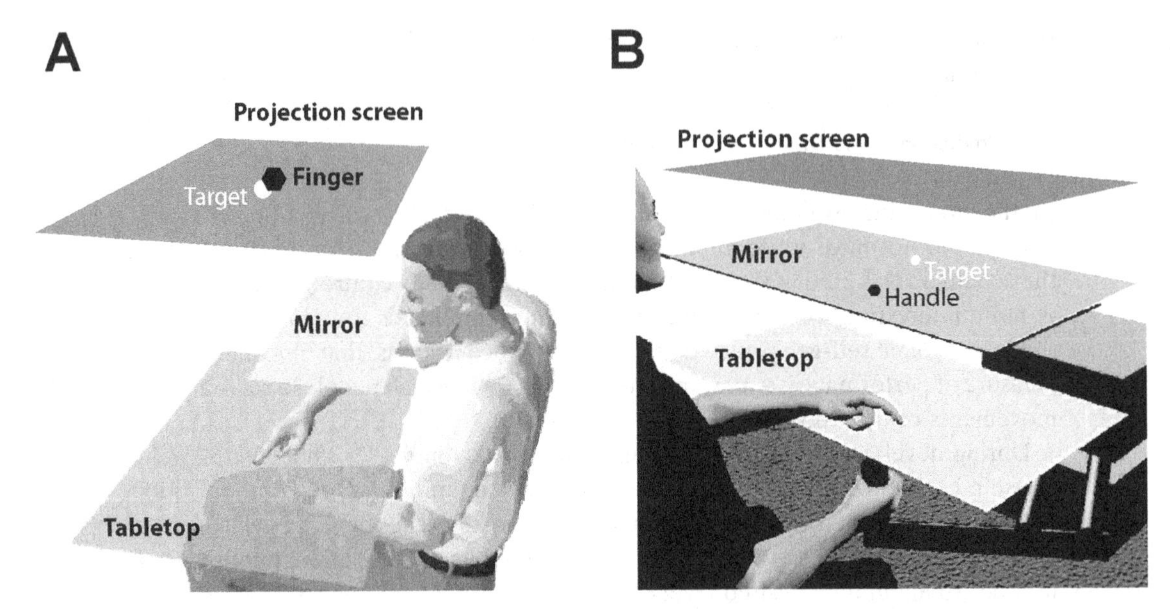

Figure 6.6
Two common paradigms to investigate mechanisms of adaptation in human motor control. (A) Visuomotor rotation paradigm. People reach with the hand above the tabletop to a target projected by a video projector and viewed through a mirror. In this way, they cannot see their hand, but they see the target in the same plane as their hand. They are provided with feedback regarding their hand position via a cursor (labeled finger on the projection screen). During the adaptation component of the experiment, the location of the projected cursor is shifted with respect to hand position, requiring the subject to alter the motor commands to the hand in order to place the cursor in the target. Occasionally, the subject is also asked to point to their right hand with their left hand. (From van Beers, Wolpert, and Haggard, 2002.) (B) Force field adaptation paradigm. People reach with their right hand while holding the handle of a robotic manipulandum. The aim is to place a cursor representing handle position inside a target. The robot produces forces that perturb the movement. Occasionally, the subject is also asked to point to their right hand with their left hand. (From Haith et al., 2008.)

with their finger at targets while wearing prism lenses that displaced the visual field laterally. When the displacement was to the left, subjects initially had errors (an overshoot) in that direction and after some practice, they learned to compensate for the visual displacement. Helmholtz observed that as soon as the prisms were removed, subjects made erroneous movements to the right of the target. This is known as an *aftereffect* of adaptation.

Nearly a century later, in the early 1960s there was renewed interest in motor adaptation because of the space program. Astronauts were preparing to leave earth, and there was concern as to whether their motor system could function in zero gravity. Could the brain adapt their motor commands so that they could function in this radically different environment? Were there methods to help speed the process of adaptation? To quantify the ability of the brain to adapt,

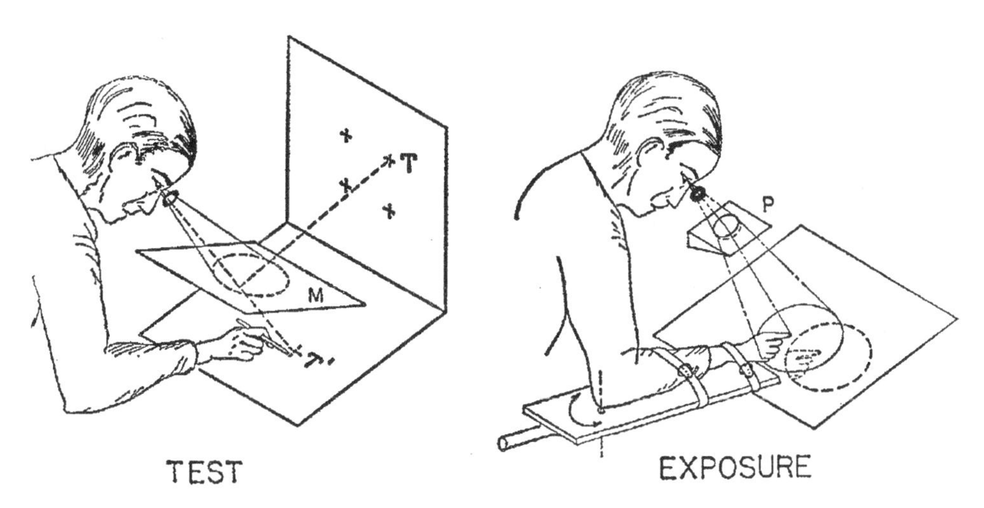

Figure 6.7
Adaptation requires sensory prediction errors. During the exposure part of the experiment, the subject's forearm was tied to a board that could move about the elbow joint. The subject either actively moved her elbow, or her elbow was passively moved by the board. In both cases, she viewed the hand through prism glasses. After a period of self-generated or passive movements, she was tested in a paradigm in which she pointed to targets without the prisms. The targets were displayed through a mirror, preventing her from seeing her hand. Only the active exposure condition resulted in aftereffects. (From Held and Freedman, 1963.)

Richard Held and Sanford Freedman (Held and Freedman, 1963) repeated Helmholtz's experiment with a new twist. They compared the performance of subjects when they actively moved their arm while viewing their finger through prism glasses, versus when they viewed their finger but their arm was passively moved for them (figure 6.7). In the experiment, the subject's arm was tied to a board that could move about the elbow. When the subject actively moved her elbow, the result was a circular motion of the finger. When the experimenters moved the board, the result was the same circular motion of the subject's finger, but now there were little or no motor commands that were generated by the subject. In both cases, the subject viewed the motion of the finger via a prism that induced a displacement in the visual feedback. After this viewing, the subject was tested in a pointing task (marked Test in figure 6.7). Held and Freedman found that in the test session, the subject showed aftereffects, but only if during the earlier session she had viewed her hand while actively moving it. She did not have aftereffects if she had viewed her hand while it was moved passively. In their words: "Although the passive-movement condition provided the eye with the same optical information that the active-movement condition did, the crucial connection between motor output and visual feedback was lacking." In our terminology, sensory prediction error was missing in the passive condition, as the subjects did not actively generate a movement, and therefore could not predict the sensory consequences.

There is a more recent example of visuomotor adaptation that provides striking evidence for the crucial role of sensory prediction errors. Pietro Mazzoni and John Krakauer (2006) had people move their wrist so that the position of the index finger was coupled with the position of a cursor on a screen. There were always eight targets on display, spanning 360 degrees. On a given trial, one of the targets would light up and the subject would move the cursor in an out-and-back trajectory, hitting the target and then returning to the center. After a baseline familiarization period (40 trials), the experimenters imposed a 45-degree counterclockwise rotation on the relationship between the cursor and finger position (early adaptation, figure 6.8A). Let us represent this perturbation as a hidden state and label it with r. Now, a motor command u that moved the hand in direction θ did not produce a cursor motion in the same direction, but in direction $\theta + r$. If we label the predicted sensory consequences $\hat{y} = \theta$ and the observed consequences $y = \theta + r + \varepsilon_y$, then there is a sensory

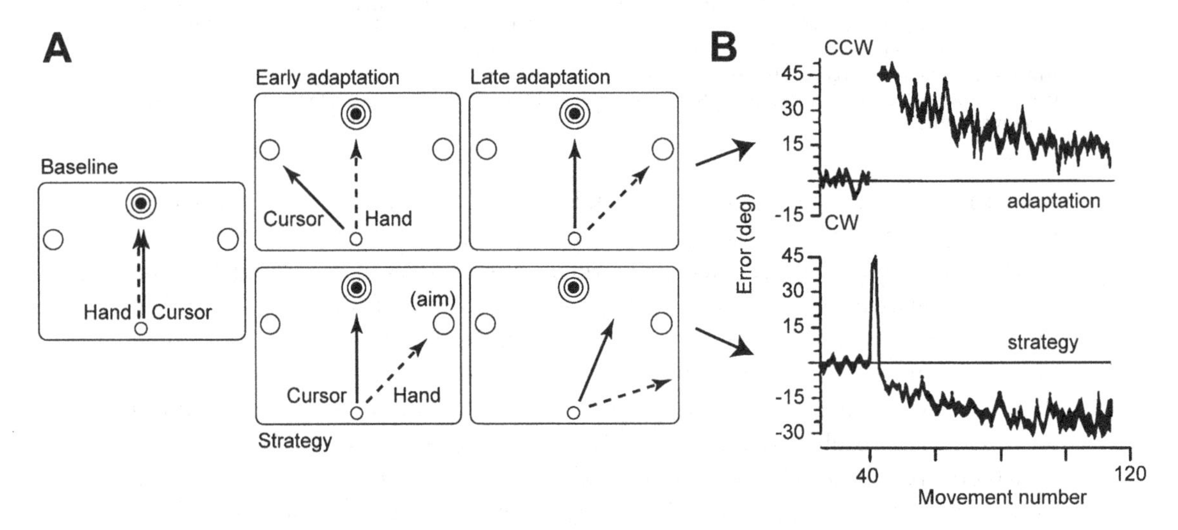

Figure 6.8
An example of learning a visuomotor rotation. (A) Subjects were asked to make an out-and-back motion with their hand so a cursor was moved to one of eight targets. In the baseline condition, hand motion and cursor motion were congruent. In the adaptation condition, a 45-degree rotation was imposed on the motion of the cursor and the hand. In the adaptation group (top two plots), the subjects gradually learned to move their hand in a way that compensated for the rotation. In the strategy group (bottom two plots), after two movements subjects were told about the perturbation and asked to simply aim to the neighboring target. (B) Endpoint errors in the adaptation and strategy groups. The strategy group immediately compensated for the endpoint errors, but paradoxically, the errors gradually grew. The rate of change of errors in the strategy and adaptation groups was the same. The rapid initial improvement is due to learning in the explicit memory system, whereas the gradual learning that follows is due to an implicit system. (From Mazzoni and Krakauer, 2006.)

prediction error $y-\hat{y}$. The objective is to use this prediction error to update an estimate for $\hat{r}$. With that estimate, for a target at direction θ^*, we can generate a motor command $u=\theta^*-\hat{r}$ to bring the cursor to the target. Indeed, after about eighty trials, in response to target at θ^* people would move their hands to θ^*-40 so the cursor would land within 5 degrees of the target (as shown in the "adaptation" subplot of figure 6.8B).

Now, Mazzoni and Krakauer tried something quite clever: They took another group of naïve subjects, and after they had experienced a couple of rotation trials, they simply told them: "Look, you made two movements that had large errors because we imposed a rotation that pushed you 45 degrees counterclockwise. You can counter the error by aiming for the neighboring clockwise target." That is, simply issue the motor command $u = \theta^* - 45$ and as a consequence, the cursor will move at direction θ and land at the target. Indeed, the subjects followed this strategy: on the very next trial, all the error dropped to zero (strategy group, figure 6.8B). However, now something very interesting happened: As the trials continued, the errors gradually grew! What's more, the rate of change in the errors in this "strategy" group was exactly the same as the rate of change in the regular adaptation paradigm.

To explain this, Mazzoni and Krakauer hypothesized that on trial 43, when the subjects in the strategy group were producing the motor commands that brought the cursor to the target, there was still a discrepancy between the predicted and observed sensory consequences of motor commands $y-\hat{y}$. This is because whereas *explicitly* they had been told of the perturbation, *implicitly* their estimate was still around zero, $\hat{r}\approx 0$. The implicit estimate (the motor system's estimate) learned from prediction error, and that learning took many trials (as shown in the adaptation group, figure 6.8B).

This experiment hints at the complexity of the problem that we are considering: there are multiple learning systems in the brain. The explicit system is typically associated with our conscious awareness, whereas the implicit system is typically associated with an unconscious process. Both can formulate internal models, learn from prediction errors, and contribute to our motor commands. The data in figure 6.8B hint that the explicit system learns extremely quickly, whereas the implicit system learns gradually. In the tasks that we will be considering in this chapter, our assumption will be that we are primarily gauging the influence of the implicit system. The main reason for this assumption is that, as we will see shortly, people with amnesia who have severe deficit in remembering explicit information nevertheless learn to alter their motor commands in the adaptation experiments. This does not mean that in healthy people, internal models learned by the explicit system play no role in adaptation. Indeed, we will return to this topic when we consider the multiple timescales of memory in the next chapter. However, for now, let us imagine that in a typical adaptation experiment involving tens to hundreds of trials, the main contributor is the implicit system.

In summary, an internal model is simply an association between motor commands and their sensory consequences. The driving force in learning an internal model is the sensory prediction error.

6.5 A Generative Model of Sensorimotor Adaptation Experiments

Whereas in the visuomotor paradigms the visual consequence of motor commands is altered but not the proprioceptive consequences, in a force field paradigm both sensory consequences are altered, resulting in multiple prediction errors. An example of the force field adaptation paradigm is shown in figure 6.9. In this experiment, Shadmehr and Mussa-Ivaldi (1994) had a volunteer reach to a target while holding a handle attached to a robotic arm (figure 6.10A). When the robot motors were turned off, the hand motion was generally straight (figure 6.9A). To induce adaptation, the robot produced a force field that pushed on the hand as a function of hand velocity, as shown in figure 6.9B. These forces perturbed the reach trajectories (figure 6.9C). With practice, the subject adapted his motor commands and produced forces on his hand that generally compensated for the robot forces, resulting in a restoration of the straight trajectories (figure 6.9D). When the forces were unexpectedly removed (in catch trials), the hand trajectory exhibited an aftereffect.

When the subjects are caught by surprise by the field, the hand trajectory is perturbed, but it eventually reaches the target, as shown in figure 6.8C. This is because the field depends only on the velocity of the hand: At the end of movement the hand is at rest, and there is no perturbing force pushing it away from the target. Reaching the target is all that subjects were asked to do. The experiment shows that subjects are compelled to adjust their motor command so as not only to reach the target, but to do so along a path that is similar to the initial unperturbed path. If asked to describe the forces they experience and what they did to compensate, they would not have a clear answer. Nevertheless, their brain alters the motor commands and attempts to bring the cursor to the target in more or less a straight line.

An important observation is that when the severely amnesic subject H.M. and other amnesic subjects were tested on this task (figure 6.10B), they learned and showed after effects during the training session, a few hours after this initial training session, and even the next day. This adaptation took place despite the fact that they could not consciously remember having seen the robot or having done the task before (Shadmehr, Brandt, and Corkin, 1998). Clearly, a crucial aspect of adaptation in the force field paradigm is via an implicit, nonconscious memory system. (It is of course possible that some explicit system was involved in the motor learning but that this explicit component itself was not remembered the next day. What is clear

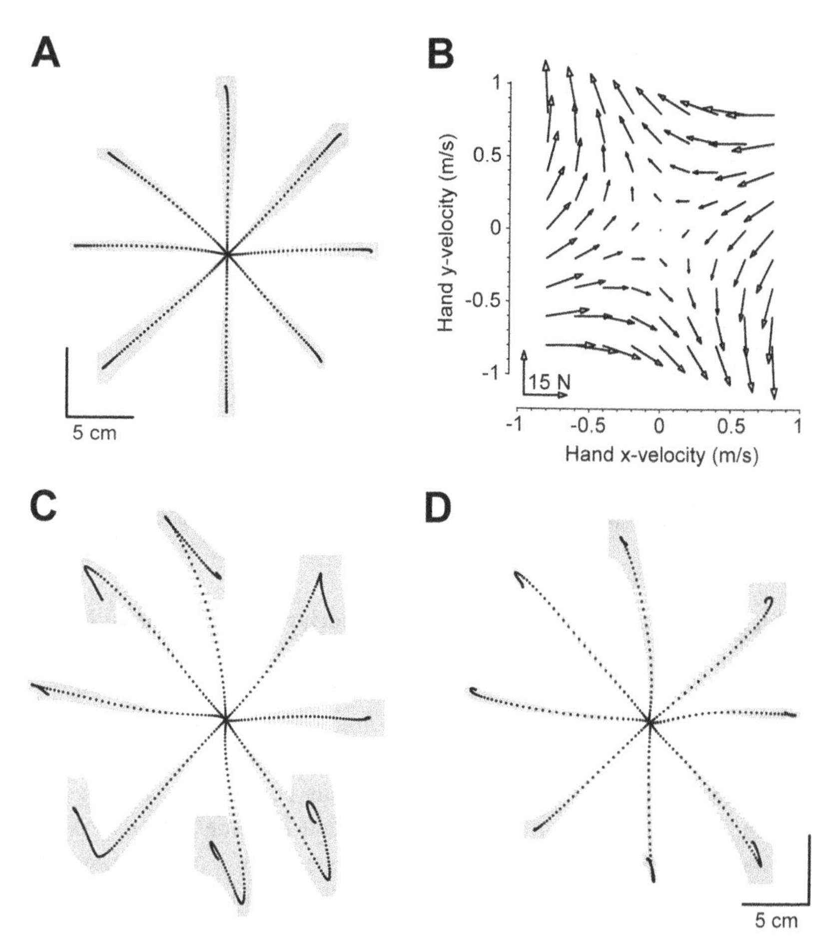

Figure 6.9

Example of adaptation to a force field. (A) Volunteers made reaching movements while holding the handle of a robotic manipulandum in order to place a cursor in a target. When the robot did not produce a force field (termed null field), hand trajectories were straight. All trajectories shown here were made without visual feedback during the movement (about one-fourth of the trials were without visual feedback). (B) The force field produced by the robot. (C) Hand trajectories early in the training protocol (first 250 trials). (D) Hand trajectories late in the training protocol (trials 750–1000). (From Shadmehr and Mussa-Ivaldi, 1994.)

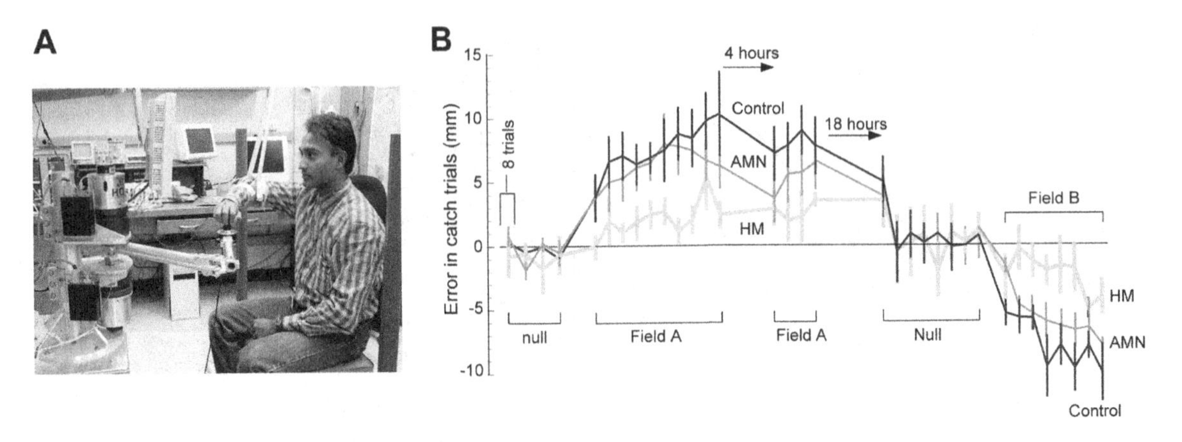

Figure 6.10
Example of adaptation to a force field. (A) Experimental setup. (B) Size of aftereffects in two groups of people: healthy controls, and amnesiacs (AMN), including subject HM. (From Shadmehr, Brandt, and Corkin, 1998.)

is that it is possible to recall motor memory without a declarative awareness of that memory.)

How do subjects adapt to the visuomotor or force field perturbations? This is a very broad question that might be answered at many levels. For example, our answer might invoke something about the molecular basis of memory in the brain regions involved in motor control. However, let us focus here on the basic computational problem: in principle, what is the problem that is being solved? As we will see in this and the following chapters, the perturbation is producing two kinds of errors: a sensory prediction error (the motor commands did not produce the expected hand trajectory in visual and proprioceptive coordinates), and a reward prediction error (the motor commands did not get us to the target in time and so we did not get rewarded for our efforts). In this chapter, we will focus on the problem of accurately predicting the sensory consequences of the motor command.

Perhaps the simplest approach is to imagine that our hand did not move as we had expected because there was a force (a hidden state) that we did not account for. Therefore, if we could estimate this force, we could accurately predict the motion that should result. To get the hand to the target, we would produce motor commands that compensate for this perturbing force. In the visuomotor paradigm, the perturbation is a bias that distorts cursor position with respect to hand position (another hidden state). If we had a correct estimate of this perturbation, we could produce motor commands that accounted for it, producing accurate movements of the cursor to the target.

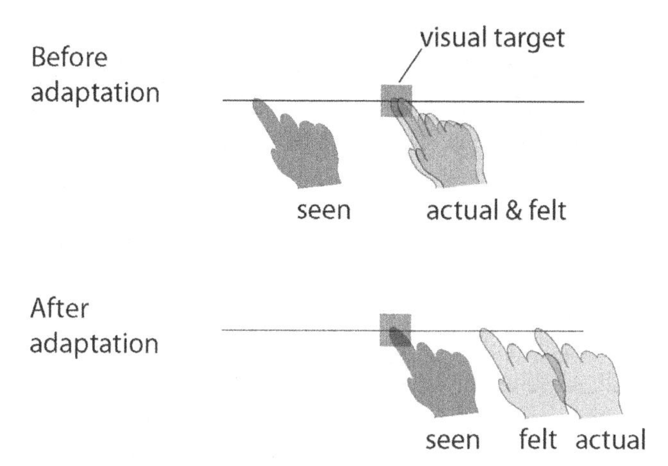

Figure 6.11
Adapting to a visuomotor perturbation induces an illusion regarding location of the trained hand. Top: Before adaptation, when the unseen right hand is placed at a target location, the estimated position of that hand corresponds to the actual location. Bottom: After adaptation to the visuomotor perturbation, the subject estimates their right hand to be somewhere between the actual position and the seen position.

To approach the problem of learning internal models, we need to specify a generative model that relates motor commands with sensory consequences. To motivate our approach, we need to consider a curious fact: People who participate in these experiments do not simply learn to produce motor commands that compensate for a perturbation. Rather, practice produces a sensory illusion.

For example, consider a paradigm in which people use their right arm to adapt to a visuomotor rotation, that is, they reach to a target and eventually learn to cancel the imposed bias on the visual feedback. It turns out that the training also alters where they think their right arm is located in space (figure 6.11). Let us explain this result in some detail. In figure 6.5, the two setups not only allow the subject to reach to a target, but the setups also allow the subject to point to the position of one hand with the other hand. Now suppose that before the reach task starts, we remove vision of both hands, have the subject move one hand to some location, and then have him point with the other hand to the location of the moved hand. People are usually pretty accurate at this kind of alignment. Now as the visuomotor adaptation experiment begins, the subject reaches to the visual target and she would see her hand (or cursor representing her hand) to the left of the target because of the bias that has been imposed by the visual display (figure 6.11, before adaptation). Trial after trial, the seen position of the hand gets closer to the target, and the actual hand position gets farther from the target (figure 6.11, after adaptation). However, if we now ask the subject to estimate her hand position, she points to somewhere a bit to the left of where it actually is (van Beers, Wolpert, and Haggard, 2002). Indeed, a similar thing happens when people adapt their reaching movements to force fields (Haith et al., 2008): If a force pushes the hand to the right, people learn to compensate for these forces but afterward believe that the hand is a little to the right of its actual position. That is, it is not just the motor commands that have adapted. Something has also changed the perception of where the hand is located in space. Adaptation seems to produce a sensory illusion.

We can account for these results by describing the process of adaptation in the framework of state estimation. A reasonable approach was suggested by Adrian Haith, Carl Jackson, Chris Miall, and Sethu Vijayakumar (2008). Their generative model is graphically described in figure 6.12. To make things easy, let us represent our problem in a one-dimensional world (so our variables are scalars). The position of the hand (or finger) on trial n, represented by variable h, depends on the motor commands u, and is observed by the brain through two independent sensors, vision and proprioception, represented by y_v and y_p. The sensory measurements

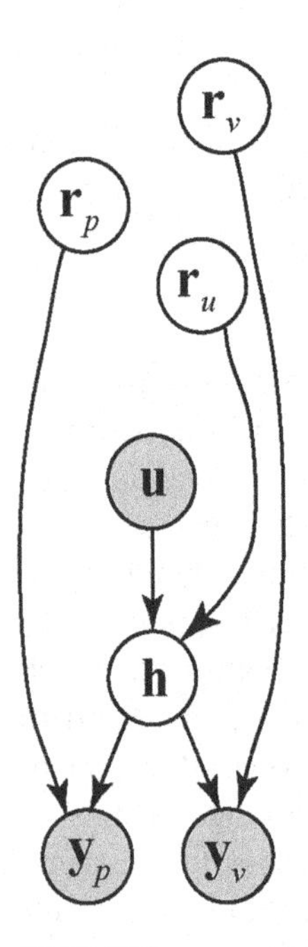

Figure 6.12
A generative model for motor adaptation (Haith et al., 2008). Motor commands u, affect the position of the hand h, which is observed by the brain through two independent sensors, vision y_v and proprioception y_p. There are three potential sources of perturbation: perturbations r_u that alter the relationship between motor commands and hand position, and perturbations r_v and r_p that alter the relationship between hand position and sensory measurements.

are affected by noise $\varepsilon_v \sim N(0, s_v)$, and $\varepsilon_p \sim N(0, s_p)$, as well as by two potential sources of perturbation r_v and r_p:

$$\begin{aligned} y_v^{(n)} &= r_v^{(n)} + h^{(n)} + \varepsilon_v \\ y_p^{(n)} &= r_p^{(n)} + h^{(n)} + \varepsilon_p \end{aligned} . \tag{6.8}$$

Motor commands produce a change in the state of the finger, and they too are subject to a potential source of perturbation r_u, and noise $\varepsilon_u \sim N(0, q_u)$. Grossly simplifying the dynamics of movements, let us represent this relationship as:

$$h^{(n+1)} = u^{(n)} + r_u^{(n)} + \varepsilon_u. \tag{6.9}$$

Equation (6.8) is a model of the sensory system: we are assuming that our two sensors, vision and proprioception, can tell us something about the position of our finger. We are also assuming that our model of how the sensors are related to the actual position of the hand may be biased and that bias is a form of perturbation. Similarly, equation (6.9) is a model of how the motor commands produce a hand position. This model can also be biased, that is, our model of the relationship between motor commands and the position of the hand may be wrong. This bias is represented as a perturbation (e.g., a force field is a perturbation that "adds" a bias to our motor commands, moving our hand in an unexpected way).

Now when we see a visual target at y_v^*, our objective is to place the visual representation of our hand y_v at the target. To do that, we need to produce the following motor command:

$$u^{(n)} = y_v^* - \hat{r}_v - \hat{r}_u. \tag{6.10}$$

That is, we produce a motor command that cancels the estimated perturbations $\hat{r}_v$ and $\hat{r}_u$. When we are asked to indicate the position of our hand by pointing to it, we are being asked about our estimate $\hat{h}$. Summarizing our model, the estimation problem is as follows: given motor command u and the sensory feedback y_v and y_p, estimate the state of the perturbations r_v, r_p, and r_u, and also estimate the position of the hand h.

Before we solve the problem mathematically, let us get an intuitive feel for what the solution should look like. The purpose of training is to help us form a better estimate of the various perturbations so that by the end of training, given a motor command u, our predictions about sensory feedback ($\hat{y}_v$ and $\hat{y}_p$) match the actual observations (y_v and y_p). Right after the visual perturbation is introduced (say $r_v = -2$, i.e., we see our hand or cursor 2 cm to the left of where it actually is), $\hat{y}_v$ will differ substantially from y_v, i.e., there will be a prediction error regarding the visual sensory feedback. However, at the same time we will have $\hat{y}_p \approx y_p$, that is, no prediction error regarding proprioceptive feedback. What can account for this pattern of prediction errors?

One possibility is that all the prediction error is due to perturbation r_v. For example, if $r_v = -2$, then by setting $\hat{r}_v = -2$ and $\hat{r}_p = \hat{r}_u = 0$, equation (6.10) will bring the visual representation of the hand to the target as well as maintain $\hat{y}_p \approx y_p$. This is one way with which we can assign credit to the various potential perturbations. However, this is not the only way. If $r_v = -2$, then by setting $\hat{r}_v = -1.5$, $\hat{r}_u = -0.5$, and $\hat{r}_p = +0.5$, once again equation (6.10) will be sufficient to make $\hat{y}_v = y_v$ and $\hat{y}_p = y_p$, that is, eliminate the prediction error. If we assign the credit this way, we interpret part of the visual perturbation in terms of a motor perturbation r_u, and part in terms of sensory perturbations r_v and r_p. Indeed, as long as $\hat{r}_v + \hat{r}_u = -2$ and $\hat{r}_p + \hat{r}_u = 0$, we have a solution to our problem. So we see that in principle, there are multiple solutions to our problem. Which kind of credit assignment is best? The credit assignment will depend on the uncertainty that we have regarding how the various hidden states (the perturbations) affected our observations.

The key idea is that despite the fact that the subject has experienced a prediction error because the experimenter introduced a bias on the visual feedback, from the point of view of the subject the optimal thing to do may not be an assignment of the error solely to a visual perturbation. The error may also be due to an incorrect internal model that associates motor commands with position of the hand, as well as sensory feedback that measures that position. To make a sensible decision, the subject must take into account the reliability of each source of information. And this can only be done based on past experience. As we will see, it is this ambiguity that can account for the fact that during adaptation to a visual perturbation, subjects acquire a sensory illusion regarding estimate of their hand position.

6.6 Accounting for Sensory Illusions during Adaptation

Let us represent the learning problem in a visuomotor adaptation experiment via the generative model shown in figure 6.12. Set vector $\mathbf{x}$ to be the states that we need to estimate $\mathbf{x} = [r_v \quad r_p \quad r_u \quad h]^T$ and vector $\mathbf{y}$ to be the sensory observations $\mathbf{y} = [y_v \quad y_p]^T$. The generative model can be written as:

$$
\begin{aligned}
\mathbf{x}^{(n)} &= A\mathbf{x}^{(n-1)} + \mathbf{b}u^{(n-1)} + \boldsymbol{\varepsilon}_x \qquad \boldsymbol{\varepsilon}_x \sim N(0,Q) \\
\mathbf{y}^{(n)} &= C\mathbf{x}^{(n)} + \boldsymbol{\varepsilon}_y \qquad \boldsymbol{\varepsilon}_y \sim N(0,R)
\end{aligned}
$$

$$
A = \begin{bmatrix} a_v & 0 & 0 & 0 \\ 0 & a_p & 0 & 0 \\ 0 & 0 & a_u & 0 \\ 0 & 0 & 1 & 0 \end{bmatrix} \quad \mathbf{b} = \begin{bmatrix} 0 \\ 0 \\ 0 \\ 1 \end{bmatrix} \quad C = \begin{bmatrix} 1 & 0 & 0 & 1 \\ 0 & 1 & 0 & 1 \end{bmatrix}. \tag{6.11}
$$

The matrix A has diagonal terms that are all less than one, implying that we believe that all perturbations are transitory and will eventually go away. To estimate the position of the hand and the various perturbations, we apply the Kalman algorithm. We start with a belief $\hat{\mathbf{x}}^{(n|n-1)}$ and uncertainty $P^{(n|n-1)}$ and produce a motor command $u^{(n-1)}$ (using equation 6.10) and predict its consequences $\hat{\mathbf{y}}^{(n)}$ (using equation 6.11). Our motor command produces a hand position (equation 6.9) and its visual and proprioceptive sensory consequences $\mathbf{y}^{(n)}$ (equation 6.8, where we apply the experimenter's visual perturbations). Using our prior uncertainty, we form the Kalman gain:

$$K^{(n)} = P^{(n|n-1)}C^T\left(CP^{(n|n-1)}C^T + R\right)^{-1}. \tag{6.12}$$

We then update our estimate of the perturbations and hand position by combining our predictions with the observations:

$$\begin{aligned} \hat{\mathbf{x}}^{(n|n)} &= \hat{\mathbf{x}}^{(n|n-1)} + K^{(n)}\left(\mathbf{y}^{(n)} - \hat{\mathbf{y}}^{(n)}\right) \\ P^{(n|n)} &= \left(I - K^{(n)}C\right)P^{(n|n-1)} \end{aligned}. \tag{6.13}$$

The final step is to form the prior uncertainty for the next movement, which is computed as:

$$P^{(n+1|n)} = AP^{(n|n)}A^T + Q. \tag{6.14}$$

Figure 6.13A shows a typical simulation result. In this simulation, all noise variances were independent and set to 1 (matrices R and Q), and a_v, a_p and a_u were set at 0.99. The visual target was set at $y_v^* = 1$. For the first ten trials, $r_v = 0$ but for the remaining trials $r_v = -2$, that is, a visual perturbation was imposed at trial 10 and beyond. All other perturbations were maintained at zero. By end of the adaptation period, the algorithm has assigned a nonzero value to $\hat{r}_p$ and $\hat{r}_u$. As a result, the estimate of current hand position $\hat{h}$ is closer to the visual target than actual hand position h (figure 6.13C). That is, there is a perceptual illusion of where the hand is located.

The reason for this illusion is that the prediction error $y_v - \hat{y}_v$ can be due to a combination of estimation errors in $\hat{r}_u$ and $\hat{r}_v$. It is possible that some of the prediction error is due to a faulty internal model that associates motor commands with hand position (misestimation of $\hat{r}_u$), while some of it is due to the model of the sensory system that measures hand position via visual feedback (misestimating $\hat{r}_v$). However, because there is never a bias in the proprioceptive feedback ($y_p - \hat{y}_p$ is zero on average), any change in $\hat{r}_u$ must be balanced with an opposite change in $\hat{r}_p$. It is the fact that $\hat{r}_p$ is not zero that produces the perceptual illusion of feeling one's hand at a location other than it actually is.

The model explains that the perceptual illusion regarding where the hand is located arises because of the relative uncertainties in the sensory and motor factors that affect the estimate of hand position. To explore this idea further, in figure 6.13B

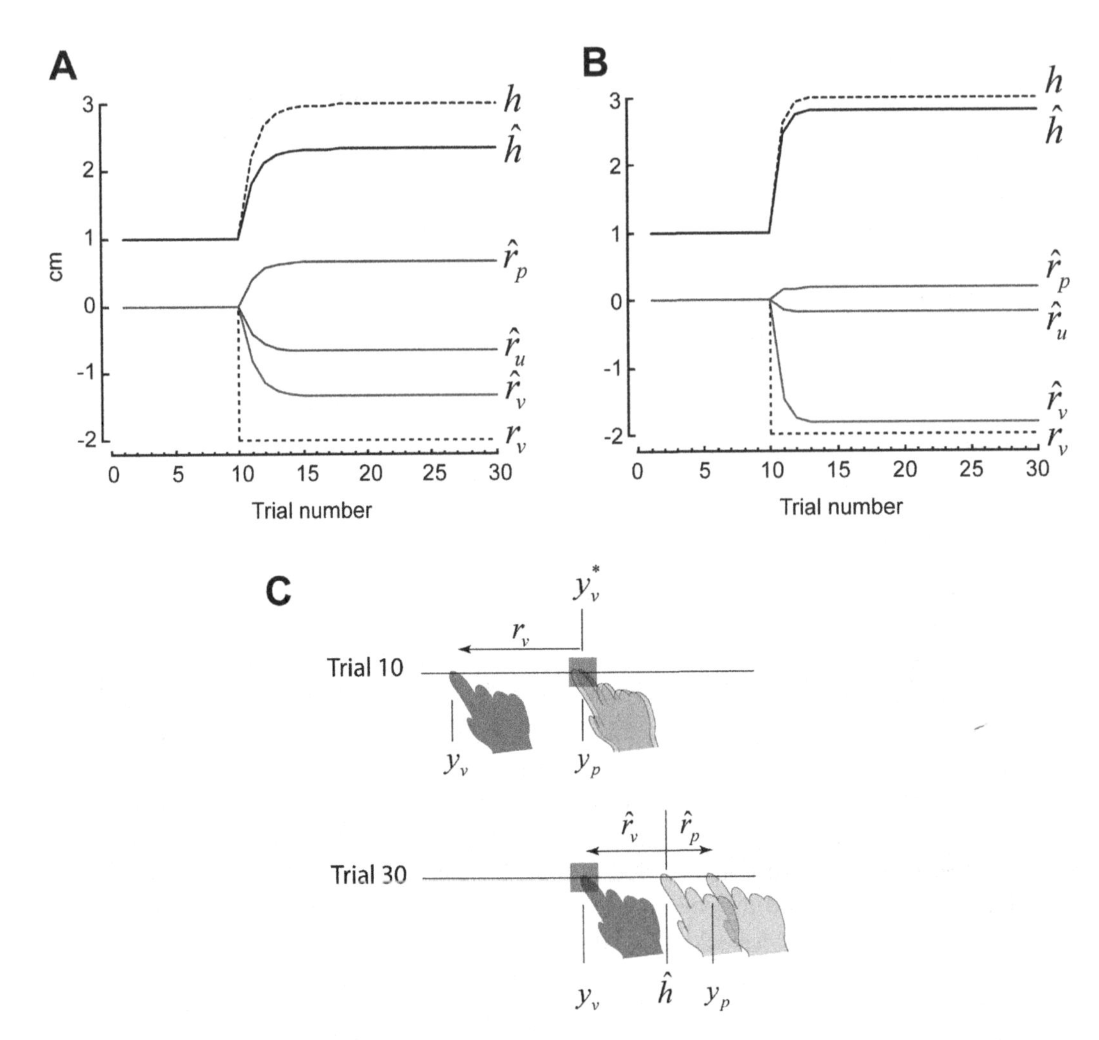

Figure 6.13
Simulations for a reaching experiment with a visuomotor displacement. On trial 10, a perturbation r_v is imposed so that the visually observed hand position y_v is 2 cm to the left of the target, as shown in part C. Trial after trial, the motor commands adapt to place the hand h so that y_v coincides with the target $y_v^* = 1$. (A) With equal uncertainties regarding the source of the perturbation, the simulation estimates the perturbation as a combination of $\hat{r}_p$, $\hat{r}_v$, and $\hat{r}_u$. Because $\hat{r}_u$ is not zero, estimate of hand position $\hat{h}$ is different than actual hand position h, as shown in part C. (B) The effect of increased state noise in the variable r_v (the noise term for this variable in matrix Q in equation 6.11). If one is most uncertain about the state of a visual perturbation, then the sensory errors are assigned almost entirely to this modality, resulting in little difference between $\hat{h}$ and h (that is, no sensory illusion). (C) Estimated hand position and sources of perturbation for (A).

we increased the state noise associated with r_v (the noise term for this variable in matrix Q in equation 6.11). Now there is little or no perceptual illusion as the algorithm assigns nearly all of the error to $\hat{r}_v$. That is, if one is very uncertain about the relationship between hand position and the visual sensory feedback, then most of the prediction error is assigned to $\hat{r}_v$. As a result, $\hat{r}_p$ remains near zero, and there is no perceptual illusion regarding hand position.

In summary, when we generate motor commands, perturbations like force fields or visuomotor rotations produce discrepancies between the predicted and observed sensory consequences. The process of adaptation involves learning an internal model that accurately predicts these sensory consequences. An important consequence of this learning is often a sensory illusion. The generative model in figure 6.12 explains the adaptation process as one of estimating the potential contributions of the various factors that could have caused the prediction error. The ambiguity in these potential causes can account for both the adaptive changes in the motor commands and the sensory illusions.

6.7 The History of Prior Actions Affects Patterns of Learning

We started this chapter with backward blocking, an example of learning from the classical conditioning literature. In that example, it seemed that learning depended not only on the prediction errors, but also the history of the previous observations that the animal had made. The Kalman model accounted for the learning by keeping a history of these observations (i.e., the covariance in the inputs) in the uncertainty matrix. Does something like that happen in human learning as well? Do we respond to prediction errors based on the history of our prior observations?

In the backward blocking experiment, it was the history of the previous inputs (light and tone) that affected the learning in the rat. We can think of these two inputs as contextual cues that the rat was trying to associate with the unpleasant shock. We imagined that the rat's objective was to find an appropriate weighting for the two cues. After a training period in which both cues were present, prediction errors during a training period in which only one cue was present affected the state associated with the absent cue. The Kalman algorithm explained that this was because in the initial training period, the history of the two cues led to a negative covariance associated with the uncertainty of the two states. This covariance affected the credit assignment of the prediction errors.

Now suppose that these two cues are contexts in which we perform an action like moving a cursor to a target. In one context, we move our arm and the cursor is associated with the motion of our hand. In another context, we move only our wrist and the cursor is associated with the motion of our index finger. The cursor that we are moving is perturbed by a visuomotor rotation of 30 degrees, and we are going to learn to compensate for this perturbation. As John Krakauer, Pietro Mazzoni, Ali Ghazizadeh, Roshni Ravindran, and Reza Shadmehr (2006) observed, people show rather peculiar behaviors during these context-dependent adaptation experiments. First, for a given magnitude of perturbation, they tend to learn much slower in the arm context than in the wrist context. That is, for some reason the wrist context is easier to learn. Second, the arm context generalizes to the wrist context: after volunteers learn the 30-degree rotation in the arm context, they are much better than naïve when tested in the same 30-degree rotation in the wrist context. However, the wrist context does not generalize to the arm context: after training in the wrist task, they are about the same as naïve in the arm task. That is, the pattern of generalization is asymmetric between these two contexts. What can account for these curious facts?

It turns out that Krakauer and colleagues could account for these observations when they formulated the problem in an estimation framework. Interestingly, they found out that the problem looked a lot like the backward blocking example that we considered earlier.

To start, consider that subjects were trained in two situations: arm-controlled cursor and wrist-controlled cursor. In each case, the cursor indicated the position of the end-effector (hand or finger). We represent this position in polar coordinates and focus only on its angular component. That is, if in trial n the end-effector angle is $e^{(n)}$ and the imposed perturbation (rotation) is $r^{(n)}$, then the computer displays the cursor at $y^{(n)}$:

$$y^{(n)} = e^{(n)} + r^{(n)}. \tag{6.15}$$

Now there are different contexts in which movements are performed. Let $\mathbf{c}^{(n)}$ be a binary vector that specifies this context and $\mathbf{w}^{(n)}$ be the weight vector that specifies the contribution of the context to the perturbation:

$$\begin{aligned} r^{(n)} &= \mathbf{c}^{(n)T}\mathbf{w}^{(n)} \\ y^{(n)} &= e^{(n)} + \mathbf{c}^{(n)T}\mathbf{w}^{(n)} + \varepsilon_y^{(n)} \qquad \varepsilon_y^{(n)} \sim N\left(0,\sigma^2\right) \end{aligned} \tag{6.16}$$

In equation (6.16), we have the "measurement" equation, specifying that what the subject observes (cursor position) is affected by some weighted contribution of the two contexts. The problem of learning is to estimate these contributions: the state that we are trying to estimate is specified by vector $\mathbf{x} = [w_1, w_2, e]^T$. The state update equation takes the form:

$$\mathbf{x}^{(n+1)} = A\mathbf{x}^{(n)} + \mathbf{b}u^{(n)} + \boldsymbol{\varepsilon}_x \qquad \boldsymbol{\varepsilon}_x \sim N(0,Q). \tag{6.17}$$

So the generative model takes the form specified by equations (6.16) and (6.17). Given a target at $y_t^{(n)}$, we rely on our prior estimate of the perturbation

$$\hat{r}^{(n)} = \mathbf{c}^{(n)T} \hat{\mathbf{w}}^{(n|n-1)} \tag{6.18}$$

to produce a motor command that compensates for the estimated perturbation:

$$u^{(n)} = y_t^{(n)} - \hat{r}^{(n)}. \tag{6.19}$$

To explain the peculiar experimental results, Krakauer and colleagues made a critical assumption. They assumed that for their tasks, the context was not defined in terms of which muscles or joints moved, but in terms of motion of body parts in extrinsic space. In the wrist context, only the hand moved and not the upper arm. However, in the arm context, both the hand and the upper arm moved. As a result, the wrist context was defined by the context vector $\mathbf{c}^{(n)} = [0 \quad 1]^T$, whereas the arm context was defined by $\mathbf{c}^{(n)} = [1 \quad 1]^T$. The basic idea then is that the two contexts are not independent, but that one (the arm context) is a superset of the other (the wrist context).

If the two contexts are viewed in this way, then the arm context is identical to the situation in our earlier classical conditioning experiment in which both light and tone were present (figure 6.4). Recall that the presence of both cues made the uncertainty matrix P acquire negative off-diagonal elements, implying that when one weight goes up, the other weight should go down. Krakauer and colleagues assumed that in the course of daily activity, most of our movements are arm movements, and few are just wrist movements in which the upper arm does not move. In effect, they assumed that the initial uncertainty matrix $P^{(1|0)}$ had negative covariance in its off-diagonal elements. With this prior uncertainty, learning will indeed be slower in the arm context than the wrist context. The reason is the negative covariance.

This idea is flushed out in a simulation. Let us consider wrist training. The experimenter sets $r^{(n)} = 30$ and asks the learner to move the cursor with the wrist. The learner assumes that the context is $\mathbf{c}^{(n)} = [0 \quad 1]^T$. Figure 6.14A shows the two components of the vector $\hat{\mathbf{w}}^{(n|n-1)}$, that is, the weight associated with the upper arm $\hat{w}_1$ and the weight associated with the wrist $\hat{w}_2$. With each trial, $\hat{w}_2$ increases toward 30 degrees. However, despite the fact that the context is wrist only, $\hat{w}_1$ becomes negative, resulting in an estimate for the whole arm $(\hat{w}_1 + \hat{w}_2)$ that is only slightly positive. Therefore, the model reproduces the result that wrist training will not have a significant impact on subsequent training with the arm (i.e., wrist training does not generalize to arm training). The prior uncertainty matrix with negative off-diagonal elements is directly responsible for this generalization pattern.

Next consider arm training. Figure 6.14B shows the simulation results when we set $r^{(n)} = 30$ and train in the arm context by setting $\mathbf{c}^{(n)} = [1 \quad 1]^T$. The observed errors produce changes in $\hat{w}_1$ and $\hat{w}_2$, but because the covariance in the uncertainty matrix

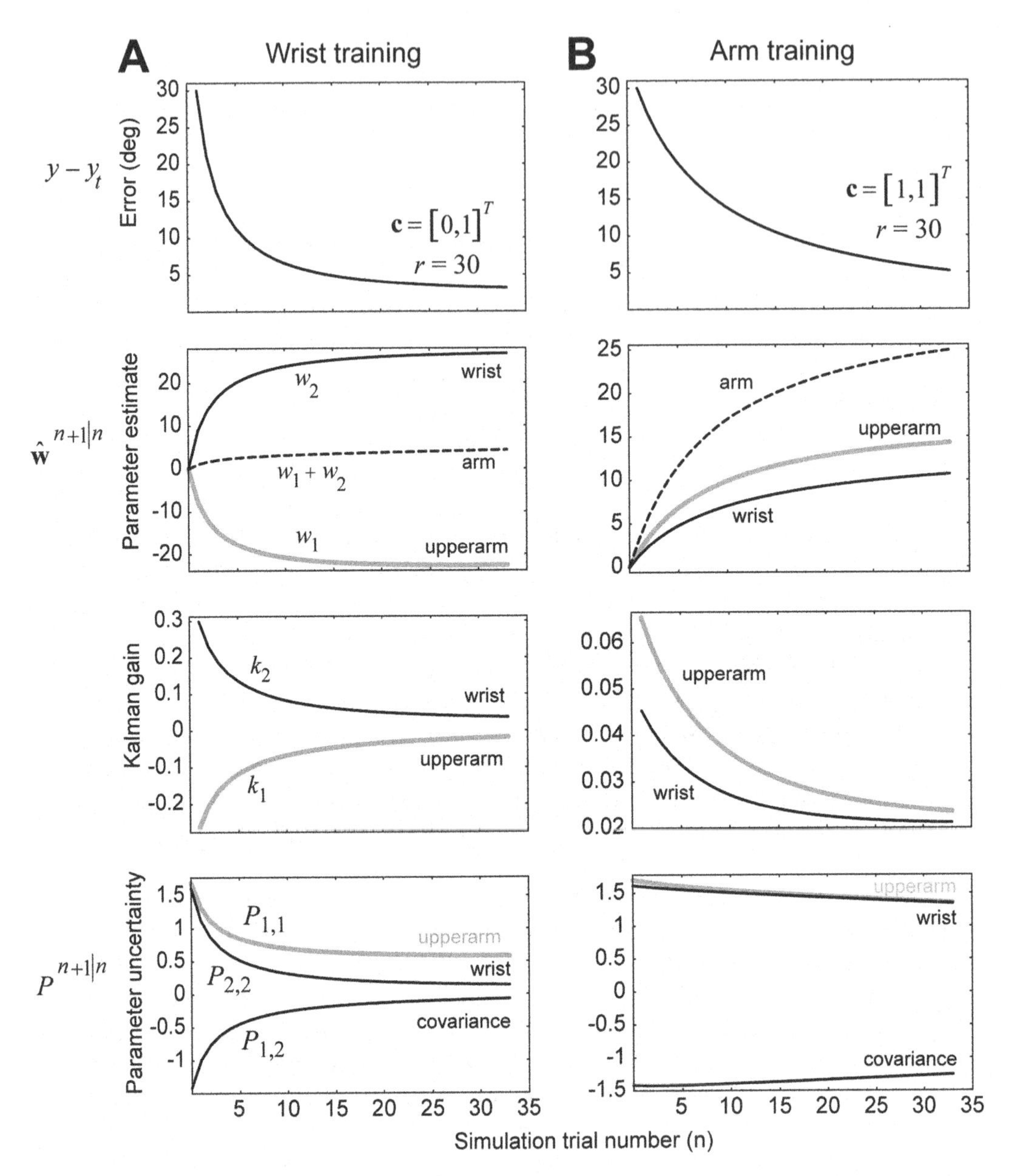

Figure 6.14
Simulation results for the wrist and arm context experiments. All simulations begin with the same initial conditions. (A) Simulation of a 30-degree rotation in the wrist context. With each trial, the estimate for the wrist increases toward 30 deg. Despite the fact that only the wrist context is present, the estimate for the upper arm becomes negative. This is because the uncertainty matrix has negative off-diagonal elements, which arises from the prior assumption that motion of the upper arm usually results in motion of the wrist. (B) Simulation of a 30-degree rotation in the arm context. Errors produce changes in the estimates of both the upper arm and the wrist, resulting in transfer to the wrist. Despite identical initial conditions, learning with the arm is slower than learning with the wrist. (From Krakauer et al., 2006.)

is negative, the sensitivity to error (Kalman gains) are much smaller in the arm context than when the task is performed in the wrist context. Consequently, the arm context is learned more slowly than the wrist context. Despite the fact that the uncertainty matrix $P^{(1|0)}$ and the initial estimate $\hat{\mathbf{w}}^{(1|0)}$ are identical in the two simulations of figure 6.14A and B, the errors decline about twice as slowly in the context of the arm as compared to the wrist. Furthermore, the same uncertainty matrix dictates a generalization from arm to wrist, as the Kalman gain is positive for both the upper arm and wrist. As a consequence, arm training results in the estimate for the wrist $\hat{w}_2$ to increases to about 10 degrees. If we now test the system in the wrist context, it has already learned much of the perturbation and will show better performance than naïve. Therefore, the simulations also reproduce the asymmetric patterns of generalization (from arm to wrist, but not from wrist to arm).

6.8 Source of the Error

Suppose that you are a tennis player and after reading the latest tennis magazine you are tempted to change your racquet. You go to the pro shop, pick up a racquet, step into the court, and hit a few balls. The results look pretty bad. The balls are flying all over the place. What is the problem? Is the problem something to do with the internal model of your own arm (for example, the muscles have gotten a bit weaker since you last played), or is the problem with the internal model of the racquet (is the center of mass different than what you are used to)? In principle, when one interacts with a novel tool and experiences an error in performance, the learning problem is one of credit assignment. Should one adjust the parameters of the internal model of their own arm, or the parameters of the novel tool?

The framework that we have built extends naturally to encompass this problem of credit assignment. The generative model of figure 6.15 shows that the parameters that we wish to estimate can be classified into parameters associated with our own body (i.e., internal parameters $\mathbf{x}_{in}$), and parameters associated with the tool (i.e., external parameters $\mathbf{x}_{ext}$). To learn optimally, the uncertainty that we have regarding each set of parameters should describe how we should divide up the error and update our estimates of the parameters.

Max Berniker and Konrad Kording (2008) used the model in figure 6.15 to explain experimental results regarding how people generalize their adaptation. Their idea was that if the brain adjusted the parameters associated with the body, then adaptation should generalize when the body is used in another configuration. For example, if I am using my right arm to hit serves with a new tennis racquet and I use the errors to update parameters associated with the internal model of my arm ($\mathbf{x}_{in}$), then after adaptation if I use my right arm and the same racquet to hit a forehand, I should be able to generalize. However, if the brain adjusted the

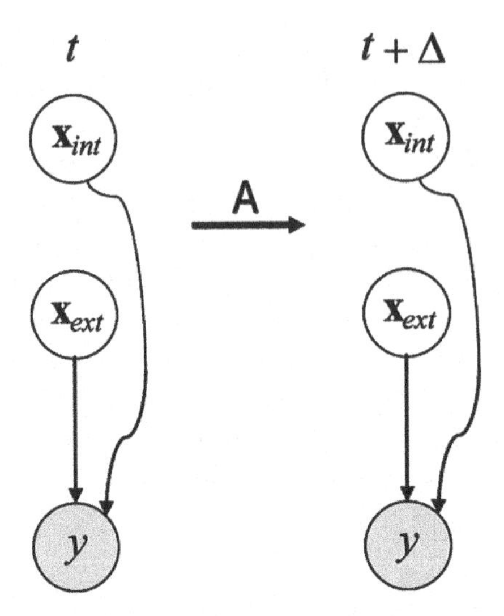

Figure 6.15
A generative model of the source of error in a tool use task. Parameters associated with the tool $\mathbf{x}_{ext}$ and the body $\mathbf{x}_{in}$ contribute to observations $\mathbf{y}$.

parameters associated with the tool, then the generalization patterns should be different: adaptation should generalize to other limbs when those limbs are holding the tool. For example, if during hitting serves with my right arm I update parameters associated with the internal model of the racquet ($\mathbf{x}_{ext}$), then I should be able to control the racquet even when I am using my left arm.

Consider the task pictured in figure 6.16. In this task, a volunteer holds the handle of a robot and reaches to a target position. The robot produces forces on the hand, perturbing its motion. If we interpret these perturbations as having something to do with the object that we are holding in hand, we can write the forces in terms of the velocity of that object:

$$\mathbf{f} = B\dot{\mathbf{x}}_o. \tag{6.20}$$

Now if we make a reaching movement and observe a difference between what we observed and what we predicted (in terms of motion of the object), we would use the prediction error to update our estimate of the parameter B. Alternatively, suppose that we interpret the perturbations as having something to do with the motion of our own body, i.e., torques $\boldsymbol{\tau}$ that depend on velocity of our arm:

$$\boldsymbol{\tau} = W\dot{\mathbf{q}}. \tag{6.21}$$

In equation (6.21), the vector $\mathbf{q}$ refers to angular configuration of the shoulder and elbow joints and $\dot{\mathbf{q}}$ refers to joint angular velocities. If we interpret the perturbations

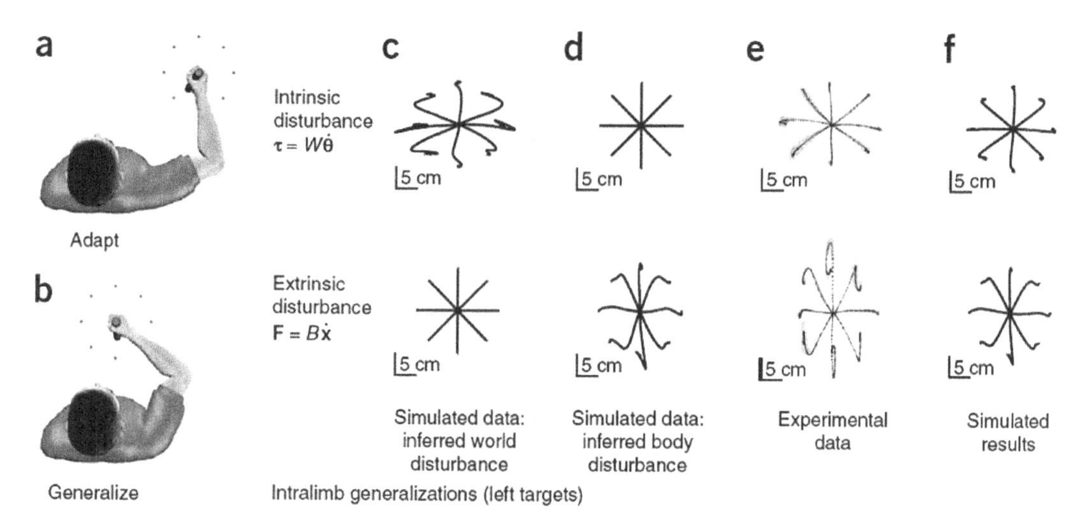

Figure 6.16
Within-limb generalization patterns after learning to reach with a novel tool. Experimental data from Shadmehr and Mussa-Ivaldi (1994). Simulation data from Berniker and Kording (2008). (a) Volunteers reached while holding the handle of a robotic arm. The training took place in the right workspace (as shown). (b) After completion of training, they were tested in the left-workspace as shown. (c) Simulation and experimental data are shown for motion in the left workspace, that is, the test of generalization. Two conditions were considered: in the intrinsic disturbance condition, the robot produced a force field that was defined in terms of motion of the subject's joint angles, i.e., $\boldsymbol{\tau} = W\dot{\boldsymbol{\theta}}$. In another condition, the robot produced a field that was defined in terms of the motion of the subject's hand, i.e., $\mathbf{f} = B\dot{\mathbf{x}}$. The simulated data here shows that if the learning assigned all the errors to the model of the tool, performance in the intrinsic disturbance would be poor, and performance in the extrinsic disturbance would be excellent. (d) The simulated data here shows that if the learning assigned all the errors to the model of the subject's own arm, performance in the intrinsic disturbance would be excellent, and performance in the extrinsic disturbance would be poor. (e) Experimental results. (f) Simulation results in which most but not all of the error assignment was to the model of the subject's own arm.

as having something to do with motion of our arm, then when we make a movement and observe a prediction error, we would be updating our estimate of the parameter W. How can we tell whether the brain is interpreting the prediction errors in terms of the external object, or the internal body?

Suppose that we learn to move the object with our hand in the right workspace (as shown in figure 6.16a), and then we test for generalization by making movements in the left workspace (as shown in figure 6.16b). In the left workspace we observe one of two different kinds of force fields. One of the fields is the same as in the right workspace when we represent it in terms of the velocity of the object. This is called an "extrinsic field." The other is the same as in the right workspace when we represent it in terms of velocity of our arm. This is called an "intrinsic field." If we learned by updating B in equation (6.20), then we should do well when our arm is in the left workspace and we encounter an extrinsic field, but do poorly if we encounter an intrinsic field. On the other hand, if we learned by updating W in equation (6.21), then we should do well in the left workspace only if we encounter an intrinsic field.

Shadmehr and Mussa-Ivaldi (1994) performed this experiment and found that in the left workspace, performance was much better for the intrinsic field than the extrinsic field. Figure 6.16e (top) shows reaching movements in the left workspace in the intrinsic field, and figure 6.16e (bottom) shows reaching movements in the left workspace in the extrinsic field. This suggests that subjects predominantly interpreted the prediction errors in terms of an internal model in equation (6.21), that is, something about their own arm. However, the subjects still learned at least a little about the dynamics of the object (i.e., equation 6.20). This conclusion was inferred because of the results of a later experiment by Sarah Criscimagna-Hemminger, Opher Donchin, Michael Gazzaniga, and Reza Shadmehr (2003). In that experiment, volunteers learned to control the robot with their right arm and then were tested with their left arm (figure 6.17). Performance on the extrinsic field during test of the left arm was better than naïve (naïve being defined as performance of people who had not previously learned any tasks with the robot), and

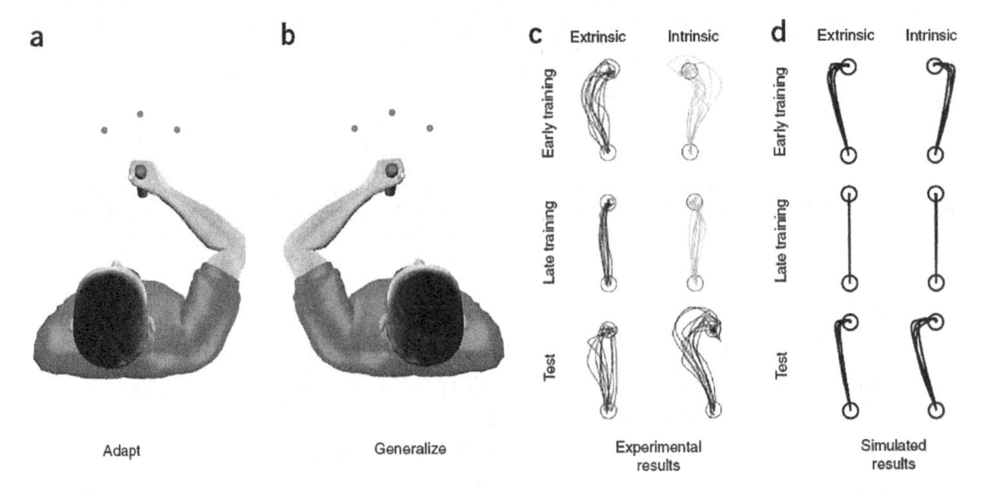

Figure 6.17
Between-limb generalization after learning to reach with a novel tool. Experimental data from Criscimagna-Hemminger et al. (2003). Simulation data from Berniker and Kording (2008). (a) Volunteers reached while holding the handle of a robotic arm. The training took place with their right arm to targets shown. (b) Testing took place with the left arm in either an intrinsic field in which the forces were defined in terms of motion of the joints, or extrinsic field in which the forces were defined in terms of motion of the hand. (c) Subjects trained with their right hand in either an extrinsic field or an intrinsic field. The early and late training data shows trajectories of the right hand. After completion of training, they were tested with the left hand on the same field. Performance was much better for the extrinsic field. (d) Simulation data from a system that assigned most of the error to the body (parameters of the right arm), but some of the error to the tool.

also better than performance of people who were given the intrinsic field on their left arm. Together, the results of within-arm generalization (figure 6.16) and between arms generalization (figure 6.17) suggested that during adaptation, the subjects mostly updated a model similar to equation (6.21), but still learned a little about the model in equation (6.20). That is, in learning to control the robot, the subjects appeared to be learning mostly in the intrinsic coordinate system of their own body, and a significant but smaller amount in the extrinsic coordinate system of the object.

To explain these results, Berniker and Kording (2008) modeled dynamics of the two-degree-of-freedom, humanlike arm using the following:

$$I(\mathbf{q})\ddot{\mathbf{q}} + C(\mathbf{q},\dot{\mathbf{q}})\dot{\mathbf{q}} + W_m\dot{\mathbf{q}} = \boldsymbol{\tau}_m + \boldsymbol{\tau}_p + \boldsymbol{\tau}_n. \tag{6.22}$$

Here, the 2×1 vector $\mathbf{q}$ represents joint angles, $I(\mathrm{q})$ is the inertia matrix, $C(\mathbf{q},\dot{\mathbf{q}})$ is the coriolis/centripetal matrix, W_m is joint viscosity, $\boldsymbol{\tau}_m$ is the commanded torque, $\boldsymbol{\tau}_p$ is a perturbation torque, and $\boldsymbol{\tau}_n$ is noise. Like earlier works (Shadmehr and Mussa-Ivaldi, 1994), Berniker and Kording assumed that the commanded torque was the sum of three components: a feedback torque that stabilized the trajectory, an internal model that compensated for inertial dynamics, and an internal model that estimated the perturbation from external sources:

$$\begin{aligned} \boldsymbol{\tau}_m &= \hat{I}\ddot{\mathbf{q}}^*(t) + \hat{C}\dot{\mathbf{q}}^*(t) + \hat{W}_m\dot{\mathbf{q}}^*(t) - \hat{\boldsymbol{\tau}}_p \\ &\quad + K_1\left(\mathbf{q}^*(t) - \mathbf{q}\right) + K_2\left(\dot{\mathbf{q}}^*(t) - \dot{\mathbf{q}}\right) \end{aligned}. \tag{6.23}$$

K_1 and K_2 represent stiffness and viscosities that stabilize the motion around desired trajectory $\mathbf{q}^*(t)$. The term $\hat{\boldsymbol{\tau}}_p$ is an estimate of the external perturbation:

$$\hat{\boldsymbol{\tau}}_p = J^T \hat{B}\dot{\mathbf{x}}. \tag{6.24}$$

The matrix J is the Jacobian that relates endpoint displacements and joint displacements. The parameters that the system tried to estimate were W_m (joint viscosity, reflecting a belief that the errors are due to internal perturbations), and B (reflecting a belief that the errors are due to external perturbations). The eight parameters were collected in vector $\mathbf{p}$. The output of the system $\mathbf{y}(t)$ was hand position and velocity $[\mathbf{x} \quad \dot{\mathbf{x}}]$. The estimate of this output was $\hat{\mathbf{y}}(\hat{\mathbf{p}},t)$. The generative model took the form:

$$\begin{aligned} p_i(t+\Delta) &= ap_i(t) + \varepsilon_i(t) \qquad & \varepsilon_i &\sim N(0,q_i) \\ \mathbf{y}(t) &= H(t)\mathbf{p} + \mathbf{w}(t) \qquad & \mathbf{w} &\sim N(0,R) \end{aligned}. \tag{6.25}$$

Here, p_i represents the elements of the vector $\mathbf{p}$ (the parameters that described the velocity dependent perturbations), and the matrix $H(t)$ is a linear approximation of the nonlinear dynamics in $\mathbf{y}(\mathbf{p},t)$. $H(t)$ was approximated as:

$$H(t)_{i,j} = \frac{\hat{\mathbf{y}}_i(\hat{\mathbf{p}},t) - \hat{\mathbf{y}}_i(\hat{\mathbf{p}} + \delta_j, t)}{\delta}. \tag{6.26}$$

In equation (6.26), the term $H_{i,j}$ refers to the element (i,j) of matrix H and δ_j is a small perturbation on the jth element of the $\mathbf{p}$ vector. The credit assignment between parameters representing internal perturbations W_m and external perturbations B depends solely on our assumption regarding the relative variance of the noise term ε_i in equation (6.25). To explain the predominately intrinsic patterns of generalization, Berniker and Kording inferred that the noise in the internal parameters W_m must be larger than external parameters B. That is, the model that explained the data was one in which the brain was more uncertain about the parameters that described the viscosity of the arm as compared to the dynamics of the object in hand. In this way, the model explained both the within-arm patterns of generalization (due to changes in the internal parameters W_m), and the across-arm patterns of generalization (due to changes in the external parameters B).

This inference may seem a little odd. How could it be that when the hand is holding a weird, unfamiliar tool (the robot), movement errors are (at least partially) associated with the internal model of the arm, and not the tool? Doesn't the brain "understand" that it is the tool that is disturbing the hand's motion? Is there another way that we can test the theory? Indeed there is. If the theoretical framework is right and indeed the brain is learning to update an internal model of its own arm, then something strange should happen: When people let go of the robot and make reaches in free air, they should have aftereffects of the previous learning. That is, despite the fact that the brain "knows" that the hand is no longer holding on to the novel tool, because it associated the errors partially to the internal model of its own arm, it should have aftereffects of the previous learning. To test for this, JoAnn Kluzik, Jörn Diedrichsen, Reza Shadmehr, and Amy Bastian (2008) trained people to reach with the robot and then tested them in two conditions (figure 6.18A). In one condition (called "free"), the experimenters removed the handle from the robot (the volunteers could see this) and then asked the volunteers to simply reach in free air. In the second condition (called "robot-null"), the handle remained on the robot, but the forces were turned off. They found that indeed, reaching movements in free air had significant aftereffects (figure 6.18B), and the size of these aftereffects in free air were about half the size of the robot-null condition. The results are fairly consistent with the theoretical framework, suggesting that about half of the errors were assigned to the internal model of the arm, and the rest assigned to the internal model of the tool.

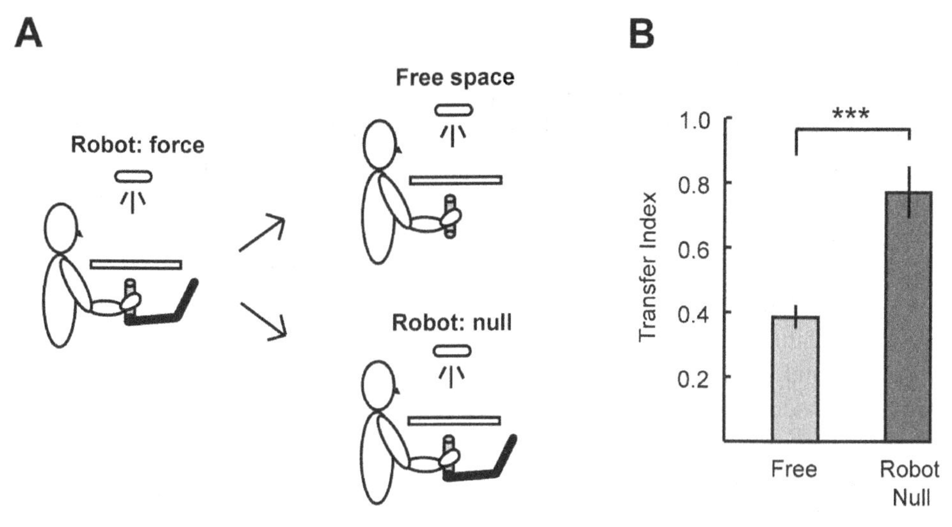

Figure 6.18
Learning to reach while holding a novel tool alters the internal model of both the subject's arm and the tool. (A) Subjects trained to make reaching movements while holding the handle of a robot. After training, one group of subjects observed as the robot was disconnected from the handle (free), while another group was simply told that the robot motor were turned off (robot null). The reason for holding on to the handle in the free condition was to ensure that the arm/wrist was kept precisely in the same configuration as in training. (B) Transfer index quantified the size of the errors in reaching in each of the two conditions with respect to the size of the error in catch trials, i.e., trials in which the force field was unexpectedly turned off. (From Kluzik et al., 2008.)

Summary

In this chapter, we considered the problem of learning to make accurate predictions. We framed this problem in terms of state estimation. We imagined that in making a prediction about a variable that we can observe, there are some sets of hidden states that affect that variable, making it so that sometimes our predictions are wrong. When we observe a prediction error, we try to assign credit to these hidden states, updating our estimate of these states so that we can improve our predictions.

In the Bayesian way of learning, this credit assignment problem is solved using methods that maximize the probability of a posterior distribution about our hidden variables. In the non-Bayesian methods of learning, as in LMS, this credit assignment is solved based on the contribution of each hidden state to the prediction (the greater the contribution, the greater the credit).

In some forms of biological learning, as in backward blocking, animals appear to learn in a way that resembles the Bayesian method and not LMS. In this example, the learner appears to keep a history of the prior observations, resulting in a credit assignment that resembles the uncertainties accumulated in the Kalman gain. These uncertainties are the covariance between the hidden states.

Sensorimotor adaptation is a widely studied example of biological learning. The key requirement for adaptation is sensory prediction error: adaptation occurs when there is a discrepancy between the predicted and observed sensory consequences of motor commands. There are two widely studied paradigms: visuomotor adaptation, in which the visual consequences of a motor command is manipulated, and force field adaptation, in which both the visual and proprioceptive consequences are manipulated.

An internal model is simply a generative model that relates motor commands with sensory consequences. The hidden states are the sensory and motor perturbations. A Bayesian process of estimation of these hidden states accounts for numerous observations in how people adapt their movements to perturbations, including sensory illusions, and asymmetric patterns of generalization.

7 Learning Faster

The title of this chapter sounds a little like a product that you might find advertised on TV in the wee hours of the night. "We will not only help you learn (your favorite task here) faster, but if you order now, we will also send you this amazing new . . ." It sounds like the promises of a snake oil salesman. But as we will see here, if we view learning in biological systems as a Bayesian state estimation process, then the mathematical framework that we developed provides us with some useful ideas on how to encourage the process of learning.

7.1 Increased Sensitivity to Prediction Errors

One of the properties of the state estimation framework is that the sensitivity to error (the Kalman gain) depends on the relative size of the state uncertainty with respect to the measurement uncertainty. Said in another way, the sensitivity to error depends on the ratio of the confidence in the state that we are trying to estimate with respect to the confidence of our measurements. For example, suppose that we are attempting to estimate state x, which is governed by the following generative model:

$$\begin{aligned} x^{(n+1)} &= ax^{(n)} + \varepsilon_x^{(n)} \quad \varepsilon_x \sim N(0,\sigma_x) \\ y^{(n)} &= x^{(n)} + \varepsilon_y^{(n)} \quad \varepsilon_y \sim N(0,\sigma_y) \end{aligned} . \tag{7.1}$$

The sensitivity to prediction error, written as our Kalman gain $k^{(n)}$, will depend on the ratio of the prior state uncertainty $p^{(n|n-1)}$ with respect to measurement uncertainty $p^{(n|n-1)} + \sigma_y$:

$$k^{(n)} = \frac{p^{(n|n-1)}}{p^{(n|n-1)} + \sigma_y} . \tag{7.2}$$

When the learner observes a prediction error but is uncertain about the sensory measurement (noise σ_y is large), he will have a relatively small learning rate. When

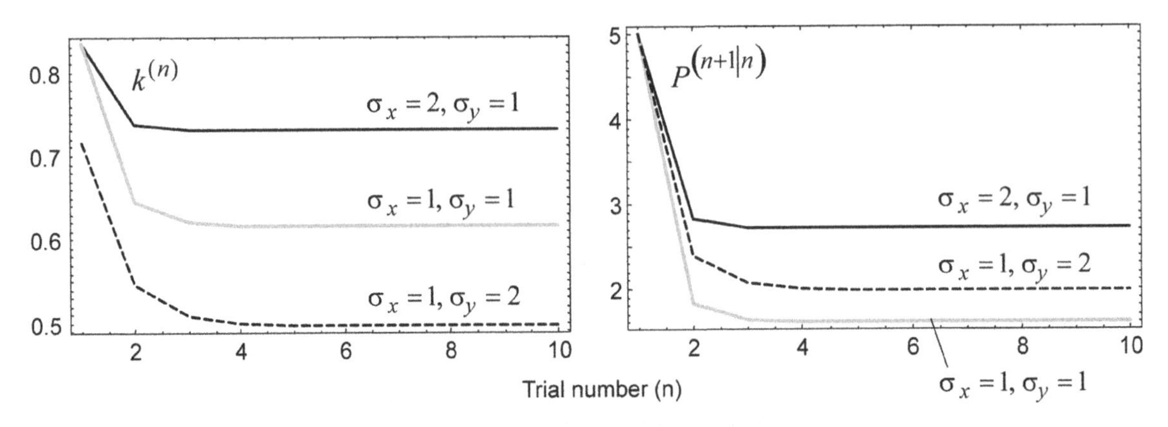

Figure 7.1
Learning rates depend on uncertainties of the learner. The left panel shows the Kalman gain, and the right panel shows the uncertainty in the state estimate for the generative model of equation 7.1. As the noise in the state update equation σ_x increase, the uncertainty in the state estimate p increases, and the sensitivity to error k increases. As the noise in the measurement σ_y increases, the uncertainty in the state estimate p increases, but the sensitivity to error k decreases.

the learner observes the same prediction error but is uncertain about his own predictions (the state uncertainty p is large), he will have a relatively large learning rate. Therefore, the state estimation framework predicts that through simple techniques, you should be able to up-regulate or down-regulate rates of adaptation.

Let us do some simulations to explore how sensory and state noise of the underlying system affects uncertainty and sensitivity to error (i.e., learning rate). Figure 7.1 shows that if the variance of the state noise σ_x were to increase (reflecting our uncertainty about how the underlying state changes with time), then state uncertainty p increases, resulting in an increase in the sensitivity to error k. Figure 7.1 also shows that if the variance of the measurement noise σ_y were to increase (reflecting our uncertainty about our observations), our state uncertainty p increases, but this increase is generally dwarfed by the increase in the measurement noise. As a result, the sensitivity to error k decreases. Therefore, if one could alter the learner's uncertainty about the state that is being estimated or her confidence about the observation, one might be able to change the learner's sensitivity to error. The learning rate should be adaptable to the conditions of the experiment.

Johannes Burge, Marc Ernst, and Martin Banks (2008) tested this idea in a simple reaching task. They had people hold a pen in hand and reach to a visual target. Visual feedback was withheld until the end of the reach (figure 7.2). The form of the visual feedback was a blurry collection of pixels in which intensities of the pixels were described by a Gaussian function. To manipulate the subject's measurement uncertainty, they changed the variance of this Gaussian. They considered two

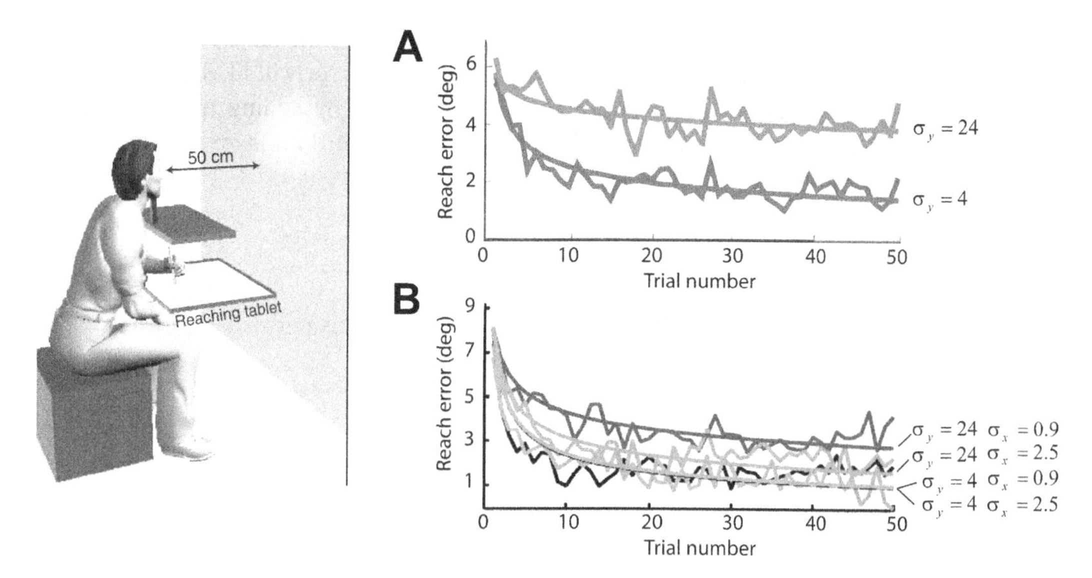

Figure 7.2
Sensitivity to prediction error can be increased when the subject is made more uncertain about the state of the perturbation. Volunteers reached to a target while a visuomotor perturbation altered their sensory feedback. (A) Performance when the measurement uncertainty was small ($\sigma_y = 4$) or large ($\sigma_y = 24$). (B) For a given measurement uncertainty, increased state uncertainty (increasing σ_x) produced a faster learning rate. (From Burge, Ernst, and Banks, 2008.)

conditions, a low measurement uncertainty condition in which the Gaussian's variance was $\sigma_y = 4$ deg, and a high uncertainty condition in which $\sigma_y = 24$ deg. Normally, the center of the Gaussian corresponded to the hand position at end of the reach. However, to produce adaptation, a bias was added to this relationship. In summary, the feedback $y^{(n)}$ on trial n was a blurry Gaussian centered at $h^{(n)} + r$, where h is hand position at end of trial n, and r is the visual perturbation, with variance σ_y. The prediction was that with the more uncertain feedback (larger σ_y), rates of learning would be reduced. Indeed, Burge, Ernst, and Banks found that the adaptation rate was slower when σ_y was larger (figure 7.2A).

You might be a little skeptical about this result, because with a really blurry feedback, there may be little motivation to adapt. For example, in figure 7.2A, the plot for σ_y may never reach zero, so it is not clear that we are really looking at a reduced sensitivity to error. We will return to this problem shortly, but for now consider a more interesting question: How do we increase the learning rate? The answer would have important practical implications because rehabilitation of people who have suffered brain injury relies on learning to recover lost function, something akin to adaptation. Finding ways to improve learning rates might result in better ways to train patients during rehabilitation.

The estimation framework predicts that if one could make the learner uncertain about the state of the perturbation, then the sensitivity to error will likely increase. Burge, Ernst, and Banks (2008) approached this problem by adding trial-to-trial variability to the perturbation, that is, by making the perturbation $r^{(n)}$ act like a random walk about a mean value r_0:

$$\begin{aligned} x^{(n)} &= x^{(n-1)} + \varepsilon_x \qquad & \varepsilon_x &\sim N(0,\sigma_x) \\ r^{(n)} &= r_0 + x^{(n)} \end{aligned} . \tag{7.3}$$

They then assumed that the subjects would be estimating this perturbation using a generative model of the usual form:

$$\begin{aligned} r^{(n)} &= r^{(n-1)} + \varepsilon_r \qquad & \varepsilon_r &\sim N(0,\sigma_r) \\ y^{(n)} &= h^{(n)} + r^{(n)} + \varepsilon_y \qquad & \varepsilon_y &\sim N(0,\sigma_y) \end{aligned} . \tag{7.4}$$

Their experimental results are plotted in figure 7.2B. When the perturbation was more variable from trial to trial, the learning was somewhat faster. For a given feedback blurriness (for example, $\sigma_y = 24$ deg), a larger variance in the random walk of the perturbation (σ_r) produced a more rapid adaptation.

Kunlin Wei and Konrad Kording (2008) also tested this idea of altering learning rates, but they approached it in a way that did not rely on fitting exponential functions to the data. Rather, their approach relied on trial-to-trial measures of sensitivity to error. They asked volunteers to move their hands through a visual target (figure 7.3) and provided each volunteer with visual feedback as the hand crossed the target area. This visual feedback represented the hand position, plus a random perturbation. To set up our generative model for this problem, suppose that on trial n, at the end of the reach the hand is at $h^{(n)}$ and the imposed perturbation is $x^{(n)}$. The computer displays its feedback at $y^{(n)}$. However, for some blocks the visual feedback is a single dot at $y^{(n)}$, while for other blocks it is a distribution of dots centered at $y^{(n)}$. For example, the feedback can be five dots drawn from a small variance distribution, or five dots drawn for a large variance distribution. The larger this variance, the larger the variance of the measurement noise σ_y in equation (7.5). Our generative model takes the form:

$$\begin{aligned} x^{(n+1)} &= x^{(n)} + \varepsilon_x \qquad & \varepsilon_x &\sim N(0,\sigma_x) \\ y^{(n)} &= h^{(n)} + x^{(n)} + \varepsilon_y \qquad & \varepsilon_y &\sim N(0,\sigma_y) \end{aligned} . \tag{7.5}$$

Notice that here the perturbation is zero mean and there is no consistent bias in the movements. Therefore, it is not possible to plot a learning curve because in fact, there is nothing to learn. However, it is still possible to measure sensitivity to error. On trial n, the computer instructs the subject to move the hand to target location $y_t^{(n)}$. To do so, the subject predicts the perturbation on this trial $\hat{x}^{(n|n-1)}$ and moves the hand to cancel that perturbation:

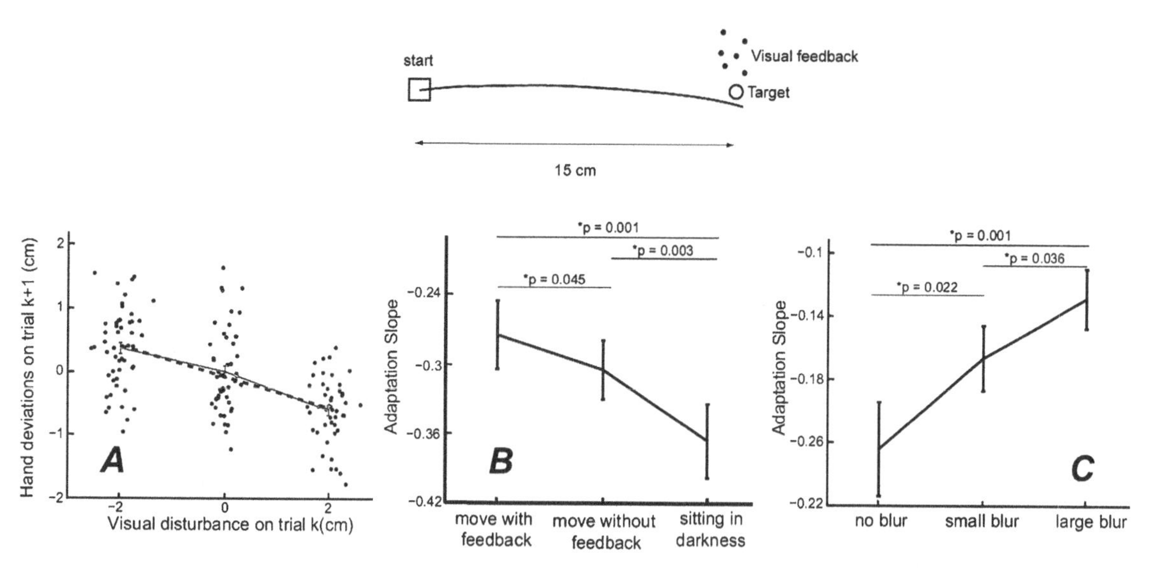

Figure 7.3
Altering the sensitivity to error through manipulation of feedback. Subjects reached to a target and feedback was provided regarding their endpoint. (A) Hand deviations on trial $k + 1$ as a function of perturbation on trial k. A positive perturbation on trial k produces a negative compensation on trial $k + 1$. The slope of the line is sensitivity to error. The more negative this slope, the greater the sensitivity to error in trial k. (B) Sensitivity to error increased after a period in which no feedback was provided after a movement, and after a period of sensory deprivation in which no movements or feedback were provided (sitting in darkness). (C) When feedback was blurred, adaptation slope became less negative, indicating a reduced sensitivity to error. (From Wei and Kording, 2008.)

$$h^{(n)} = y_t^{(n)} - \hat{x}^{(n|n-1)}. \tag{7.6}$$

The computer provides feedback to the subject by displaying a cursor at position $y^{(n)}$. The subject observes an error between the feedback $y^{(n)}$ and the predicted feedback $h^{(n)} + \hat{x}^{(n|n-1)}$ and updates his estimate of the perturbation:

$$\hat{x}^{(n|n)} = \hat{x}^{(n|n-1)} + k^{(n)}\left(y^{(n)} - h^{(n)} - \hat{x}^{(n|n-1)}\right). \tag{7.7}$$

Wei and Kording tested the idea that increasing observation noise σ_y (uncertainty about the measurement) would reduce the learning rate $k^{(n)}$. For example, in figure 7.3A we see that when the vertical perturbation on trial n was zero, vertical position of the hand on the trial $n + 1$ had a zero mean. This means that on average, subjects did not alter their movements on trial $n + 1$ if on trial n their movement was not perturbed. Once a –2 cm visual perturbation was imposed on trial n, on trial $n + 1$ the distribution of hand positions had a slightly positive mean. That is, the brain responded to the errors in trial n by changing behavior on the next trial (to compensate for the –2 cm perturbation, on trial $n + 1$ the subject moved their hand to +0.4). However, we are particularly interested in the sensitivity to error—that is, the learning rate $k^{(n)}$. The sensitivity to error is the slope of the line in figure 7.3A. To see this, consider that hand position on trial $n + 1$ is simply

$$\begin{aligned}
h^{(n+1)} &= y_t^{(n+1)} - \hat{x}^{(n+1|n)} \\
&= y_t^{(n+1)} - \hat{x}^{(n|n-1)} - k^{(n)}\left(y^{(n)} - h^{(n)} - \hat{x}^{(n|n-1)}\right) \\
&= y_t^{(n+1)} + h^{(n)} - y_t^{(n)} - k^{(n)}\left(y^{(n)} - y_t^{(n)}\right) \\
h^{(n+1)} - h^{(n)} &= -k^{(n)}\left(y^{(n)} - y_t^{(n)}\right)
\end{aligned} \quad (7.8)$$

The steeper this slope becomes (more negative), the larger the sensitivity to error $k^{(n)}$. Wei and Kording noted that as the subject's measurement uncertainty increased (single dot, five dots with a small distribution, five dots with a large distribution), sensitivity to error became smaller (the slope became more shallow; figure 7.3C). Therefore, when the feedback that subjects received about their actions became more uncertain and noisy, sensitivity to prediction errors declined and people learned less from their errors.

The estimation framework's key prediction is that the learners will exhibit increased sensitivity to error when they are made uncertain about their own predictions. Wei and Kording attempted to increase this prior uncertainty in two ways. First, they had their subjects make a minute of reaching movements without any visual feedback. They argued that in effect, this increased the magnitude of the noise variance σ_x in equation (7.5). In the subsequent test period, they provided visual feedback regarding end of the reach but on random trials they added a disturbance. They measured the change in performance on trial $n + 1$ as a function of the error on trial n. Indeed, they found a slightly larger sensitivity to error than in a control condition in which the prior movements had feedback (figure 7.3B). They next tried a different way to increase the subject's uncertainty about their state: they had the subjects sit quietly for a minute in darkness. This time passage can be argued to increase one's uncertainty about the state of one's hand as it removes both visual feedback as well as velocity dependent proprioceptive feedback that occurs when one makes movements. Indeed, the sensitivity to error in the subsequent test trials was even larger than in the control condition (figure 7.3B). (However, the experiment may also need another control: having subjects sitting still with full vision of the hand. The effect observed by Wei and Kording could reasonably be attributed to the period of inactivity reducing the reliability of motor commands for some time immediately following the pause.)

In summary, if we view learning in a Bayesian framework, we can make certain predictions about how to speed up or slow down the process of learning. Speeding up learning can occur if the learner becomes more sensitive to prediction errors, which can be achieved by making the learner more uncertain about her own predictions. Slowing down learning can occur if the learner's sensory measurements become less reliable.

7.2 Modulation of Forgetting Rates

In theory, another way to manipulate sensitivity to error is via our belief regarding how the state that we are trying to estimate changes from trial to trial. Suppose our generative model takes the form:

$$\begin{aligned} x^{(n+1)} &= ax^{(n)} + \varepsilon_x \qquad \varepsilon_x \sim N(0,\sigma_x) \\ y^{(n)} &= x^{(n)} + \varepsilon_y \qquad \varepsilon_y \sim N(0,\sigma_y) \end{aligned} . \tag{7.9}$$

If the state x that we are trying to estimate is basically constant from one trial to the next, that is, $a \approx 1$, then what we learn in one trial should be retained for the next trial. That is, the posterior estimate $\hat{x}^{(n|n)}$ in trial n should be "retained" and become the prior estimate $\hat{x}^{(n+1|n)}$ in trial n+1:

$$\hat{x}^{(n+1|n)} = a\hat{x}^{(n|n)}. \tag{7.10}$$

However, if we believe that the state x is uncorrelated from one trial to the next, that is, $a \approx 0$, then what we learn in one trial should be forgotten by the next trial. In a sense, if we believe that $a \approx 0$, then the state that we are trying to estimate is pure noise, and whatever prediction errors we observe on a given trial should be forgotten as we form our prior for the next trial.

Do people reduce their sensitivity to prediction error when the perturbations are pure noise? Do they increase their sensitivity when the perturbations are correlated? To answer this question, we need a way to estimate the subject's sensitivity to error. There are, in principle, two things that we can measure on each trial: the subject's prediction $\hat{y}^{(n)}$ and the subject's prediction error $y^{(n)} - \hat{y}^{(n)}$. If we assume that the $\hat{y}^{(n)}$ reflects the subject's prior, then the change in the subject's predictions from one trial to the next is a function of the prediction error, with a sensitivity that is proportional to the learning rate k:

$$\begin{aligned} \hat{y}^{(n)} &= \hat{x}^{(n|n-1)} \\ \hat{x}^{(n+1|n)} &= a\hat{x}^{(n|n-1)} + ak^{(n)}\left(y^{(n)} - \hat{y}^{(n)}\right) \\ \hat{y}^{(n+1)} &= \hat{x}^{(n+1|n)} \\ \hat{y}^{(n+1)} - \hat{y}^{(n)} &= (1-a)\hat{x}^{(n|n-1)} + ak^{(n)}\left(y^{(n)} - \hat{y}^{(n)}\right) \end{aligned} \tag{7.11}$$

If the subject has learned that $a \approx 0$, then the sensitivity to error $ak^{(n)}$ should also go to zero. In effect, if the learner is optimal, then she will stop learning when the perturbations are simply noise with no correlations from trial to trial.

Maurice Smith and Reza Shadmehr (2004) tested this idea by having people reach while holding a robotic arm that produced force perturbations. The force perturbation $f^{(n)}$ on trial n was described by a discrete function in which the correlation

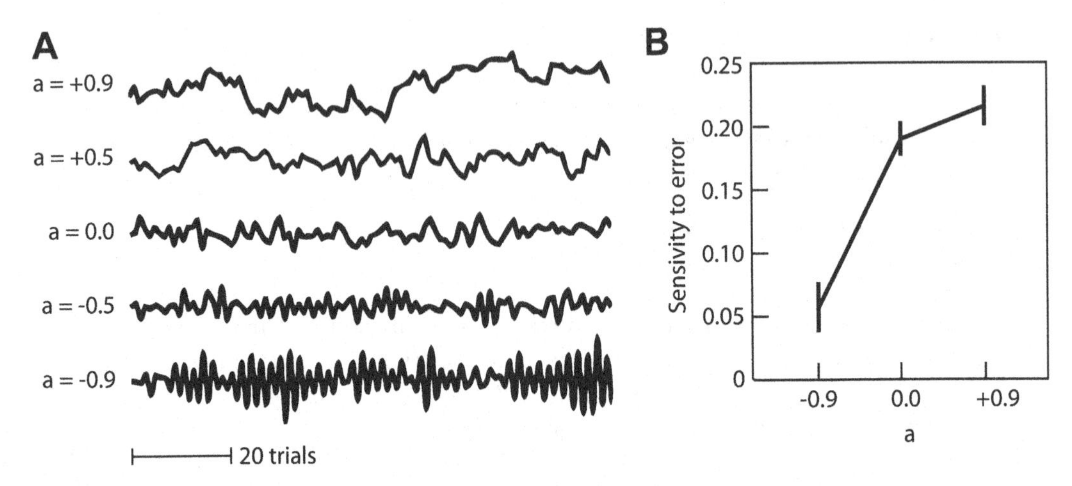

Figure 7.4
Altering the sensitivity to error through manipulation of the state update equation. Subjects held the handle of a robotic arm and made reaching movements to a target. The robot perturbed the hand with a force field. (A) Force perturbations are plotted as a function of trial number. These forces were generated via equation (7.12). (B) As the autocorrelation parameter a increased, sensitivity to error increased. (From Smith and Shadmehr, 2004.)

between one trial to the next was controlled within each block of training by parameter a:

$$f^{(n+1)} = af^{(n)} + \varepsilon \qquad \varepsilon \sim N(0, p^2). \tag{7.12}$$

In figure 7.4A we have examples of the perturbations as a function of trial. When $a = +0.9$, the perturbation on trial n were positively correlated with trial $n + 1$. This correlation disappeared when $a = 0$, and became negative when $a = -0.9$. The authors estimated the change in the motor output $\hat{y}^{(n+1)} - \hat{y}^{(n)}$ as a function of the error in the previous trial (similar to figure 7.3A). This sensitivity to error is plotted in figure 7.4B. When $a = 0$, the sensitivity to error was around 0.19, meaning that subjects compensated for about 19 percent of their prediction errors. When $a = +0.9$, sensitivity to error was about 0.22. When $a = -0.9$, the sensitivity to error was around 0.06. Therefore, the apparent learning rate was modulated by the statistics of the perturbation.

However, this last experiment unmasks a troubling feature of our attempt to couch biological learning in the framework of state estimation. Note that when $a = -0.9$, the theoretical prediction is that the Kalman gain will have the opposite sign as to when $a = +0.9$, but equal magnitude. That is, negatively correlated perturbations are just as predictable (and therefore learnable) as positively corrected ones. However, what we see in figure 7.4B is that people tend to shut down their learning system when they are presented with negatively correlated perturbations. What is happening here?

The theoretical problem that we are facing is that in optimal estimation (as in the Kalman framework), we need to have knowledge of the generative model, that is, the model that describes how the hidden states that we are trying to estimate are related to the data that we are observing. In other words, we need to have an internal model of the data and its relationship to our actions. But in reality, no one gives us this generative model. We do not have the very thing that we need in order to do optimal estimation. For example, no one provides the learner in figure 7.4, or in any of our examples thus far, the generative model that is producing the perturbations.

One way to view the results of the experiment in figure 7.4 is to assume that the brain is estimating both the parameter a of the generative model and the state of the perturbation $f^{(n)}$. If indeed the brain was attempting to estimate the parameter a, then what we should see is a modulation of the forgetting rates from trial to trial. For example, suppose that the perturbations are governed by equation (7.12) in which $a \approx 1$, and suppose that you as a learner have an accurate estimate of this parameter (because of prior learning or experience with the perturbations). Now suppose that through some artificial means during ten trials in a row whatever predictions you make, you observe zero prediction error:

$$y^{(n)} - \hat{y}^{(n)} = 0. \tag{7.13}$$

Let us call these *error clamp* trials because the prediction error is clamped to zero. If in your generative model $a \approx 1$, then you should exhibit little or no change in your motor output during these ten error-clamp trials. That is, you should show very little forgetting. On the other hand, if the perturbations are governed by $a \approx 0$, and you have an accurate internal model of this generative process, then your motor output should show complete forgetting during these same error-clamp trials.

This way of thinking allows us to arrive at an interesting idea: decay of our memories (forgetting rate) should be—at least in part—a reflection of our internal model of the environment in which we acquired that memory. We should retain what we learned in environments that changed slowly, but forget what we learned in environments that changed rapidly. In the motor learning literature there are a number of reports demonstrating that the brain retains a recently acquired memory better if that memory was acquired as a consequence of a gradual rather than a sudden change in the environment (Kagerer et al., 1997; Klassen et al., 2005; Michel et al., 2007). For example, adaptation to prism glasses can produce long-lasting motor memories if the visual distortion is introduced gradually (Hatada et al., 2006), but it tends to decay quickly if the distortion is introduced suddenly (Martin et al., 1996). Adaptation to gradual perturbations of gait produces longer-lasting aftereffects than adaptation to rapidly changing perturbations (Reisman et al., 2007). Why should the rate of change in the environment influence decay rates of motor memory?

In principle, if an environment changes slowly it might lead to the implicit belief that the changes are likely to be lasting, which in turn should influence the retention properties of the memory. On the other hand, if the environment changes suddenly, one should still adapt to the changes, but the speed of the change might imply that whatever is learned should be rapidly forgotten because the changes in the environment are unlikely to be sustained. This view implies that the long-term statistics of the experience, which presumably leads to formation of a more accurate generative model, that is, an estimate of parameter a in equation (7.12), should have a strong effect on the "forgetting" rate associate with the state estimate.

Vincent Huang and Reza Shadmehr (2009) found that when people reached in an environment in which perturbations changed rapidly, motor output adapted rapidly but then decayed rapidly in a post-adaptation period in which performance errors were eliminated (error-clamp trials). On the other hand, in an environment that changed gradually motor output adapted gradually but then decayed gradually. They used this behavioral assay to test whether prior experience with a different rate of environmental change would affect forgetting rates of the motor output. The idea was that through experience with a sequence of perturbations, the brain builds a generative model of those perturbations. Part of this generative model is the parameter a in equation (7.12). If this parameter indicates that the perturbations are changing slowly, then one should remember what one has learned in the subsequent error clamp trials.

To test this idea, volunteers were provided with a prior experience with a perturbation sequence, and then tested on a second sequence. For example, group 1 initially trained with a perturbation sequence that was steplike, while group 3 trained with a gradual perturbation sequence (figure 7.5A, "learn" period). After this brief period of training, a very long period of washout trials was presented for which there were no perturbations. Following the washout, the subjects once again were exposed to a sequence of perturbations. In this relearning block, the perturbations for both groups 1 and 3 were gradual. Retention was then assayed in the following error-clamp block.

Huang and Shadmehr found that the initial training in the learning block biased the forgetting rates in the relearning block. People who had initially observed a rapidly changing sequence of perturbations subsequently had rapid forgetting after the gradual relearning (figure 7.5B, group 1). People who had initially observed a gradually changing sequence of perturbations subsequently had slow forgetting after the same gradual relearning (figure 7.5B, group 3). Similarly, subjects that had prior experience with a rapidly changing perturbation (learn block, group 4) exhibited more rapid forgetting in the error-clamp period that followed relearning (group 4, figure 7.5C) as compared to subjects whose prior experience was with a gradually changing perturbation (groups 2 and 5, figure 7.5C). Therefore, prior experience with

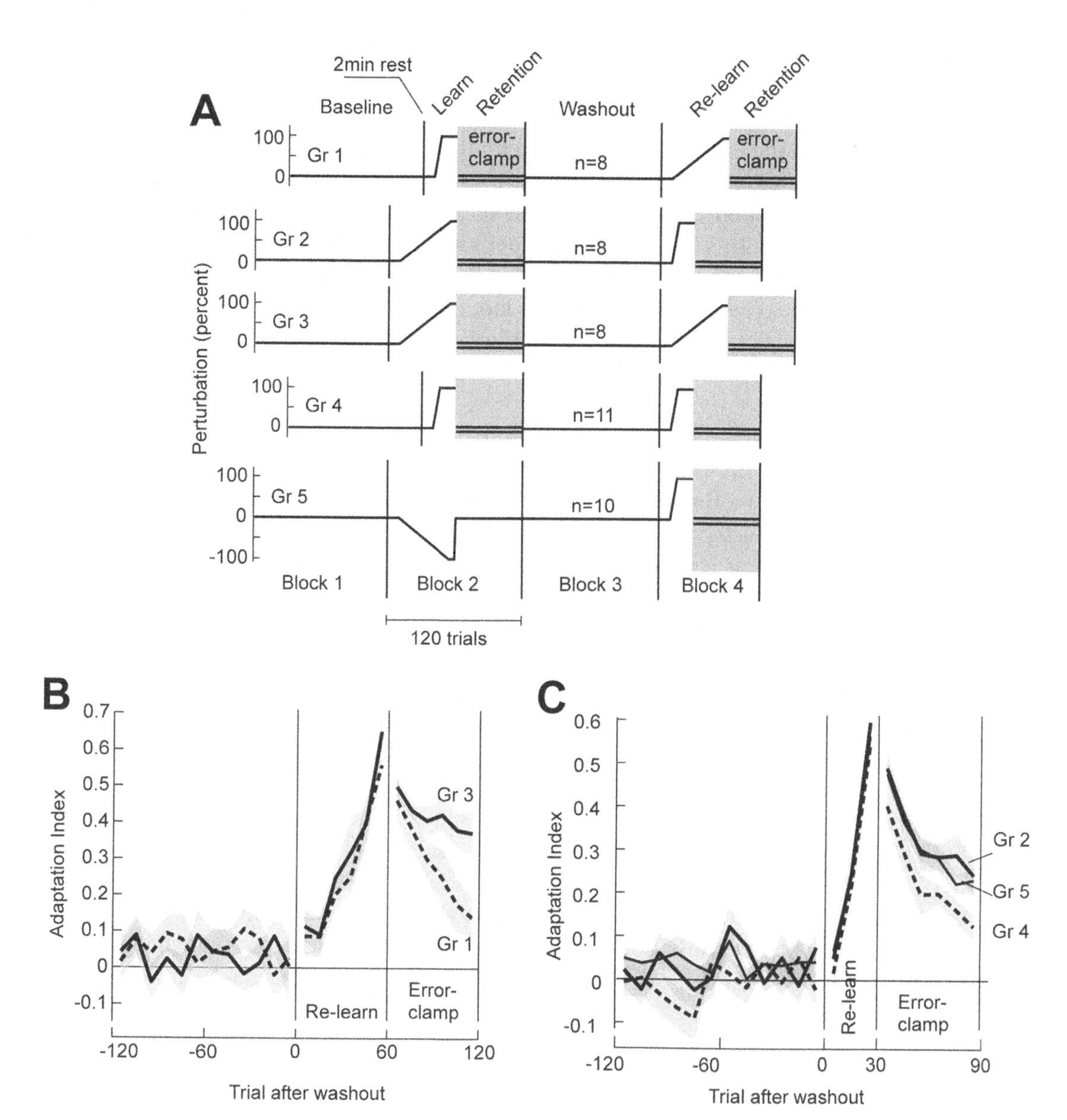

Figure 7.5
Retention of motor memory as a reflection of the statistics of the perturbations. (A) On each trial, subjects reached to a target and were perturbed by a force that either increased abruptly (as a function of trial) or gradually. The gray region indicates error-clamp trials, that is, trials in which errors in movement were artificially clamped to zero, enabling measurement of motor output while minimizing error dependent learning. In block 1, no perturbations were applied. In block 2, learning was followed by a measure of retention in error-clamp trials. In block 3, no perturbations were applied (termed washout). In block 4, relearning took place followed by a measure of retention in error-clamp trials. (B) In block 3 (relearning), groups 1 and 3 both experienced a gradually increasing force. However, the prior experience of group 3 was with the same gradually increasing force, while the prior experience of group 1 was with a rapidly increasing force. In error-clamp trials, force output of group 3 gradually declined, while output of group 1 rapidly declined. (C) In block 3, groups 2, 4, and 5 experienced a rapidly increasing force. However, the prior experience of both groups 2 and 5 was with a gradually increasing force. In error-clamp trials, force output of groups 2 and 5 gradually declined. (From Huang and Shadmehr, 2009.)

a rapid change in the environment biased retention of memories acquired later in response to a gradual change in the same environment, increasing forgetting rates. Similarly, prior experience in an environment that changed gradually enhanced retention in response to a rapid change in that same environment.

These results suggest that as our brain is presented with a prediction error, it tries to learn a generative model. This generative model is a description of how the state of the perturbation is related to the observations or measurements. It then uses this generative model to optimally estimate the state of the perturbation.

Summary

If we view learning in a Bayesian state estimation framework, we can make certain predictions about how to speed up or slow down the process of learning. Learning will speed up if the learner becomes more sensitive to prediction errors, which can be achieved by making the learner more uncertain about her own predictions. Learning will slow down if the learner's sensory measurements become less reliable.

To make a prediction, the learner relies on her prior estimate. The prediction error allows the learner to form a posterior estimate. However, the prior belief on the next trial depends on the generative model and how it dictates trial-to-trial change in the hidden states. In effect, the generative model's trial-to-trial change is a forgetting rate. Presumably, biological learning is not only a process of state estimation, but also a process in which the brain learns the structure of the generative model, that is, how the hidden states change from one trial to the next. This can account for the fact that retention, as measured by decay of motor output in error-clamp trials, is slower after perturbations that change gradually versus those that change rapidly.

8 The Multiple Timescales of Memory

Despite the modest success of our simple framework for learning, it turns out that there is something seriously wrong with the models that we have considered so far in our discussion of learning and memory. They fail to account for two fundamental properties of memory: *savings* and *spontaneous recovery*. Let us explain these two features of memory with some examples.

8.1 Savings and Spontaneous Recovery of Memory

A few years ago, Yoshiko Kojima, Yoshiki Iwamoto, and Kaoru Yoshida (2004) took a monkey to a darkened room and measured its eye movements (saccades) in response to display of visual targets. Usually, when a target is shown at 10 degrees with respect to current fixation, the animal makes a 10-degree saccade to foveate the target. If we define *saccadic gain* as the ratio between the displacement of the eyes (about 10 degrees) and the displacement of the target (also 10 degrees), then in the healthy monkey (and healthy person) the saccadic gain is approximately one. However, Kojima and colleagues were interested in having the monkey learn to change this gain. To do this, they tested the monkey in a gain adaptation paradigm (McLaughlin, 1967). In the experiment, a target is shown at 10 degrees, but as soon as the eyes begin moving toward it, the 10-degree target is extinguished and a new target is shown at 13.5 degrees. Saccades are so fast (peak speed of greater than 400 deg/s) and so brief (movements complete within 60 ms) that the brain cannot use any sensory feedback during the saccade to help control it. In fact, one is effectively blind during a saccade. So in response to a 10-degree target, the monkey makes a 10-degree saccade, observes the endpoint error, and follows this with a second saccade. As the trials continue, the brain learns to increase the saccadic gain so that in response to the 10-degree target it makes a larger than 10-degree saccade. This is called gain-up adaptation.

Kojima and her mentors performed a variant of a classic experiment called *extinction*. They initially trained the monkey on the gain-up task. They then followed this

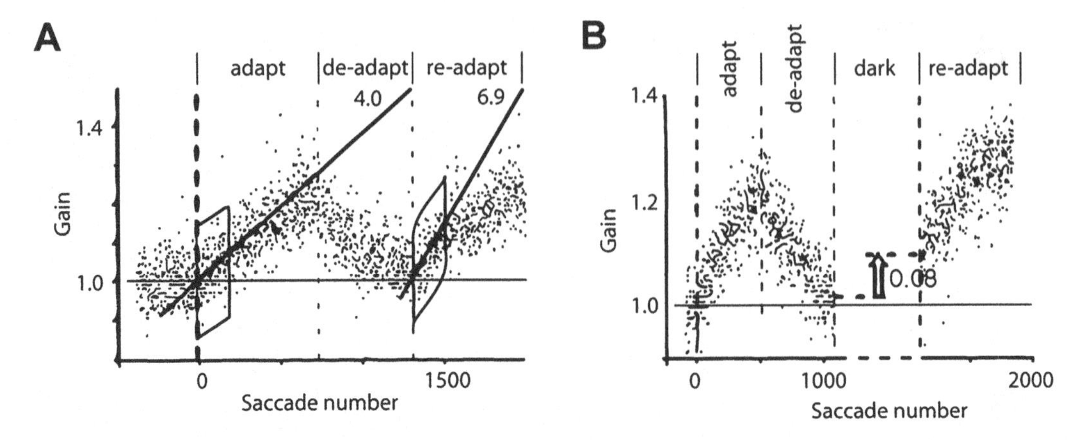

Figure 8.1
Savings and spontaneous recovery of motor memory. (A) A monkey made saccades to a target at 10 degrees. In the adaptation period, the 10-degree target was displaced to 13.5 degrees as the saccade initiated, resulting in an endpoint error. Slowly, the monkey learned to respond to the 10 degrees by making a larger-than-10-degree saccade. That is, the gain of the saccade increased. Saccadic gain is defined as the ratio between the displacement of the eyes and the displacement of the target. In the de-adapt period, the 10-degree target was displaced to 6.5 degrees. The de-adapt period continued until the saccade gain returned back to baseline. However, upon readaptation, the rate of relearning was faster than learning during the initial adaptation period. This phenomenon is called savings. The slope of the first 150 trials is noted by the line. (B) Adapt and de-adapt periods were followed by thirty minutes of darkness in which the animal made saccades but had no stimuli or other visual information. In the readapt period, saccade gain appeared to spontaneously change from the end of the de-adapt period. This phenomenon is called spontaneous recovery. (From Kojima, Iwamoto, and Yoshida, 2004.)

training with a gain-down task in which the 10-degree target was extinguished and a new target was shown at 6.5 degrees. This is called extinction training because the objective is to counter the initial adaptation with a reverse-adaptation protocol until the gain is returned back to baseline (i.e., gain of one). They observed that in the gain-up trials, the gain of saccades increased, and in the following gain-down trials, the gain returned back to near baseline (figure 8.1A). The rate of adaptation in the gain-up trials was assayed by fitting a line to the initial 150 trials (this is noted by the slope of the line in figure 8.1A, which is about 4.0). The training in the gain-up session was followed by training in a gain-down session until the gain had returned to one. That is, in the de-adaptation session the performance of the animal returned to baseline. Was the motor memory also at baseline, that is, had the de-adaptation session wiped out the memory that was acquired due to the prior adaptation session? To answer this question, Kojima et al. once again presented the animal with the gain-up paradigm (labeled readapt in figure 8.1A). They observed that the animal relearned faster (the slope of the line is now 6.9). This faster relearning is evidence of *savings*. Savings refers to the fact that despite return of behavior to a baseline condition, some component of the animal's memory still remembered the earlier gain-up training, producing a faster relearning. Therefore, the gain-down training did not completely wash out the previous gain-up training.

In the next set of experiments (figure 8.1B), Kojima and her mentors once again trained the monkey in the gain-up paradigm and followed this with a gain-down paradigm until behavior had returned to baseline. At this point they simply turned the lights off in the room and left the monkey alone for thirty minutes. When they returned and retested the monkey in the gain-up paradigm, they noticed that the gain had suddenly increased. This is an example of *spontaneous recovery* of motor memory. With passage of time after extinction, the initial memory of adaptation seemed to spontaneously recover.

Another example of spontaneous recovery is in a form of classical conditioning in honeybees. Nicola Stollhoff, Randolf Menzel, and Dorothea Eisenhardt (2005) placed a bee in a test chamber and presented it with an odor (oil from a carnation flower) for five seconds. Two seconds after the odor onset, they touched the antenna of the bee with a toothpick that was moistened with sucrose. Training consisted of three such odor (conditioned stimulus, CS) and sucrose (unconditioned stimulus, US) pairings. They measured whether the bee extended its proboscis between the onset of the CS and the presentation of the US (a positive score indicates that the bee performed this act). In figure 8.2 we see that on the first trial, none of the bees anticipated the sugar, but on the second trial about 60 percent of the bees put out their proboscis after they sensed the fragrance of the carnation flower. This ratio reached 85 percent on the third trial. The next day they brought the bee back to the test chamber and presented it with five trials in which the odor was present, but the antenna was not touched (CS-only trials). During this period of extinction training, the percent of bees that responded to the odor gradually dropped, so that by the fifth trial, only about 15 percent of the bees extended their proboscis in anticipation of the sugar. Effectively, performance had returned to near baseline. Finally, they brought back some of the bees at one hour, two hours, and so on after completion of the CS-only trials for a test of retention. They observed that with each passing hour, the percentage of bees that extended their proboscis increased, reaching about 50 percent by twenty-four hours. Therefore, with passage of time after extinction, the initial memory that associated the odor with the sucrose seemed to recover spontaneously.

Savings and spontaneous recovery are two fundamental properties of biological learning. Savings suggests that errors that reverse the direction of learning in a behavioral sense may not reverse the state of memory in a neural sense, as evidenced by the observation that relearning after extinction may be faster than original learning. This implies that de-adaptation may not destroy the memory that was acquired during adaptation. Spontaneous recovery suggests that errors that

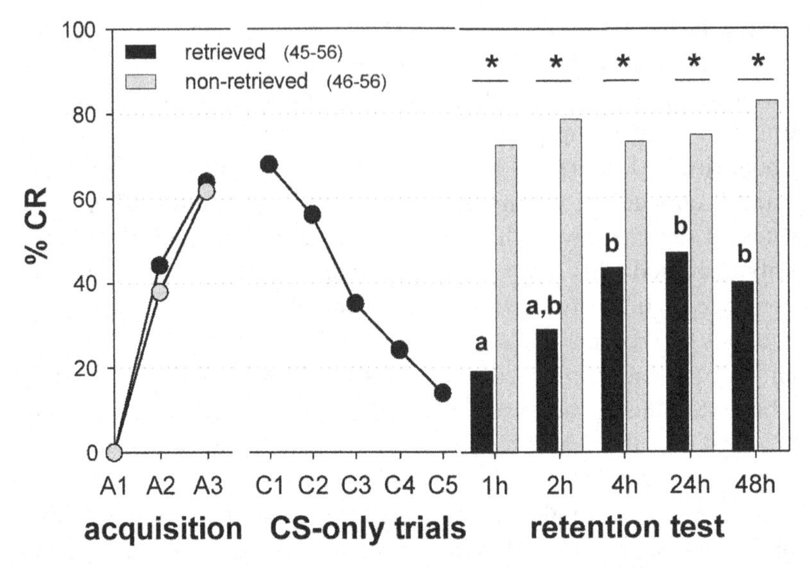

Figure 8.2
Acquisition, extinction, and spontaneous recovery of memory in a honeybee. In the three acquisition trials, bees were provided with an odor (CS, conditioned stimulus), and sugar (US, unconditioned stimulus). The y-axis records the percentage of animals that extended their proboscis in response to the odor (percent conditioned response). The next day, some of the bees were provided with five extinction trials (CS-only). At time periods thereafter (retention test), different bees were tested for CS-only. The black bars denote the performance of animals that did not receive extinction training. In the hours that followed extinction training, animals exhibited spontaneous recovery of the initial learning. (From Stollhoff, Menzel, and Eisenhardt, 2005.)

reverse the direction of learning produce a new memory (call it extinction memory, e.g., CS associated with no US), that competes with the original adaptation memory (CS associated with US). With the passage of time after the training episode that produced extinction, the memory acquired during the adaptation training appears to be reexpressed.

8.2 Two-State Model of Learning

Maurice Smith, Ali Ghazizadeh, and Reza Shadmehr (2006) proposed a simple mathematical model of learning to account for spontaneous recovery and savings. Imagine that when our brain encounters a prediction error, the error affects multiple learning systems, each one with its own internal state: some that are highly sensitive to error but have poor retention, and others that are poorly responsive to error but have strong retention. The key new ideas were to (1) represent learning as a problem in state estimation in which there were multiple hidden states and (2) associate a different timescale of forgetting to each state.

To illustrate the idea, suppose that we make a prediction $\hat{y}^{(n)}$. Say that this prediction is the sum of two internal states, one a "fast" state $x_f^{(n)}$ and the other a "slow" state $x_s^{(n)}$:

$$\hat{y}^{(n)} = x_f^{(n)} + x_s^{(n)}. \tag{8.1}$$

Both the fast and the slow states learn from the prediction error $\tilde{y}^{(n)} = y^{(n)} - \hat{y}^{(n)}$, but from one trial to the next partially forget what they have learned. The forgetting in each state is specified by parameter a, and sensitivity to error is specified by parameter b. The important assumptions are that the fast system's learning rate b_f is larger than the slow system's learning rate b_s, while the fast system's retention rate a_f is smaller than the slow state's retention rate a_s:

$$\begin{aligned} x_f^{(n+1)} &= a_f x_f^{(n)} + b_f\left(y^{(n)} - \hat{y}^{(n)}\right) \\ x_s^{(n+1)} &= a_s x_s^{(n)} + b_s\left(y^{(n)} - \hat{y}^{(n)}\right) \\ & 0 < a_f < a_s < 1 \\ & 0 < b_s < b_f < 1 \end{aligned} \tag{8.2}$$

Let us use this model to consider the experiment by Kojima and colleagues. In that experiment (figure 8.1A), the animal adapted to a perturbation, and this period of adaptation was followed by a period of de-adaptation until performance returned to baseline. The de-adaptation was followed by readaptation, during which the animal exhibited savings, i.e., a faster relearning. Figure 8.3A shows a simulation of this scheme. A perturbation of magnitude one is imposed for a few hundred trials (adaptation period) and then the perturbation reverses sign (de-adaptation period) until performance returns to baseline. At the onset of the adaptation period the prediction errors $\tilde{y} = y - \hat{y}$ are large, causing rapid changes in the fast state x_f. As the training continues, $\tilde{y}$ becomes smaller and the forgetting term a_f in the fast state becomes relatively more important. As a result, near the end of the adaptation period most of the prediction $\hat{y}$ is due to the slow state x_s. Now in the de-adaptation period, the prediction errors are once again very large, causing a rapid change in the fast state. By the end of the de-adaptation period, the prediction $\hat{y}$ is at zero, that is, the system appears to be at baseline. Yet, this is only because the fast and slow states are in competition, effectively canceling each other. Now the de-adaptation period is followed by another period of readaptation. In this relearning period, the system adapts faster than before, exhibiting savings (compare dashed line with the solid line in the lower subplot of figure 8.3A). The reason for the faster relearning is that the prediction errors during the de-adaptation period hardly affect the slow state. Savings is due to resistance of the slow state to change, enhancing the ability of the system to return to its previously adapted state in the readaptation period.

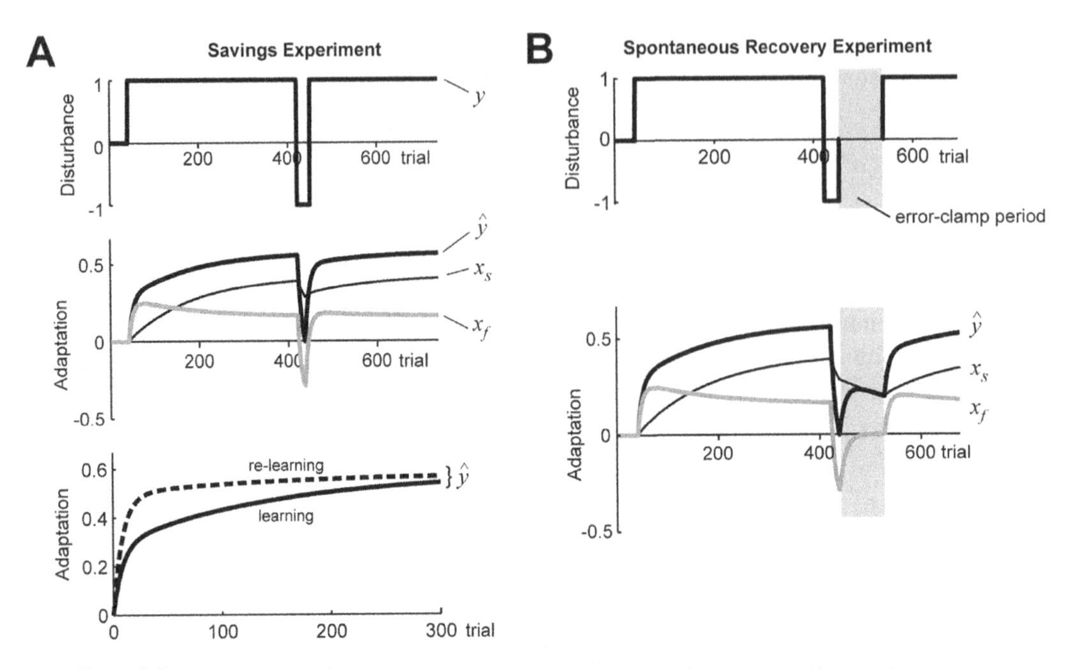

Figure 8.3
Simulation of a two-state system in learning from prediction error (equations 8.1 and 8.2). (A) Savings experiment. The disturbance $y^{(n)}$ induces adaptation, which initially causes a large change in the fast state. With more trials, the slow state rises while the fast state falls (as the errors decline and forgetting becomes a stronger factor than prediction error). In the de-adaptation period, the training continues until $\hat{y}$ returns to zero. The fast state rapidly changes, setting up a competition between the fast and slow states. In readaptation, the slow state is already in an adapted position, which makes relearning faster than original learning (dashed line vs. solid line, bottom figure). (B) Spontaneous recovery experiment. Adaptation is followed by de-adaptation until performance returns to baseline. During the error-clamp period, prediction errors are set to zero on every trial. During this period the fast state rapidly decays while the slow state slowly decays. The sum of the two produces recovery of the output $\hat{y}$ to near a previously adapted state despite the fact that there are no errors to learn from.

The basic idea of this two-state model is that during the brief de-adaptation period, the system's predictions can change rapidly and return to baseline, but not through complete unlearning of what it has learned in the past. Rather, a fast state of memory is put in competition with a slow state. The system's output $\hat{y}$ exhibits washout only because the two states happen to balance each other.

While there are some similarities between the simulations of extinction (figure 8.3A) and the behavioral data (figure 8.1A), there are also some critical differences. In both the simulations and the behavioral data, the de-adaptation period continues until behavior has reached baseline. In the behavioral data, however, the length of de-adaptation training is fairly similar to the length of adaptation training (about 700 trials of adaptation followed by about 500 trials of de-adaptation), whereas in the simulations the adaptation period is much longer than the de-adaptation period (about 400 trials of adaptation followed by 20 trials of de-adaptation). As Eric Zarahn and colleagues pointed out (Zarahn et al., 2008), the two-state model would show complete unlearning if the de-adaptation period was as long as the adaptation period. This is inconsistent with the behavioral data, and something that we will return to later in this chapter.

Now let us consider the spontaneous recovery phenomenon. In Kojima's experiment (figure 8.1B), the monkey was left alone in the dark for thirty minutes. When it was once again asked to make a saccade, the monkey made hypermetric saccades, as it had initially learned to make in the gain-up paradigm. What happened during this period of darkness? The monkey surely moved its eyes, but it had little or no sensory feedback to tell it if the saccades that it made were accurate. In effect, darkness reduced the monkey's ability to measure sensory consequences of its motor commands. Let us imagine that this is equivalent to making movements, but having no errors to learn from, that is, the "darkness" period is equivalent to setting the error term to zero:

$$y^{(n)} - \hat{y}^{(n)} = 0. \tag{8.3}$$

(Having zero error is not the same as having no sensory feedback, but this is a useful initial step. We will return to the question of no sensory feedback once we recast our problem using a generative model.) So in effect, the "dark" period is like an error-clamp period in which movements are made, but there is zero error on each trial.

In the readaptation period that comes after the error-clamp period, the model exhibits a jump in performance with respect to end of the de-adaptation period (figure 8.3B). Therefore, the model accounts for the Kojima data in figure 8.1B by suggesting that during the "dark" period, the animal made saccades but had no visual errors to learn from. During this "dark" period, the fast state rapidly declined, the slow state slowly declined, and the sum of the two produce the spontaneous recovery. The states decline toward zero because there were no errors to learn from. In this way, changes in the memory states were only due to the forgetting terms a_f and a_s. The rapid decay of the fast state but the gradual decay of the slow state made $\hat{y}$ rise up from zero, resulting in spontaneous recovery of the previously learned behavior.

The strongest prediction of the model is regarding how the behavior should evolve during the error-clamp period (figure 8.3B): There should be a very rapid rise of output followed by a gradual decline. A number of recent experiments have directly tested this prediction by measuring motor output during error-clamp trials. In a saccade task, Vince Ethier, David Zee, and Reza Shadmehr (2008) examined saccadic gain changes in people (figure 8.4A). As before, adaptation was followed by de-adaptation until gain had returned to near baseline. The training was then

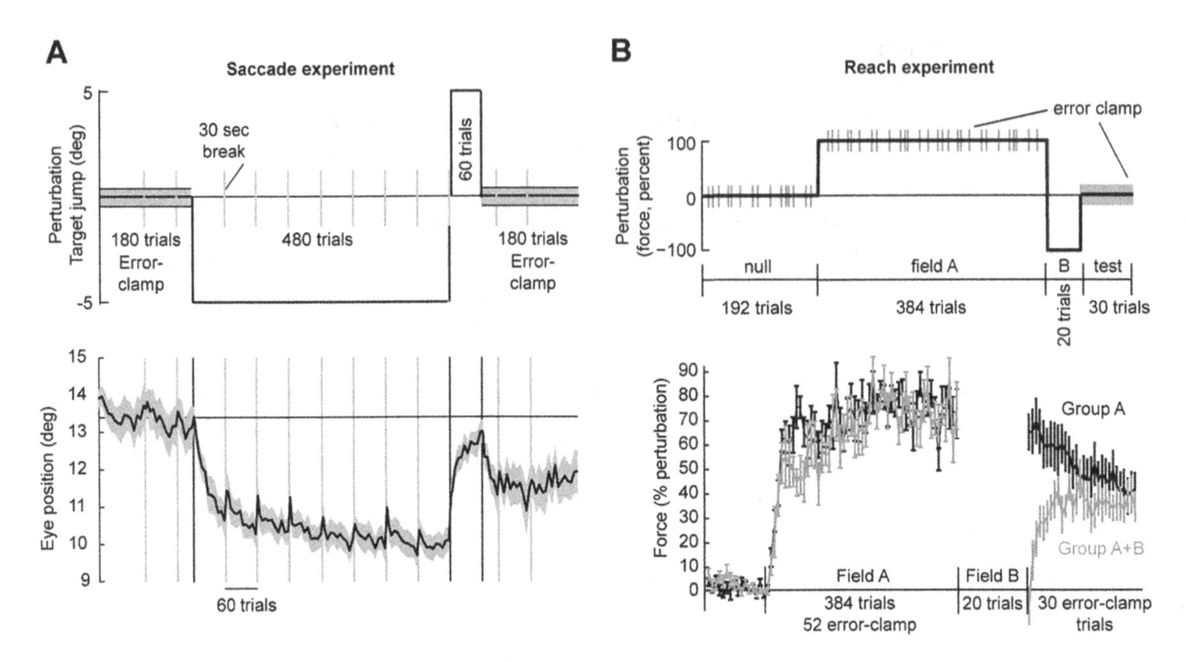

Figure 8.4
Spontaneous recovery following adapt/de-adapt training. (A) Saccade experiment. Top figure shows the state of the perturbation, and the bottom figure shows motor output (eye displacement during each saccade). Target appeared at 15 degrees. Adaptation began with a gain-down block. Saccade amplitudes are displayed in the lower plot. Note the rapid initial adaptation that was followed by gradual adaptation. Also note the forgetting that took place at set breaks (gray vertical lines). De-adaptation (gain-up block) returned motor output back to near baseline. During the subsequent error-clamp trials, motor output returned toward the previously adapted state. (From Ethier, Zee, and Shadmehr, 2008.) (B) Reach experiment. People held the handle of a robotic arm and reached to a target. The robot perturbed the arm perpendicular to its direction of motion with a force field. With practice, people learned to expect a force perturbation and compensated. The gray vertical lines indicate error-clamp trials, trials in which the robot produced a stiff channel that guided the hand straight to the target, removing error and allowing the experimenters to measure how hard subject pushed against the wall. This was a proxy for the motor output (plotted in the bottom figure). One group adapted to field A and was then presented with a block of error-clamp trials (group A). Another group adapted to field A and then was presented with B until performance returned to baseline, at which point a block of error-clamp trials was presented (group A + B). Motor output of A + B group spontaneously recovered from baseline and converged to output of group A, that is, the effect of field B completely washed out within thirty trials after de-adaptation. (From Criscimagna-Hemminger and Shadmehr, 2008.)

followed by a block of error-clamp trials. Each error-clamp trial began with a visual target. As soon as the saccade was initiated, the target was turned off. After completion of the saccade the target reappeared at the current location of fixation, effectively minimizing endpoint errors. As the model had predicted, saccade gain after the de-adaptation period exhibited spontaneous recovery: a rapid change from baseline followed by a gradual change (figure 8.4A, the final period of error-clamp trials).

The same pattern appeared in adaptation of reaching movements (Smith, Ghazizadeh, and Shadmehr, 2006; Criscimagna-Hemminger and Shadmehr, 2008). For example, Sarah Criscimagna-Hemminger and Shadmehr (2008) asked volunteers to hold the handle of a robotic arm and reach to a target. Forces that were perpendicular to the direction of motion perturbed the reach. With training, the brain learned to predict and compensate for these forces. To measure the motor output that represented this prediction, in some trials the robot produced a virtual channel consisting of two stiff walls that surrounded a straight line between the start and the target positions (Scheidt et al., 2000). In these error-clamp trials, it was possible to measure both the motor output (i.e., what the brain predicted about the perturbation, reflected in the forces that the subject produced against channel walls), and minimize any errors in the movement. In the baseline trials before imposition of the perturbations, people simply moved their hand straight to the target, producing little or no forces against the channel walls (figure 8.4B, "null" condition). Two groups of subjects were then considered. One group experienced only a field that perturbed the hand to one direction (field A). Another group was first exposed to A, and then to field B. Field B had a perturbation that was equal but opposite in direction to field A. Effectively, in the A + B group the experiment replicated the adapt/de-adapt paradigm that we saw earlier in the saccades. In group A, a block of error-clamp trials immediately followed training in A. In group A + B, a block of error-clamp trials followed training in B. While group A showed a motor output that gradually declined from the adapted state, group A + B demonstrated spontaneous recovery, that is, motor output rose from baseline toward the previously adapted levels. However, something interesting happened: Spontaneous recovery of motor output was so strong in the A + B group that motor output converged to the A group. This could only happen if de-adaptation training produced *little or no unlearning* of previous adaptation. That is, it appeared that de-adaptation training instantiated an independent fast memory, and this fast memory rapidly faded in the subsequent error-clamp trials.

The spontaneous recovery patterns in figure 8.4B suggest that a long period of training is sufficient to produce a slow motor memory that is resistant to subsequent performance errors (e.g., in field B). When the field changes and induces large performance errors, the errors do not erase the memory of A, but rather install a competing memory of B. The general implication of this work is that once a memory has been acquired, all further learning may be instantiation of competing memories.

8.3 Timescales of Memory as a Consequence of Adapting to a Changing Body

While the two-state model of equation (8.2) is certainly more capable of accounting for the experimental data than a single-state model, we need to ask why the nervous system should learn in this way. That is, what might have given rise to this kind of learning? What is this kind of learning good for?

The main new idea in this section is that animals exhibit patterns of learning and forgetting, that is, the multiple timescales that we alluded to earlier, because these timescales describe perturbations that naturally affect their own body. There are many natural perturbations that can affect the body, some with rapid timescales and others with slower timescales. Perhaps when we perform a task and observe an error, our brain tries to estimate the source of the error with the same generative model that represents natural perturbation to the body.

To illustrate the point, let us suppose that you are designing the control mechanism of an autonomous robot. You recognize that the input–output properties of the motors in various limbs will change with use and with passage of time. For example, with repeated use over a short period, a motor may experience heating and change its response transiently until it cools. On the other hand, with repeated use over a long period of time the motors may decline in function due to wear and tear. Both of these conditions will produce movement errors—that is, the inputs to the motor will not produce the expected sensory feedback. The errors will require your controller to adapt and send updated commands to the motors to produce the desired actions. However, your controller should interpret these errors differently: Errors that have a fast time scale should result in rapid adaptive changes, but should be quickly forgotten (because a heating motor will cool simply by passage of time). Errors that persist for extended periods of time should result in slow adaptive changes that should be remembered (because a worn-out motor will not fix itself no matter how long we wait). Now if a mischievous child were to attach a small weight to one of the arms, the sudden drop in performance of the arm might be interpreted as a motor heating up and therefore produce a rapid adaptation followed by rapid forgetting if the arm is held still for a while (during which time we would expect the motor to cool, removing the potential source of performance error). In this way, we might view the problem of learning as one of credit assignment to various timescales of memory. To solve the credit assignment problem, we need a generative model that describes the likely sources of perturbations and their timescales.

Konrad Kording, Joshua Tenenbaum, and Reza Shadmehr (2007) proposed that our nervous system faces similar problems in controlling our body. Properties of our muscles change due to a variety of disturbances, such as fatigue, disease, exercise, and development. The states of these disturbances affect the motor gain, that is, the ratio of movement magnitude relative to the input signal. States of disturbances unfold over a wide range of timescales. Therefore, when the nervous system observes an error in performance, it faces a credit assignment problem: Given that there are many possible perturbation timescales that could have caused the error, which is the most likely? It seems reasonable that the solution to the credit assignment problem should dictate the temporal properties of the resulting memory. That is, adaptation in response to things that are likely to be permanent (slow states) should be remembered, while adaptation in response to things that appear transient (fast states) should be forgotten.

Now this kind of thinking does not explain all forms of perturbations. For example, a sudden injury may take a long time to heal. In this case, a perturbation that comes on rapidly does not go away rapidly. Similarly, certain low frequency events like a car accident or childbirth leave memories that are never forgotten. In these scenarios, there are neuromodulators that are released due to the heightened emotional state of the learner, making memories "stick." But for the motor system, the most likely source of a natural perturbation is fatigue. For example, you decide to walk up a few flights of stairs. Step after step, the leg muscles fatigue, altering their force production capabilities. To maintain your stride, the brain must keep an estimate of the state of the muscles so that it can deliver the right amount of activation. This state estimate should change rapidly as the leg muscles fatigue. However, when you arrive at your floor and get to the lab and sit down, the fatigue state of the muscles naturally fades away. Your estimate of the state of the muscles should also rapidly fade away, without requiring errors to drive changes in your estimate. You do not want a brain that assumes that the muscles are still fatigued after sitting for ten minutes just because during that period you have not taken a step and therefore had not had the opportunity to observe a prediction error. Perhaps a generative model that describes dynamics of these natural, internally driven sources of perturbation in our motor system acts as a generative model for external perturbations as well. That is, the way animals adapt to artificially induced perturbations on their movements may be based on a generative model of perturbations that affect their own bodies.

To make our generative model, we will make two assumptions. First, we will assume that there are many sources of perturbations that can affect our performance, with some perturbation states changing rapidly, while others changing slowly. Second, we will assume that states that change rapidly are affected by greater noise and variability than those that change slowly.

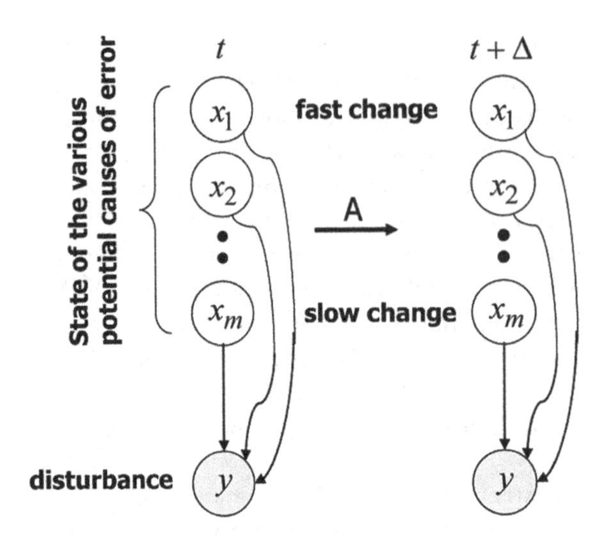

Figure 8.5
A graphical representation of the generative model of equation (8.4).

Such a generative model might look something like figure 8.5. We represent the state of each potential source of perturbation by x_i. Some of these perturbation states decay quickly with time, while others decay slowly. This is reflected in the elements of the diagonal state transition matrix A. When we make an observation y, we see the combined effects of all the states:

$$\begin{aligned} \mathbf{x}^{(n+1)} &= A\mathbf{x}^{(n)} + \boldsymbol{\varepsilon} \\ y^{(n)} &= \mathbf{c}^T \mathbf{x}^{(n)} + \eta \end{aligned} . \tag{8.4}$$

To note the different timescales of change for each state x_i, we can manipulate the elements of the matrix A to reflect our assumption that perturbation state x_1 decays more rapidly than perturbation state x_2, etc.

$$0 < A_{1,1} < A_{2,2} < \cdots < A_{m,m} < 1.$$

(The subscripts in the above expression refer to the elements of the matrix.) For simplicity, we will assume that A is diagonal. Furthermore, it seems rational that we should be most uncertain about states that change most quickly, an idea that we can incorporate by making the fast states in equation (8.4) have the most amount of noise.

$$\begin{aligned} &\boldsymbol{\varepsilon} \sim N(\mathbf{0}, Q) \\ &Q = \begin{bmatrix} \sigma_1^2 & 0 & 0 \\ 0 & \ddots & 0 \\ 0 & 0 & \sigma_m^2 \end{bmatrix} \qquad \sigma_1^2 > \cdots > \sigma_m^2 \end{aligned} . \tag{8.5}$$

Finally, while we cannot observe these states individually, what we can observe is their combined effects; so we set vector **c** in equation (8.4) to have elements that are all one:

$$\mathbf{c} = [1 \quad 1 \quad \cdots \quad 1]^T.$$

In summary, when we observe a disturbance in our performance, our job is to estimate which state is the most likely candidate. Is the disturbance due to a fast changing state? In that case, we should learn a lot from our performance error (because we are least certain about the fast states), but also quickly forget it (because we assume that these states quickly decay, e.g., $A_{1,1} \ll 1$). Is the error due to a slow changing state? In that case, we should change these states by a relatively small amount (because we are most certain about the slow states), but remember what we learned (because we assume that these states decay slowly).

With this generative model, Kording, Tenenbaum, and Shadmehr used the Kalman algorithm to make predictions about learning and memory. Their specific model had thirty hidden states, with timescales that were distributed exponentially between 2 and 10^5 trials. The simulation result for the spontaneous recovery experiment is shown in figure 8.6A. To simulate the darkness period, rather than setting errors to zero, the sensory noise η in equation (8.4) was set to near infinite variance. As a consequence, during the darkness period the Kalman gain became zero, and the performance was dictated by the terms in matrix A.

Because the timescales ranged from very short to very long, the model was also able to replicate results from a long-term adaptation experiment. An example of a long-term adaptation experiment is shown in figure 8.6B. In this experiment, Rick Robinson, Robijanto Soetedjo, and Christopher Noto (2006) trained a monkey on a gain-down saccade adaptation protocol for twenty-two days. During each training session the gain decreased, but some of this learning was forgotten during the time between sessions (figure 8.6B). On the twenty-third day and beyond, the targets were no longer moved during the saccades, encouraging the animal to return the saccadic gain back to one (called the washout period). Interestingly, between the sessions in the washout period the forgetting now reversed directions. That is, whereas during adaptation forgetting was toward a gain of one, now during washout forgetting was toward a gain of 0.5.

The simulations with the generative model of equation (8.4) produced a similar result (figure 8.6C). The simulations explained that during initial days of gain-down adaptation, most of the changes were in the fast states (they became negative, where zero indicates a baseline condition). These states exhibited forgetting between the sessions, and so behavior reverted back toward gain of one between training days. By the end of training on day 22, only the slowest of the slow states had changed (became negative) and there was little or no contribution to behavior from the faster

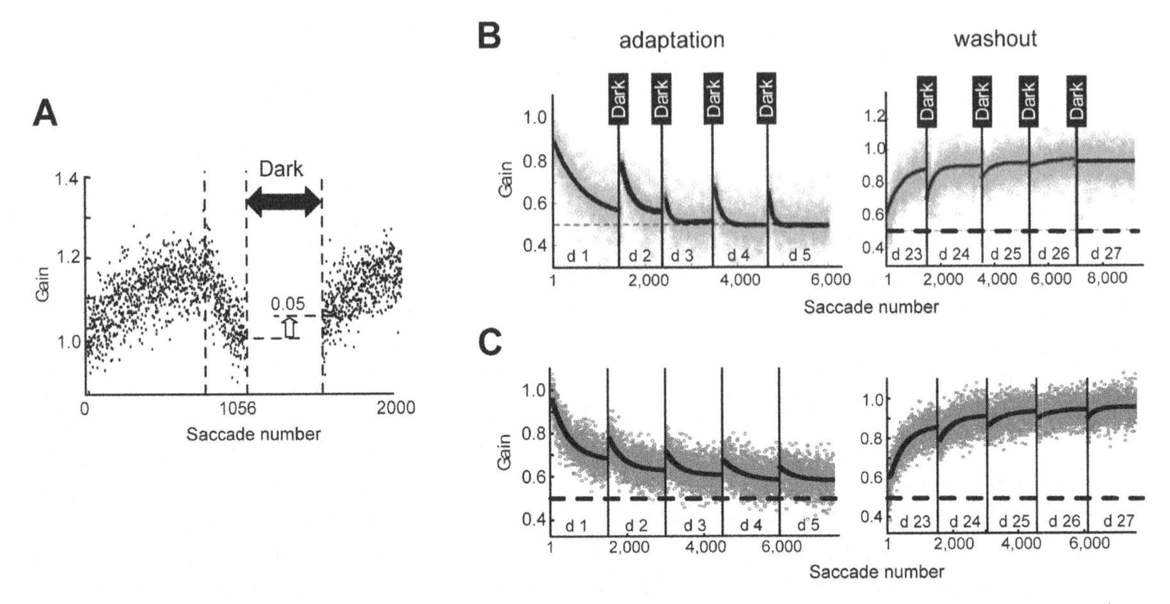

Figure 8.6
Simulations of the model in equation (8.4). (A) Simulations results for the spontaneous recovery experiment (figure 8.1B). (B) Experimental data for a multiday experiment. A monkey trained in a gain-down saccade adaptation paradigm for twenty-two days (data for five days are shown in the adaptation subplot). Between days the animal was kept in the dark. Note that the forgetting is toward gain of one. On day 23 and beyond, the learning was washout (target remained stationary, encouraging the gain to return to one). Note that during washout, the forgetting is toward a gain of 0.5. (From Robinson, Gordon, and Gordon, 2006.) (C) Simulation results for the model of equation (8.4). (From Kording, Tenenbaum, and Shadmehr, 2007.)

states (which were all near zero). When the washout trials begin on day 23, all the changes were once again in the fast states (which now became positive). Between the training days, the fast states once again forgot back to zero, but because the slow states had not returned to baseline, between-day forgetting in the washout session appeared to be in the opposite direction of adaptation.

The reason why the simulations needed thirty states rather than two is because of the very long length of the training protocol. Whereas from day 3 to day 22 there was little change in performance (as shown by the generally flat lines in figure 8.6B), in the model there were significant changes as the performance shifted from a reliance on some of the faster states in the early days of training to the slowest states as the weeks wore on. The effect of this very long-term training could only be recorded by states that changed on a similar long-term timescale. As a result, very long-term training produced memories that decayed extremely slowly.

In summary, when adaptation is followed by de-adaptation, during the following period of error-clamp trials there is a spontaneous recovery of the adapted state.

This pattern is consistent with a state estimation process in which our brain assumes that the observations that it makes are driven by a generative model in which multiple hidden states contribute; some states change slowly with time, while others change rapidly.

8.4 Passive and Active Metastates of Memory

When you learn to use a tool (for example, the robot in figure 6.10A), the motor memory can be reactivated when you return a day later and are asked to use it again. For example, if during the experience with the robot you felt a force that pushed your hand to the right on day 1, upon your return on day 2 on your very first movement you will produce a motor output that pushes the robot handle to the left, in apparent anticipation of the force field you experienced yesterday. However, your experience with the robot on day 1 was rather brief, perhaps an hour long. For the rest of the day you continued to make reaching movements without the tool. Presumably, the motor memory that you acquired regarding the novel tool during those brief training trials was "deactivated" for the rest of the day, and then "reactivated" when on day 2 you needed to move the handle of the robot again. Therefore, it seems rational that motor memory should exist in two metastates: an active metastate in which the motor memory is being used to perform an action, and a passive metastate in which the memory is not engaged.

Until now, we have not considered the dissociation between these two hypothetical metastates of memory. When we talked about forgetting—for example, the period termed "retention" in figure 7.5A—we referred to an experiment in which motor output declined during error-clamp trials. This forgetting referred to decay in an activated memory, as the subjects in that experiment were using the tool as we measured their motor output. However, forgetting in the usual sense refers to changes in memory as a function of time, that is, changes to a passive memory. For example, if you learn to hit a tennis serve as a result of an hour of instruction, you may not be able to reproduce the same performance the next day when you pick up a racquet. This reduction in performance is presumably a forgetting that affects the memory when it is in a passive metastate.

We can represent the active and passive properties of memory as metastates: When the memory is activated, the rules that describe the state transitions may be different from when it is not activated. It is easy to see that forgetting in the active memory, that is, state changes as defined in error-clamp trials, cannot be the same as time-dependent forgetting when the memory is passive. For example, consider the decline in motor output for group 1 in figure 7.5B. These subjects learned to move the robot to a single direction for sixty trials, and were then placed in a block of sixty error-clamp trials. Their motor output declined to near baseline during these sixty postadaptation trials. The amount of time it takes to perform these trials is about 2–3 minutes. Experiments show that if we waited three months after a comparable period of initial training, subjects would show savings (Shadmehr and Brashers-Krug, 1997). That is, the decline in the active metastate of the memory is not the same as the decline that takes place during the same period of time in a passive metastate of memory.

Returning to the state estimation framework, this implies that the state transition matrix for the active metastate A_a must differ from the state transition matrix for the passive metastate A_p.

$$\begin{aligned} &\text{active meta-state} \begin{cases} \mathbf{x}^{(t+\Delta)} = A_a \mathbf{x}^{(t)} + \boldsymbol{\varepsilon}_a \\ y^{(t)} = \mathbf{c}^T \mathbf{x}^{(t)} + \eta_a \end{cases} \\ &\text{passive meta-state} \left\{ \mathbf{x}^{(t+\Delta)} = A_p \mathbf{x}^{(t)} + \boldsymbol{\varepsilon}_p \right. \end{aligned} \quad (8.6)$$

In the active metastate, we can make observations and use prediction errors (differences between $y^{(t)}$ and $\hat{y}^{(t)}$) to update our estimate of the perturbation state. In the passive metastate, the states continue to change but we cannot make observations. (We made the superscripts in equation 8.6 in terms of time, rather than trial, so that the matrices A_a and A_p could be compared. In effect, we will assume that a "trial" takes a constant amount of time, and use this calibration to represent changes in the states of the passive metastate.)

Sarah Criscimagna-Hemminger and Reza Shadmehr (2008) performed an experiment that estimated A_a and A_p. They assumed a two-state model of motor memory (fast and slow states) and trained people in a reaching task with a robot that perturbed the limb with force fields. They had two groups of subjects: group A, which trained for hundreds of trials on field A, and group A + B, which had the same training in A but then briefly was trained in the opposite field B until motor output returned to baseline. We can assume that for group A, most of the memory by the end of the training is of the slow variety, whereas for Group A + B, there is an equal amount of slow and fast memories (canceling each other, reflected in the fact that by end of training in B, motor output was at baseline). They then waited a variable period of time and measured performance in error-clamp trials. They observed that at 0 and 2 minutes after training, motor output of the A + B group rose from zero and converged onto the motor output of the A group (figure 8.7A). That is, there was a spontaneous recovery, indicating resistance of the slow state to change. This implies that at 0 and 2 minutes after acquisition of B, the "fast" memory (of B) rapidly decayed within thirty trials. However, at 10 minutes and beyond, motor output of the A + B group no longer started at zero, but at a significantly higher value. Importantly, the performance of A and A + B groups no longer converged within thirty trials. That is, within a few minutes after acquisition, the fast memory of B no longer decayed away within thirty trials. It had somehow gained resistance to the error-clamp trials.

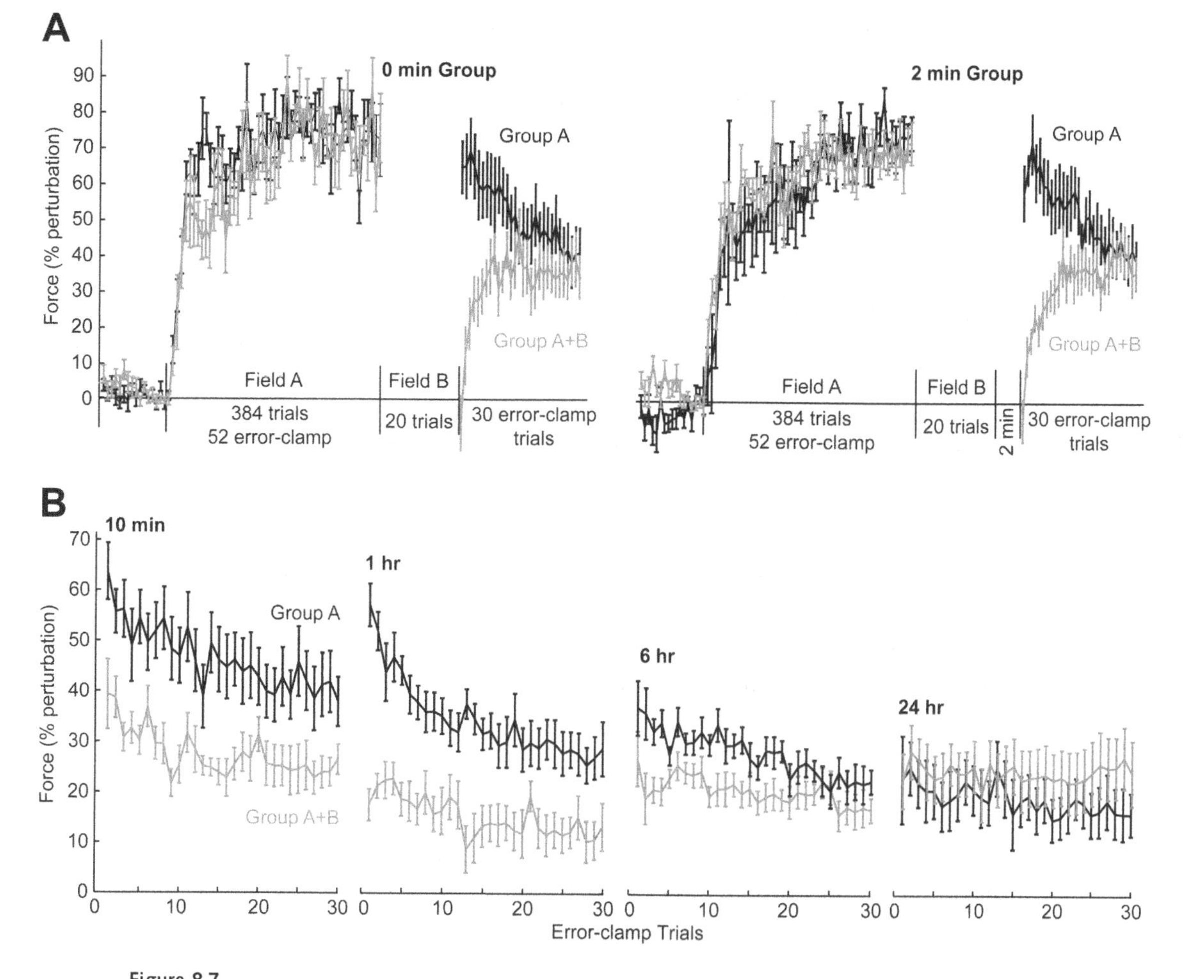

Figure 8.7
Effect of time passage on the spontaneous recovery of motor memory. The task was a reaching movement while holding the handle of a robotic arm, as in figure 8.4B. The data describes force produced by subjects during reaching movements as a percentage of the ideal force perturbation. All data are measurements in error-clamp trials. Group A trained on field A for 384 trials and was then presented with 30 error-clamp trials at a particular time after completion of training. Group A + B trained in field A and then immediately for a brief period in field B (field B had forces that were opposite that of field A). (A) The 0 min group: immediately after completion of training in A or A + B, performance was assayed in error-clamp trials. Group A + B exhibited spontaneous recovery. In the 2 min group, two minutes of delay was added to the end of completion of A or A + B training before performance was assayed in error-clamp trials. Similar to the 0 min interval, the 2 min interval also showed spontaneous recovery that started near zero and rose to meet the A only group. (B) As the interval between completion of training and error-clamp trials increased, the pattern of spontaneous recovery changed. The changed pattern suggested that whereas at 0 and 2 min after acquisition, the memory of B was fragile and declined to zero within 30 trials, with further passage of time some memory of B decayed away but the rest gained resistance and no longer decayed to zero within the same number of trials. (From Criscimagna-Hemminger and Shadmehr, 2008.)

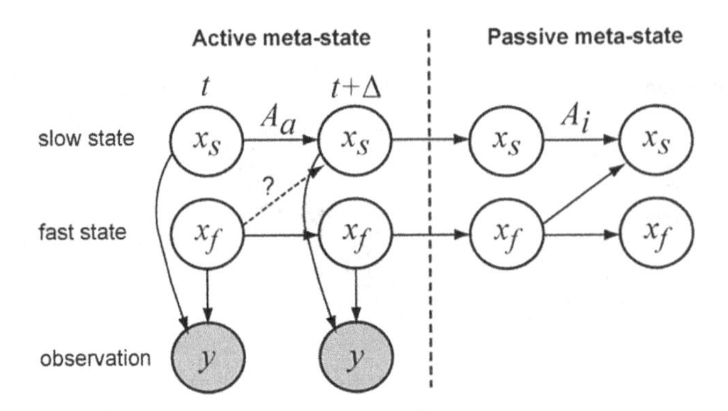

Figure 8.8
A model of the effect of motor memory under passive (the task is not being performed) and active (the task is being performed) processes, as in equation (8.6). In the passive metastate, some of the fast state transitions into a slow state.

The resulting generative model that fitted the data well is shown in figure 8.8. The state transition matrix in the active metastate had values of $A_a = [0.984, 0; 0, 0.855]$. The state transition matrix in the passive metastate had values of $A_p = [0.9999, 0.0043; 0, 0.983]$, where $\Delta = 6$ sec for a typical intertrial interval. (The matrix elements may look similar, but keep in mind that six hours has 3,600 "trials," so these numbers are raised to very large powers and a small difference matters.) The passage of time in the active state was far more destructive to the state of memory than the same passage of time in the passive metastate. That is, there was more forgetting in the active metastate. Furthermore, note that A_p had a nonzero off-diagonal term. This indicates that in the passive metastate, some of the fast state became a slow state with passage of time. This transition of the fast into a slow state during the time when the memory was not activated is an example of a transformation that makes it resistant to change, a phenomenon that is called consolidation.

In summary, these results suggest that error-driven learning not only engaged processes that adapt with multiple timescales, but that once practice ends and the memory is placed in a passive state, passage of time transforms some of the fast states into slower states that resist change when the task again resumes and the memory is reactivated.

8.5 Protection of Motor Memories

When a period of adaptation training is followed by de-adaptation, does the brain protect the memory that was acquired during adaptation, or do the prediction errors during de-adaptation destroy the previously acquired memory, producing effective unlearning? This simple question has been surprisingly difficult to answer. While saccade experiments in monkeys such as those shown in figure 8.1, and classical conditioning experiments such as those shown in figure 8.2, would suggest that adaptation memories are protected during de-adaptation, there is also significant data that does not agree with this view.

For example, Graham Caithness and colleagues (2004) trained volunteers in a reaching task that for some groups involved visuomotor rotation and for other groups involved force fields that perturbed motion of the arm. They adapted the volunteers to a perturbation for n trials, and then reversed the perturbation and continued the de-adaptation training for an equal number of n trials. When they retested the volunteers in the original perturbation, they found that performance was no different than naïve—that is, there was no evidence of savings. They write, "When people successively encounter opposing transformations (A then B) of the same type (e.g., visuomotor rotations or force fields), memories related to A are reactivated and then modified while adapting to B." They concluded that memory of adaptation was unprotected during de-adaptation, resulting in catastrophic destruction.

A year later, John Krakauer and colleagues (2005) repeated some of the experiment performed by Caithness and colleagues and found different results. They trained volunteers in a reaching task that perturbed the relationship between hand position and cursor position via a visuomotor rotation. They observed that when adaptation to a perturbation was followed by adaptation to the reverse-perturbation, in some conditions the relearning of the initial perturbation was faster (these conditions involved either a large number of trials in which subjects adapted to the perturbation, or passage of time between initial adaptation and the de-adaptation episode). Because the experiment uncovered evidence for savings, they write, "The persistence of interference across long intervals . . . is not definitive proof of erasure of initial learning." Based on their results the authors concluded that the inability to show savings in previous experiments was not due to destruction of the memory, but an inability of the nervous system to express that memory.

The published data demonstrates that in experiments in which adaptation is followed by de-adaptation, some researchers have found evidence of savings while others have not. Indeed, there is currently no consensus on the question of whether memories that are acquired during adaptation are protected during de-adaptation. One possibility is that savings, as defined by a faster relearning, is a weak assay of memory. Is there a different way to assay motor memory?

A recent experiment approached the issue from a different perspective. The idea was to use the error-clamp trials as a way to probe the contents of motor memory. Recall that in the error-clamp trials, movements are constrained so that no matter the motor output, there is no performance error and the subject is rewarded for her behavior. We saw earlier that when a long period of adaptation is followed by a brief period of de-adaptation, motor output in the following error-clamp trials is a mixture of both the memory acquired during adaptation and the memory acquired during de-adaptation (figure 8.4). For example, when people train for ~400 trials to reach in force field A, and then are trained for 20 trials in the opposite force field B, the motor output in the following error-clamp trials starts at zero but then rapidly rises toward *A* (figure 8.4B). This suggests that during the error-clamp trials the brain expresses both the memory of A and the memory of B (B rapidly decays, while A gradually decays, resulting in the phenomena of spontaneous recovery). Sarah Pekny, Sarah Criscimagna-Hemminger, and Reza Shadmehr trained people so that a long period of exposure to B preceded an equally long period of exposure to A (Pekny et al. 2011). They then briefly reexposed the subjects to B. This paradigm was termed BAb. They argued that if A destroys B, then the brief exposure to B after A should not produce savings, and more importantly, the pattern of spontaneous recovery that is expressed in the error-clamp trials show be the same in BAb and Ab. If, on the other hand, the memory of B is protected during adaptation to A, then the motor output in the error-clamp trials should show evidence for B in the BAb group as compared to the Ab group.

The results of this experiment are shown in figure 8.9. When reexposed to B, the BAb group relearned slightly faster than the Ab group (middle column, figure 8.9B). More importantly, the motor output of the BAb group in the error-clamp trials is clearly biased toward field B as compared to the Ab group. Because of the large difference between Ab and BAb groups in the error-clamp trials, it seems likely that in the BAb group the memory of B was not destroyed by the subsequent training in A. Similar results were obtained when training in B was followed by training in a null field (Bnb group vs. nb group, left column of figure 8.9). The Bnb group displayed a motor output in the error-clamp trials that was biased toward field B. Finally, when training in B was done gradually (right column, figure 8.9A), the resulting memory for B was even stronger, producing a large bias in the error-clamp trials toward B (right column, figure 8.9C). Taken together, the results of this experiment suggest that a long period of adaptation produces a memory that is at least partially protected from the large errors that could produce unlearning during de-adaptation. That is, the performance changes that take place during de-adaptation appear to be due to learning of a new memory.

But if this is the case, how does the brain decide when to spawn a new memory? One possibility is that sudden, large prediction errors are interpreted by the brain as a change in context, resulting in engagement of new resources to learn from prediction errors, rather than unlearn the previously acquired memory. Jeong-Yoon Lee and Nicolas Schweighofer (2009) proposed a model of adaptation in which motor output was supported by a single fast adaptive state (represented by scalar

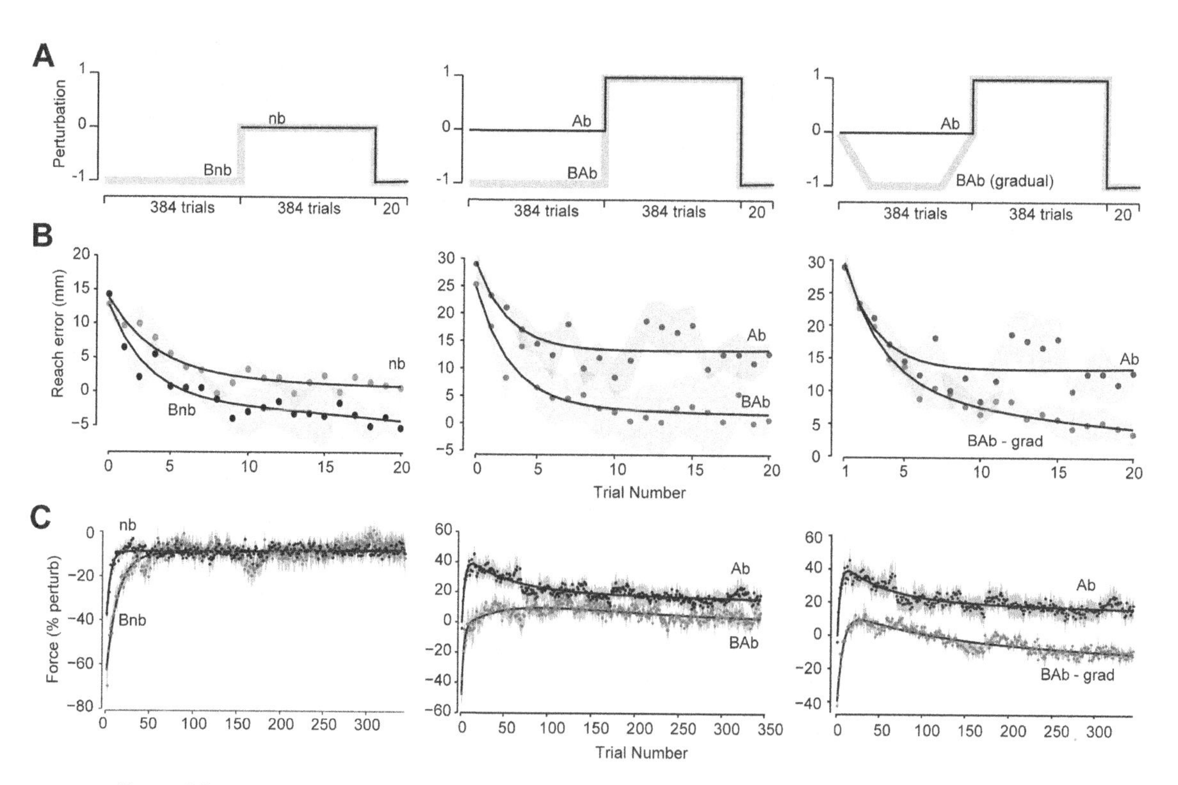

Figure 8.9
De-adaptation does not erase adaptation memories, as evidence by patterns of spontaneous recovery. The task was reaching in a force field. (A) Perturbation patterns. Left column: null followed by a brief period of adaptation to field B (termed nb), vs. training that starts with B, then washout, then brief readaptation to B (termed Bnb). Middle column: Ab vs. BAb. Right column: Ab vs. gradual BAb. (B) Performance during the twenty trials of B (relearning). There is a trend toward faster relearning for groups that had previously been exposed to B. However, this improved performance is marginal. (C) Following twenty trials of B, all movements were in error-clamp trials. In all cases, the prior training in B is reflected in the more negative forces that subjects produce in the error-clamp trials. During error-clamp trials, motor output is a mixture of A and B, with the strongest evidence for B exhibited in the gradual condition.

variable x_f), and multiple slow adaptive states (represented by a vector variable $\mathbf{x}_s$). The appropriate slow state was selected based on context $\mathbf{c}$:

$$\begin{aligned}
\hat{y}^{(n)} &= x_f^{(n)} + \mathbf{c}^{(n)T}\mathbf{x}_s^{(n)} \\
x_f^{(n+1)} &= a_f x_f^{(n)} + b_f\left(y^{(n)} - \hat{y}^{(n)}\right) \\
\mathbf{x}_s^{(n+1)} &= A_s\mathbf{x}_s^{(n)} + b_s\left(y^{(n)} - \hat{y}^{(n)}\right)\mathbf{c}^{(n)}
\end{aligned} \quad . \tag{8.7}$$

This model predicts that whereas the fast state is completely erasable (because it is not context selectable), the slow states can be afforded protection if the context changes. The lack of context dependence of fast states accounts for the fact that if training in A precedes training in B, performance in B is worse than naïve, a phenomenon called *anterograde interference*. Yet, the model can account for savings because context allows protection of the slow states. Lee and Schweighofer (2009) did not describe a method with which one might select the correct context. The mechanisms of spawning new memories and contextual selection of old ones remains poorly understood.

8.6 Multiple Timescales of Memory in the Cerebellum

The adaptation tasks that we have considered here are known to rely on the cerebellum. For example, damage to the cerebellum makes people and other animals generally impaired in their ability to adapt to perturbations in saccadic eye movements (Straube et al., 2001; Golla et al., 2008) or reaching movements (Smith and Shadmehr, 2004; Rabe et al., 2009; Criscimagna-Hemminger, Bastian, and Shadmehr, 2010). Are the multiple timescales of memory, savings, and spontaneous recovery reflected in the neural activity of the cerebellum? Unfortunately, relatively little is known regarding the neural changes in the cerebellum during saccade and reach adaptation. However, classical conditioning also exhibits savings and spontaneous recovery, and for one form of classical conditioning there is substantial data on the role that the cerebellum plays in encoding the memory. Here we will briefly review this data, as it sheds light on the neural basis of the fast and slow processes that potentially support learning.

Mike Mauk and his students have been systematically identifying the neural basis of a form of classical conditioning in rabbits. In a typical experiment, a rabbit is presented with a tone for 550 ms (the conditioned stimulus, CS). The rabbit does not respond to this tone. Next, they apply a small current to a region around one eye of the rabbit for 50 ms at the end of the tone (the unconditioned stimulus, US). In response to the mild shock, the rabbit closes its eyes. With repeated trials, the rabbit learns to associate the CS with the US and closes its eyes at around 500 ms, right before the US is given. This change in behavior is mediated via changes in two places in the cerebellum: the cerebellar cortex, and the interpositus nucleus (one of the deep cerebellar nuclei). In the cerebellar cortex, the tone (CS) is conveyed via activity that reaches the Purkinje cells via mossy fibers (figure 8.10). The US is conveyed via activity that reaches the Purkinje cells via climbing fibers from the inferior olive. The coincidence of these two signals produces reduction in the strength of the mossy fiber-Purkinje cell synapse, inducing plasticity in the cerebellar cortex. In the cerebellar nuclei, the changes in the Purkinje cell discharge combine with input from the mossy fibers to produce plasticity in the interpositus nucleus.

Tatsuya Ohyama and Mike Mauk (2001) performed an experiment that suggests that the cerebellar cortical plasticity takes place fairly rapidly after start of training, whereas interpositus nucleus plasticity takes place slowly and follows the changes

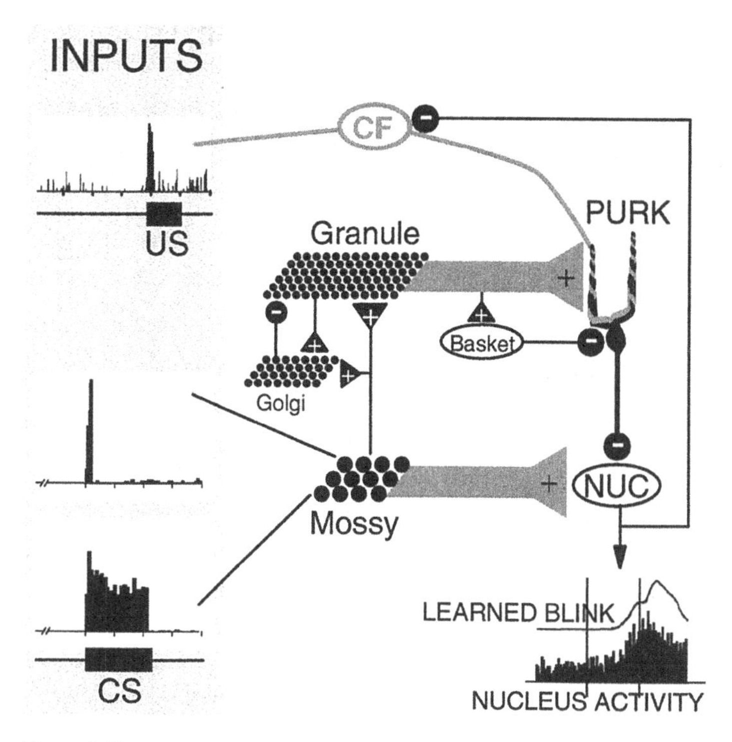

Figure 8.10
Circuitry of the cerebellum and its relationship to eyelid classical conditioning. During the task, mossy fibers convey information from the CS. This input affects granule cells, and ultimately the Purkinje cells, which inhibit the cells in the deep cerebellar nucleus. Climbing fibers convey information about the US. Learning is driven by changes in two locations: the synapse between granule cell and Purkinje cell, and the synapse between the mossy fiber and deep cerebellar nucleus cell. (From Medina, Garcia, and Mauk, 2001.)

in the cerebellar cortex. In their experiment, in phase 1 the rabbits were trained briefly with a long-duration CS. The training was too brief to produce a learned eye-blink response. However, Ohyama and Mauk posited that synaptic plasticity took place in the cerebellar cortex (reduction in the granule cell-Purkinje cell synaptic strength, figure 8.11B), but because there was still no plasticity at the nucleus, this change in the cerebellar cortex could not be expressed in terms of change in behavior. In phase 2, the rabbits were trained for a many trials with a short-duration CS. This training produced robust eye-blink response to the CS, which was posited to include plasticity in both the cerebellar cortex and the nucleus (reduction in the granule cell-Purkinje cells synaptic strength, increase in the mossy fiber-nucleus cell strength, figure 8.11C). In the test period, the animals were presented with a very long duration CS. The model predicted that the animals would produce a double blink (figure 8.11D), and this was indeed confirmed in the behavioral data. The

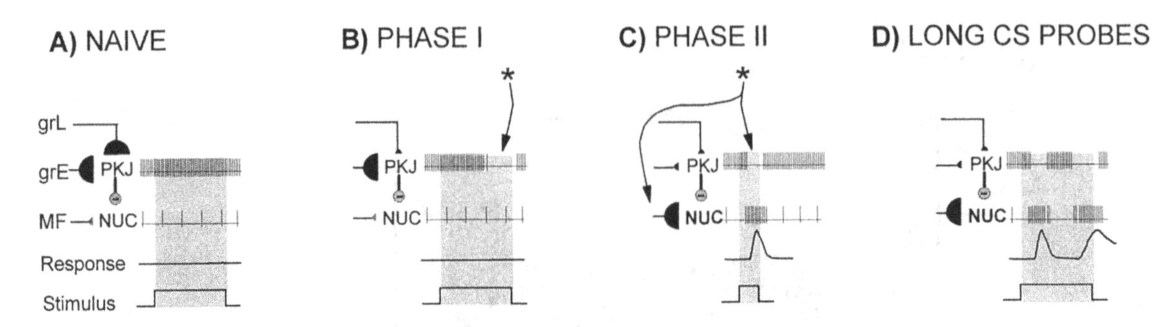

Figure 8.11
A schematic representation of the two timescales of learning in the cerebellum. (A) In the naïve animal, Purkinje cells activity is not affected by the stimulus (CS). (B) In phase 1, a few trials of long-duration CS is presented with US. There is no behavioral response. Yet, the Purkinje cells show a change in their response (reduced activity near offset of CS). (C) In phase 2, many trials of short-duration CS is presented with US. There is robust behavioral response. This response is due to plasticity in the granule cell-Purkinje cells synapse and mossy fiber-nucleus cell synapse. (D) In the CS probe trials, a long duration CS is presented without US. The behavioral response is a double blink. (From Ohyama and Mauk, 2001.)

double blink response suggests that the brief training (to the long duration CS, figure 8.11B) was sufficient to produce fast adaptation in the cerebellar cortex, which could subsequently be expressed when adaptation took place in the cerebellar nucleus. Based on these data, it would appear that the neural basis of fast and slow adaptive processes may be distinct, with a fastlike process residing in the cerebellar cortex and a slowlike process residing in the cerebellar nucleus.

To investigate the neural basis of savings after extinction, Javier Medina, Keith Garcia, and Mike Mauk (2001) trained their rabbits for five days, ensuring that they produced a robust eye-blink response to the CS (A5, figure 8.12). At this point, they injected a drug (picrotoxin, PTX) into the deep nucleus that was an antagonist to GABA, the neurotransmitter that Purkinje cells release to communicate with the cells in the deep nuclei. This effectively disconnected the cerebellar cortex from the cerebellar nucleus. They noticed that the rabbit still produced a response to the tone. However, this response was at the start of the tone and not near its end (PTX on A5, figure 8.12). Therefore, there was a memory in the deep cerebellar nucleus that had formed due to training, but this memory did not have the temporal specificity that was present when the whole network was intact. They now trained the rabbits for forty-five days of CS alone. That is, they performed extinction training. After fifteen days of extinction training, the rabbits no longer responded to the CS (E15, figure 8.12). However, when they injected the GABA antagonist into the cerebellar nucleus, the animals once again started responding to the tone. That is, despite the extinction of behavior, a component of the original CS-US memory was present in the deep nucleus. This component appeared to be unaffected by the fifteen days of extinction training. After forty-five days of extinction training, still a small amount

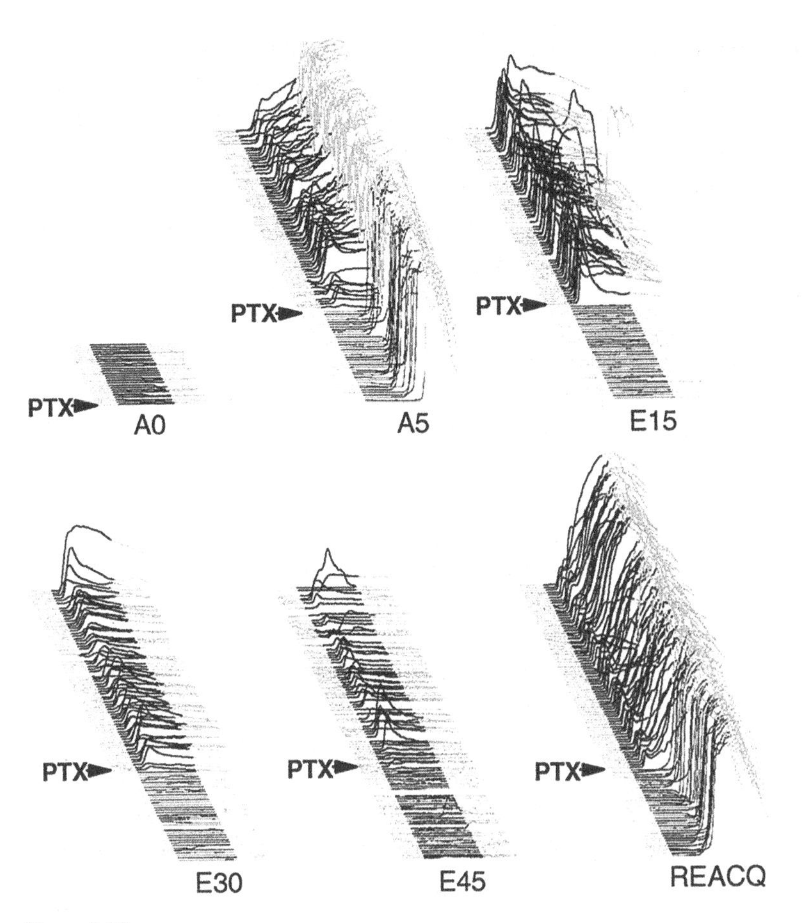

Figure 8.12
Eyelid traces throughout acquisition and extinction for one animal. A0: before start of training. A5: Fifth day of CS-US training. E15: Fifteenth day of extinction training (CS only). The location of the black arrow indicates the time during the session when picrotoxin (PTX) was infused into the cerebellar nucleus. The black portion of each trace represents the time when the CS was present. (From Medina, Garcia, and Mauk, 2001.)

of response to CS remained in the cerebellar nucleus (E45, figure 8.12). After these forty-five days of extinction, Medina and colleagues retrained the animals with the CS-US pair and observed that relearning was faster than naïve (REACQ, figure 8.12). The improvement in relearning with respect to naïve was correlated with the magnitude of the response that was present in the cerebellar nucleus, that is, the greater the memory in the nucleus, the greater the behavioral measure of savings.

Together, these results demonstrate that the multiple timescales of memory during classical conditioning are partly due to different rates of learning in the cerebellar cortex and nuclei. The faster rate of learning appears in the cerebellar cortex. When adaptation is followed by extinction, the memory in the cerebellar nucleus shows very strong resistance to change, and appears to be responsible for the savings that is exhibited during relearning.

Summary

When people and other animals adapt their movements to a perturbation, removal or reversal of that perturbation may return behavior to baseline, but this does not imply that the change in the direction of errors erases the acquired memory. The evidence for this comes from savings (faster relearning), and spontaneous recovery of the acquired memory in the time period after reversal of the perturbation. One way to account for these behaviors is to assume that the learner responds to the prediction errors with adaptive processes that have multiple timescales. Some of the processes have fast timescales, learning rapidly from error but also exhibiting rapid forgetting. Some of the processes have slow timescales, learning gradually from error but exhibiting strong retention and resistance to forgetting. Sudden large errors in performance may act as a cue that context of the task has changed, facilitating spawning of new memories and protection of old ones. Much of the ability to adapt our movements to perturbations is due to the cerebellum. The learning in this structure exhibits a faster rate of plasticity in the cerebellar cortex than the cerebellar nuclei. During extinction, the slowly acquired memories in the nuclei exhibit strong resistance to change so that despite return of behavior to baseline, the memory in the cerebellar nucleus is maintained and appears to form the basis for savings during relearning.

9 Building Generative Models: Structural Learning and Identification of the Learner

Suppose that it is true that when the brain generates a motor command, it also predicts the sensory consequences. The state estimation framework that we developed in the last few chapters suggests that when the actual sensory measurements arrive, the brain should combine what it predicted with what it measured to form an estimate of the state of the body (and whatever else contributes to things that it can sense). By doing so, it can estimate this state better than if it were to rely on the sensors alone. For example, we can estimate the location of a visual stimulus better if we can predict its position as well as see it, as compared to if we can only see it but not predict it (Vaziri, Diedrichsen, and Shadmehr, 2006). However, this last statement is true only if the predictions that the brain makes are unbiased.

Being an unbiased predictor is actually rather difficult, because our body and the objects we interact with have dynamics: Inputs (motor commands) and outputs (sensory measurements) are not related via some static mapping. Rather, dynamics implies that state of the system changes both as a function of time (passive dynamics) and as a function of the input (active dynamics). What is more, the relationship between inputs and outputs can change (for example, muscles can fatigue), making it necessary for us to adjust our predictions. We saw that one way our brain can learn to make better predictions is to assume a generative model that describes the relationship between inputs and outputs via some hidden states (for example, the fatigue state of the muscle can be one of these states). The prediction errors can guide the process of estimating these hidden states, thereby improving the predictions.

The elephant in the room is the generative model itself. Where does the brain get such a model? That is, how does our brain go about finding (presumably, learning) a model that has a *structure* or topology that can in principle represent the relationship between inputs (e.g., motor commands) and observations (sensory measurements)? To predict accurately the sensory consequences of a motor command, we need to be able to have a model that can approximate behavior of a dynamical system. We are no longer talking about the hidden states of some generic model, but rather the topology of the model itself. How do we discover this topology, the structure of the dynamical system that we wish to control? The problem that our brain faces is one of system identification.

Consider the problem of learning to control motion of a table tennis paddle. It would be useful if during practice, we learned a model that had a structure with hidden states that could also help us with learning control of a tennis racket. This should be possible because the two objects are rigid-body inertial systems with dynamics that are similar in structure. If we could somehow discover this structure while playing table tennis, it could vastly speed up our learning of tennis. In this chapter, we will consider this problem of structural learning.

How is structural learning different than the learning that we had considered in the last few chapters? The problem of learning an internal model, a model that can accurately predict sensory consequence of movements, can be approached from two perspectives. In the perspective that we had considered in the last few chapters (which we called the state-estimation perspective), we started with a specific model that could in principle represent the relationship between inputs and outputs, and then estimated the hidden states of this model based on our prediction errors. We did not consider where such a model with these specific hidden states might come from. In the perspective that we will consider here, the problem is to discover this structure from the relationships (input–outputs) that we observe. Structural learning relies on the long-term history of the motor commands and their sensory consequences.

9.1 Structure of Dynamics for Two Example Systems

Suppose that you want to build a model that can help you predict motion of your arm. Because you tend to hold different kinds of objects in your hand (and these objects change the arm's dynamics), you want this model to be flexible enough so that it can readily predict motion in different scenarios. Let us sketch what this model might look like.

For simplicity, let us assume that the arm will move in the horizontal plane. The relationship between torques at the shoulder τ_1 and elbow τ_2 and motion (joint angular position q_1 and q_2, and their derivatives) for the system shown in figure 9.1A is:

$$\begin{aligned}\begin{bmatrix}\tau_1\\ \tau_2\end{bmatrix} &= \begin{bmatrix} a_1+2a_3\cos(q_2)+a_2 & a_2+a_3\cos(q_2)\\ a_2+a_3\cos(q_2) & a_2\end{bmatrix}\begin{bmatrix}\ddot{q}_1\\ \ddot{q}_2\end{bmatrix}\\ &\quad+\begin{bmatrix}-a_3\dot{q}_2\sin(q_2) & -a_3(\dot{q}_1+\dot{q}_2)\sin(q_2)\\ a_3\sin(q_2)\dot{q}_1 & 0\end{bmatrix}\begin{bmatrix}\dot{q}_1\\ \dot{q}_2\end{bmatrix}.\end{aligned} \tag{9.1}$$

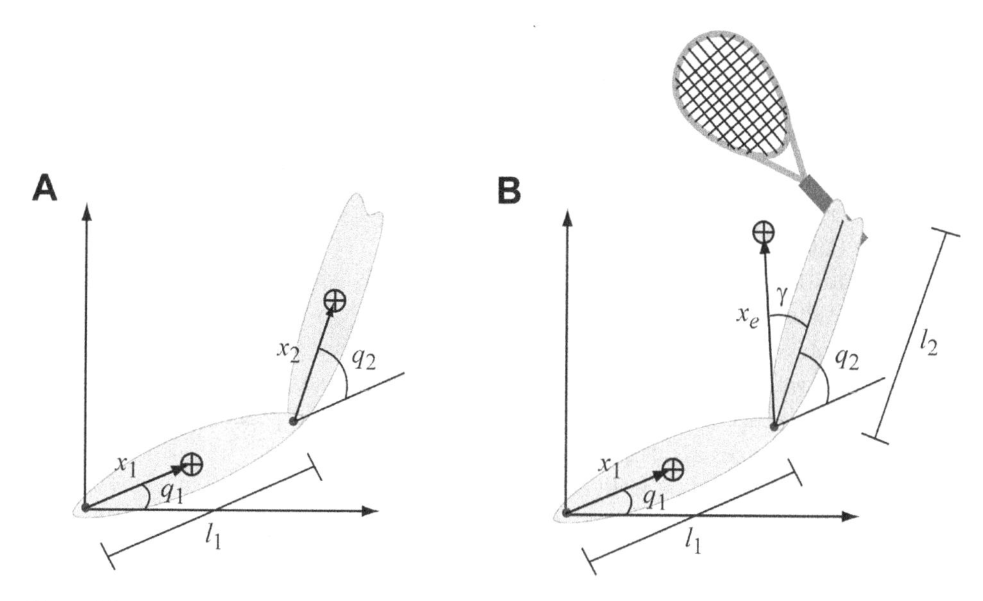

Figure 9.1
A schematic of an arm moving in the horizontal plane with and without a tennis racquet. The circled plus signs indicate center of mass of each link of the arm.

(If you are interested in how to derive these equations, check out the Introduction to Dynamics lecture notes at http://www.shadmehrlab.org/book/dynamics.pdf.) Equation (9.1) is called *inverse dynamics* because it maps states to forces. Equation (9.1) is basically the familiar $F = m\ddot{x}$ but written for the complicated system of figure 9.1, in coordinates of torques and joint rotations. *Forward dynamics* is the map from forces to change in states. In equation (9.1), the parameters a_i are constants that depend on the mass properties of the arm:

$$\begin{aligned} a_1 &= m_1 |x_1|^2 + I_1 + m_2 l_1^2 \\ a_2 &= m_2 |x_2|^2 + I_2 \\ a_3 &= m_2 l_1 |x_2| \end{aligned} , \tag{9.2}$$

where m_1 and m_2 are masses of the upper arm and forearm, $|x_1|$ and $|x_2|$ are lengths of the vector that points to the center of the mass of each segment, I_1 and I_2 are the moments of inertia of each segment, and l_1 is the length of the upper arm. We can rewrite equation (9.1) using vector notation:

$$\boldsymbol{\tau} = H(\mathbf{a},\mathbf{q})\ddot{\mathbf{q}} + \mathbf{c}(\mathbf{a},\mathbf{q},\dot{\mathbf{q}}). \tag{9.3}$$

Matrix H represents inertia of the arm. Inertia depends on parameter $\mathbf{a}$ (some combination of mass and link lengths of the arm, as specified in equation 9.2), and position of the arm $\mathbf{q}$. Vector $\mathbf{c}$ represents the centripetal and coriolis forces. These forces also depend on parameter $\mathbf{a}$, as well as position and velocity of the arm. A fundamental characteristic of equation (9.3) is that the torques that are due to acceleration are linearly separable from forces that are due to velocity. A second important characteristic is that the parameter $\mathbf{a}$ appears linearly, that is, torques are a linear function of the parameter that might change if we were to hold different objects in our hand.

To see this last point, let us consider what happens when you hold an object in your hand. When you hold an object in your hand, such as a cup of coffee, the addition of this mass increases m_2, I_2, and $|x_2|$. This means that the equations of motion for your arm when you are holding a cup has the same structure as when you are not holding the cup, with the only difference being in the parameters a_i. That is, holding a cup changes $\mathbf{a}$, but nothing else.

What happens when you hold a tennis racket? Unlike the coffee cup, the racket shifts the center of mass off the forearm (figure 9.1B), resulting in the following equations of motion:

$$\begin{bmatrix}\tau_1\\ \tau_2\end{bmatrix}=\begin{bmatrix}H_{11} & H_{12}\\ H_{21} & H_{22}\end{bmatrix}\begin{bmatrix}\ddot{q}_1\\ \ddot{q}_2\end{bmatrix}+\begin{bmatrix}-h\dot{q}_2 & -h(\dot{q}_1+\dot{q}_2)\\ h\dot{q}_1 & 0\end{bmatrix}\begin{bmatrix}\dot{q}_1\\ \dot{q}_2\end{bmatrix}, \tag{9.4}$$

where

$$\begin{aligned}H_{11}&=a_1+2a_3\cos(q_2)+2a_4\sin(q_2)\\ H_{12}&=H_{21}=a_2+a_3\cos(q_2)+a_4\sin(q_2)\\ H_{22}&=a_2\\ h&=a_3\sin(q_2)-a_4\cos(q_2)\end{aligned} \tag{9.5}$$

and

$$\begin{aligned}a_1&=m_1|x_1|^2+I_1+I_e+m_e|x_e|^2+m_e l_1^2\\ a_2&=m_e|x_e|^2+I_e\\ a_3&=m_e l_1|x_e|\cos(\phi)\\ a_4&=m_e l_1|x_e|\sin(\phi)\end{aligned} \tag{9.6}$$

Notice that switching from a cup (roughly a point mass) to a racket adds one parameter to our equation, but it maintains the fact that forces that depend on acceleration and velocity remain separable. Now if we were to switch from holding a tennis racket to a table-tennis paddle (or any other rigid object), once again all that changes are the parameters a_i, with no change in the structure of our relationship in equation (9.4). In sum, we see that physics implies regularity in the relationship between motion and forces. In particular, for our arm the forces that are produced due to acceleration and velocity are linearly separable. Furthermore, regardless of the rigid object that we may hold in our hand, the forces remain linear in terms of parameters representing masses, lengths, and so on.

Now suppose that we knew this structure and wanted to use that information to estimate the parameters a_i. From equation (9.3), we have:

$$\begin{aligned}\ddot{\mathbf{q}} &= -H^{-1}(\mathbf{a},\mathbf{q})\mathbf{c}(\mathbf{a},\mathbf{q},\dot{\mathbf{q}}) + H^{-1}(\mathbf{a},\mathbf{q})\boldsymbol{\tau} \\ &= G(\mathbf{a},\mathbf{q},\dot{\mathbf{q}},\boldsymbol{\tau})\end{aligned} \tag{9.7}$$

Equation (9.7) is parameterized by a small number of unknowns, the elements a_i of the vector $\mathbf{a}$. These are some of the hidden states of the system, which we can estimate from sensory observations. To use our state estimation framework, we need to linearize the relationship between sensory observations and the states that we wish to estimate. Let us label the states of our system as the column vector $\mathbf{x} = [\mathbf{q}, \dot{\mathbf{q}}, \mathbf{a}]$. Using equation (9.7), we can approximate $\ddot{\mathbf{q}}$ around our current estimate $\hat{\mathbf{x}}$:

$$\ddot{\mathbf{q}} = G(\hat{\mathbf{x}},\boldsymbol{\tau}) + \left.\frac{dG}{d\mathbf{x}}\right|_{\hat{\mathbf{x}}}(\mathbf{x}-\hat{\mathbf{x}}), \tag{9.8}$$

and then write the state update equation as a linear function of current state $\mathbf{x}(t)$ and some constant terms:

$$\begin{aligned}\mathbf{q}(t+\Delta) &= \mathbf{q}(t) + \dot{\mathbf{q}}(t)\Delta \\ \dot{\mathbf{q}}(t+\Delta) &= \dot{\mathbf{q}}(t) + \left.\frac{dG}{d\mathbf{x}}\right|_{\hat{\mathbf{x}}}\mathbf{x}(t)\Delta - \left.\frac{dG}{d\mathbf{x}}\right|_{\hat{\mathbf{x}}}\hat{\mathbf{x}}\Delta + G(\hat{\mathbf{x}},\boldsymbol{\tau})\Delta. \\ \mathbf{a}(t+\Delta) &= A\mathbf{a}(t) + \boldsymbol{\varepsilon}_a\end{aligned} \tag{9.9}$$

The terms on the second line of equation (9.9) are simply an expansion of velocity in terms of acceleration, that is, $\dot{\mathbf{q}}(t+\Delta) = \dot{\mathbf{q}}(t) + \ddot{\mathbf{q}}(t)\Delta$. The last line in equation (9.9) is the state equation regarding the parameters $\mathbf{a}$, which includes state noise. The measurement equation is:

$$\mathbf{y}(t) = L\mathbf{x}(t) + \boldsymbol{\varepsilon}_y. \tag{9.10}$$

In equation (9.10), the matrix L indicates the state variables that we can observe (typically position and velocity). Having approximated our nonlinear system with a linear equation, we can now use state estimation techniques to form an estimate $\hat{\mathbf{x}}$, which includes the parameter that we are looking for, $\hat{\mathbf{a}}$.

The point is that there exists a model that has the appropriate structure for the dynamics that we wish to approximate. In this model, the problem of representing dynamics of different objects that we might hold in our hand reduces to changes in the space spanned by the vector $\mathbf{a}$. Within this four-dimensional space we can represent arm dynamics for a large range of objects.

9.2 Evidence for Learning a Structural Model

Daniel Braun, Ad Aertsen, Daniel Wolpert, and Carsten Mehring (2009) argued that if during learning people adjust not just the parameters of a default generative model but learn a new one, then this new model should help them with learning of tasks that are also supported by the same structure (but perhaps with different parameter values). The idea is that if you learn to ride one kind of bike (say a mountain bike), you should be able to rapidly learn to ride a race bike. If you learn to play table tennis, it should also help you learn tennis. To test for this structure specific facilitation, they exposed a group of subjects to a long period of visuomotor adaptation during reaching. The caveat was that from trial to trial, the perturbations were uniformly distributed between –90 degrees and +90 degrees. If learning were merely an updating of parameter values for an existing model, then random perturbations should produce no sustained change in these parameter values (because the mean perturbation is zero). After this period of random rotation perturbations, they presented subjects with a constant +60-degree rotation. They found that compared to naïve subjects, the people with the prior exposure to the random rotations were much better in learning the constant rotation (figure 9.2, random rotation group). Next, they considered prior

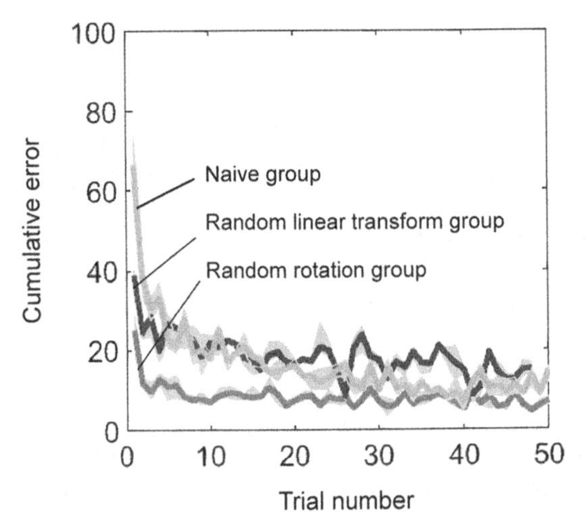

Figure 9.2
Subjects trained in a long sequence of trials in which a rotation was imposed on the motion of a cursor. The trial-to-trial distribution of the rotation perturbation was random. After this period of training, performance was measured in a sequence of trials in which the perturbation was a constant +60-degree rotation. Their data is shown in the preceding plot in the Random rotation group. Performance was significantly better than a naïve group that had prior training in trials for which there were no perturbations. Performance was also better than another group in which prior training was in a perturbation that had rotation, shearing, and scaling (labeled as Random linear transform). (Figure from Braun et al., 2009.)

training that had random errors, but not errors of the rotation class. In this group, the random errors were generated in a long set of trials in which the movements were transformed by a rotation, a shearing, and a scaling (a linear transformation). This group performed significantly worse than the group that had prior training in the rotation perturbation (figure 9.2, random linear transform group).

Therefore, the prior exposure to a rotation perturbation, despite being random and unlearnable, appeared to significantly improve learning rates for a member of the same perturbation class. This is consistent with the idea that during exposure to the rotation class of perturbations, people learned the structure of the perturbation, despite being unable to learn the specific parameter of that perturbation, and this structure aided their learning when they subsequently were exposed to a constant rotation.

9.3 Nonuniqueness of the Structure

The problem that we have been discussing is closely related to the general system identification problem where inputs $\mathbf{u}^{(1)}, \mathbf{u}^{(2)}, \ldots$ are given to a system and outputs $\mathbf{y}^{(1)}, \mathbf{y}^{(2)}, \ldots$ are measured and the intention is to identify the dynamics of that system. We do not know how to solve this problem in general, but we can solve it if we assume that the system that we are trying to identify has a linear structure. Our objective here is to show the computations that are an essential part of finding the correct structure.

We will assume that dynamics of the system that we are trying to model is of the form shown below, with hidden states $\mathbf{x}$ that are of unknown dimensionality, and with unknown matrices A, B, C, and D:

$$\begin{aligned} \mathbf{x}^{(n+1)} &= A\mathbf{x}^{(n)} + B\mathbf{u}^{(n)} + \boldsymbol{\varepsilon}_x^{(n)} \qquad \boldsymbol{\varepsilon}_x \sim N(0,Q) \\ \mathbf{y}^{(n)} &= C\mathbf{x}^{(n)} + D\mathbf{u}^{(n)} + \boldsymbol{\varepsilon}_y^{(n)} \qquad \boldsymbol{\varepsilon}_y \sim N(0,R) \end{aligned}.$$

The objective is to find the structure, that is, matrices A, B, C, and D, and noise properties Q and R, that describes the relationship between the sequence of inputs $\mathbf{u}^{(1)}, \mathbf{u}^{(2)}, \ldots$ that we gave to the system and the sequence of measurements $\mathbf{y}^{(1)}, \mathbf{y}^{(2)}, \ldots$ that we made from the system. We will identify the generative model that produced the data that we observed.

An elegant technique for identifying the system in equation (9.8) is via *subspace analysis* (van Overschee and De Moor, 1996). Let us begin with the deterministic problem in which our system is without noise. That is, let us suppose that the structure that we wish to identify has the following form:

$$\begin{aligned} \mathbf{x}^{(n+1)} &= A\mathbf{x}^{(n)} + B\mathbf{u}^{(n)} \\ \mathbf{y}^{(n)} &= C\mathbf{x}^{(n)} + D\mathbf{u}^{(n)} \end{aligned}. \tag{9.11}$$

We will provide a sequence of inputs $\mathbf{u}^{(1)}, \mathbf{u}^{(2)}, \ldots$ to this system, and measure outputs $\mathbf{y}^{(1)}, \mathbf{y}^{(2)}, \ldots$. The parameters that we are searching for are A, B, C, D, and $\mathbf{x}^{(1)}$ (the initial conditions). However, there are an infinite number of parameters that can give us this exact input–output sequence. That is, there is no unique solution to our problem. To see this, from equation (9.11) we can write:

$$\begin{bmatrix} \mathbf{x}^{(2)} & \mathbf{x}^{(3)} & \cdots & \mathbf{x}^{(p+1)} \\ \mathbf{y}^{(1)} & \mathbf{y}^{(2)} & \cdots & \mathbf{y}^{(p)} \end{bmatrix} = \begin{bmatrix} A & B \\ C & D \end{bmatrix} \begin{bmatrix} \mathbf{x}^{(1)} & \mathbf{x}^{(2)} & \cdots & \mathbf{x}^{(p)} \\ \mathbf{u}^{(1)} & \mathbf{u}^{(2)} & \cdots & \mathbf{u}^{(p)} \end{bmatrix}. \tag{9.12}$$

In equation (9.12), we have arranged the vectors and matrices to form new matrices. For example, the matrix on the left side of equation (9.12) is composed of columns with elements that are specified by column vectors $\mathbf{x}$ and $\mathbf{y}$. If we now multiply the state equation by an arbitrary but invertible matrix T, we have:

$$\begin{bmatrix} T & 0 \\ 0 & I \end{bmatrix} \begin{bmatrix} \mathbf{x}^{(2)} & \mathbf{x}^{(3)} & \cdots & \mathbf{x}^{(p+1)} \\ \mathbf{y}^{(1)} & \mathbf{y}^{(2)} & \cdots & \mathbf{y}^{(p)} \end{bmatrix} = \begin{bmatrix} T & 0 \\ 0 & I \end{bmatrix} \begin{bmatrix} A & B \\ C & D \end{bmatrix} \begin{bmatrix} \mathbf{x}^{(1)} & \mathbf{x}^{(2)} & \cdots & \mathbf{x}^{(p)} \\ \mathbf{u}^{(1)} & \mathbf{u}^{(2)} & \cdots & \mathbf{u}^{(p)} \end{bmatrix}. \tag{9.13}$$

In equation (9.13), the matrix I is an identity matrix of appropriate size. Simplifying equation (9.13), we have:

$$\begin{bmatrix} T\mathbf{x}^{(2)} & T\mathbf{x}^{(3)} & \cdots & T\mathbf{x}^{(p+1)} \\ \mathbf{y}^{(1)} & \mathbf{y}^{(2)} & \cdots & \mathbf{y}^{(p)} \end{bmatrix} = \begin{bmatrix} TA & TB \\ C & D \end{bmatrix} \begin{bmatrix} \mathbf{x}^{(1)} & \mathbf{x}^{(2)} & \cdots & \mathbf{x}^{(p)} \\ \mathbf{u}^{(1)} & \mathbf{u}^{(2)} & \cdots & \mathbf{u}^{(p)} \end{bmatrix}. \tag{9.14}$$

Now let us represent the right side of equation (9.14) in the same transformed space of the left side:

$$\begin{bmatrix} T\mathbf{x}^{(2)} & T\mathbf{x}^{(3)} & \cdots \\ \mathbf{y}^{(1)} & \mathbf{y}^{(2)} & \cdots \end{bmatrix} = \begin{bmatrix} TA & TB \\ C & D \end{bmatrix} \begin{bmatrix} T^{-1} & 0 \\ 0 & I \end{bmatrix} \begin{bmatrix} T & 0 \\ 0 & I \end{bmatrix} \begin{bmatrix} \mathbf{x}^{(1)} & \mathbf{x}^{(2)} & \cdots \\ \mathbf{u}^{(1)} & \mathbf{u}^{(2)} & \cdots \end{bmatrix}. \tag{9.15}$$

Rearranging the above equation provides us with the crucial observation that the same input–output data can be generated with very different parameters TAT^{-1}, TB, CT^{-1}, and D, and states $T\mathbf{x}$:

$$\begin{bmatrix} T\mathbf{x}^{(2)} & T\mathbf{x}^{(3)} & \cdots & T\mathbf{x}^{(p+1)} \\ \mathbf{y}^{(1)} & \mathbf{y}^{(2)} & \cdots & \mathbf{y}^{(p)} \end{bmatrix} = \begin{bmatrix} TAT^{-1} & TB \\ CT^{-1} & D \end{bmatrix} \begin{bmatrix} T\mathbf{x}^{(1)} & T\mathbf{x}^{(2)} & \cdots & T\mathbf{x}^{(p)} \\ \mathbf{u}^{(1)} & \mathbf{u}^{(2)} & \cdots & \mathbf{u}^{(p)} \end{bmatrix}. \tag{9.16}$$

In comparing equation (9.16) with equation (9.12) we arrive at two important ideas. First, given an input–output sequence of data, it is not possible to estimate a unique set of parameters for equation (9.11) because there are an infinite number of equally valid candidates. Therefore, we need to change our objective and settle for finding one set of parameters among this infinite set. The second insight is that we can find a parameter set if we could estimate the state sequence $\mathbf{x}$ in any arbitrary transformed space $T\mathbf{x}$. As we will show, when we pick this transformed space carefully, the problem lends itself to a closed form solution.

9.4 Subspace Method: Intuitive Ideas

In equation (9.11), output $\mathbf{y}$ is a linear function of the state $\mathbf{x}$ and input $\mathbf{u}$. If we could somehow remove the effects of the sequence of inputs $\mathbf{u}^{(1)},\mathbf{u}^{(2)},\ldots,\mathbf{u}^{(p)}$ from the sequence of outputs $\mathbf{y}^{(1)},\mathbf{y}^{(2)},\ldots,\mathbf{y}^{(p)}$, we would be left with a sequence that is a linear transformation on the sequence of states, that is, $C\mathbf{x}^{(1)},C\mathbf{x}^{(2)},\ldots,C\mathbf{x}^{(p)}$. If we had a linear transformation of the sequence of states, our problem would be trivial, as we could then recover the parameters that produced those states from equation (9.16). The subspace method is a geometric technique that precisely accomplishes this goal (van Overschee and De Moor, 1996).

When we project vector $\mathbf{a}$ onto $\mathbf{b}$, we get a vector in the direction of $\mathbf{b}$ with the magnitude of $\|\mathbf{a}\|\cos(\alpha)$, where $\|\mathbf{a}\|$ represents the length of vector $\mathbf{a}$ and α is the angle between $\mathbf{a}$ and $\mathbf{b}$. Using the expression $\mathbf{a/b}$ to represent this projection, we have:

$$\begin{gathered}
\mathbf{a}/\mathbf{b} \equiv \|\mathbf{a}\|\cos(\alpha)\frac{\mathbf{b}}{\|\mathbf{b}\|} \\
\cos(\alpha) = \frac{\mathbf{a}^T\mathbf{b}}{\|\mathbf{a}\|\|\mathbf{b}\|} \\
\mathbf{a}/\mathbf{b} = \frac{\mathbf{a}^T\mathbf{b}}{\|\mathbf{b}\|\|\mathbf{b}\|}\mathbf{b} = \mathbf{a}^T\mathbf{b}\left(\mathbf{b}^T\mathbf{b}\right)^{-1}\mathbf{b}
\end{gathered} \quad . \tag{9.17}$$

Now suppose that we have an arbitrary matrix A of size 3×3.

$$A = \begin{bmatrix} a_{11} & a_{12} & a_{13} \\ a_{21} & a_{22} & a_{23} \\ a_{31} & a_{32} & a_{33} \end{bmatrix} = \begin{bmatrix} \mathbf{a}_1^T \\ \mathbf{a}_2^T \\ \mathbf{a}_3^T \end{bmatrix}. \tag{9.18}$$

Each row of this matrix is a vector in 3D space. If the row vectors are linearly independent (meaning that the rank of A is 3), then the row vectors can serve as a basis set to span 3D space. That is, we can construct any 3D vector as a linear combination of the row vectors of A. Now suppose we have another matrix B in which the row vectors are also in 3D space, but the rank is 2.

$$B = \begin{bmatrix} b_{11} & b_{12} & b_{13} \\ b_{21} & b_{22} & b_{23} \end{bmatrix} = \begin{bmatrix} \mathbf{b}_1^T \\ \mathbf{b}_2^T \end{bmatrix}. \tag{9.19}$$

In this case, the space spanned by the row vectors of B is a plane. This plane defines a subspace of the space spanned by row vectors of A. When we project matrix A onto B, we are projecting the row vectors of A onto the subspace spanned by the row vectors of B. This is shown in figure 9.3A. To project vector $\mathbf{a}$ onto matrix B, we have:

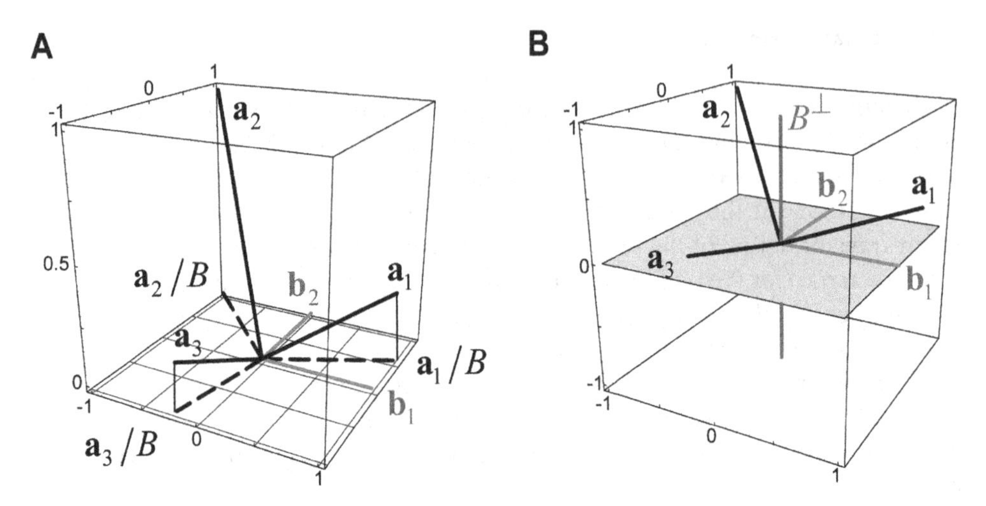

Figure 9.3
Projecting a vector onto the subspace spanned by the row vectors of a matrix. (A) When we project matrix A onto B, we are projecting the row vectors of A (equation 9.18) onto the subspace spanned by the row vectors of B (equation 9.19). (B) We use the term $B^{\perp}$ to define the space that is perpendicular to the space spanned by row vectors of B. Therefore, $B^{\perp}$ is simply a line in this example.

$$\mathbf{a}/B \equiv B^T\left(BB^T\right)^{-1} B\mathbf{a}. \tag{9.20}$$

The result of equation (9.20) is a column vector. To represent it as a row vector we transpose it:

$$(\mathbf{a}/B)^T = \mathbf{a}^T B^T\left(BB^T\right)^{-1} B.$$

To project matrix A onto matrix B we have:

$$A/B = \begin{bmatrix} (\mathbf{a}_1/B)^T \\ (\mathbf{a}_2/B)^T \\ (\mathbf{a}_3/B)^T \end{bmatrix} = AB^T\left(BB^T\right)^{-1} B. \tag{9.21}$$

An important problem for us is to find the space that is perpendicular to the space defined by a matrix. The space spanned by the row vectors of B is simply a plane. Therefore, the space perpendicular to this plane is simply a line (figure 9.3B). Let us use the term $B^{\perp}$ to refer to the space perpendicular to the row vectors of B. Projecting A onto $B^{\perp}$ we have:

$$A/B^{\perp} = A\left(I - B^T\left(BB^T\right)^{-1} B\right). \tag{9.22}$$

It then follows that $B/B^{\perp} = 0$.

Returning to our problem, in equation (9.11) we see that vector $\mathbf{y}^{(n)}$ is a linear combination of vectors $\mathbf{x}^{(n)}$ and $\mathbf{u}^{(n)}$, which means that vectors $\mathbf{x}^{(n)}$ and $\mathbf{u}^{(n)}$ are the bases for vector $\mathbf{y}^{(n)}$. By projecting the vector $\mathbf{y}^{(n)}$ onto a space perpendicular to $\mathbf{u}^{(n)}$, we will be left with a vector in the subspace spanned by $\mathbf{x}^{(n)}$. We will not recover $\mathbf{x}^{(n)}$ as a result of this projection, but we will recover a vector proportional to $\mathbf{x}^{(n)}$. As we noted in equation (9.16), our problem has no unique solution anyway, so we have no desire to recover $\mathbf{x}^{(n)}$. All we need to do is recover a vector proportional to it. If we do so, we can produce a generative model that precisely replicates the inputs and outputs of the system in question.

9.5 Subspace Analysis

Our plan of attack is as follows. Suppose we construct matrices U and Y so that we have a compact way to represent the history of the inputs that we gave and outputs that we measured. We know that each row of matrix Y lives in the space spanned by the row vectors in X (the history of hidden states, which is unknown to us), and U (the history of inputs that we gave). We will project Y onto the subspace perpendicular to U, written as $U^{\perp}$. By doing so, we will get rid of the contribution of U, leaving only the components of Y that live in the subspace described by X. Therefore, by projecting Y onto $U^{\perp}$, we will end up with a new matrix that is precisely equal to CX. Because C is a constant matrix, we will in fact recover the history of the hidden states up to a constant "multiple."

The first step is to arrange the input and output data in what is called a *Hankel matrix* as follows:

$$Y_{1|i} \equiv \begin{bmatrix} \mathbf{y}^{(1)} & \mathbf{y}^{(2)} & \cdots & \mathbf{y}^{(j)} \\ \mathbf{y}^{(2)} & \mathbf{y}^{(3)} & \cdots & \mathbf{y}^{(j+1)} \\ \vdots & \vdots & & \vdots \\ \mathbf{y}^{(i)} & \mathbf{y}^{(i+1)} & \cdots & \mathbf{y}^{(i+j-1)} \end{bmatrix} \qquad U_{1|i} \equiv \begin{bmatrix} \mathbf{u}^{(1)} & \mathbf{u}^{(2)} & \cdots & \mathbf{u}^{(j)} \\ \mathbf{u}^{(2)} & \mathbf{u}^{(3)} & \cdots & \mathbf{u}^{(j+1)} \\ \vdots & \vdots & & \vdots \\ \mathbf{u}^{(i)} & \mathbf{u}^{(i+1)} & \cdots & \mathbf{u}^{(i+j-1)} \end{bmatrix}. \tag{9.23}$$

In these matrices, $i \ll j$, that is, the number of rows is much smaller than the number of columns. If we now label the (unknown) state sequence as matrix X,

$$X_i \equiv \begin{bmatrix} \mathbf{x}^{(i)} & \mathbf{x}^{(i+1)} & \cdots & \mathbf{x}^{(i+j-1)} \end{bmatrix}, \tag{9.24}$$

we see that by projecting the row vectors of Y onto the subspace spanned by $U^{\perp}$, we can recover matrix CX. That is, we can recover the subspace spanned by the state vectors. To show this, let us write the output matrix Y as a linear function of states and inputs:

$$Y_{1|i}=\begin{bmatrix} C\mathbf{x}^{(1)}+D\mathbf{u}^{(1)} & C\mathbf{x}^{(2)}+D\mathbf{u}^{(2)} & \cdots \\ CA\mathbf{x}^{(1)}+CB\mathbf{u}^{(1)}+D\mathbf{u}^{(2)} & CA\mathbf{x}^{(2)}+CB\mathbf{u}^{(2)}+D\mathbf{u}^{(3)} & \cdots \\ CA^2\mathbf{x}^{(1)}+CAB\mathbf{u}^{(1)}+CB\mathbf{u}^{(2)}+D\mathbf{u}^{(3)} & CA^2\mathbf{x}^{(2)}+CAB\mathbf{u}^{(2)}+CB\mathbf{u}^{(3)}+D\mathbf{u}^{(4)} & \cdots \\ \vdots & \vdots & \ddots \end{bmatrix}.$$

We will now attempt to write the history of measurements $Y_{1|i}$ as a linear function of history of states X_i and history of input $U_{1|i}$. If we define matrices Γ_i and H_i as follows:

$$\Gamma_i \equiv \begin{bmatrix} C \\ CA \\ \vdots \\ CA^{i-1} \end{bmatrix}, \quad H_i \equiv \begin{bmatrix} D & 0 & 0 & \cdots & 0 \\ CB & D & 0 & \cdots & 0 \\ CAB & CB & D & & 0 \\ \vdots & \vdots & & \ddots & \vdots \\ CA^{i-2}B & CA^{i-3}B & \cdots & CB & D \end{bmatrix} \tag{9.25}$$

then we can write:

$$Y_{1|i}=\Gamma_i X_1 + H_i U_{1|i}. \tag{9.26}$$

Equation (9.26) represents a shorthand way of writing the relationship between the history of inputs, the history of hidden states, and the history of observations. It is useful to use a similar notation to write the relationship between the history of states and the history of inputs. To do so, we note the following:

$$\begin{aligned} \mathbf{x}^{(2)} &= A\mathbf{x}^{(1)}+B\mathbf{u}^{(1)} \\ \mathbf{x}^{(3)} &= A^2\mathbf{x}^{(1)}+AB\mathbf{u}^{(1)}+B\mathbf{u}^{(2)} \\ \mathbf{x}^{(4)} &= A^3\mathbf{x}^{(1)}+A^2B\mathbf{u}^{(1)}+AB\mathbf{u}^{(2)}+B\mathbf{u}^{(3)} \\ \mathbf{x}^{(i+1)} &= A^i\mathbf{x}^{(1)}+A^{i-1}B\mathbf{u}^{(1)}+A^{i-2}B\mathbf{u}^{(2)}+\cdots+B\mathbf{u}^{(i)} \end{aligned} \tag{9.27}$$

If we now define matrix Δ_i as follows:

$$\Delta_i \equiv \begin{bmatrix} A^{i-1}B & A^{i-2}B & \cdots & B \end{bmatrix}, \tag{9.28}$$

we can write the state update equation as:

$$X_{i+1}=A^i X_1+\Delta_i U_{1|i}. \tag{9.29}$$

We begin our campaign with equation (9.26). Rearranging this equation, we have:

$$X_1=\Gamma_i^* Y_{1|i}-\Gamma_i^* H_i U_{1|i}. \tag{9.30}$$

The superscript * in equation (9.30) indicates a pseudo-inverse. Inserting the preceding representation into equation (9.29), we have:

$$X_{i+1} = A^i\Gamma_i^* Y_{1|i} - A^i\Gamma_i^* H_i U_{1|i} + \Delta_i U_{1|i}. \tag{9.31}$$

In this equation, we know $Y_{1|i}$ and $U_{1|i}$ but nothing else. Let us rewrite equation (9.31) in terms of things that we know and things that we do not know. If we label things that we know with matrix $W_{1|i}$ as follows:

$$W_{1|i} \equiv \begin{bmatrix} U_{1|i} \\ Y_{1|i} \end{bmatrix}, \tag{9.32}$$

and then define matrix L_i as follows:

$$L_i \equiv \begin{bmatrix} \Delta_i - A^i\Gamma_i^* H_i & A^i\Gamma_i^* \end{bmatrix}, \tag{9.33}$$

we can now write the history of the states as a linear function of things that we know:

$$X_{i+1} = L_i W_{1|i}. \tag{9.34}$$

From equation (9.26) we have:

$$Y_{i+1|2i} = \Gamma_i X_{i+1} + H_i U_{i+1|2i}. \tag{9.35}$$

Inserting equation (9.34), we have:

$$Y_{i+1|2i} = \Gamma_i L_i W_{1|i} + H_i U_{i+1|2i}. \tag{9.36}$$

Now we project our history of observations $Y_{i+1|2i}$ onto the subspace perpendicular to our history of inputs $U_{i+1|2i}^{\perp}$:

$$Y_{i+1|2i}/U_{i+1|2i}^{\perp} = \Gamma_i L_i W_{1|i}/U_{i+1|2i}^{\perp} + H_i U_{i+1|2i}/U_{i+1|2i}^{\perp}. \tag{9.37}$$

Because the second term on the right side of equation (9.37) is zero, we have:

$$Y_{i+1|2i}/U_{i+1|2i}^{\perp} = \Gamma_i L_i W_{1|i}/U_{i+1|2i}^{\perp}. \tag{9.38}$$

Note that in the preceding equations, we know everything except $\Gamma_i L_i$. Let us rearrange this equation and put our known quantities on the right side and the unknowns on the left side:

$$\Gamma_i L_i = \left[Y_{i+1|2i}/U_{i+1|2i}^{\perp}\right]\left[W_{1|i}/U_{i+1|2i}^{\perp}\right]^*. \tag{9.39}$$

Now if we simply multiply both sides with another known quantity, $W_{1|i}$, we have:

$$\Gamma_i L_i W_{1|i} = \left[Y_{i+1|2i}/U_{i+1|2i}^{\perp}\right]\left[W_{1|i}/U_{i+1|2i}^{\perp}\right]^* W_{1|i}. \tag{9.40}$$

The left side of this equation can be simplified via equation (9.34):

$$\Gamma_i X_{i+1} = \left[Y_{i+1|2i}/U^{\perp}_{i+1|2i}\right]\left[W_{1|i}/U^{\perp}_{i+1|2i}\right]^* W_{1|i}. \tag{9.41}$$

The right side of equation (9.41) includes quantities that are all known to us, and so we can compute them. Let us label this matrix as O_{i+1}:

$$O_{i+1} \equiv \left[Y_{i+1|2i}/U^{\perp}_{i+1|2i}\right]\left[W_{1|i}/U^{\perp}_{i+1|2i}\right]^* W_{1|i}. \tag{9.42}$$

The term on the left side of equation (9.41) is simply a linear transformation of the states that we are looking for:

$$O_{i+1} = \Gamma_i X_{i+1} = \begin{bmatrix} C \\ CA \\ \vdots \\ CA^{i-1} \end{bmatrix} \begin{bmatrix} \mathbf{x}^{(i+1)} & \mathbf{x}^{(i+2)} & \cdots & \mathbf{x}^{(i+j)} \end{bmatrix}. \tag{9.43}$$

At this point we can compute the matrix O_{i+1}. Our final step is to factor it so that we recover a matrix $\hat{X}_{i+1}$. Our estimate $\hat{X}_{i+1}$ will not be equal to X_{i+1}, but it will be related to it by a linear transformation. If we do a singular value decomposition of matrix O_{i+1}, we have:

$$\begin{aligned} O_{i+1} &= PSV \\ &= \begin{bmatrix} P_1 \\ P_2 \\ \vdots \\ P_i \end{bmatrix} \begin{bmatrix} s_1 & 0 & \cdots & 0 \\ 0 & s_2 & \cdots & 0 \\ \vdots & \vdots & \ddots & \vdots \\ 0 & 0 & \cdots & s_n \end{bmatrix} \begin{bmatrix} \mathbf{v}^{(1)} & \mathbf{v}^{(2)} & \cdots & \mathbf{v}^{(j)} \end{bmatrix}^* \end{aligned} \tag{9.44}$$

In this decomposition, the dimensions of matrix P_i will be the same as matrix C, which is the same as the dimensions of matrices CA, CA^2, and so on. The dimensions of the matrix S are $n \times n$, where n is the dimension of the state vector $\mathbf{x}$. The crucial idea is that the number of singular values associated with matrix O_{i+1} is the size of the state vector that we are seeking. While we cannot recover the state matrix X_{i+1}, we can recover something that is a linear transformation of it. To see this, say that matrix T is an arbitrary and unknown invertible matrix of appropriate size. The singular value decomposition of O_{i+1} can be written as:

$$\Gamma_i X_{i+1} = PSV = PS^{1/2}TT^{-1}S^{1/2}V. \tag{9.45}$$

In equation (9.45), the term $S^{1/2}$ is the square root of the matrix S, which in this case is simply a matrix with the square root of each diagonal term. The state matrix X_{i+1} that we are looking for is related to the singular value decomposition matrices as follows:

$$X_{i+1} = T^{-1}S^{1/2}V. \tag{9.46}$$

We do not know matrix T. So if we simply set our state estimate as:

$$\hat{X}_{i+1} = S^{1/2}V, \tag{9.47}$$

then our state estimate will be a linear transformation on the actual states:

$$\hat{X}_{i+1} = TX_{i+1}. \tag{9.48}$$

Having found an estimate of the hidden states $\hat{X}_{i+1}$, the problem of finding the parameters of the system (matrices A, B, C, D) is now trivial. If we define the matrix V without its last column as:

$$V| \equiv \begin{bmatrix} \mathbf{v}^{(1)} & \mathbf{v}^{(2)} & \cdots & \mathbf{v}^{(j-1)} \end{bmatrix} \tag{9.49}$$

and the matrix V without its first column as:

$$|V \equiv \begin{bmatrix} \mathbf{v}^{(2)} & \mathbf{v}^{(3)} & \cdots & \mathbf{v}^{(j)} \end{bmatrix} \tag{9.50}$$

and similarly for state estimates as:

$$\begin{aligned} \hat{X}_{i+2}\big| &= S^{1/2} \ |V \\ \hat{X}_{i+1}\big| &= S^{1/2} \ V| \end{aligned} \tag{9.51}$$

and similarly for input and output matrices as:

$$\begin{aligned} Y_{i+1}| &\equiv \begin{bmatrix} \mathbf{y}^{(i+1)} & \mathbf{y}^{(i+2)} & \cdots & \mathbf{y}^{(j+i-1)} \end{bmatrix} \\ U_{i+1}| &\equiv \begin{bmatrix} \mathbf{u}^{(i+1)} & \mathbf{u}^{(i+2)} & \cdots & \mathbf{u}^{(j+i-1)} \end{bmatrix}, \end{aligned} \tag{9.52}$$

we will have a simple linear equation:

$$\begin{bmatrix} \hat{X}_{i+2}| \\ Y_{i+1}| \end{bmatrix} = \begin{bmatrix} \hat{A} & \hat{B} \\ \hat{C} & \hat{D} \end{bmatrix} \begin{bmatrix} \hat{X}_{i+1}| \\ U_{i+1}| \end{bmatrix} \tag{9.53}$$

to solve for the unknown parameters A, B, C, D:

$$\begin{bmatrix} \hat{A} & \hat{B} \\ \hat{C} & \hat{D} \end{bmatrix} = \begin{bmatrix} \hat{X}_{i+2}| \\ Y_{i+1}| \end{bmatrix} \begin{bmatrix} \hat{X}_{i+1}| \\ U_{i+1}| \end{bmatrix}^{*}. \tag{9.54}$$

We can use the following to find the initial state:

$$\hat{\mathbf{x}}^{(1)} = \hat{A}^{-i}\hat{\mathbf{x}}^{(i+1)} - \hat{A}^{-1}\hat{B}\mathbf{u}^{(1)} - \hat{A}^{-2}\hat{B}\mathbf{u}^{(2)} - \cdots - \hat{A}^{-i}\hat{B}\mathbf{u}^{(i)}. \tag{9.55}$$

Now the important point to note is that while our system with parameters $\hat{A}$, $\hat{B}$, $\hat{C}$, and $\hat{D}$ will be indistinguishable from the original system with parameters A, B, C, D (in terms of input–output behavior), our estimates are not the same value as the system parameters. They are related by an unknown linear transformation as defined by matrix T:

$$\begin{bmatrix} \hat{A} & \hat{B} \\ \hat{C} & \hat{D} \end{bmatrix} = \begin{bmatrix} TAT^{-1} & TB \\ CT^{-1} & D \end{bmatrix}. \tag{9.56}$$

In summary, the problem of structural learning is that of describing a dynamical system that in principle can accurately predict the sensory consequences of motor commands, that is, learn the structure of a forward model. Even for a linear system without noise, this problem has no unique solution. However, we can find one particular solution by projecting the history of our measurements upon a subspace that is perpendicular to the space defined by the history of our motor commands. Therefore, a fundamental part of structuring learning is to keep a history of the motor commands so that one can build this subspace. Once the measurements (sensory observations) are projected onto this subspace, the result is a linear transformation of the hidden states of the system that we wish to model.

9.6 Examples

Consider a system with the following dynamics:

$$\begin{aligned} x^{(n+1)} &= 0.75x^{(n)} + 0.3u^{(n)} \\ y^{(n)} &= 0.5x^{(n)} \\ x^{(1)} &= 0.0 \end{aligned} \quad . \tag{9.57}$$

Suppose that we give a sequence of square-wave inputs $\{u^{(1)}, u^{(2)}, \ldots, u^{(p)}\}$ to this system, as shown in the top row of figure 9.4A, and observe the sequence of outputs $\{y^{(1)}, y^{(2)}, \ldots, y^{(p)}\}$, as shown in the bottom row of figure 9.4A. From this input/output data, we form the $Y_{1|i}$ and $U_{1|i}$ matrices in equation (9.23) (in our example here, we set $i = 4$), and then the matrix O_{i+1} from equation (9.42). The singular value decomposition of this matrix provides a single singular value, which specifies that the size of the state vector is 1, and the following estimate of the structure for this system:

$$\begin{matrix} \hat{A} = 0.75 & \hat{B} = -0.09 \\ \hat{C} = -1.7 & \hat{D} = 0 \end{matrix} \quad \hat{x}^{(1)} = 0. \tag{9.58}$$

Notice that we did not recover the system parameters. Despite this, the system with the parameters in equation (9.58) is identical to the one in equation (9.57). For example, if we give the input $\{u^{(1)}, u^{(2)}, \ldots, u^{(p)}\}$ to our system with parameters in equation (9.58), we find $\sum_n \left(\hat{y}^{(n)} - y^{(n)}\right)^2 = 0$.

The system in equation (9.57) was relatively simple in that there was only a single state and our measurement on each trial was proportional to that

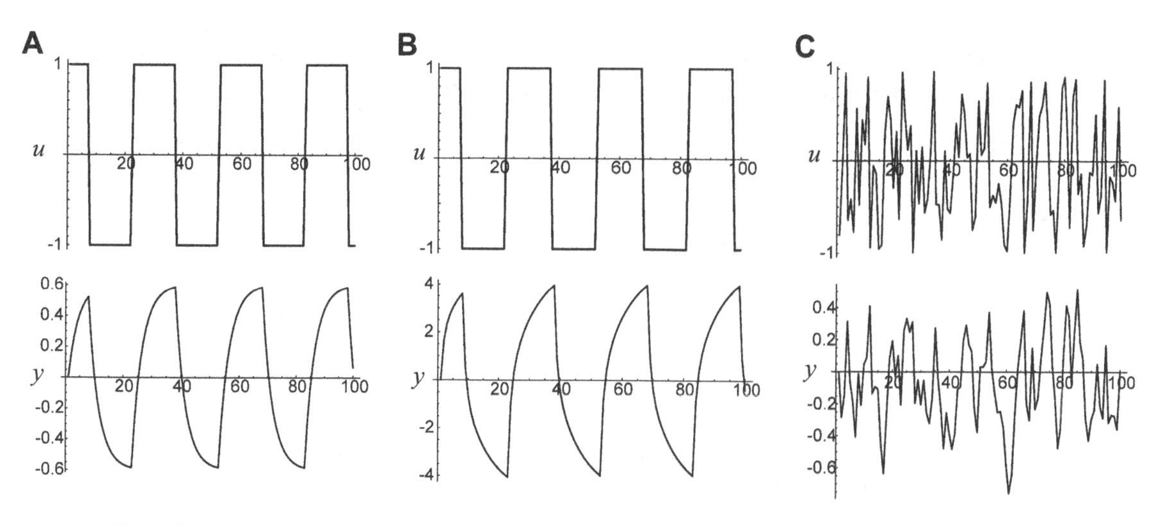

Figure 9.4
Dynamics of some sample systems. (A) System of equation (9.57), driven by inputs $\{u^{(1)},u^{(2)},\ldots,u^{(p)}\}$ that are shown and producing outputs $\{y^{(1)},y^{(2)},\cdots,y^{(p)}\}$. (B) System of equation (9.59), driven by inputs that are shown and producing the plotted outputs. (C) System of equation (9.62), driven by random noise inputs that are shown and producing the plotted outputs.

state. To make our problem more interesting, suppose that we have a system with many states, and that we can only observe the sum of these states:

$$\mathbf{x}^{(n+1)} = \begin{bmatrix} 0.9 & 0 & 0 \\ 0 & 0.5 & 0 \\ 0 & 0 & 0.2 \end{bmatrix} \mathbf{x}^{(n)} + \begin{bmatrix} 0.3 \\ 0.5 \\ 0.8 \end{bmatrix} u^{(n)} \quad . \qquad (9.59)$$
$$y^{(n)} = [1 \quad 1 \quad 1]\mathbf{x}^{(n)}$$
$$\mathbf{x}^{(1)} = [0.1 \quad -0.1 \quad 0]^T$$

The inputs to this system and the outputs are shown in figure 9.4B. Superficially, the response in figure 9.4B is similar to that in figure 9.4A, except perhaps for a slower time-constant. However, when we run the algorithm (we set $i = 5$ in equation 9.23), we see that the matrix O_{i+1} has 3 singular values, and we arrive at the following estimate for the structure of the system:

$$\begin{bmatrix} \hat{A} & \hat{B} \\ \hat{C} & \hat{D} \end{bmatrix} = \begin{bmatrix} \begin{bmatrix} 0.659 & -0.303 & 0.014 \\ -0.260 & 0.549 & 0.134 \\ 0.070 & 0.191 & 0.393 \end{bmatrix} & \begin{bmatrix} -0.302 \\ -0.155 \\ -0.063 \end{bmatrix} \\ [-4.53 \quad -1.47 \quad 0.073] & 0 \end{bmatrix} \qquad \hat{\mathbf{x}}^{(1)} = \begin{bmatrix} -0.011 \\ 0.0354 \\ 0.0375 \end{bmatrix}. \qquad (9.60)$$

In equation (9.60), $\hat{A}$ is a 3×3 matrix, $\hat{B}$ is 3×1, $\hat{C}$ is 1×3, and $\hat{D}$ is a scalar. Our estimate produces an exact match to the measured data: $\sum_n \left(\hat{y}^{(n)} - y^{(n)}\right)^2 = 0$.

Finally, let us consider a system that is driven by random inputs $u^{(n)}$ rather than a square wave. An example is shown in figure 9.4C. The dynamics of this system are as follows:

$$\begin{aligned} \mathbf{x}^{(k+1)} &= \begin{bmatrix} 0.9 & 0 \\ 0 & 0.5 \end{bmatrix} \mathbf{x}^{(k)} + \begin{bmatrix} 0.1 \\ 0.3 \end{bmatrix} u^{(k)} \\ y^{(k)} &= \begin{bmatrix} 1 & 1 \end{bmatrix} \mathbf{x}^{(k)} \\ \mathbf{x}^{(1)} &= \begin{bmatrix} 0.1 & -0.1 \end{bmatrix}^T \end{aligned} \quad . \tag{9.61}$$

Subspace analysis uncovers two singular values, and it provides the following estimate for the structure of this system:

$$\begin{bmatrix} \hat{A} & \hat{B} \\ \hat{C} & \hat{D} \end{bmatrix} = \begin{bmatrix} 0.657 & -0.241 & -0.267 \\ -0.158 & 0.743 & -0.029 \\ 1.46 & 0.319 & 0 \end{bmatrix} \qquad \hat{\mathbf{x}}^{(1)} = \begin{bmatrix} 0.028 \\ -0.127 \end{bmatrix}. \tag{9.62}$$

The resulting output of the estimated system exactly matches the measured data.

The systems that we arrived at in equation (9.60) and (9.62) may produce the same outputs for a sequence of inputs as that in equation (9.59) and (9.61), but they are much harder to interpret. For example, the system in equation (9.59) has three states, each that decays with a different rate, with no interaction between these states (i.e., the matrix A is diagonal). Furthermore, the observation y is simply the sum of these states, in which the states are weighted equally. In contrast, our estimate in equation (9.60) is a system with rather complicated interactions between the states, with observation y that weights the states unequally. It seems hard to believe that these two systems are really the same. To show that they are, let us find the matrix T in equation (9.56) that transforms one set of parameters into the other. The key point is the relationship between A and $\hat{A}$:

$$\hat{A} = TAT^{-1}. \tag{9.63}$$

In our original system in equation (9.59), the matrix A was diagonal. How do we factor $\hat{A}$ into the form shown in equation (9.61) such that the matrix A is diagonal? We proceed by finding the eigenvectors and eigenvalues of matrix $\hat{A}$. Suppose that $\hat{A}$ has eigenvalues $\lambda_1, \lambda_2, \cdots, \lambda_k$, and eigenvectors:

$$\mathbf{q}_1 = \begin{bmatrix} q_{11} \\ \vdots \\ q_{1k} \end{bmatrix}, \mathbf{q}_2 = \begin{bmatrix} q_{21} \\ \vdots \\ q_{2k} \end{bmatrix}, \cdots \tag{9.64}$$

Let us arrange the eigenvectors and values in matrix form:

$$Q \equiv [\mathbf{q}_1 \quad \mathbf{q}_2 \quad \cdots \quad \mathbf{q}_k]$$
$$L \equiv \begin{bmatrix} \lambda_1 & 0 & \cdots & 0 \\ 0 & \lambda_2 & \cdots & 0 \\ \vdots & \vdots & \ddots & \vdots \\ 0 & 0 & \cdots & \lambda_k \end{bmatrix}. \tag{9.65}$$

Multiplying the matrix $\hat{A}$ by its eigenvectors gives us:

$$\begin{aligned} \hat{A}Q &= \left[\hat{A}\mathbf{q}_1 \quad \hat{A}\mathbf{q}_2 \quad \cdots \quad \hat{A}\mathbf{q}_k\right] \\ &= [\lambda_1\mathbf{q}_1 \quad \lambda_2\mathbf{q}_2 \quad \cdots \quad \lambda_k\mathbf{q}_k] \end{aligned}. \tag{9.66}$$

We can factor the matrix $\hat{A}Q$ as follows:

$$\begin{aligned} \hat{A}Q &= \begin{bmatrix} \lambda_1 q_{11} & \lambda_2 q_{21} & \cdots & \lambda_k q_{k1} \\ \vdots & \vdots & \ddots & \vdots \\ \lambda_1 q_{1k} & \lambda_2 q_{2k} & \cdots & \lambda_k q_{kk} \end{bmatrix} \\ &= \begin{bmatrix} q_{11} & q_{21} & \cdots & q_{k1} \\ \vdots & \vdots & \ddots & \vdots \\ q_{1k} & q_{2k} & \cdots & q_{kk} \end{bmatrix} \begin{bmatrix} \lambda_1 & 0 & \cdots & 0 \\ 0 & \lambda_2 & \cdots & 0 \\ \vdots & \vdots & \ddots & \vdots \\ 0 & 0 & \cdots & \lambda_k \end{bmatrix}, \end{aligned} \tag{9.67}$$

which we can summarize as:

$$\hat{A}Q = QL. \tag{9.68}$$

We finally arrive at a factored form of $\hat{A}$ in terms of a diagonal matrix L:

$$\hat{A} = QLQ^{-1}. \tag{9.69}$$

So if we set $T = Q$, where Q is the matrix containing the eigenvectors of $\hat{A}$, we can transform our estimated parameter $\hat{A}$ to a form that is diagonal, producing another equivalent system to the one that originally produced the input/output data. Note that once again we cannot find the parameters of the original system, but we can find parameters that make a system that is indistinguishable from the original one from the point of view of dynamics.

9.7 Estimating the Noise

In the case that our system has noise, in the form:

$$\begin{aligned} \mathbf{x}^{(n+1)} &= A\mathbf{x}^{(n)} + B\mathbf{u}^{(n)} + \boldsymbol{\varepsilon}_x^{(n)} \qquad \boldsymbol{\varepsilon}_x \sim N(0,Q) \\ \mathbf{y}^{(n)} &= C\mathbf{x}^{(n)} + D\mathbf{u}^{(n)} + \boldsymbol{\varepsilon}_y^{(n)} \qquad \boldsymbol{\varepsilon}_y \sim N(0,R) \end{aligned}, \tag{9.70}$$

the key step of determining the size of the hidden states by examining the number of singular values in O_{i+1} will need to be slightly modified. In principle, the number of singular values will be equal to i in equation (9.23). That is, once the system has noise, we can no longer identify with absolute certainty the size of the hidden state vector. In practice, the singular values need to be examined and hopefully most will be rather small and can be disregarded. Once the parameters $\hat{A}$, $\hat{B}$, $\hat{C}$, and $\hat{D}$ are estimated, the noise variance Q and R can be computed from the residuals in the fit. If we define the residual in the estimate of state as in equation (9.71), then state noise is the variance of this estimate:

$$\begin{aligned}\tilde{\mathbf{x}}^{(n)} &= \hat{\mathbf{x}}^{(n)} - \hat{A}\hat{\mathbf{x}}^{(n-1)} - \hat{B}\mathbf{u}^{(n-1)} \\ Q &= E\left[(\tilde{\mathbf{x}} - E[\tilde{\mathbf{x}}])(\tilde{\mathbf{x}} - E[\tilde{\mathbf{x}}])^T\right]\end{aligned} \quad (9.71)$$

Similarly, measurement noise is the variance of the residual in the estimate of output:

$$\begin{aligned}\tilde{\mathbf{y}}^{(n)} &= \mathbf{y}^{(n)} - \hat{\mathbf{y}}^{(n)} \\ R &= E\left[(\tilde{\mathbf{y}} - E[\tilde{\mathbf{y}}])(\tilde{\mathbf{y}} - E[\tilde{\mathbf{y}}])^T\right]\end{aligned} \quad (9.72)$$

9.8 Identifying the Structure of the Learner

There is a practical application for the framework that we just developed: We can use it to model the learner. For example, say that we have collected some data from a subject during some behavioral task in which she made movements and adapted to a perturbation. We imagine that our subject has a state on each trial, and based on this state she programs a motor command (which we can record based on their movements). As a consequence of their motor command, our subject makes an observation (e.g., the limb did not move in the predicted direction), and then learns from the resulting prediction error. Our data set consists of a series of trials in which we have given perturbations and a series of movements (or predictions) that the subject has made. We can represent the learner as a dynamical system in which her states change as a consequence of her prediction errors. We are interested in quantifying the timescales of these states, and sensitivity of each state to a prediction error.

Let us apply this idea to a simple task: saccade adaptation. A subject is provided with a target, makes a saccade to that target, and during that saccade the target is moved to a new location. As the saccade terminates, the subject observes that the target is not on the fovea, that is, there is a prediction error. Trial after trial, the subject learns to alter the magnitude of the saccade to minimize this error. Suppose that the state of the learner can be represented with vector $\mathbf{x}$ (of unknown dimensionality). The state is affected by three factors: visual error $\tilde{y}$ at end of a saccade, passage of time between trials, and Gaussian noise $\boldsymbol{\varepsilon}_x$. If we assume that the intertrial interval is constant, then we can write the change in states of the learner as:

$$\mathbf{x}^{(n+1)} = A\mathbf{x}^{(n)} + \mathbf{b}\tilde{y}^{(n)} + \boldsymbol{\varepsilon}_x. \tag{9.73}$$

In this equation, the matrix A specifies how the states will change from trial n to $n+1$ because of passage of time, and the vector $\mathbf{b}$ specifies how the states will change because of the error observed on trial n. We cannot directly observe the states, but can measure saccade amplitude on trial n as $y^{(n)}$. Let us assume that saccade amplitude $y^{(n)}$ is affected by target location $p^{(n)}$, some inherent bias that the subject may have y_b, the state of the subject $\mathbf{x}^{(n)}$, plus noise ε_y inherent in the execution of the movement. This is written as:

$$y^{(n)} = p^{(n)} - y_b + \mathbf{c}^T\mathbf{x}^{(n)} + \varepsilon_y. \tag{9.74}$$

In equation (9.74), the vector $\mathbf{c}$ specifies the relative weight of each state in influencing the saccade amplitude.

Like any system identification problem, we want to give "inputs" to our learner and then measure her behavior. These inputs are in the form of a perturbation—that is, we will move the target $p^{(n)}$ by amount $u^{(n)}$ during the saccade, so that when the eye movement completes, there will be some endpoint errors. The error on that trial will be:

$$\begin{aligned} \tilde{y}^{(n)} &= p^{(n)} + u^{(n)} - y^{(n)} - y_b \\ &= u^{(n)} - c^T\mathbf{x}^{(n)} - \varepsilon_y \end{aligned}. \tag{9.75}$$

Inserting equation (9.75) into equation (9.73) produces our state space model of the saccade adaptation task:

$$\begin{aligned} \mathbf{x}^{(n+1)} &= \left(A - \mathbf{b}\mathbf{c}^T\right)\mathbf{x}^{(n)} + \mathbf{b}\left(u^{(n)} - \varepsilon_y\right) + \boldsymbol{\varepsilon}_x & \boldsymbol{\varepsilon}_x &\sim N(0,Q) \\ y^{(n)} &= p^{(n)} - y_b + \mathbf{c}^T\mathbf{x}^{(n)} + \varepsilon_y & \varepsilon_y &\sim N(0,r) \end{aligned}. \tag{9.76}$$

In a typical experiment, for each subject we give a sequence of targets $p^{(n)}$ and on each trial displace that target during the saccade by amount $u^{(n)}$. We then measure the saccade amplitude $y^{(n)}$. Our objective is to find the structure of the learner, i.e., parameters A, $\mathbf{b}$, and $\mathbf{c}$ in equation (9.76).

Vincent Ethier, David Zee, and Reza Shadmehr (2008) performed a saccade adaptation experiment and then used subspace analysis to estimate the structure of the learner from trial-to-trial movement data. In the experiment, they considered two kinds of trials: perturbation trials in which the target of the saccade was moved during the saccade, and *error-clamp* trials in which errors were eliminated by moving the target so that it was located on the fovea at the endpoint of the saccade. In adaptation trials, $u^{(n)}$ was the intrasaccadic target displacement. In error-clamp trials,

$$u^{(n)} = y^{(n)} - p^{(n)} + y_b. \tag{9.77}$$

The paradigm is summarized in figure 9.5A. The experiment began with error-clamp trials, was then followed by a gain-down session, then a gain-up session ("extinction"), and finally error-clamp trials. The objective was to unmask the multiple timescales of memory and test for "spontaneous recovery." In summary, the mathematical problem consisted in finding the parameters of the dynamical system in equation (9.76), given a sequence of inputs $u^{(n)}$ (target displacements) and measurements $y^{(n)}$ (saccade amplitudes).

The averaged saccade amplitudes produced by a group of subjects is shown in figure 9.5B. These averaged data were analyzed using subspace methods and parameters for the resulting system were identified. Using these parameters, a state estimation procedure (Kalman filter) was then used to estimate the two hidden states of the system. The model's performance is shown in figure 9.5C and the estimated states of the system (the fast and slow states) are plotted in figure 9.5D. The fast state shows rapid adaptation at set starts and forgetting at set breaks. Of interest is the fact that the slow state shows no forgetting at set breaks, and little or no unlearning during extinction.

A particularly useful way to visualize the parameters of the model is one in which the states are assumed to be independent, that is, the transition matrix $A - \mathbf{b}\mathbf{c}^T$ is diagonal. This would produce a time constant of forgetting for each state. Suppose that we represent the state equation in continuous time:

$$\dot{\mathbf{x}}(t) = A_c\mathbf{x}(t) + \mathbf{b}_c u(t) + \pi_x, \tag{9.78}$$

where $A_c = \Delta^{-1}(A - \mathbf{b}\mathbf{c}^T - I)$, $\mathbf{b}_c = \mathbf{b}\Delta^{-1}$, and Δ is the intertrial interval, set to 1250 ms. If we represent vector $\mathbf{x}$ as $[x_f, x_s]^T$, that is, the fast and slow states, then λ_s and λ_f refer to the time constant of the solution to this differential equation. The fit to the group data produced a decay time constant of $\lambda_f = 28$ sec for the fast state and $\lambda_s =$ 7 min for the slow state. Error sensitivity of the fast state was 18 times larger for the fast state, i.e., $b_f/b_s = 18$. Finally, saccade amplitudes relied almost twice as much on the slow than the fast state: $c_s/c_f = 1.7$.

9.9 Expectation Maximization (EM)

Among people who study control theory, the subspace approach is well known and often used for system identification. In the machine learning community,

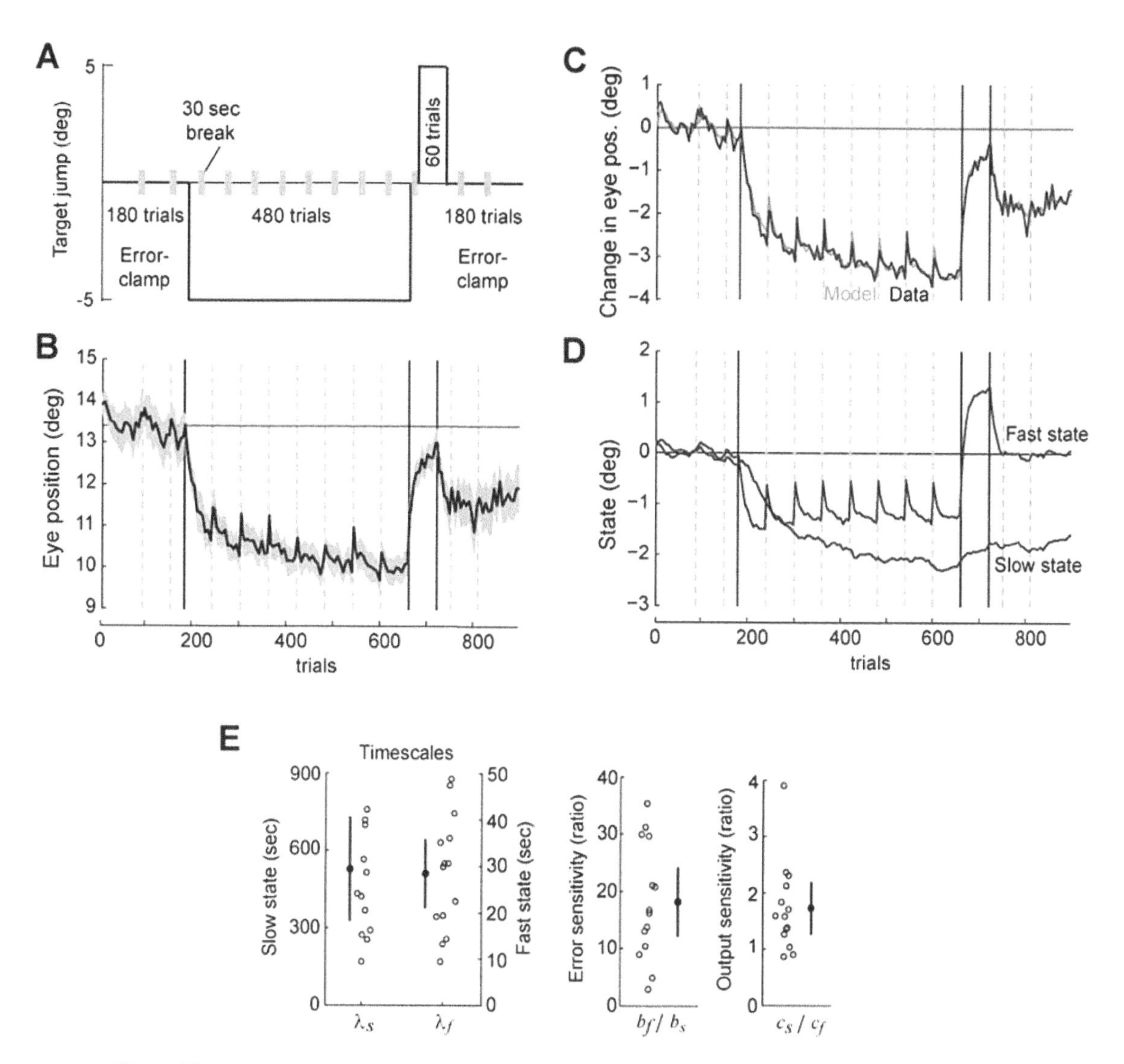

Figure 9.5
Identifying the structure of learners in a saccade adaptation experiment. (A) Experimental protocol. The experiment began with error-clamp trials, was followed by a gain-down session, followed by a gain-up session ("extinction") and finally error-clamp trials. (B) The averaged saccade amplitudes produced by a group of subjects. (C) Model result. Subspace analysis was used to identify the structure of the system, and then a Kalman filter was used to give a running estimate of the two hidden states. (D) Estimate of the two hidden states. (E) The time constants of the two states (forgetting rates), the error sensitivity of the two states, and the contribution of the two states to output. The values for fits to individual subjects are shown, as is the mean of that distribution. (From Ethier, Zee, and Shadmehr, 2008.)

however, this approach is less used. Instead, an algorithm called Expectation Maximization (EM) (Shumway and Stoffer, 1982) is often employed for solving the problem of system identification (Ghahramani and Hinton, 1996). Here, we briefly introduce EM as it has been used effectively to represent the human learner in terms of a linear state-space model (Cheng and Sabes, 2006).

Given some sequence of inputs $\mathbf{u}^{(1)},\mathbf{u}^{(2)},\cdots$ and a sequence of measurements $\mathbf{y}^{(1)},\mathbf{y}^{(2)},\cdots$, our objective is to find the structure (matrices A, B, C, etc.) of the linear system:

$$\begin{aligned} \mathbf{x}^{(n+1)} &= A\mathbf{x}^{(n)} + B\mathbf{u}^{(n)} + \boldsymbol{\varepsilon}_x^{(n)} \qquad & \boldsymbol{\varepsilon}_x \sim N(0,Q) \\ \mathbf{y}^{(n)} &= C\mathbf{x}^{(n)} + \boldsymbol{\varepsilon}_y^{(n)} \qquad & \boldsymbol{\varepsilon}_y \sim N(0,R) \end{aligned} \quad (9.79)$$

Unlike subspace analysis, here we will assume that we know the dimensionality of the hidden states $\mathbf{x}$. There are two kinds of unknowns in our model: the structural quantities $\theta = \{A,B,C,Q,R,\hat{\mathbf{x}}^{(0|0)},P^{(0|0)}\}$ (where the last two terms are the prior estimate of state and its variance), and the hidden states $\hat{\mathbf{x}}^{(1)},\hat{\mathbf{x}}^{(2)},\cdots$. If we knew the structure θ, then finding the hidden states would be easy (via the Kalman filter, for example). If we knew the hidden states, then finding the structural quantities would be easy (via maximizing a likelihood measure of the observe data). EM proceeds by performing these two steps in sequence, and repeats until the parameters converge.

The starting point in EM is to describe the expected complete log-likelihood. In our problem, this is the joint probability of the hidden states $\{\mathbf{x}\}_1^N = \{\mathbf{x}^{(1)},\mathbf{x}^{(2)},\cdots,\mathbf{x}^{(N)}\}$ and the observed quantities $\{\mathbf{y}\}_1^N = \{\mathbf{y}^{(1)},\mathbf{y}^{(2)},\cdots,\mathbf{y}^{(N)}\}$, given the inputs $\mathbf{u}^{(1)},\mathbf{u}^{(2)},\cdots$ and structural parameters θ. The expected complete log-likelihood that we are after is defined as:

$$\mathbb{Q} \equiv E[\log p(\{\mathbf{x}\},\{\mathbf{y}\}|\theta,\{\mathbf{u}\})]. \quad (9.80)$$

In the E step, we fix θ and try to maximize the expected complete log-likelihood by setting expected value of our states $\{\mathbf{x}\}_1^N$ to their posterior probabilities (done via a Kalman filter). In the M step, we fix the expected value of our states $\{\mathbf{x}\}_1^N$ and try to maximize the expected complete log-likelihood by estimating the parameters θ.

To compute the expected complete log-likelihood, let us start with the following equality:

$$p\left(\mathbf{x}^{(0)},\mathbf{x}^{(1)},\mathbf{y}^{(1)}\middle|\mathbf{u}^{(0)}\right) = p\left(\mathbf{y}^{(1)}\middle|\mathbf{x}^{(1)},\mathbf{x}^{(0)},\mathbf{u}^{(0)}\right)p\left(\mathbf{x}^{(1)},\mathbf{x}^{(0)}\middle|\mathbf{u}^{(0)}\right). \quad (9.81)$$

Using equation (9.79), this can be simplified to:

$$p\left(\mathbf{x}^{(0)},\mathbf{x}^{(1)},\mathbf{y}^{(1)}\middle|\mathbf{u}^{(0)}\right) = p\left(\mathbf{y}^{(1)}\middle|\mathbf{x}^{(1)}\right)p\left(\mathbf{x}^{(1)}\middle|\mathbf{x}^{(0)},\mathbf{u}^{(0)}\right)p\left(\mathbf{x}^{(0)}\right). \quad (9.82)$$

Similarly, we have:

$$\begin{aligned} p\left(\mathbf{x}^{(0)},\mathbf{x}^{(1)},\mathbf{x}^{(2)},\mathbf{y}^{(1)},\mathbf{y}^{(2)}\middle|\mathbf{u}^{(1)},\mathbf{u}^{(0)}\right) &= p\left(\mathbf{y}^{(1)},\mathbf{y}^{(2)}\middle|\{\mathbf{x}\}_0^2,\{\mathbf{u}\}_0^1\right)p\left(\{\mathbf{x}\}_0^2\middle|\{\mathbf{u}\}_0^1\right) \\ &= p\left(\mathbf{y}^{(2)}\middle|\mathbf{y}^{(1)},\{\mathbf{x}\}_0^2,\{\mathbf{u}\}_0^1\right)p\left(\mathbf{y}^{(1)}\middle|\{\mathbf{x}\}_0^2,\{\mathbf{u}\}_0^1\right) \\ &\quad\times p\left(\mathbf{x}^{(2)}\middle|\mathbf{x}^{(1)},\mathbf{x}^{(0)},\{\mathbf{u}\}_0^1\right)p\left(\mathbf{x}^{(1)},\mathbf{x}^{(0)}\middle|\{\mathbf{u}\}_0^1\right) \\ &= p\left(\mathbf{y}^{(2)}\middle|\mathbf{x}^{(2)}\right)p\left(\mathbf{y}^{(1)}\middle|\mathbf{x}^{(1)}\right) \\ &\quad\times p\left(\mathbf{x}^{(2)}\middle|\mathbf{x}^{(1)},\mathbf{u}^{(1)}\right)p\left(\mathbf{x}^{(1)}\middle|\mathbf{x}^{(0)},\mathbf{u}^{(0)}\right)p\left(\mathbf{x}^{(0)}\right) \end{aligned} \quad . \qquad (9.83)$$

So we can conclude that:

$$p\left(\{\mathbf{x}\}_0^N,\{\mathbf{y}\}_1^N\middle|\theta,\{\mathbf{u}\}_1^N\right)=\left[\prod_{n=1}^N p\left(\mathbf{y}^{(n)}\middle|\mathbf{x}^{(n)}\right)\right]\left[\prod_{n=1}^N p\left(\mathbf{x}^{(n)}\middle|\mathbf{x}^{(n-1)},\mathbf{u}^{(n-1)}\right)\right]p\left(\mathbf{x}^{(0)}\right). \qquad (9.84)$$

In equation (9.84), we have the following normal distributions:

$$\begin{aligned} p\left(\mathbf{y}^{(n)}\middle|\mathbf{x}^{(n)}\right) &= N\left(C\mathbf{x}^{(n)},R\right) \\ p\left(\mathbf{x}^{(n+1)}\middle|\mathbf{x}^{(n)},\mathbf{u}^{(n)}\right) &= N\left(A\mathbf{x}^{(n)}+B\mathbf{u}^{(n)},Q\right) \end{aligned} . \qquad (9.85)$$

Next, we find the log of the expression in equation (9.84). This is our complete log-likelihood:

$$\begin{aligned} \log p\left(\{\mathbf{x}\}_0^N,\{\mathbf{y}\}_1^N\middle|\theta,\{\mathbf{u}\}_1^N\right) &= -\sum_{n=1}^N \frac{1}{2}\left(\mathbf{y}^{(n)}-C\mathbf{x}^{(n)}\right)^T R^{-1}\left(\mathbf{y}^{(n)}-C\mathbf{x}^{(n)}\right) \\ &\quad -\sum_{n=1}^N \frac{1}{2}\left(\mathbf{x}^{(n)}-A\mathbf{x}^{(n-1)}-B\mathbf{u}^{(n-1)}\right)^T Q^{-1}\left(\mathbf{x}^{(n)}-A\mathbf{x}^{(n-1)}-B\mathbf{u}^{(n-1)}\right) \\ &\quad -\frac{1}{2}\left(\mathbf{x}^{(0)}-\hat{\mathbf{x}}^{(0)}\right)^T P_0^{-1}\left(\mathbf{x}^{(0)}-\hat{\mathbf{x}}^{(0)}\right) \\ &\quad -\frac{N}{2}\log|R|-\frac{N}{2}\log|Q|-\frac{1}{2}\log|P_0|+\text{const} \end{aligned} . \qquad (9.86)$$

In the E step, we fix the parameters θ and find the state estimate $\{\hat{\mathbf{x}}\}_1^N$ that maximizes the expected value of the quantity in equation (9.86). The state estimate is typically the Kalman estimate, or better yet the estimate that depends both on the past and future observations of $\mathbf{y}$: this is called smoothing (Anderson and Moore, 1979), in which

$$\hat{\mathbf{x}}^{(n)} \equiv E\left[\mathbf{x}^{(n)}\middle|\{\mathbf{y}\}_1^N\right]. \qquad (9.87)$$

In the M step, we fix the state estimate $\{\hat{\mathbf{x}}\}_1^N$, and find the parameters θ that maximize the complete log-likelihood. To do so, we find the derivative of the expected value of the quantity in equation (9.86) with respect to parameters θ, set it to zero, and then use the derivative to find the estimate of these parameters. To show how to do this, let us label the complete log-likelihood expression in equation (9.86) by the shorthand l_c. We have:

$$\frac{dl_c}{dC} = \sum_{n=1}^{N} R^{-1}\mathbf{y}^{(n)}\mathbf{x}^{(n)T} - R^{-1}C\mathbf{x}^{(n)}\mathbf{x}^{(n)T} = 0$$
$$E\left[\frac{dl_c}{dC}\right] = \sum_{n=1}^{N} R^{-1}\mathbf{y}^{(n)}\hat{\mathbf{x}}^{(n)T} - R^{-1}CP^{(n)} = 0 \tag{9.88}$$

In equation (9.88), the matrix $P^{(n)}$ refers to the variance of the state estimate, $P^{(n)} \equiv E\left[\mathbf{x}^{(n)}\mathbf{x}^{(n)T} \mid \{\mathbf{y}\}_1^N\right]$. Solving for C, we have:

$$C_{new} = \left[\sum_{n=1}^{N} \mathbf{y}^{(n)}\hat{\mathbf{x}}^{(n)T}\right]\left[\sum_{n=1}^{N} P^{(n)}\right]^{-1}. \tag{9.89}$$

To find a new estimate for R, it is convenient to find the derivative of the log-likelihood with respect to R^{-1}:

$$\frac{dl_c}{dR^{-1}} = \frac{N}{2}R + \sum_{n=1}^{N} C\mathbf{x}^{(n)}\mathbf{y}^{(n)T} - \frac{1}{2}C\mathbf{x}^{(n)}\mathbf{x}^{(n)T}C^T - \frac{1}{2}\mathbf{y}^{(n)}\mathbf{y}^{(n)T}. \tag{9.90}$$

Setting the above quantity to zero gives us the new estimate for R:

$$R_{new} = \frac{1}{N}\left(\sum_{n=1}^{N} CP^{(n)}C^T + \mathbf{y}^{(n)}\mathbf{y}^{(n)T} - 2C\hat{\mathbf{x}}^{(n)}\mathbf{y}^{(n)T}\right). \tag{9.91}$$

For parameter A we have:

$$\frac{dl_c}{dA} = \sum_{n=1}^{N} Q^{-1}\mathbf{x}^{(n)}\mathbf{x}^{(n-1)T} - Q^{-1}A\mathbf{x}^{(n-1)}\mathbf{x}^{(n-1)T} - Q^{-1}B\mathbf{u}^{(n-1)}\mathbf{x}^{(n-1)T}. \tag{9.92}$$

The new estimate for A becomes:

$$A_{new} = \left(\sum_{n=1}^{N} P^{(n,n-1)} - B\mathbf{u}^{(n-1)}\hat{\mathbf{x}}^{(n-1)T}\right)\left(\sum_{n=1}^{N} P^{(n-1)}\right)^{-1}. \tag{9.93}$$

For parameter B we have:

$$\frac{dl_c}{dB} = \sum_{n=1}^{N} Q^{-1}\mathbf{x}^{(n)}\mathbf{u}^{(n-1)T} - Q^{-1}A\mathbf{x}^{(n-1)}\mathbf{u}^{(n-1)T} - Q^{-1}B\mathbf{u}^{(n-1)}\mathbf{u}^{(n-1)T}. \tag{9.94}$$

The new estimate for B becomes:

$$B_{new} = \left(\sum_{n=1}^{N} \hat{\mathbf{x}}^{(n)}\mathbf{u}^{(n-1)T} - A\hat{\mathbf{x}}^{(n-1)}\mathbf{u}^{(n-1)T}\right)\left(\sum_{n=1}^{N} \mathbf{u}^{(n-1)}\mathbf{u}^{(n-1)T}\right)^{-1}. \tag{9.95}$$

Finally, for parameter Q we have:

$$\begin{aligned}\frac{dl_c}{dQ^{-1}} = \frac{N}{2}Q + \sum_{n=1}^{N} \mathbf{x}^{(n)}\mathbf{x}^{(n)T} + A\mathbf{x}^{(n-1)}\mathbf{x}^{(n)T} + B\mathbf{u}^{(n-1)}\mathbf{x}^{(n)T} \\ - A\mathbf{x}^{(n-1)}\mathbf{x}^{(n-1)T}A^T - A\mathbf{x}^{(n-1)}\mathbf{u}^{(n-1)T}B^T - \frac{1}{2}B\mathbf{u}^{(n-1)}\mathbf{u}^{(n-1)T}B^T\end{aligned} \quad (9.96)$$

The new estimate for Q becomes:

$$\begin{aligned}Q_{new} = \frac{1}{N}\sum_{n=1}^{N} P^{(n)} - 2AP^{(n-1,n)} - 2B\mathbf{u}^{(n-1)}\hat{\mathbf{x}}^{(n)T} + 2AP^{(n-1)}A^T \\ +2A\hat{\mathbf{x}}^{(n-1)}\mathbf{u}^{(n-1)T}B^T + B\mathbf{u}^{(n-1)}\mathbf{u}^{(n-1)T}B^T\end{aligned} \quad (9.97)$$

Summary

In building a model that can predict things, one can begin with some generic structure and then optimally fit the parameters of this structure to the observed data. This is the problem of state estimation, something that we considered in the last few chapters. However, a generic structure often has the disadvantage of being poorly suited to the task at hand, resulting in poor generalization, and slow learning. It would be useful to build a new structure or model topology that is specific to the data that one observes. This would give one the capability to learn to ride a small bike at childhood, and then generalize to larger bikes as one grows older.

The mathematical problem is one of finding the structure of a stochastic dynamical system that can, in principle, be given a sequence of inputs and produce the sequence of outputs that match that observed from the real system. For linear stochastic systems, a closed form solution exists; this is called subspace analysis. This approach relies on the fact that each observation $\mathbf{y}^{(n)}$ is a linear combination of states $\mathbf{x}^{(n)}$ and inputs $\mathbf{u}^{(n)}$, which means that vectors $\mathbf{x}^{(n)}$ and $\mathbf{u}^{(n)}$ are the bases for vector $\mathbf{y}^{(n)}$. By projecting the vector $\mathbf{y}^{(n)}$ onto a vector perpendicular to $\mathbf{u}^{(n)}$, one is left with a vector in the subspace spanned by $\mathbf{x}^{(n)}$ and is proportional to $\mathbf{x}^{(n)}$. Recovery of this subspace in which the hidden states $\mathbf{x}^{(n)}$ reside allows one to find the structure of the dynamical system.

An application of this formalism is in modeling biological system and how they learn. The idea is to give some inputs to the learner, and observe their behavior, and then use that input–output relationship to discover the structure of the learner. An example of this is in saccade adaptation, in which it was found that in a typical single session experiment, the learner is well represented via a fast system that decays with a time constant of about

25 seconds and is highly sensitive to errors, and a slow system that shows very little if any decay and is also an order of magnitude less sensitive to error.

An alternate approach to estimating the structure of a linear dynamical system is via Expectation Maximization. In this procedure, one begins by assuming a structure and then finds the hidden states. Next, one uses these estimates of the hidden states and finds the structural quantities.

10 Costs and Rewards of Motor Commands

If we were to summarize the concept of internal models thus far, it would go something like this: (1) You begin with a prior belief regarding the state of your body and the environment; (2) you generate some motor command; (3) you predict the resulting change in the state of your body and the environment; (4) you predict the information that should be conveyed by your sensory organs; (5) you combine the sensory feedback with your predictions; and (6) you update your estimate of the state. Internal models are the computational maps that allow you to make the predictions in steps 3 and 4: the prediction about the change caused by your motor commands in the state of your body and the environment, and the prediction about the sensory information that should be provided by your sensory afferents. These predictions provide a source of information that is independent of the actual sensory feedback. If your internal models are unbiased, by combining their predictions with the actual sensory feedback you can produce an estimate of state that is better than either source of information alone. Therefore, a fundamental advantage of internal models is that it can improve your ability to sense the world around you.

However, our brain would be nothing more than a fancy surveillance system if all it could do was estimate the state of the environment accurately. We have a central nervous system in order to *do things*, namely, generate motor commands and perform actions. An often-cited example is the sea squirt, a cucumber-shaped organism that in its larval form actively swims and is equipped with a brainlike structure of three hundred cells (Llinas, 2001). It uses these cells to search for a suitable surface. Once it finds a surface, it buries its head, absorbs most of its brain, and never swims again. So if the purpose of having a nervous system is to move, how does one generate useful motor commands? It turns out that if we have an internal model of our body and the environment, then we can generate motor commands that change the state of our body and the environment to one that is more advantageous for us. That is, the ability to predict the consequences of our motor commands is crucial for generating movements so that we can get *reward*. Without accurate internal models, we would have little ability to perform movements that provide us with rewarding outcomes.

In this chapter and the next one, we will begin considering the problem of how to produce useful motor commands—that is, motor commands that can change the state of our body to one that is more rewarding or valuable. We will imagine that there is something of value that we would like to have, and that this object is at some position. It is better to receive reward sooner rather than later, so there will be a cost associated with the time it takes to arrive at this valuable state. We will have to exert some effort in terms of motor commands to get to this rewarding state, and therefore we will have to balance the need for acquiring reward with costs that we are willing to pay in terms of motor commands (or effort). Together, the reward that we expect to get when we arrive at the goal state, and the effort that we expect to expend to arrive at that state, form a cost function. Our objective will be to find a *control policy* that, given any current state, produces the "best" motor commands in the sense that the motor commands will bring us to the goal state while minimizing a cost. Internal models allow us to predict how motor commands will change our state, and therefore they are critical for forming control policies. Our main theoretical tool will be optimal control theory, a framework in which we will describe the cost of movements. Our ultimate objective is to produce a theory that can not only account for the regularity that exists in movements of people but can also explain why movement patterns change with aging, development, and disease.

10.1 Voluntary Eye Movements

As you are reading this book, your brain shifts your gaze from one point to another, rapidly moving your eyes. Each movement is a saccade that positions your eyes so that the fovea can sample the visual space. The fovea has densely packed neurons that provide an exquisitely fine resolution of the visual scene. It is like a very high-resolution camera. However, the rest of our retina does not have nearly the resolution as the fovea. As we move away from the fovea on the retina, the density of the neurons drops exponentially, and as a result, the visual acuity drops exponentially (figure 10.1). It is clear that we make saccades in order to place the image of interest on our fovea. But why don't we have an eye that has uniformly high-resolution capabilities? That way, we wouldn't have to make saccades. Why is it that our retina is endowed with only one region of high resolution?

In his book *Movements of the Eyes*, Richard Carpenter writes:

> The reason why the whole visual field is not provided with the luxury grade of vision is presumably [because] the dense packing of the receptors and the one-to-one connections to the optic nerve would lead to the latter becoming unmanageably large. For example, if our entire retina were of the high-quality type, the cross-sectional area of our optic nerve would have to increase by a factor of over two hundred, and no doubt the size of the blind spot would also have to increase proportionately. (1988, p. 7)

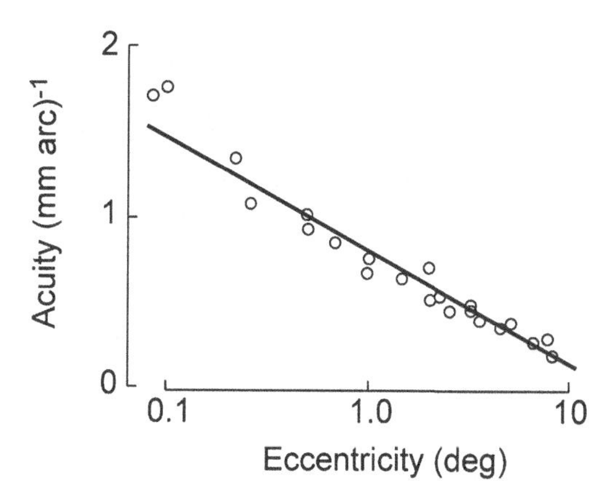

Figure 10.1
Spatial acuity as a function of distance from the center of fovea. The line represents the function $y = 0.8(1 - \log_{10}x)$, indicating that acuity declines exponentially as a function of distance from the fovea. (Plot constructed from Carpenter, 1988, using data from Green, 1970, and Jacobs, 1979).

So it would seem that having the very high-resolution area limited to a small part of the human retina allows us to have a relatively small blind spot, as well as a manageable optic nerve, but at the cost of having to constantly move the eyes so to place the image of interest on our fovea. In fact, voluntary eye movements occur only in vertebrates that have a fovea (Walls, 1962).

In making these saccades, our brain solves two kinds of problems. First, it selects where to look, and next, it programs the motor commands that move the eyes to that location. Let us consider the problem of where to look. It has long been recognized that the scan sequence, that is, the locations that people fixate when they are given an image, is not random. Alfred Yarbus (1967), a Russian physicist turned psychologist, did pioneering work on this; he invented new methods to record eye movements and wrote the results of his dissertation, which took about twenty years to complete, in a book that was published in 1965 and was translated to English two years later. In thinking about how we direct movements of our eyes to examine a picture, he wrote: "It may be seen to some people that when we examine an object we must trace its outlines with our eye and, by analogy with tactile sensation, 'palpate' the object. Others may consider that, when looking at a picture, we scan the whole of its surface more or less uniformly with our eyes." However, Yarbus performed experiments that showed that motion of the eyes was neither like the motion of the hand in examining a surface, nor uniform like a scanning beam that you find on a copy machine. For example, he presented to his subjects a painting by Ivan Shishkin, *Morning in the Pine Forest*, in which four black bears are playing on a fallen tree (figure 10.2). He imagined that people might look more at parts of the image that had a lot of detail, such as the intricate branches on the trees. He thought

Figure 10.2
Saccade scan paths for viewing of images. Top image: Reproduction of Ivan Shishkin's painting *Morning in the Pine Forest*. The records of the eye movements were made during free examination of the picture over two minutes. Middle image: Shishkin's painting *Countess Mordvinova's Forest*. The records were made during ten minutes of free examination. Bottom image: Petrushavich's sculpture *My Child*. The records were made during two minutes of free examination. (Data reproduced from Yarbus, 1967).

that perhaps people might look more at colors that they had previously identified to be their favorites, but they did not seem to do that. Instead, he found that people tended to move their eyes so that the image of one of the bears fell on the fovea (that is, they fixated one of the bears). When he presented another painting by Shishkin, *Countess Mordvinova's Forest*, people tended to fixate the hunter, which is only a small figure in a rather large and highly detailed image. For the Petrushavich sculpture *My Child*, the fixations were centered on the faces of the mother and child (especially the eyes), and the hands. From these experiments, Yarbus concluded:

> All records of the eye movements show that neither bright nor dark elements of a picture attract the observer's attention unless they give essential and useful information. . . . When examining complex objects, the human eye fixates mainly on certain elements of these objects. Analysis of the eye-movement records show that the elements attracting attention contain . . . information useful and essential for perception.

Written more formally, the idea is that our brain continuously assigns a value to every part of the visible space, forming a priority or salience map (Gottlieb, Kusunoki, and Goldberg, 1998; Fecteau and Munoz, 2006). Each saccade is a movement with which the brain directs the fovea to a region where currently, the value is highest. Therefore, the reason we make a saccade is to place the currently most valuable part of the image on our fovea. The landscape of this value function, that is, the position of the currently most valuable part of the image, would of course depend on the task: if the *Morning in the Pine Forest* painting is shown and the subject is asked to look for a shoe, they are likely to direct their eyes to the underbrush. Generalizing from this, we can speculate that the reason we move is that our motor commands are expected to change the state of our body (or state of the environment) to something that is more advantageous or rewarding. Some seven hundred years ago, the Persian poet Hafez expressed this thought elegantly:

> Because of our wisdom, we will travel far for love,
> As all movement is really a sign of thirst,
> And speaking really says "I am hungry to know you."

10.2 Expected Reward Discounts the Cost of the Motor Commands

Perhaps when we are looking at a visual scene, the decision of where to move our eyes is specified by the peaks and valleys of a landscape of values or expected rewards. For example, in viewing a scene consisting of a face and nonface objects, we are naturally drawn to the face region first and spend a longer time looking at the face compared to the rest of the scene (Cerf et al., 2008). Benjamin Hayden, Purak Parikh, Robert Deaner, and Michael Platt (2007) found that the opportunity to look at another person of opposite sex is a valued commodity. They found that this was especially true for men, for whom physical attractiveness of faces of women was one dimension along which value increased rapidly. Indeed, face images activate the reward circuitry of the brain (Bray and O'Doherty, 2007; Kampe et al., 2001). The implication is that faces have an intrinsically higher value than other images. This accounts for the fact that images of attractive faces play a prominent role in advertising.

Until recently, it was thought that the motor commands that the brain generates to move the eyes to a given location are generally independent of the value that the brain assigns to that location. Saccades are so short in duration (50–70 ms), and the eyes move with such a high velocity (400–500 deg/s), that the motor commands that move the eyes during a saccade were thought to be invariant programs: given a desired displacement of the eyes, the program played out the motor commands. Indeed, during a saccade the motion of the eyes tends to be stereotypical, exhibiting a fairly consistent relationship between amplitude, duration, and velocity. Yarbus (1967) noted that as saccade amplitudes increased, durations also increased (figure 10.3A). As better recording techniques developed, his observations regarding the approximately linear relationship between saccade duration and amplitude were confirmed (figure 10.3B). In addition, it also became apparent that during a saccade, the eye velocity had a symmetric profile for small amplitudes, but an elongated profile for larger amplitudes (figure 10.3C and D) (Bahill, Clark, and Stark, 1975; Collewijn, Erkelens, and Steinman, 1988).

This regularity is important because in principle, each person could have chosen a different trajectory of moving their eyes: For a given displacement, one subject might have moved their eyes more slowly than the other, with a peak velocity that arrived earlier or later. The fact that we see regularity (in healthy people) is exciting because it encourages us to ask why the brain should control the movements in this way. In this chapter and the next, we build a theory that can account for this regularity. Along the way, we will need to establish two ideas: that movements are directed toward rewarding states, and that this reward seems to discount the effort it takes to produce that movement. We will attempt to account for the regularity in movements by claiming that the specific way that we move is the best that we could move, given that our goal is to maximize the rate at which we receive reward.

Let us establish the idea that saccade kinematics are affected by the reward that is expected at the end of the movement. We will do this by reviewing data from a few experiments. Yoriko Takikawa, Reiko Kawagoe, Hideaki Itoh, Hiroyuki Nakahara, and Okihide Hikosaka (2002) trained monkeys to make saccades to a remembered target location (as shown in figure 10.4A). The targets would appear in one of four positions. In each block of trials only one of the four targets was associated with a juice reward. That is, the monkey would get juice after making a saccade to

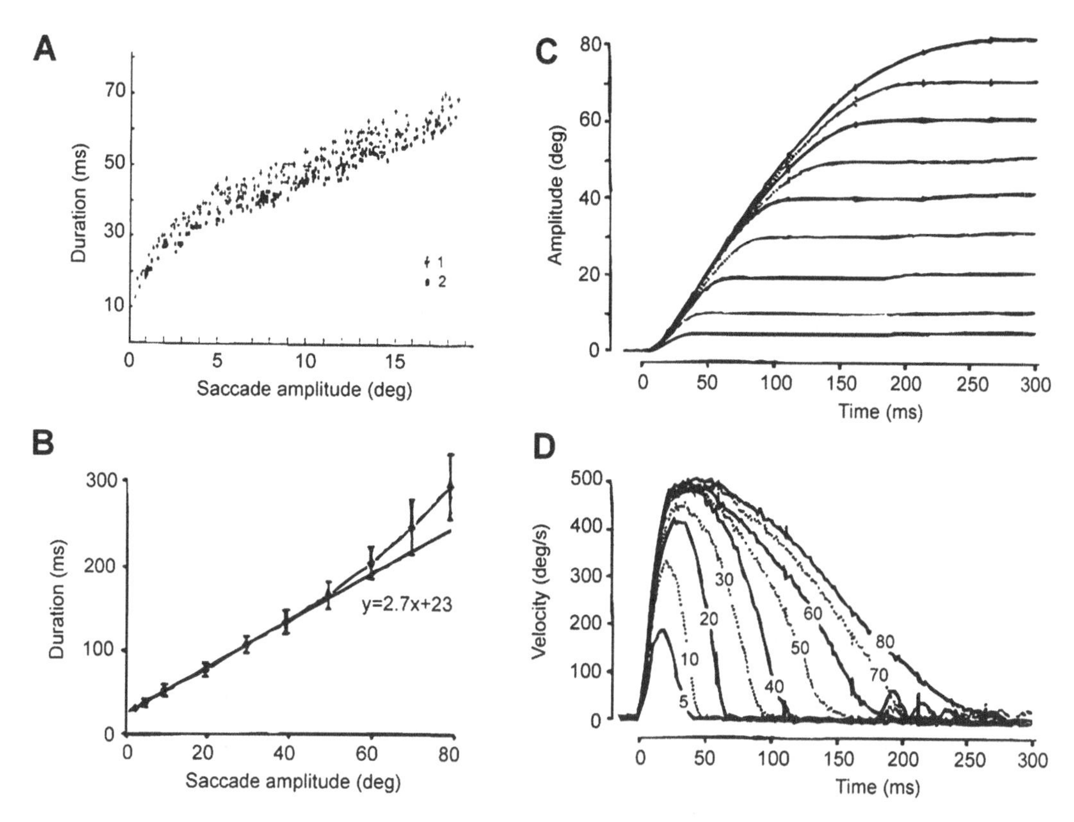

Figure 10.3
Kinematic characteristics of human saccades. (A) Duration and amplitude relationship for small to medium size saccades of two subjects (from Yarbus, 1967). (B) Duration and amplitude relationship for a larger range of saccade amplitudes from three subjects. The error bars are 1SD. Note that the variance of the saccade durations increases with saccade amplitude. (C, D) Amplitude and velocity properties of saccades for one subjects (average of four trials). (Panels B–D from Collewijn et al., 1988.)

this baited target. The monkey would nevertheless have to make saccades to the other targets as well (so that the trials would proceed to the baited target), but would not get a reward for the unbaited targets. Examples of the resulting eye positions and velocities are shown in figure 10.4B. The eye velocity is higher, and latency (reaction time) is shorter, when the target is associated with juice as compared to when it is not (figure 10.4C). In fact, when the monkey has to make a saccade to a nonrewarding target, the movement has a long duration and the velocity profile occasionally has multiple peaks. Therefore, the relative reward that was associated with the stimulus affected the velocity and latency of the movement. In effect, in programming the motor commands that moved the eyes, the brain was more willing to "spend" the motor commands if the result of the movement was some juice.

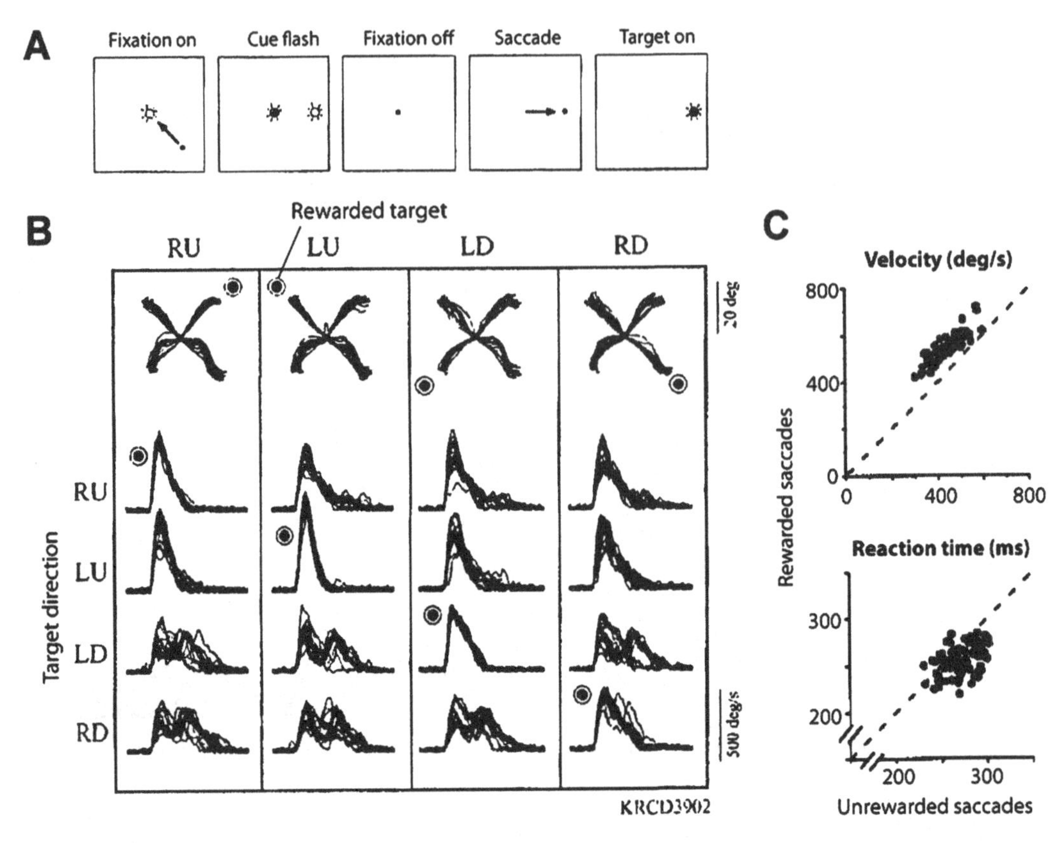

Figure 10.4
Expected reward affects control of movements. (A) The task for the monkey is to saccade to the remembered location of a visual target that can appear in one of four locations (LU: left upper, RD: right down, etc.). However, only one target location in a given set is rewarded. (B) The location of the rewarded target is identified by the label of the column. The top row indicates the rewarded target location in each set (filled circle). The bottom four rows show saccade speed to each target location under each reward condition. When the target is rewarded, saccades to that location have a higher speed, smaller duration, and less variability. (C) Comparison of saccade peak velocities and latencies for rewarded and unrewarded movements. Peak velocities are higher and latencies are shorter for movements that are expected to be rewarded. (From Takikawa et al., 2002.)

This reward dependent variability in saccade velocities was not limited to targets that were baited with juice. For example, consider a scenario in which there are some objects on a table: a cup of tea, a newspaper, a sugar bowl, and some utensils. Suppose you fixate these objects one at a time, making saccades from one to another. Now you decide to reach for the sugar spoon. As you saccade to the spoon, your eye velocity will be higher and the duration of the saccade will be shorter than if you were to make the same amplitude saccade to the spoon but without the intention of reaching for it. That is, saccades that fixate the goal of a reaching movement are faster than saccades that fixate that same image outside the context of reaching (Epelboim et al., 1997; van Donkelaar, Siu, and Walterschied, 2004; Snyder et al., 2002). Perhaps the value of the stimulus that affords the saccade (e.g., the spoon) is higher when that object is also the stimulus that simultaneously affords a reaching movement.

In the laboratory, we like to present the same visual stimulus (often a simple LED) repeatedly, asking the volunteers to fixate it over and over. Just as you might be less interested in looking at a (particularly boring) picture for the fiftieth time, making a repeating sequence of saccades to a small point of light might also reduce your interest regarding the visual target. Indeed, repeatedly making saccades to the same visual stimulus produces eye movements with smaller velocities and longer durations (Montagnini and Chelazzi, 2005; Chen-Harris et al., 2008; Golla et al., 2008). An example of this is shown in figure 10.5A, in which subjects made 60 saccades from one target to another (target LEDs were positioned 15 degrees apart and displayed at a rate of about once per 1.5 sec). The targets appeared in a short, fixed sequence that repeated many times in each set. Haiyin Chen-Harris, Wilsaan Joiner, Vincent Ethier, David Zee, and Reza Shadmehr (2008) noted that saccade velocities dropped as the set proceeded, and then recovered after a short break in which the subjects simply closed their eyes. In almost lockstep fashion, durations increased during each set and then dropped after the short break (figure 10.5B). Saccade amplitudes, however, remained fairly stable and showed little correlation to changes in duration and velocity (figure 10.5C). In other words, the eye arrived at the target, yet with a high velocity and short duration at the start of each set, and with a lower velocity and longer duration at the end of each set. Were these changes due to some form of fatigue in the muscles of the eyes?

To test for this, Minnan Xu-Wilson, Haiyin Chen-Harris, David Zee, and Reza Shadmehr (2009) repeated this experiment but now suddenly changed the sequence of stimuli in the middle of a set. The idea was that if the effect on velocities was due to a muscular or neural fatigue, then not much should happen if one makes a small change in the sequence of targets. However, if the effect was due to devaluation of the stimuli because of their predictability, then a sudden but small change in the sequence should produce recovery. That is, novelty might produce an increase in

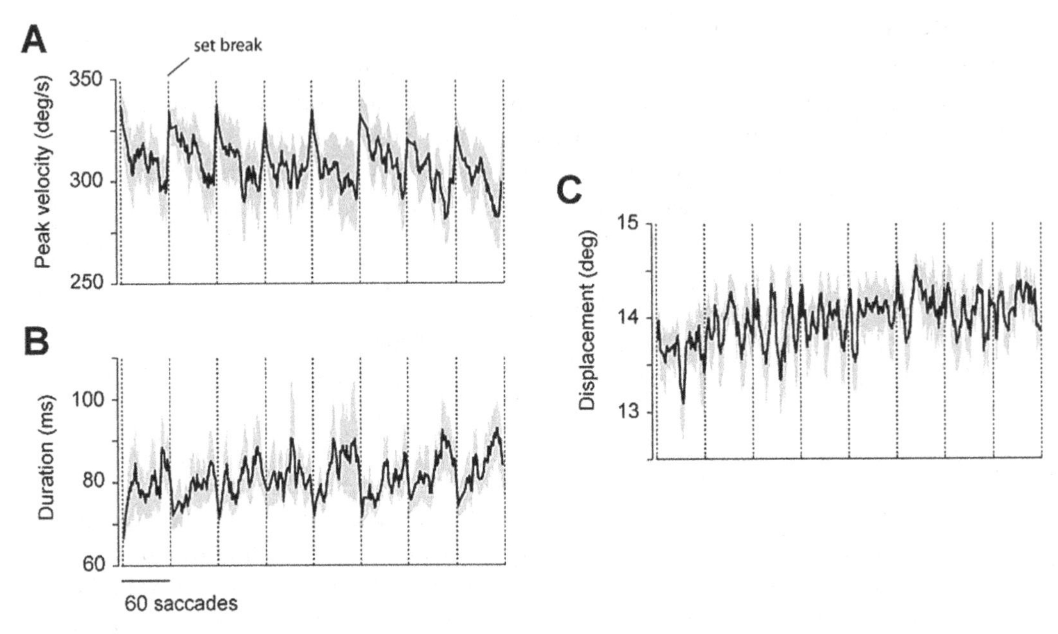

Figure 10.5
Repeatedly making saccades to the same target reduces saccade velocities. (A) In a darkened room, volunteers made saccades to targets that appeared in a short, fixed sequence, at a rate of about one saccade per 1.5 seconds. The figure shows saccade velocities for targets that appeared at 15 degrees displacement. Set breaks were about thirty seconds in duration. (B) Saccade durations increased as velocities decreased. (C) Eye displacement showed little or no correlation to changes in duration and velocity. (From Xu-Wilson et al., 2009.)

value, and therefore an increase in velocities. They found that, on average, during a sequence of 60 saccades the velocities dropped by around 15 percent (figure 10.6B, "set break"). This decrease was not due to fatigue of the oculomotor muscles or circuitry, because when the sequence of targets suddenly changed in the middle of a set, the velocities once again recovered in the very next saccade (figure 10.6B, "sequence change"). These results suggest but certainly do not prove that when a visual stimulus repeats, the internal value that the brain associates with it declines, resulting in a relative increase in the cost of the motor commands, and therefore reduced saccade velocities.

Together, these results suggest that certain manipulations (food, repetition, etc.) alter the implicit value that the brain assigns to the target of the saccade, and that in turn affect the motor commands that move the eyes. To test this idea more directly, Minnan Xu-Wilson, David Zee, and Reza Shadmehr (2009) asked people to make saccades to foveate a point of light in a dimly lit room. Instead of giving people reward in terms of juice or money, the experiment provided them with an image of a face, an inverted face, an object, or simply noise. The experiment began

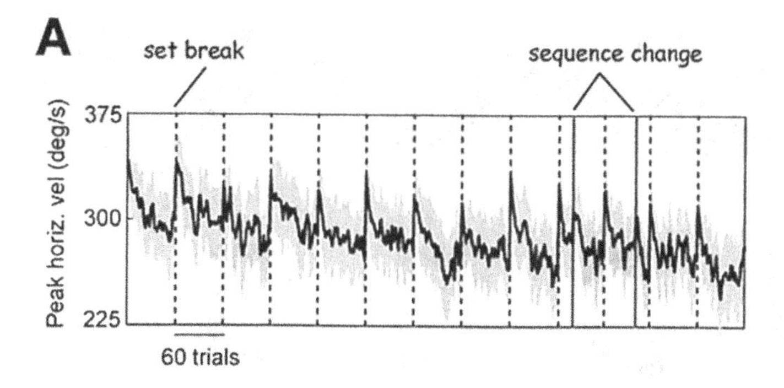

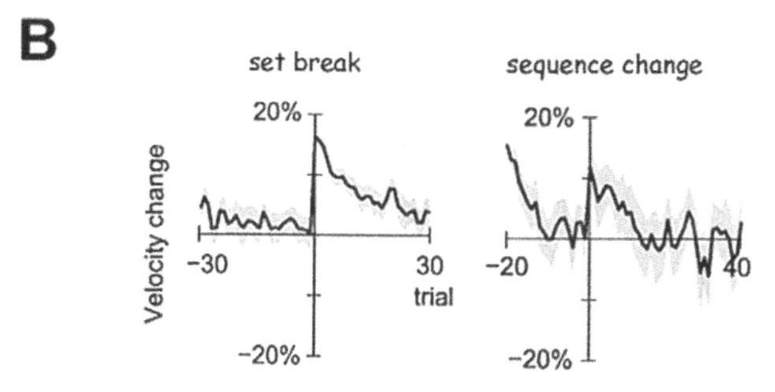

Figure 10.6
Changes in saccade velocities may be related to predictability of the stimulus. (A) In a darkened room, volunteers made saccades to targets that appeared in a short, fixed sequence. The figure shows saccade velocities for targets that appeared at 15 degrees displacement. In the tenth and eleventh sets, the target sequence suddenly changed. The dashed lines indicate brief set breaks. (B) The data were aligned to the saccade following set breaks and to the saccade following sequence change. (From Xu-Wilson et al., 2009.)

with fixation of a point of light, and then flash of an image to one side of fixation (figure 10.7A). The eyes had to maintain fixation and could not move to look at the image (they of course could identify the image as being face or not, but could not saccade to it). After the image disappeared, the light moved, causing the subject to make a saccade. After completion of the saccade, the subjects were provided with the image that had been previously flashed (figure 10.7B). In this way, subjects made a saccade in anticipation of being "rewarded" by an image. The results showed that saccades that were made in anticipation of seeing the image of a face were faster than inverted faces, objects, and noise (figure 10.7D). The implication is that because faces have an intrinsically higher value than other images, this increased value discounts the motor commands, resulting in faster saccades.

In summary, we perform saccades to change the state of our eyes. Usually, this makes it so that a highly valued part of the visual scene falls on our fovea. Therefore, the motor commands that move the eyes during a saccade change the state of the

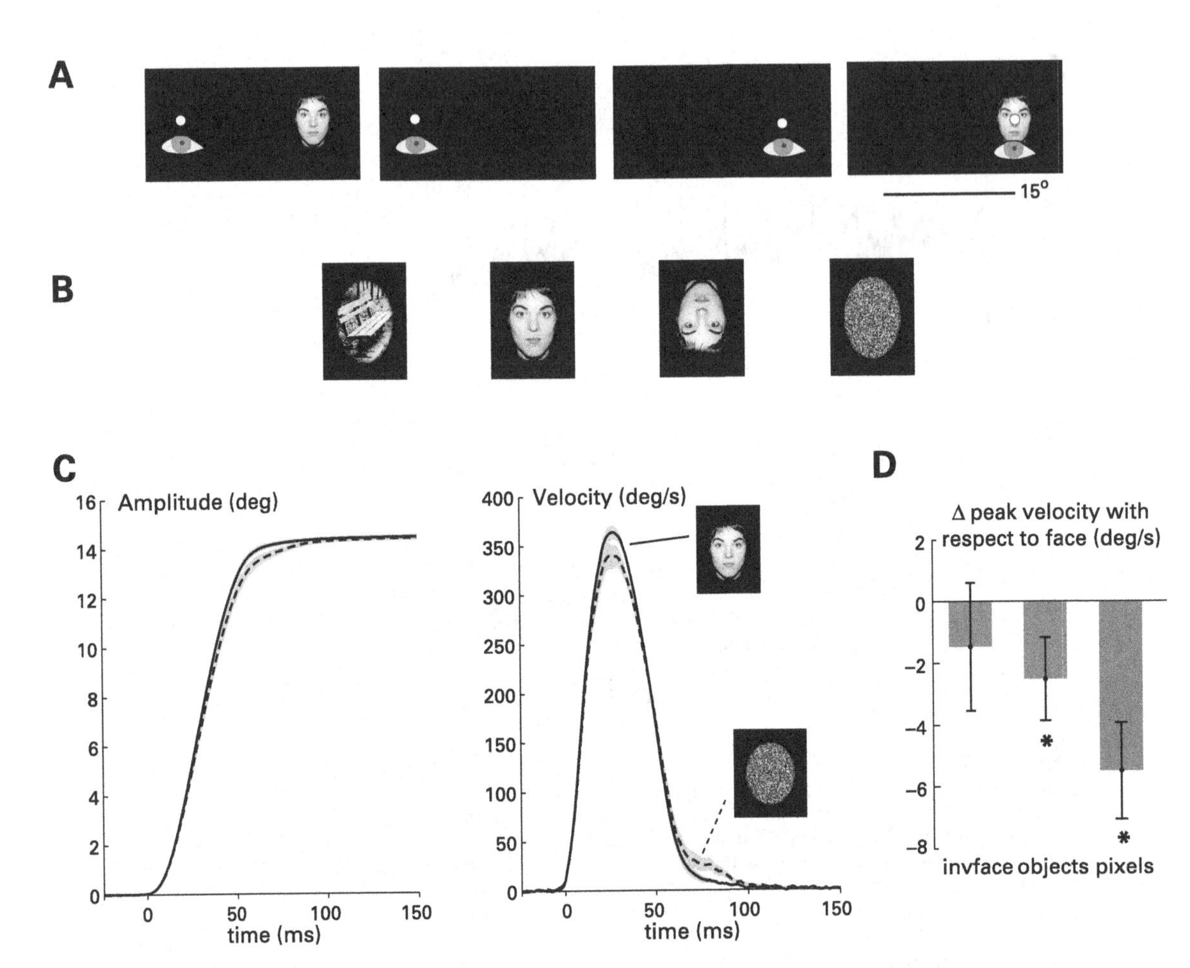

Figure 10.7
The intrinsic value of visual information affects saccade velocities. (A) Study protocol. Volunteers fixated an LED in a darkened room while the picture was flashed at 15 degrees. Subjects maintained fixation until the LED moved. After completion of a saccade, the volunteers were showed the same picture. Therefore, saccades were made in anticipation of viewing a face, an object, inverted face, or noise. (B) Examples of pictures used in the experiment. (C) Subjects made saccades with higher velocities in anticipation of seeing a face. (D) Relative decline in peak speed of saccades with respect to face. (From Xu-Wilson et al., 2009.)

eyes to one that is more valuable. If we consider saccades as an example of the simplest and most stereotypical movements that our brain controls, we see that even here there is no "hard wired" sequence of motor commands that alters the state of our body. Rather, it appears that the relative gain that we expect from the change in state (the expected value of the final state versus the value of the current state) affects the strength of motor commands that our brain generates to produce this change in state.

10.3 Movement Vigor and Encoding of Reward

Writing instruments are one of the most common tools that we use in our daily lives. A striking feature of damage to the human striatum is micrographia, an impairment of writing where letters become very small and writing speed becomes slow. This condition is most common in degenerative diseases of the basal ganglia like Parkinson's disease (Van Gemmert, Teulings, and Stelmach, 2001). However, it can also occur with focal lesions. Figure 10.8 provides an example of micrographia in a patient who suffered an ischemic stroke in the right basal ganglia, in the head of the caudate nucleus (Barbarulo et al., 2007). When he was asked to copy a four- or eight-letter string of characters with his left hand, he wrote much smaller than with his right hand.

Micrographia reflects an abnormal choice of speed and amplitude and is one manifestation of generalized slowing of movement, called bradykinesia.

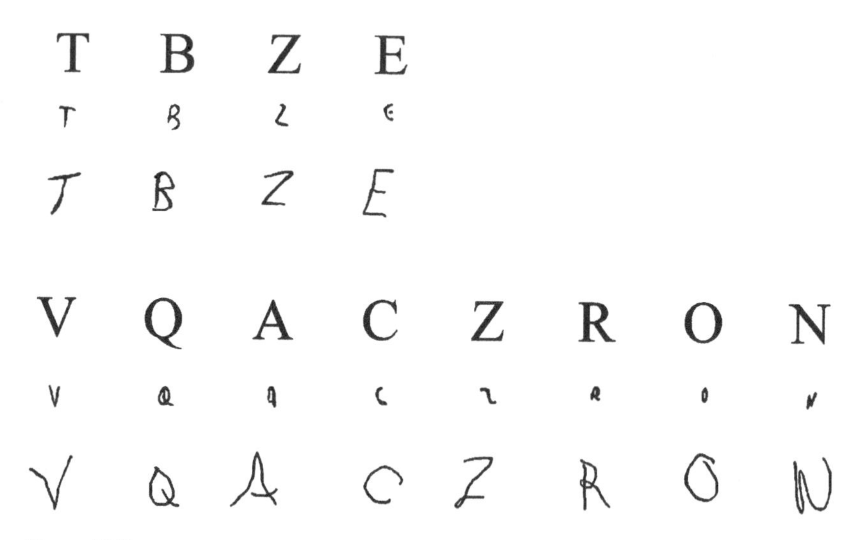

Figure 10.8
Writing ability of patient FF, who suffered a lesion in the right caudate nucleus. Four- and eight-letter string copying (models on the upper lines) by the right (middle lines) and the left hand (lower lines). Micrographia was evident only with the right hand. (From Barbarulo et al., 2007.)

Bradykinesia is most prevalent in Parkinson's disease (PD), a disease that arises from loss of neurons in the substantia nigra (a part of the basal ganglia). These neurons release the neurotransmitter dopamine in various parts of the brain, and their activity is strongly tied to expectation of reward (Fiorillo, Tobler, and Schultz, 2003). For example, when animals are presented with a visual stimulus that predicts the amount and probability of reward (juice), after a period of training the dopaminergic cells in the substantia nigra fire in proportion to the value of the stimulus (amount of reward times its probability). That is, the greater expected value of a stimulus, the greater the release of dopamine.

There is evidence that if an animal has to perform a movement to get reward, in conditions for which the movement is expected to be rewarded a larger dopamine response occurs, and this response coincides with a generally faster movement. Tomas Ljungberg, Paul Apicella, and Wolfram Schultz (1992) trained a monkey to put its hand on a table and wait for opening of a small door that hid a food box (figure 10.9). In some block of trials the box occasionally contained a food reward (called the intermittent task). In the other block of trials, the box always contained a food reward (called the full task). In all trials after the door opened the monkey made a saccade to the box, and when the box was full, the monkey also reached and picked up the food. Interestingly, in the intermittent task in trials in which the box

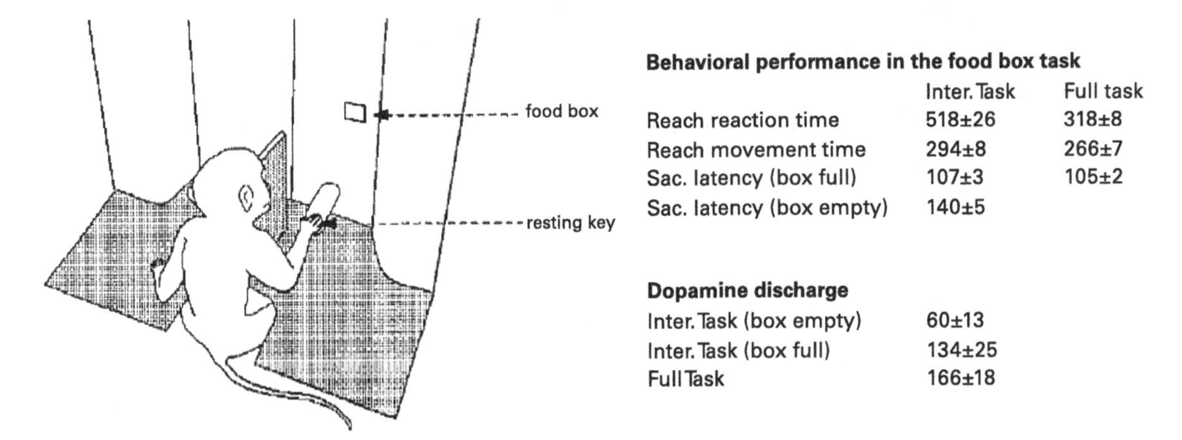

Behavioral performance in the food box task

	Inter. Task	Full task
Reach reaction time	518±26	318±8
Reach movement time	294±8	266±7
Sac. latency (box full)	107±3	105±2
Sac. latency (box empty)	140±5	

Dopamine discharge

Inter. Task (box empty)	60±13
Inter. Task (box full)	134±25
Full Task	166±18

Figure 10.9
Behavioral and dopamine response to a food reward. In animal faces a wall on which there is a food box. The door of the box opens rapidly upward, and the movement is visible and audible to the animals. The animal usually makes a saccade to the box. In the intermittent task, the box occasionally contains food, whereas in the full task, the box always contains food. When the box contains food, the animal reaches into the box and collects the reward. The behavioral performance table shows values in units of ms with ±SE. Reaction and movement times are significantly longer in the intermittent than in the full task. Saccade latencies are significantly longer in trials without food in the box. The neural discharge table shows magnitude of dopamine firing after door opening in a standard time window. (From Ljungberg, Apicella, and Schultz, 1992.)

held a piece of food the saccades had a shorter latency (107 ms, on average, vs. 140 ms in the trials in which the box was empty). After the door opened, dopamine neurons fired, and the firing rate was higher in the trials in which the box contained food than in those when it was empty. In these trials in which the box contained food, the reach was a bit slower and had a longer reaction time as compared to the reaches that were performed in the full task. The discharge of dopamine neurons was also smaller in the intermittent task with the full box versus the full task with the full box. Therefore, dopamine response seems to indicate some aspect of the value associated with a stimulus, and on the occasion in which a movement is performed, greater dopamine release often coincides with a more brisk movement (faster, with a shorter reaction time).

In PD, the dopaminergic cells in the substantia nigra gradually die, which presumably results in a reduced valuation of stimuli. Recently, Pietro Mazzoni, Anna Hristova, and John Krakauer (2007) put forth the idea that in PD, bradykinesia is not due to some inherent inability of these patients to make accurate and fast movements, but it is due to a general devaluation of the state changes caused by the movements, which in turn increases an internal measure of motor costs. They write, "The motor system has its own motivation circuit, which operates analogously to but separately from explicit motivation. . . . We propose that striatal dopamine energizes action in a more literal sense, namely by assigning a value to the energetic cost of moving" (p. 7115). In effect, they suggested that in PD, there is a high cost associated with motor commands because the stimuli that afford movements have an unusually low value assigned to them by the brain.

To test their idea, they asked subjects to make accurate reaching movements of specified speeds without visual feedback of their hand. Just like healthy controls, PD patients produced movements that had greater endpoint variance when they moved faster. However, these faster movements were simply less probable than in healthy controls. When comparing the performance of PD to that of control subjects, the PD patients demonstrated normal spatial accuracy in each condition, but required more trials than controls to accumulate the required number of movements in each speed range. Therefore, the PD patients had an increased reluctance to execute movements requiring greater effort, in spite of comparable spatial accuracy with healthy controls. Perhaps the devaluation of the visual stimulus that was produced by loss of dopamine resulted in an unusually large motor costs.

10.4 Motor Costs

Let us suppose that the purpose of generating a motor command $\mathbf{u}$ is to change our state $\mathbf{x}$ to one that is more rewarding or valuable. Perhaps the specific motor commands that our brain programs depend on a balance between the expected change in value of state, and the expected effort that it takes to produce the motor commands that cause that change in state. That is, there may be a cost to the motor commands, preventing us from taking actions, unless those actions are discounted by the expected increase in the value of our states. What might be the nature of this motor cost?

Let us imagine a task in which we provide reward for performing an action, and then see how the action that is performed changes as the cost of the motor commands changes. Suppose that you were to have a subject hold onto a rigid bar as shown in figure 10.10 (in this case, the subject is a monkey) and provide a task in which the objective is to pull up or down, push left or right, so that a force vector is produced. The handle is attached to a force transducer, and you will reward the subject based on the difference between the goal force $\mathbf{f}_g$ and the force produced $\mathbf{f}$. Now there are lots of muscles that act on the wrist joint, and each has a pulling direction specified by a unit length vector $\mathbf{p}_i$. Let us assume that when muscle i is activated by amount $u_i \geq 0$, it will produce a force $\mathbf{f}_i$ such that:

$$\mathbf{f}_i = u_i \mathbf{p}_i. \tag{10.1}$$

For the sake of argument, assume that there are eight muscles that act on the wrist, with pulling directions shown in figure 10.10A. The resulting force produced by activation vector $\mathbf{u} = [u_1, \cdots, u_8]^T$ is:

$$\mathbf{f} = P\mathbf{u}, \tag{10.2}$$

where $P = [\mathbf{p}_1, \cdots, \mathbf{p}_8]$. Given a goal force $\mathbf{f}_g$, how should these muscles be activated? Let us describe a cost function that penalizes the difference between $\mathbf{f}_g$ and $\mathbf{f}$, as well as activations $\mathbf{u}$:

$$J = (\mathbf{f}_g - P\mathbf{u})^T (\mathbf{f}_g - P\mathbf{u}) + \lambda \sum_i u_i^m. \tag{10.3}$$

We want to see how the patterns of muscle activations change when we change the motor costs (i.e., the parameters λ and m). We set $|\mathbf{f}_g| = 1$ and vary its direction along a circle and minimize the above cost for various values of λ and m, arriving at $\mathbf{u}^* = [u_1^*, \cdots, u_8^*]^T$. For example, for a goal force along direction θ, the optimum activation u_1^* is plotted in figure 10.10A as a vector with magnitude u_1^* along direction θ. In figure 10.10B, each muscle's activation $u_i^*(\theta)$ is normalized with respect to the maximum activation for that muscle, resulting in a "tuning function." The residual force magnitude $|\mathbf{f}_g - \mathbf{f}|$ is plotted in figure 10.10C. We note that the muscles are not necessarily most active in their pulling directions. For example, muscles with pulling directions along 270 and 340 degrees are maximally active to force directions other than their pulling directions. We also note that as λ changes, there is little change in the shape of the tuning functions (the effect is mainly on the residuals). Most importantly, as m increases, the tuning functions become broad. That is, when m is small, the muscles do not "share" the burden of producing force, but rather specialize in

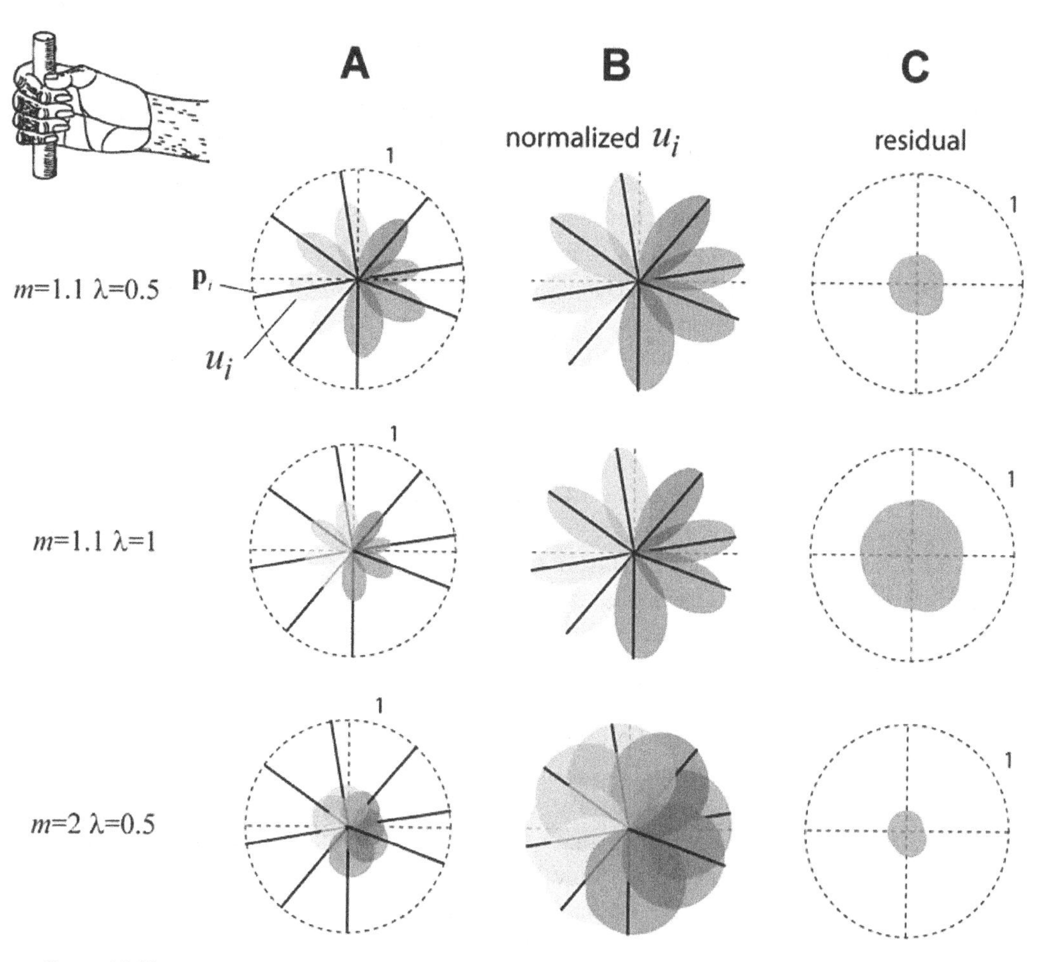

Figure 10.10
Minimization of the cost specified by equation (10.3) with the muscle model of equation (10.2). Broad tuning curves result from increased motor costs. (A) Muscle activation patterns. (B) Normalized activation patterns, that is, tuning curves. (C) Force residuals.

their own pulling direction and do little in other directions. In contrast, as m increases, the muscles share the burden, becoming active in many directions. The interesting idea is that the shape of the tuning functions of muscles may be a useful way to infer the nature of the motor costs used by the brain.

The idea that the shape of the muscle tuning functions may be a reflection of the motor cost was put forth by Andrew Fagg, Ashwin Shah, and Andrew Barto (2002). They considered the EMG data that was recorded by Donna Hoffman and Peter Strick (1999) from the wrist muscles of a monkey who was instructed to produce force in various directions (figure 10.11A). Hoffman and Strick had noted that muscle tuning functions were broad, cosine-like, and that several muscles were maximally activated in directions that differed significantly from their pulling directions. (They had established the pulling direction of each muscle by electrically stimulating that muscle and noting the force produced at the handle.) For example, the tuning functions for muscle ECRB (extensor carpi radialis brevis), ECRL (extensor carpi radialis longus), ECU (extensor carpi ulnaris), and FCR (flexor carpi radialis) are shown in figure 10.11A. In this figure, the pulling direction of the muscle is noted by the solid line, and the arrow indicates the direction of maximum activation for that muscle during the voluntary force production task. For example, note that for ECRB the pulling direction (as defined by artificial stimulation of the muscle) is about 45 degrees apart from the peak of the tuning function (as defined by voluntary activation of that muscle). The simulation results for increasing values of m are shown in figure 10.11B, and for increasing values of various λ are shown in figure 10.11C. A reasonable fit to the data is at $m = 2$ and $\lambda = 0.5$, in which the simulations reproduce both the broad tuning and the mismatch between the pulling direction and peak of the tuning function.

Broad tuning functions are a fundamental characteristic of muscles and neurons in the primate motor system. For example, in the primary motor cortex, neurons exhibit broad, cosine-like tuning functions similar to those found in muscles (Georgopoulos, Schwartz, and Kettner, 1986). To see why increasing m in equation (10.3) broadens the shape of the tuning functions, consider a simplified scenario in which we have two muscles that pull nearly in the same direction. When $m \approx 1$, producing a total force of 1N can be done optimally when one of the muscles produces 1N and the other is inactive. However, when $m = 2$, the smallest motor cost is attained when each muscle produces a force of 0.5N. As Fagg, Shah, and Barto note, "Under the total squared muscle activation criterion, both muscles would be activated to equal levels, rather than activating one and not the other."

Quadratic motor costs (i.e., $m = 2$) discourage specialization, and encourage cooperation. The broad tuning of muscles and neurons may arise in part as a reflection of a cost structure in which the motor commands are penalized in a quadratic fashion.

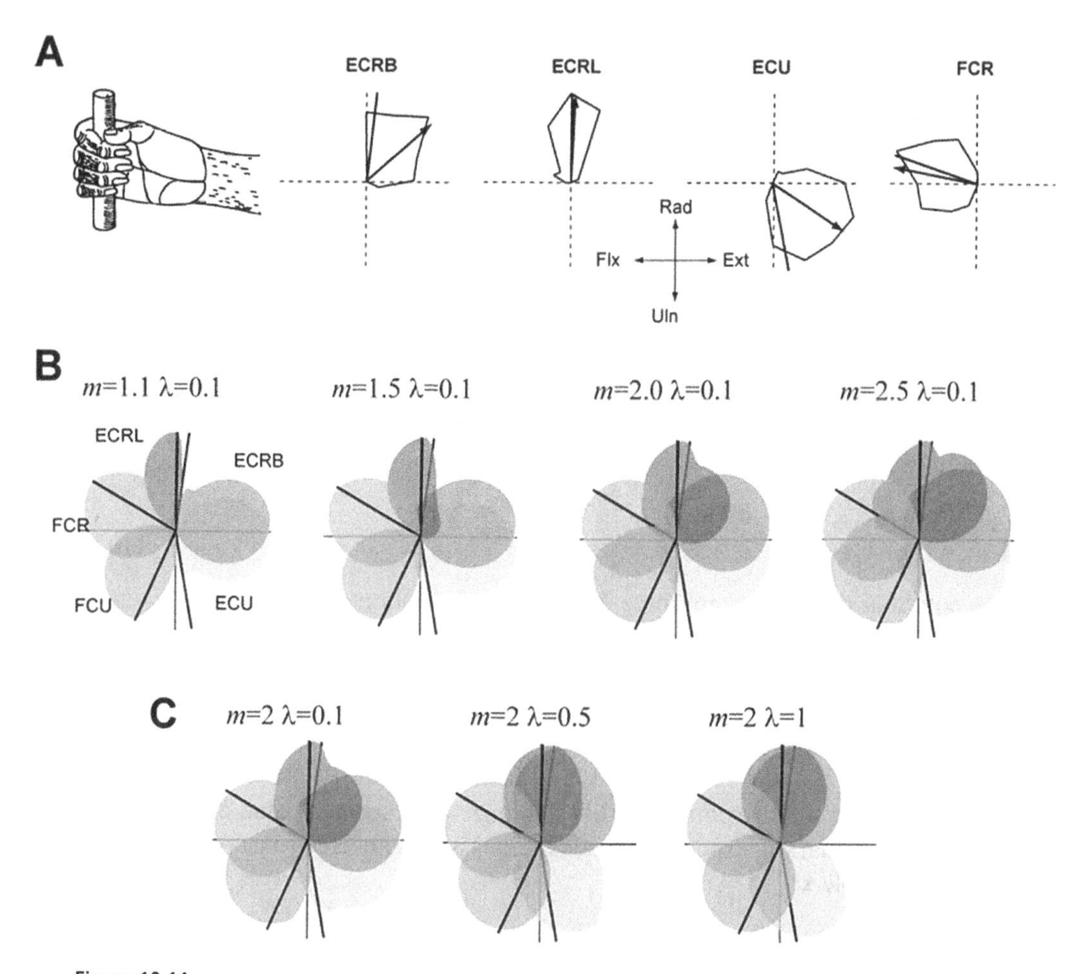

Figure 10.11
Wrist activation patterns during an isometric force production task. (A) Normalized EMG activation patterns for four wrist muscles: ECRB (extensor carpi radialis brevis), ECRL (extensor carpi radialis longus), ECU (extensor carpi ulnaris), and FCR (flexor carpi radialis). The line without an arrow indicates the pulling direction of the muscle when it is electrically stimulated, that is, the direction of force that the muscle produces when it is stimulated. The line with an arrow indicates preferred direction of the muscle, that is, direction of force for which the muscle is maximally activated during voluntary recruitment. (Data from Fagg, Shah, and Barto, 2002.) (B, C) Muscle tuning curves that minimize the cost specified by equation (10.3).

10.5 Motor Noise and Variability in Performance

A significant problem with our approach is that we have not considered the fact that when muscles are activated, they produce forces that have stochastic properties. This stochastic behavior will produce variability in the force, potentially affecting the ability to acquire reward (the first term in equation 10.3). Let us show that when we consider the noise properties of muscles, we can no longer infer that their broad tuning is a reflection of motor costs (the second term in equation 10.3). Rather, as we will see, the broad tuning can also be associated with maximizing reward.

Recall that the noise properties of muscles are not Gaussian, but signal dependent. In particular, the force generated by motor commands has signal dependent noise properties (figure 4.19). We can incorporate this fact into equation (10.1):

$$\mathbf{f}_i = u_i \mathbf{p}_i (1 + k_i \phi_i) \qquad \phi_i \sim N(0,1). \tag{10.4}$$

We note that the variance of force increases as a function of the signal u_i:

$$\text{var}[\mathbf{f}_i] = u_i^2 k_i^2 \mathbf{p}_i \mathbf{p}_i^T. \tag{10.5}$$

And so the standard deviation of force increases linearly with u_i, with slope k_i. The cost J in equation (10.3) is of course a scalar, but it now becomes a stochastic variable due to the properties of the first term in equation (10.3). The best that we can do is to find the motor commands that minimize the expected value of this cost. A useful identity that will expedite our calculations is the expected value of a "squared" random variable $\mathbf{x}$:

$$E[\mathbf{x}^T A \mathbf{x}] = E[\mathbf{x}]^T A E[\mathbf{x}] + tr[A\, \text{var}[\mathbf{x}]]. \tag{10.6}$$

The second term in equation (10.6) is the trace operator. Noting that $\mathbf{f} = \sum_i \mathbf{f}_i$, the expected value of our cost is:

$$E[J] = E\left[\mathbf{f}_g - \sum_i \mathbf{f}_i\right]^T E\left[\mathbf{f}_g - \sum_i \mathbf{f}_i\right] + tr\left[\text{var}\left[\mathbf{f}_g - \sum_i \mathbf{f}_i\right]\right] + \lambda \sum_i u_i^m. \tag{10.7}$$

The second term in equation (10.7) is:

$$\begin{aligned} tr\left[\text{var}\left[\mathbf{f}_g - \sum_i \mathbf{f}_i\right]\right] &= tr\left[\sum_i u_i^2 k_i^2 \mathbf{p}_i \mathbf{p}_i^T\right] \\ &= \sum_i u_i^2 k_i^2 \mathbf{p}_i^T \mathbf{p}_i \end{aligned} \tag{10.8}$$

The expected value of our cost becomes:

$$E[J] = (\mathbf{f}_g - P\mathbf{u})^T (\mathbf{f}_g - P\mathbf{u}) + \sum_i u_i^2 k_i^2 \mathbf{p}_i^T \mathbf{p}_i + \lambda \sum_i u_i^m. \tag{10.9}$$

When we compare equation (10.9) with equation (10.3), we see that when the motor commands have noise properties that are signal dependent, in effect we are adding a quadratic motor cost to our cost function (the second term in equation 10.9). As we increase the motor commands, we also increase the variance of these commands, affecting the variance of the output force, the task relevant variable. Because we are rewarded based on a measure of accuracy in our force, increasing the motor commands will cost us in terms of accuracy, and this cost appears as a squared term on the motor commands. (This would not be the case if the noise process were Gaussian, that is, a process in which noise variance was independent of the mean of the signal.)

For example, when the system is noiseless ($k_i = 0$ in equation 10.4), setting $m \approx 1$ produces narrowly tuned muscle activation patterns (figure 10.12A). However, when the same system has signal dependent noise ($k_i = 1$ in equation 10.4), setting $m \approx 1$ now produces broadly tuned muscle activation patterns (figure 10.12B). Therefore, the broadly tuned activation patterns (for example, the patterns observed in the wrist muscle in figure 10.10) are not necessarily a result of a quadratic motor cost. Rather, the same broadly tuning

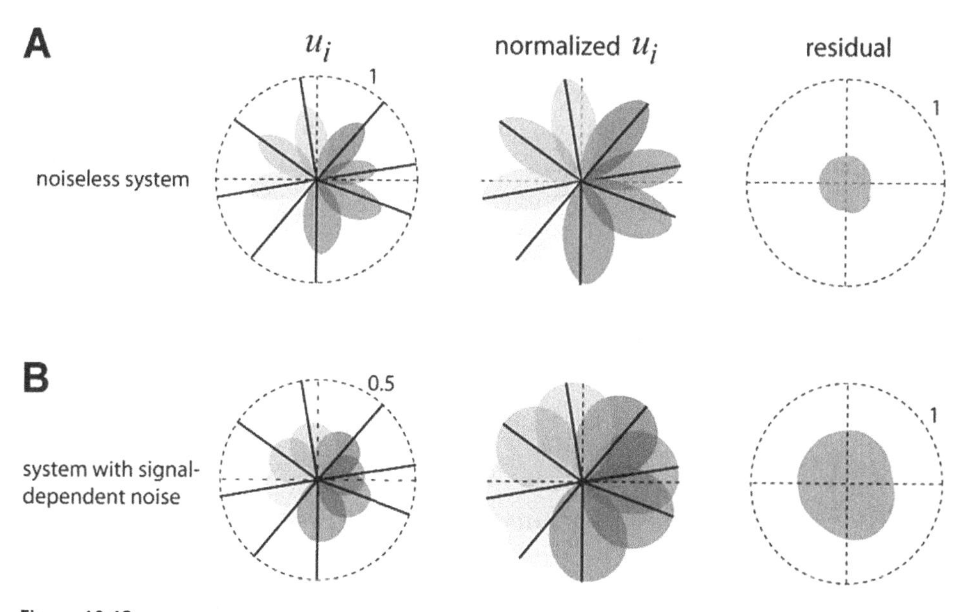

Figure 10.12
Motor commands that suffer from signal dependent noise result in broad tuning functions of muscles. (A) The noiseless system. The pattern of motor commands that minimize equation (10.9) with the constraint given in equation (10.4), with $k_i = 0$, $m = 1.1$, and $\lambda = 0.5$. (B) A system with signal dependent noise. The pattern of motor commands that minimize equation (10.9) with the constraint given in equation (10.4), with $k_i = 1$, $m = 1.1$, and $\lambda = 0.5$. The format of this figure is similar to figure 10.8.

can result when the system suffers from signal dependent noise. This is because the motor noise affects accuracy, and reward is often a function of accuracy. The problem that we face is that noise in the motor commands affects our accuracy, which in turn affects our ability to get reward. We need a way to dissociate motor costs from the effects that the motor commands have on accuracy.

10.6 Maximizing Performance While Minimizing Effort

How can we dissociate between costs that truly penalize effort and costs that penalize the inaccuracy that results from their noise? We need to know the noise properties of the system that generates force. With this information, we can estimate how much this noise affects the cost function, and then see if the system still penalizes the motor costs over and above those that arise from signal dependent noise properties of the system.

The first experiment that dissociated performance variability versus effort costs was performed by Ian O'Sullivan, Etienne Burdet, and Jörn Diedrichsen (2009). In the task, volunteers were asked to put one finger of the right hand on a force transducer, and one finger of the left hand on another force transducer (figure 10.12). They pushed with the two fingers (either the index or the little finger on each hand) so that a cursor whose position represented the sum of the two forces was placed in a target area during a seven-second period. The subjects were rewarded inversely proportional to the mean squared error of force during that period. Let us assume that the displacement of the cursor due to the force produced by a finger on the right hand is:

$$x_i \sim N\left(u_i, k_i^2 u_i^2\right). \tag{10.10}$$

Further suppose that there are no motor costs, and the only objective is to minimize the squared error between the sum of the forces produced by the two fingers and the goal force:

$$\begin{aligned} J &= E\left[\left(x_i + x_j - g\right)^2\right] \\ &= E\left[\left(x_i + x_j - g\right)\right]^2 + \mathrm{var}\left[\left(x_i + x_j - g\right)\right]. \\ &= \left(u_i + u_j - g\right)^2 + k_i^2 u_i^2 + k_j^2 u_j^2 \end{aligned} \tag{10.11}$$

Minimizing this cost results in the following optimal motor commands:

$$\begin{aligned} \frac{dJ}{du_i} &= 2\left(u_i + u_j - g\right) + 2k_i^2 u_i = 0 \qquad & u_i^* &= \frac{g - u_j}{1 + k_i^2} \\ \frac{dJ}{du_j} &= 2\left(u_i + u_j - g\right) + 2k_j^2 u_i = 0 \qquad & u_j^* &= \frac{g - u_i}{1 + k_j^2} \end{aligned} . \tag{10.12}$$

Simplifying the first equation in equation (10.12), we have:

$$u_i^* = \frac{g - \frac{g - u_i}{1 + k_j^2}}{1 + k_i^2} = \frac{g k_j^2 + u_i}{(1 + k_i^2)(1 + k_j^2)}.$$

We can now view the optimal motor commands by describing the percentage of the force produced by one hand as a function of the total force by the two hands:

$$\frac{u_i^*}{u_i^* + u_j^*} = \frac{k_j^2}{k_i^2 + k_j^2}. \tag{10.13}$$

The result in equation (10.13) implies that if the noise on the left finger is greater than that on the right, then the right hand should contribute more to the total force. The implication is that if the cost function depends only on acquisition of reward (equation 10.11), then the subject should divide the total force based on the ratio of the noise on each finger. O'Sullivan, Burdet, and Diedrichsen measured the noise properties of each finger on each hand using the same procedure as in the main task, arriving at an estimate of the rate at which standard deviation of the force for each finger increased as a function of mean force (data for a typical subject is shown in figure 10.13A). Using this estimate of k for each of the two fingers of each hand, they then computed the optimal ratio of forces and compared it to the actually measured ratio. If the selection of forces on the right and left hand depended solely on maximizing reward, then the data should fall on the diagonal line of figure 10.13B. The actual data did not agree well with this prediction.

As an alternative, O'Sullivan, Burdet, and Diedrichsen considered a cost that penalized not only a measure of inaccuracy in the task relevant variable but also a measure of motor costs:

$$J = vE\left[(x_i + x_j - g)^2\right] + \lambda(u_i^2 + u_j^2) + \mu\left(u_i^2/MVC_i^2 + u_j^2/MVC_j^2\right). \tag{10.14}$$

The term MVC_i refers to the maximum voluntary contraction of muscle i. In equation (10.14), motor commands are penalized in terms of their absolute value, and also in terms of their ratio with respect to MVC. Simplifying the preceding equation, we have:

$$\begin{aligned} J &= v(u_i + u_j - g)^2 + \left(vk_i^2 + \lambda + \mu/MVC_i^2\right)u_i^2 + \left(vk_j^2 + \lambda + \mu/MVC_j^2\right)u_j^2 \\ &= v(u_i + u_j - g)^2 + a_i u_i^2 + a_j u_j^2 \end{aligned}, \tag{10.15}$$

where $a_i = vk_i^2 + \lambda + \mu/MVC_i^2$. Minimizing this cost, we have:

$$\begin{aligned} \frac{dJ}{du_i} &= 2v(u_i + u_j - g) + 2a_i u_i \\ u_i^* &= -\frac{v(u_j - g)}{(v - a_i)} = \frac{vga_j}{-va_i - va_j + a_i a_j} \end{aligned}. \tag{10.16}$$

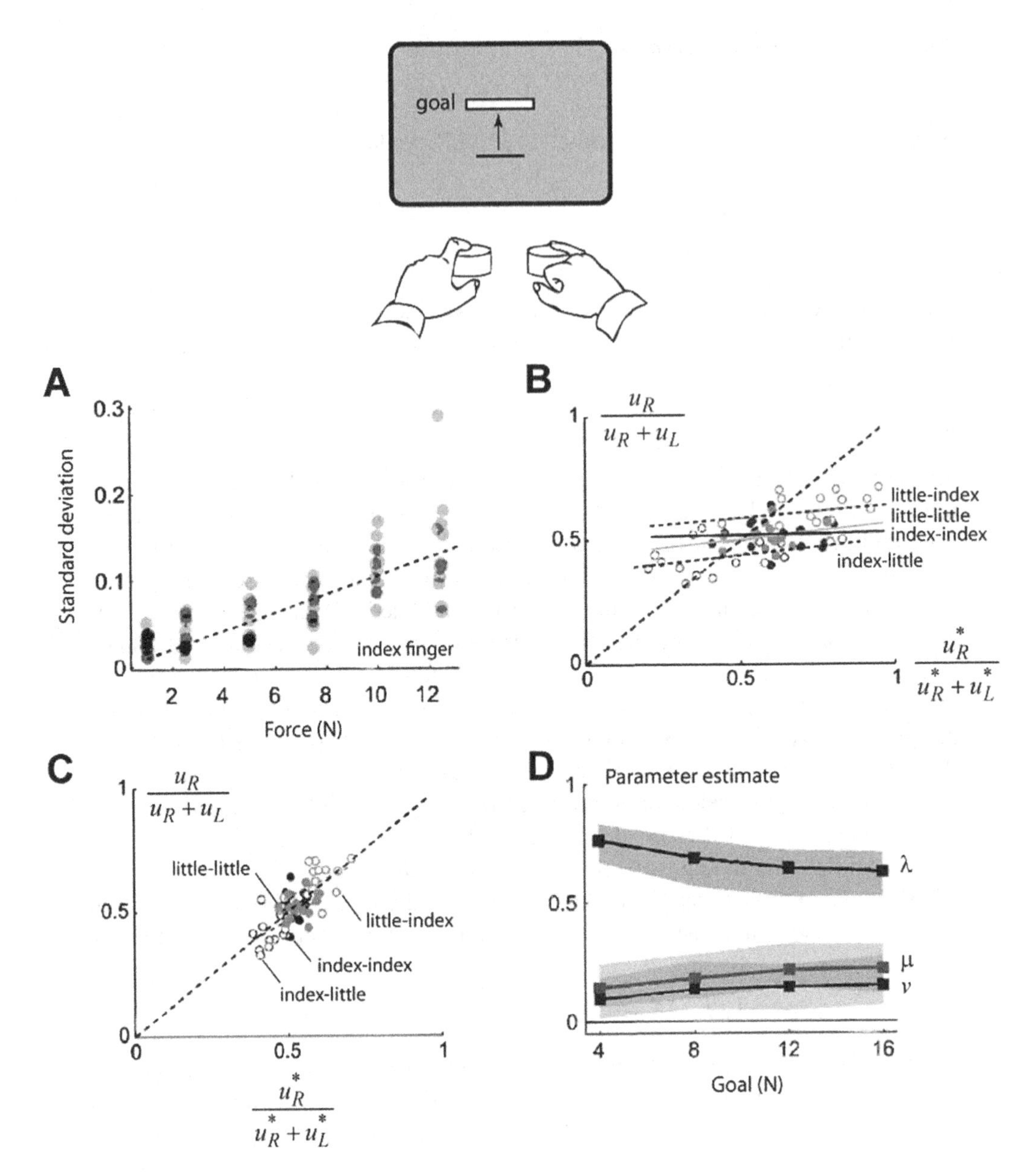

Figure 10.13
Volunteers were asked to push on two force transducers so that the sum of the two forces, as displayed by a cursor, was equal to a goal. They were rewarded inversely proportional to the mean squared error of the force. (A) When subjects were asked to perform the task with a single finger, the standard deviation of the force (as shown for one subject) increased as a function of the mean force. This exemplifies the signal dependent noise properties of the motor commands. (B) If the cost minimized by the subjects depended only on a measure of accuracy (i.e., reward), then the optimal motor commands for each condition and each hand/finger combination should be along the dashed line. (C) When the costs explicitly penalize motor commands, as well as inaccuracy (equation 10.14), there is a better fit between the predicted motor commands and the actually observed values. (D) Parameter estimate for the cost function in equation (10.14). (From (O'Sullivan, Burdet, and Diedrichsen, 2009.)

Once again, we can view the optimal motor commands by describing the percentage of the force produced by one hand as a function of the total force by the two hands:

$$\frac{u_i^*}{u_i^* + u_j^*} = \frac{a_j}{a_j + a_i}. \tag{10.17}$$

The authors then fit the model (equation 10.17) to the data (actual ratio of the forces) and found the two free parameters. (There were only two free parameters because of the constraint that the sum of the three parameters $\nu + \lambda + \mu$ should equal one.) Performance of the fitted model and the parameter values are shown in figure 10.13C and D. The term that penalized the quadratic motor costs had the largest value.

Therefore, in this simple bimanual force production task, the distribution of forces between the two hands was not done to maximize accuracy (despite the fact that maximizing accuracy was the criterion for which the subjects were rewarded). Rather, the cost of maximizing accuracy was discounted by a cost that penalized the motor commands, that is, a motor cost. The most costs appeared to be approximately associated with the squared force produced by each finger.

10.7 Motor Costs during a Movement

Most of the experiments that we have considered thus far were isometric. Motor costs also appear to influence planning of voluntary movements. Our first example is a simple task described by Yoji Uno, Mitsuo Kawato, and Ryoji Suzuki (1989) (shown in figure 10.14). The objective for the volunteers was to reach from one point to another. In one condition, the subject holds a lightweight tool that moves freely in air. In a second condition, the tool is attached to a spring that pulls the hand to the right. Without the spring, people reach in a straight line. However, once the spring is attached, the straight path incurs substantially more motor costs than a curved path. The curved path is the one that subjects choose.

In our second example, the task is to move one's hand from one point to another in a given amount of time (450 ms), but now instead of a spring, there is a velocity dependent force field that pushes the hand perpendicular to its direction of motion (Shadmehr and Mussa-Ivaldi, 1994). The field is shown as a sequence of arrows in figure 10.15A. We might consider two kinds of cost functions here: one that minimizes a measure of kinematics (e.g., a measure of smoothness like squared jerk of the hand), and one that attempts to bring the hand to the target while minimizing a measure of effort (e.g., squared force). Before the field is imposed, the motion that minimizes both kinds of costs is simply a straight line with a bell-shaped velocity profile. However, when the field is imposed, the solution for the kinematic cost remains a straight line, while the solution for the effort cost is no longer a straight

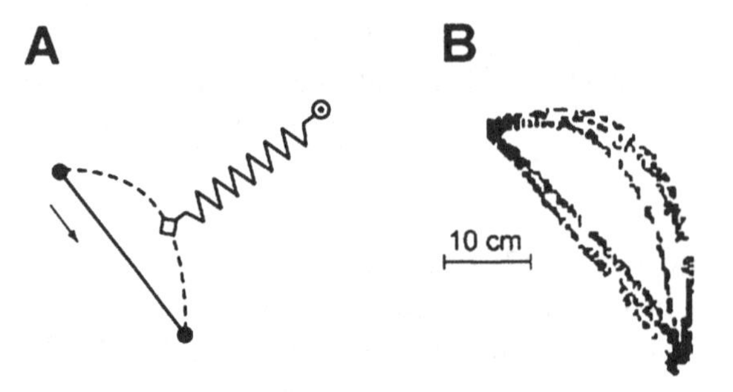

Figure 10.14
Motor costs affect reach trajectories. (A) The task is to reach from one point to another. In one condition, the reach takes place in free space (straight line). In another condition, a spring is attached to the hand. In this experiment, the maximum force produced by the spring was 10.4N and the minimum was 3.3N. (B) Data from four reaching movements of a single subject. The subject chooses to move the hand along an arc. (Data redrawn from Uno, Kawato, and Suzuki, 1989.)

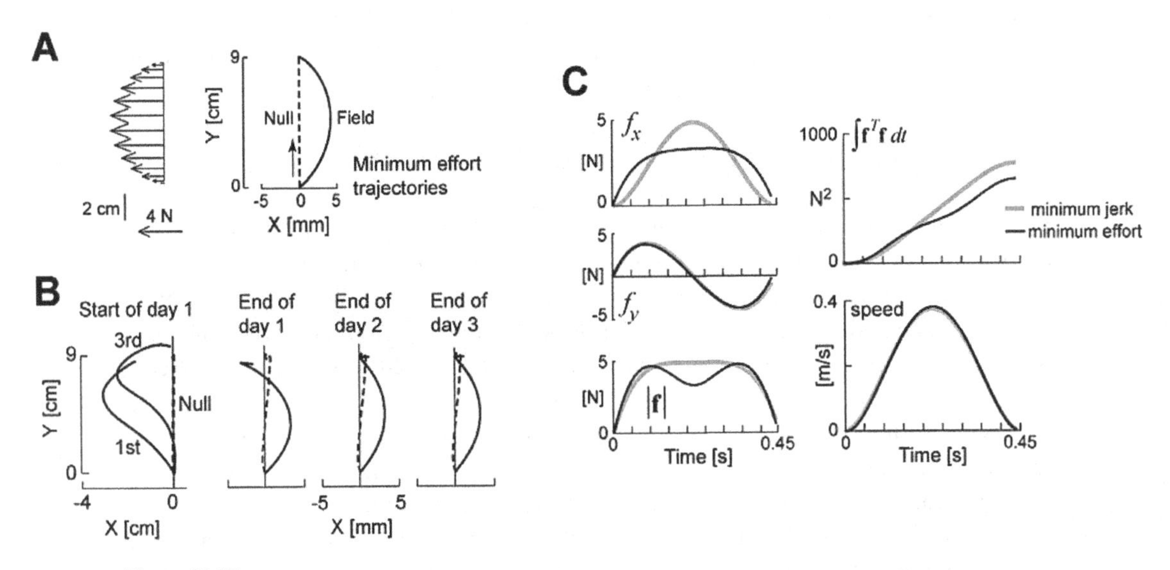

Figure 10.15
During adaptation, the motor commands attempt to minimize motor costs. (A) A velocity dependent force field pushes the hand perpendicular to its direction of motion. For example, for an upward movement the forces push the hand to the left. The motion that minimizes a motor costs composed of the sum of squared forces is not a straight line, but one that has a curvature to the right. The data show hand paths for a typical subject at start of training on day 1, and then end of training each day. Except for the first and third trials, all other trajectories are the average of fifty trials. (B) A rationale for why a curved movement is of lower cost. The curves show simulation results on forces $\mathbf{f} = [f_x \quad f_y]^T$ that the controller produces, and speed of movement $\sqrt{\dot{x}^2 + \dot{y}^2}$, in an optimal control scenario in which the objective is to minimize total motor costs $J = \int_0^p \mathbf{f}^T \mathbf{f} dt$, and in a scenario in which the objective is to minimize squared jerk $J = \int_0^p (\dddot{x}^2 + \dddot{y}^2) dt$. (Data from Izawa et al., 2008).

line. Jun Izawa, Tushar Rane, Opher Donchin, and Reza Shadmehr (2008) showed that if the field pushes the hand to the left, the policy that produces the least effort is one that moves the hand slightly to the right of a straight line, resulting in a curved movement that appears to overcompensate for the forces (figure 10.15A). To test this idea, they had subjects train in the field for a long period (three sessions over three days). They found that with training, subjects did not return their hand trajectories to a straight line. Rather, they produced a curved trajectory (figure 10.15B). To see the rationale for this behavior, figure 10.15C plots the forces produced by a policy that minimizes effort, and compares it to forces that must be produced if a mass is moving along a minimum-jerk trajectory. By moving the hand along a curved path, the controller produces less total squared force: It overcompensates early into the movement when the field is weak but undercompensates at peak speed when the field is strongest. Therefore, the curved path that appears to overcompensate for the forces actually requires less total force than a straight trajectory. If we define effort as the sum of squared force, then this path is the one that minimizes an effort cost.

These examples suggest that movement trajectories are a result of motor commands that produce changes in state that attempt to meet task goals (e.g., bring the hand to the target), while minimizing some measure of effort (e.g., cumulative squared force). The task goals describe how we will be rewarded for our efforts, while task costs describe how much we will work to acquire this reward.

Summary

We move, perhaps, because our motor commands are expected to change the state of our body (or state of the environment) to something that is more advantageous or rewarding. For example, each saccade is a movement with which the brain directs the fovea to a region where currently, the value is highest. Images of faces tend to have higher value for us humans than most other images, which is why we tend to first look at the face in a portrait.

The relative gain that we expect from the change in state (the expected value of the final state vs. the value of the current state) affects the strength of motor commands that our brain generates to produce this change in state. For example, saccades that are made to stimuli that are expected to be more rewarding have higher speed. That is, the expected reward discounts the motor commands. Diseases that affect the reward prediction mechanism of the brain produce motor disorders that maybe understood in terms of unusually small or large discounting of motor commands.

Discounting of motor commands implies that the brain maintains an implicit cost associated with generating a movement. To explore the nature of these motor costs, a number of studies have considered how the brain solves redundancies in which many muscles contribute to generating a force. Under an assumption of noiseless or Gaussian noise process, penalizing motor commands tends to discourage specialization of muscles and encourage cooperation. The implication is that the broad tuning of muscles and neurons are a reflection of a cost structure in which the motor commands are penalized.

However, motor commands tend to be affected by a noise process that is signal-dependent, in which the standard deviation of the signal increases with the mean. In this more biologically plausible scenario, the same broad tuning of muscles and neurons arises not from a motor cost, but a cost associated with minimizing variance (that is, a task cost that attempts to maximize reward). Recent experiments, however, suggest that motor commands are penalized over and above the cost associated with task performance. That is, our motor commands are due to a cost structure that includes two components: attain the greatest reward possible, while being as lazy as possible. In the next chapter we will consider the third component of this cost: time.

11 Cost of Time in Motor Control

Perhaps in no area of neuroscience is the cooperation between theory and data more fruitful than in the study of movements. The reason has much to do with the fact that movements of people and other animals tend to be highly reproducible. For example, reaching movements, eye movements, and walking each have distinct features that are present in all healthy people. A teenager may actively choose to dress very differently from his parents, but he will not exhibit walking, reaching, or eye movements that are very different. Indeed, a persistent difference in movements is often a sign of a neurological disorder. Our aim here is to describe some of this regularity and attempt to build a theory that may account for it.

Over the last twenty-five years, a large body of experimental and computational work has been directed toward offering theories that attempt to explain why there is regularity in our movements. The approaches are reminiscent of physics and its earliest attempts to explain regularity in motion of celestial objects. As with physics, the search in the field of motor control has been for normative laws, or cost functions, that govern our behavior. Here, we will focus on a theory that assumes that the purpose of movements is to achieve a rewarding state at a minimum effort. That is, we are going to assume that we make movements in order to change the state of our body to something more valuable, and effort is the cost that we pay to acquire reward.

Consider the behavior of the eyes during saccadic eye movements. When we are looking at an image, the brain shifts our gaze from location to location with saccades. During each saccade, the eyes move in a stereotypical trajectory, with temporal and spatial characteristics that are described in figure 10.3. These characteristics include a nearly linear relationship between saccade amplitude and duration (figure 10.3B) and a symmetric trajectory of eye velocities for small amplitude saccades but asymmetric velocities during large amplitude saccades (figure 10.3D). However, these characteristics are variable and the variability is not entirely due to random noise. For example, when we make a saccade to look at a face, the eyes move slightly faster and arrive at the target slightly sooner than if the goal of our saccade was to foveate another kind of image (figure 10.7). When monkeys are rewarded with juice for moving their eyes to a location, their saccade has a shorter duration and higher peak speed than when the same location is unrewarded (figure 10.4). Therefore, while saccades show certain patterns in their temporal and spatial properties, the variability in these patterns appears to have something to do with the value of the stimulus that affords the movement. Possibly, the expected reward alters the motor commands that the brain programs to make a movement. In particular, movements seem to be faster (higher speed) and have a shorter reaction time (figure 10.4C) when the value of the stimulus is higher.

These results hint that the subjective value of a state that we wish to attain by our movements somehow discounts the motor commands, making the movement faster and with a shorter duration. In a sense, by spending a greater effort, the brain produces motor commands that attain a valuable stimulus sooner. Time, even in the range of milliseconds, seems to matter because it provides the reward sooner. These ideas lead us toward a cost for movements: It seems that reward is something that is good and we should get as much of it as possible, but effort is something that is bad and we should spend as little of it as possible. A reward that is attained sooner is better than the same reward if it is attained later. By mathematically describing these ideas, a framework will be constructed that can account for the specific saccade durations and velocities of healthy humans. More interestingly, we will be able to consider changes in the reward system of the brain due to development, aging, and disease, and try to make sense of changes that occur in control of movements.

11.1 Temporal Discounting of Reward

Economists and psychologists have noted that humans (and other animals) tend to discount reward as a function of time. An example of this is an experiment by Joel Myerson and Leonard Green (1995) in which undergraduate students were asked to make a series of hypothetical decisions between monetary rewards. They were shown one card that specified how much they would be paid now, and another card that specified a larger amount that they would be paid if they waited some amount of time. Based on the choice that each subject made, the authors constructed a function that estimated how value of the reward declined as a function of time (in this case, years). For example, say that a subject picks at roughly 50 percent probability between $350 now and $1,000 in five years. This would suggest that for this person, $350 now is equal to $1,000 in five years, which would imply that $1,000 loses its value as a function of time. Myerson and Green considered two kinds of functions to describe this temporal discounting of value: hyperbolic and exponential.

$$V_e(p) = \alpha \exp(-\beta p)$$

$$V_h(p) = \frac{\alpha}{1+\beta p}$$

The variable p represents the time delay to reward acquisition. The exponential function is one that is typically used in economics and is called a discounted utility model. It assumes that future rewards are discounted because of the risk involved in waiting for them. The hyperbolic function is one that has been used by psychologists, and its rationale is based on empirical grounds: It fits the data better (figure 11.1A). (However, as we will see later, there may be a good reason for why animals use hyperbolic discount functions, since they may be trying to maximize reward per unit of time.) In particular, as can be seen in figure 11.1A, exponential discounting underestimates the value of reward at longer time delays. People generally do not discount future reward as steeply as is predicted by an exponential—the hyperbolic form of the discount function fits the data better.

Interestingly, the basic form of the temporal discount function remains hyperbolic when we go from time delay of years to time delay of seconds. Koji Jimura and

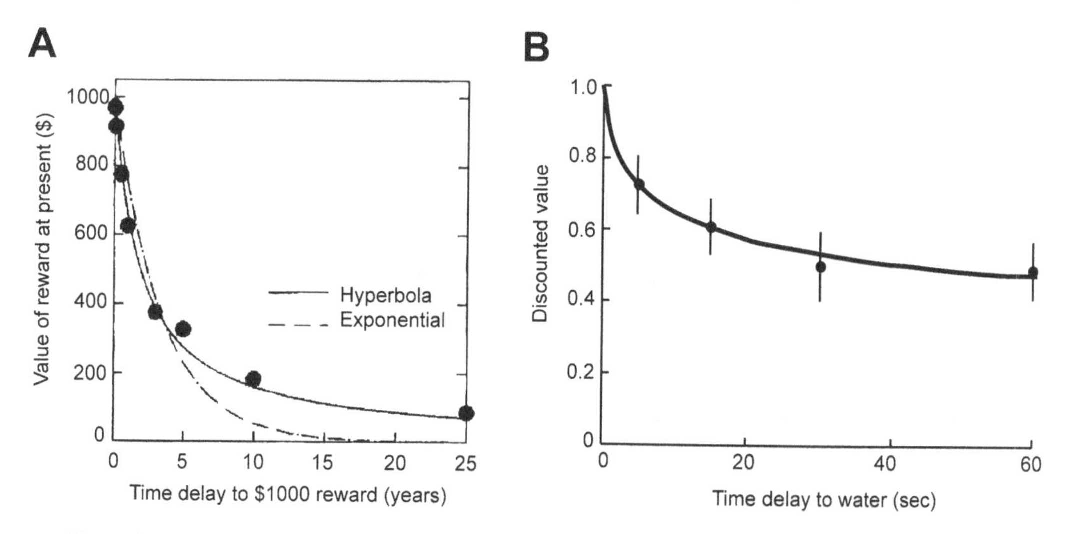

Figure 11.1
Temporal discounting of reward in people. (A) Value of $1,000 as a function of time to acquisition of the money. For example, on average, undergraduate students would accept $400 now or $1,000 at four years at equal probability. The best fits to the data via a hyperbolic function or an exponential function are shown. Exponential functions tend to fit poorly particularly for long delays. (Data from Myerson and Green, 1995.) (B) Temporal discounting of water for thirsty undergraduate students. The volunteers were given a choice between a small amount of water now and a larger amount some seconds later. The points show subjective equality. For example, students would take about 8 ml of water now than wait 30 s for 16 ml. (Data from Jimura et al., 2009).

colleagues (2009) asked undergraduates not to drink anything for a few hours and then gave them a choice between a small amount of water now and a larger amount some seconds later. They found that the temporal discount function was a hyperbola, with a rate $\beta = 0.09$ per second (figure 11.1B). Shunsuke Kobayashi and Wolfram Schultz (2008) trained thirsty monkeys to choose between a small amount of juice now and a greater amount some seconds later and found a temporal discount rate of $\beta = 0.31$ per second for one monkey and $\beta = 0.17$ per second for another monkey (figure 11.2A). The larger value of β indicates that the subject is more *impulsive*, that is, would rather take a smaller amount of reward now than wait for the larger amount. Kenway Louie and Paul Glimcher (2010) also trained thirsty monkeys to choose between a small amount of juice now and a greater amount some seconds later and found a temporal discount rate of $\beta = 0.16$ per second for one monkey and $\beta = 0.04$ per second for the another (figure 11.2C and D). Leonard Green and colleagues (2004) trained pigeons to choose between a small amount of food now or a greater amount a few seconds later and found a temporal discount rate of $\beta = 0.62$ per second. Although it is difficult to compare these results directly because the experiments are somewhat different, there are two basic ideas that emerge from this vast literature: (1) animals tend to discount reward hyperbolically as a function of time, and (2) while there are intersubject differences within species (e.g., some monkeys are more impulsive than others), there are also large differences between species, with pigeons generally more impulsive than rhesus monkeys and rhesus monkeys more impulsive than humans.

A hyperbolic function of time is not only a good fit to choices that people and other animals make regarding the temporal discounting of valuable commodities, it is also a good fit to discharge of dopamine cells in the brain of monkeys that have been trained to associate visual stimuli with delayed reward. Shunsuke Kobayashi and Wolfram Schultz (2008) trained two monkeys to watch a monitor upon which visual stimuli appeared. In a given trial, one stimulus was followed by a short delay and then a small amount of liquid. In another trial, another stimulus was followed by a longer delay and the same amount of liquid. They then recorded from dopaminergic cells in the substantia nigra (a part of the basal ganglia) and found that in response to the stimulus that predicted reward at short delay, discharge was high. In contrast, in response to the stimulus that predicted the same amount of reward after a long delay, discharge was low. This response declined approximately hyperbolically as a function of delay (figure 11.2B). Intriguingly, for the monkey that was more impulsive (monkey A in figure 11.2A), the dopamine function was slightly steeper (monkey A in figure 11.2B). Therefore, dopaminergic cells tend to give a short burst of discharge in response to a stimulus that predicts future reward, and the magnitude of this burst declines hyperbolically as a function of the expected delay to reward.

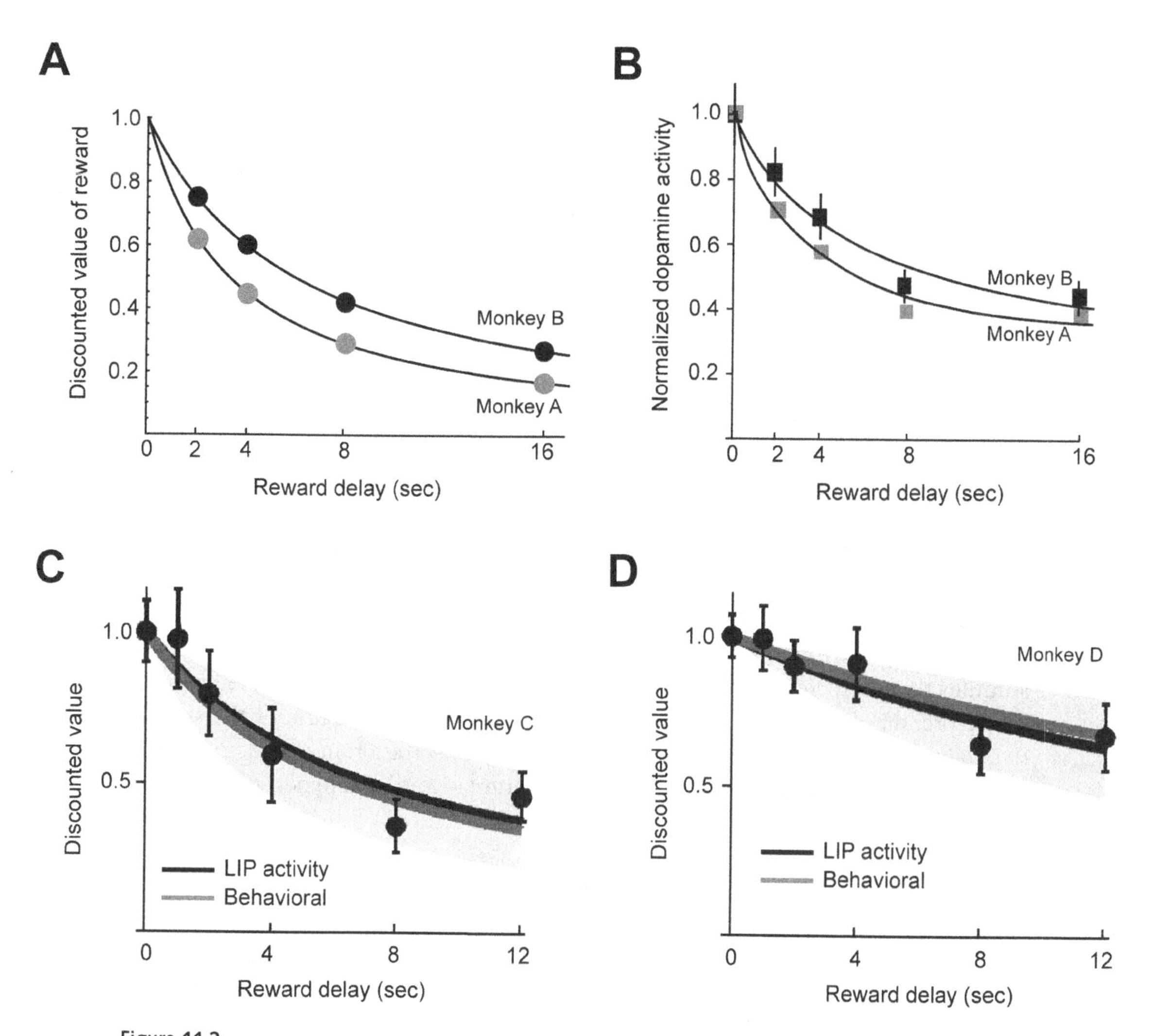

Figure 11.2
Behavioral and neural correlates of temporal discounting of reward in rhesus monkeys. (A) Monkeys were trained to fixate a center dot on a monitor while two visual stimuli appeared on the left and right side of the fixation point. One stimulus was associated with a small reward that would be available now, and the other stimulus was associated with a larger reward that would be available some seconds later. The animal made a saccade to one of the stimuli to indicate its choice. The data summarizes the choices made by the animals via a temporal discounting function. (Data from Kobayashi and Schultz, 2008.) (B) Discharge of dopaminergic cells in the substantia nigra in response to stimuli that were paired with a given amount of reward at some time in the future. In the task, the animals were trained on five visual stimuli, each of which was paired with a constant amount of reward delivered immediately, at 2 s, 4 s, etc. The plot shows the response of the population of dopamine cells to presentation of the visual stimulus. (Data from Kobayashi and Schultz, 2008.) (C) Behavioral temporal discounting of liquid reward for a monkey, along with normalized activity of cells in the LIP area of the same monkey during the decision-making task. The animal made a choice between small reward now versus a larger reward that would be available some seconds later. The animals indicated its choice by making a saccade to the visual stimulus associated with that reward/delay. In the period before the saccade, LIP cells that had their receptive field on the eventual choice responded by discharging proportional to the subjective value of the stimulus. (Data from Louie and Glimcher, 2010.) (D) Same as in (C), but for a different monkey.

This hyperbolic encoding of future reward is also present in the response of neurons in the lateral intraparietal area (LIP), an area of the brain that mediates sensorimotor transformations for control of eye movements. Cells in this area respond to stimuli in a selective region of the visual space (called a receptive field), firing before a saccade to that location. Louie and Glimcher (2010) trained monkeys in a task in which two stimuli would appear (red and green LED), and the animal had the choice of making a saccade to one of the stimuli. In a given block of trials, the red stimulus was associated with an immediate reward (small magnitude), and the green stimulus was associated with a delayed reward (larger magnitude). The animal would saccade to indicate the choice of which reward it wanted. By manipulating the magnitude of the delayed reward the authors constructed the behavioral temporal discount function for each monkey (figure 11.2C and D). Monkey C was more impulsive than monkey D. Next, the authors recorded from cells in LIP. When a stimulus appeared in a cell's receptive field, the cell responded with a firing rate that increased with the subjective stimulus value (figure 11.2C and D). That is, for the more impulsive monkey (monkey C), LIP cells tended to fire much less for a stimulus that predicted a 12 s delayed reward as compared to an immediate reward. Therefore, the discharge in LIP before the saccade was a reflection of the subjective value of the upcoming movement. It appears that the brain encodes the economic value of a given stimulus that promises future reward as a hyperbolic function of time.

11.2 Hyperbolic vs. Exponential Discounting of Reward

We only have a few data points in each subplot of figures 11.1 and 11.2, and they are often missing error bars. Why has previous work preferred to use hyperbolas to represent temporal discounting of reward vs. exponentials? The reason has to do with a basic property of decision making: change of mind.

Suppose that on Friday morning you are informed that there will be a test in one of your classes on Monday. Later that morning a friend mentions that she is planning to go out to dinner and dancing on Saturday evening, saying that it would be great if you could come. You are faced with a choice: stay home on Saturday and study for the test or go out with friends. Suppose that the value that you associate with passing your test is greater than the value that you assign to going out with friends. Indeed, on Friday morning you evaluate your two options and decide to stay home on Saturday. However, as Saturday evening approaches, you change your mind and decide to go out with friends. Let us show that hyperbolic discounting of reward can explain this behavior, whereas exponential discounting cannot.

In figure 11.3 we have plotted the temporal discounting of two rewarding events that will take place in the future. One event has a greater value but will take place

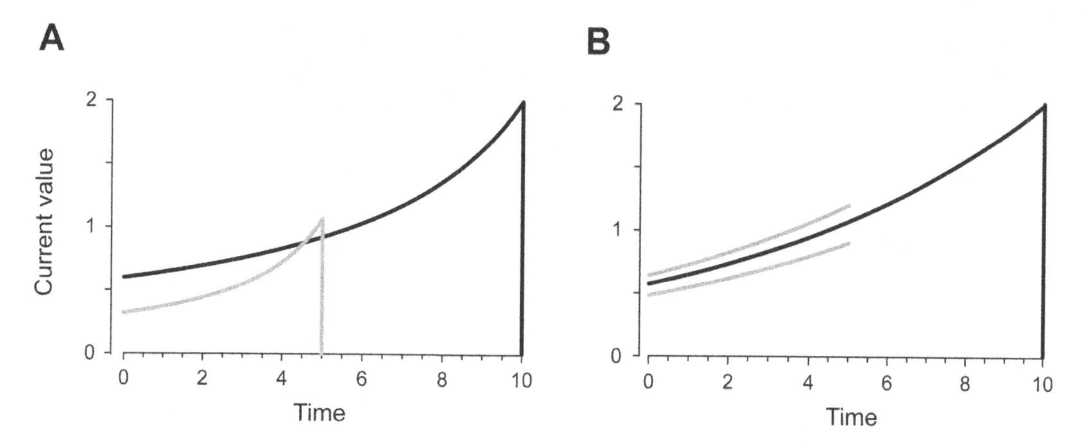

Figure 11.3
Change of mind and hyperbolic vs. exponential discounting of reward. (A) Hyperbolic discounting. At time zero, the discounted value associated with the later event (black line) is higher than the sooner event (gray line). Therefore, at time zero you pick the later event. However, as time passes and we get closer to the time of the sooner event, its discounted value becomes greater than the later event. This produces a change of mind. (B) Exponential discounting. If at time zero the discounted value of the sooner event is less than the later event, its value will remain smaller of the two choices at all times, preventing a change of mind.

later (passing your test); the other event has a smaller value but will take place sooner (going out with friends). The two rewarding events are discounted in time at the same rate. So all that matters is the value that you assign to the events at the time that they will occur. Consider a hyperbolic form of temporal discounting of these two events (figure 11.3A). At time zero (Friday morning) you evaluate your choices and pick the later event (passing your test, black line), as this has a higher current value. However, as you get closer to the time of the sooner event (going out with friends), its value becomes greater than the later event (gray line crosses the black line). On Saturday evening you change your mind and decide to go out with friends. However, if you discounted reward exponentially (figure 11.3B), the two functions would never cross and you would not change your mind. The fact that people and other animals do occasionally change their mind is inconsistent with exponential temporal discounting of reward.

11.3 A Cost for Movements

Let us now return to the problem of making movements. Suppose that the objective of any movement is to place our body in a state that is more rewarding. Further suppose that the value of the state that we wish to acquire declines hyperbolically with time, meaning that it is better to get to this valuable state sooner rather than later. But how much better is it to get there sooner? Well, that depends on our subjective value of the stimulus and our temporal discount function. However, in order to acquire this state we will have to spend some effort. Usually, the greatest effort will be associated with acquiring the rewarding state sooner (move quickly), and the smaller effort will be associated with acquiring the rewarding state later (move slowly). Putting these ideas together, we get a natural balance between effortful behavior and maximizing reward: the movement that we perform is one that maximizes reward while minimizing effort.

Here, we will construct a simple model of the eye's dynamics, produce a saccade that minimizes a cost, and explore the influence of temporal discounting on the movement kinematics. Our objective is to ask why eye trajectories have their specific regularities, and why a change in the value associated with the target of the movement might alter this trajectory. As we will see, there appears to be a link between how the brain temporally discounts reward, and the trajectory that the eye follows during a saccade. The basic idea is that the motor commands that move our body are a reflection of an economic decision regarding reward and effort.

Suppose that in order to make a movement the brain solves the following problem: generate motor commands to acquire as much reward as possible, while expending as little effort as possible. Suppose that at time t, the state of our eye is described by vector $\mathbf{x}(t)$ (representing position, velocity, etc.), our motor commands are $u(t)$, and our target is a stimulus at position g (with respect to the fovea). Our brain assigns some value α to the visual stimulus that happens to be the target of our saccade. The reward is acquired when the image of the target is on the fovea, which will require a movement that will take time. The crucial question is how passage of time should affect the reward value of the target. That is, how much more valuable is it for us to place the face on our fovea within 50 ms vs. 100 ms? Reza Shadmehr, Jean-Jacques Orban de Xivry, Minnan Xu-Wilson, and Ting-Yu Shih (2010) proposed that the same temporal discount function that described devaluation of reward in timescales of years and seconds may also describe this devaluation in timescale of milliseconds. That is, the motor system will incur a cost for delaying the acquisition of reward because of the time p that it takes to place the valuable image on the fovea:

$$J_p = \alpha\left(1 - \frac{1}{1+\beta p}\right). \tag{11.1}$$

Therefore, the longer it takes to get the target on the fovea, the larger the loss of reward value. More important, note that the cost depends on the initial value α that we assign to the stimulus, and the rate β with which we temporally discount this value. If the stimulus was interesting (e.g., face of an attractive person), we incur a greater cost for the duration of the movement than if the stimulus was uninteresting (e.g., an image made of random noise). Similarly, if one subject is more impulsive than another (i.e., a larger β), then the passage of a given amount of time will cost more for the impulsive subject.

In order to move the eyes, we will have to spend some effort in terms of motor commands. We have little information about how the brain represents motor costs, but in the previous chapter we saw some evidence that this cost is approximately a quadratic function of force. For example, Fagg, Shah, and Barto (2002) suggested that the width of the tuning function of muscles that act on the wrist could be accounted for if one assumes that the brain activates the muscles in such as way as to minimize a cost composed of the sum of the squared force contribution of each muscle (figure 10.11). Similarly, O'Sullivan, Burdet, and Diedrichsen (2009) estimated that in a task in which fingers of the two hands cooperate to produce a goal force, the brain generated a force in each finger so to minimize the sum of squared forces (figure 10.14). Based on this admittedly limited body of evidence, the best that we can currently do is to define motor costs during a movement as the sum of squared muscle activations:

$$J_u = \lambda \int_0^p u^2(t)dt. \tag{11.2}$$

The parameter λ specifies the relative cost that we incur for the motor commands. Finally, when our movement ends at time $t = p$, the target at position g should be on the fovea. This constitutes an accuracy cost, and it is mathematically convenient to represent it as a quadratic function:

$$J_x = \tau(x(p) - g)^2. \tag{11.3}$$

In equation (11.3), the term $x(p)$ represents the position of our eye at the end of the movement. The parameter τ specifies the relative cost that we incur for being inaccurate.

In summary, we assume that in performing a movement, the brain attempts to produce motor commands that minimize a cost composed of accuracy cost, effort cost, and temporal cost:

$$J = J_x + J_u + J_p. \tag{11.4}$$

Our first objective is to ask whether a hyperbolic temporal cost can account for the kinematics (i.e., duration, velocity, etc.) of saccades. Our second objective is to ask whether this temporal cost is related to reward processing in the brain. These objectives require solving an optimal control problem in which equation (11.4) serves as a cost function. The crucial prediction of the theory is that there should be specific changes in saccade durations and velocities due to changes in the reward discounting function (equation 11.1), for example, due to changes in stimulus reward value α or temporal discounting rate β.

11.4 Optimal Control of Eye Movements

In this section we will describe a very simply model of the human eye plant (i.e., dynamics of the eyes), and then find a set of motor commands that bring the eyes to the target while minimizing a cost (equation 11.4). Our objective will be to find the "best movement" that can be performed in order to be as lazy as possible (least effort), while getting as much reward as possible (least devaluation).

In our one-dimensional model of the eye in figure 11.4A, there are two elastic elements that pull the eye in each direction and a viscous element that resists this motion. Suppose that the stiffness of each spring is $k/2$. The force in the bottom spring is $\frac{k}{2}x$ and the force in the top spring is $\frac{k}{2}(x_0 - x)$. The dynamics of this system is:

$$m\ddot{x} = -\frac{k}{2}x - b\dot{x} + f + \frac{k}{2}(x_0 - x),$$

which can be simplified to:

$$m\ddot{x} = -k\left(x - \frac{x_0}{2}\right) - b\dot{x} + f. \tag{11.5}$$

If we redefine x so that we measure it from $\frac{x_0}{2}$, then the equivalent system is shown in figure 11.4B, where the equilibrium point of the spring is at $x = 0$. As a result, the dynamics of our system becomes:

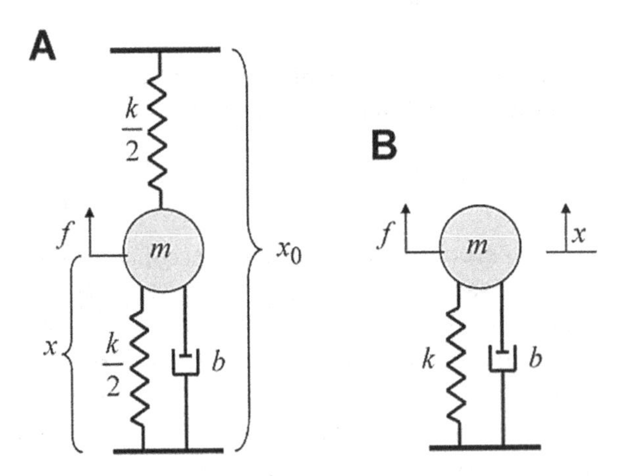

Figure 11.4
A one-dimensional model of the eye plant. Models in (A) and (B) are equivalent.

$$m\ddot{x} = -kx - b\dot{x} + f. \tag{11.6}$$

Solving for acceleration, we have:

$$\ddot{x} = -\frac{k}{m}x - \frac{b}{m}\dot{x} + \frac{1}{m}f. \tag{11.7}$$

The term f represents the active forces produced by the muscles. A very simple muscle model is one that transforms "activations" u into force:

$$\alpha_1 \frac{df}{dt} + \alpha_2 f = u. \tag{11.8}$$

We can rewrite this third-order dynamics as a sequence of first-order differential equations. Let us define $x_1 \equiv x$, $x_2 \equiv \dot{x}$, and $x_3 \equiv f$. We have:

$$\begin{bmatrix} \dot{x}_1 \\ \dot{x}_2 \\ \dot{x}_3 \end{bmatrix} = \begin{bmatrix} 0 & 1 & 0 \\ -\frac{k}{m} & -\frac{b}{m} & \frac{1}{m} \\ 0 & 0 & -\frac{\alpha_2}{\alpha_1} \end{bmatrix} \begin{bmatrix} x_1 \\ x_2 \\ x_3 \end{bmatrix} + \begin{bmatrix} 0 \\ 0 \\ \frac{1}{\alpha_1} \end{bmatrix} u. \tag{11.9}$$

David Robinson and colleagues (Keller, 1973; Robinson, Gordon, and Gordon, 1986) perturbed the human eyes, measured the mechanical response, and found that the system could be approximated via the linear system of equation (11.9) with time constants of 224, 13, and 4 ms. (In equation (11.9), if we set $k = 1$, then $b = \tau_1 + \tau_2$ and $m = \tau_1\tau_2$, where $\tau_1 = 0.224$ and $\tau_2 = 0.013$. If we set $\alpha_2 = 1$, then $\alpha_1 = 0.004$.) Our goal is to find the motor commands $u(t)$ that move the eyes so that we get as much reward as possible (i.e., minimize the temporal discounting of reward), while being as lazy and accurate as possible (i.e., minimize the effort and accuracy costs). To accomplish our goal, we will reformulate the dynamics from continuous to discrete time and then solve the optimal control problem.

Relating discrete and continuous representations of linear systems can be readily done using Euler's approximation. Suppose the continuous representation is of the form:

$$\dot{\mathbf{x}}(t) = A_c\mathbf{x}(t) + \mathbf{b}_c u(t). \tag{11.10}$$

We have:

$$\dot{\mathbf{x}}(t) \approx \frac{\mathbf{x}(t+\Delta t) - \mathbf{x}(t)}{\Delta t}. \tag{11.11}$$

From equation (11.11), we have:

$$\begin{aligned}\mathbf{x}(t+\Delta t) &\approx \dot{\mathbf{x}}(t)\Delta t+\mathbf{x}(t)\\ &\approx (A_c\mathbf{x}(t)+\mathbf{b}_c u(t))\Delta t+\mathbf{x}(t).\\ &\approx (I+A_c\Delta t)\mathbf{x}(t)+\mathbf{b}_c u(t)\Delta t\end{aligned} \tag{11.12}$$

Therefore, the discrete and continuous representations are related as follows:

$$\begin{aligned}\mathbf{x}^{(k+1)} &= A_d\mathbf{x}^{(k)}+\mathbf{b}_d u^{(k)}\\ A_d &\approx (I+A_c\Delta t)\\ \mathbf{b}_d &\approx \mathbf{b}_c\Delta t\end{aligned} \tag{11.13}$$

(Equation (11.13) describes an approximate relationship between discrete and continuous forms. One can also find an exact relationship. The technique that provides an exact solution relies on matrix exponentials. We have provided a tutorial on the subject in the web-based supplementary materials that accompany this text.) Our objective is to find a sequence of motor commands $\mathbf{u}_h=\left[u^{(0)},u^{(1)},\cdots,u^{(p-1)}\right]^T$ to minimize a cost of the form:

$$J=\left(\mathbf{y}^{(p)}-\mathbf{r}\right)^T T\left(\mathbf{y}^{(p)}-\mathbf{r}\right)+\mathbf{u}_h^T L\mathbf{u}_h+\alpha\left(1-\frac{1}{1+\beta p}\right), \tag{11.14}$$

where:

$$T\equiv\begin{pmatrix} v_1 & 0 & 0\\ 0 & v_2 & 0\\ 0 & 0 & v_3\end{pmatrix}\quad \mathbf{r}\equiv\begin{bmatrix} g\\ 0\\ 0\end{bmatrix}\quad L\equiv\begin{pmatrix}\lambda^{(0)} & 0 & 0 & 0\\ 0 & \lambda^{(1)} & 0 & 0\\ 0 & 0 & \ddots & 0\\ 0 & \cdots & 0 & \lambda^{(p-1)}\end{pmatrix}$$

and

$$\begin{aligned}\mathbf{x}^{(k+1)} &= A\mathbf{x}^{(k)}+\mathbf{b}u^{(k)}\\ \mathbf{y}^{(k)} &= C\mathbf{x}^{(k)}\end{aligned} \tag{11.15}$$

In our simulations, we set the terms C and L to be the identity matrix. The first term in equation (11.14) enforces our desire to have endpoint accuracy. Because accuracy matters only at the end of the movement, we penalize the squared difference between the state of the eye at movement end and the goal state (the goal state includes stimulus position, as well as zero velocity and acceleration). The second term penalizes effort. The third term enforces our desire to get to the target as soon as possible in order to minimize the temporal discounting of reward. The terms α and β describe how reward is valued and discounted in time. As we will see, different populations (e.g., children vs. adults, monkeys vs. humans) discount reward in different ways. Furthermore, certain diseases (e.g., schizophrenia and Parkinson's disease) alter how the brain processes reward. Equation (11.14) predicts that there should be a correlation between this discounting of reward and eye movements in these different populations.

Notice that our model of the eye plant has no noise in it. We will solve the optimal control problem for a system without noise, and then solve it again with noise. We will find that if the system has signal dependent noise, the optimum set of motor commands are quite different than if it has no noise. We will see that the trajectory of the modeled eye with signal dependent noise looks quite similar to actual eye movements, but without this noise, the trajectory is quite different from the actual eye movement.

To minimize the cost in equation (11.14), we will make a critical assumption: that the motor commands during a saccade do not have access to sensory feedback regarding state of the eye. Indeed, typical saccades are too brief for visual feedback to influence saccade trajectory (a movement ends after about 100 ms). Furthermore, proprioceptive signals from the eyes do not play a significant role in controlling saccade trajectories (Keller and Robinson, 1971; Guthrie, Porter, and Sparks, 1983). Technically, this means that we have an open-loop optimal control problem that we are trying to solve.

In order to minimize the cost in equation (11.14), given the constraint in equation (11.15), we will divide our problem into two parts: first, we select an arbitrary length of time p and find the optimal set of motor commands $\mathbf{u}_h^*(p)$ that minimize equation (11.14). We repeat this for all possible p. Second, given that we produce the optimal motor commands $\mathbf{u}_h^*(p)$, we search the space of p for the one movement time that provides us with the minimum total cost J.

Step 1: The Optimal Motor Commands for a Given Movement Duration

It is useful to write the history of states and how they relate to the history of motor commands. We note that the state $\mathbf{x}$ at time step k is related to the history of motor commands as follows:

$$\begin{aligned}
\mathbf{x}^{(1)} &= A\mathbf{x}^{(0)} + \mathbf{b}u^{(0)} \\
\mathbf{x}^{(2)} &= A\mathbf{x}^{(1)} + \mathbf{b}u^{(1)} = A^2\mathbf{x}^{(0)} + A\mathbf{b}u^{(0)} + \mathbf{b}u^{(1)} \\
\mathbf{x}^{(3)} &= A\mathbf{x}^{(2)} + \mathbf{b}u^{(2)} = A^3\mathbf{x}^{(0)} + A^2\mathbf{b}u^{(0)} + A\mathbf{b}u^{(1)} + \mathbf{b}u^{(2)}. \\
\mathbf{x}^{(k)} &= A^k\mathbf{x}^{(0)} + \sum_{j=0}^{k-1} A^{k-1-j}\mathbf{b}u^{(j)}
\end{aligned} \tag{11.16}$$

Therefore, the state at the end of our movement is:

$$\mathbf{x}^{(p)} = A^p\mathbf{x}^{(0)} + F\Gamma(\mathbf{u}_h + \boldsymbol{\varepsilon}_h), \tag{11.17}$$

where:

$$F \equiv \begin{bmatrix} A^{p-1} & A^{p-2} & A^{p-3} & \cdots & I \end{bmatrix} \quad \Gamma \equiv \begin{bmatrix} \mathbf{b} & 0 & \cdots & 0 \\ 0 & \mathbf{b} & 0 & 0 \\ \vdots & 0 & \ddots & \vdots \\ 0 & 0 & \cdots & \mathbf{b} \end{bmatrix}. \tag{11.18}$$

From equation (11.14), we have:

$$J(p) = \mathbf{x}^{(p)T} C^T T C \mathbf{x}^{(p)} - 2\mathbf{x}^{(p)T} C^T T \mathbf{r} + \mathbf{r}^T T \mathbf{r} + \mathbf{u}_h^T L \mathbf{u}_h + \alpha\left(1 - \frac{1}{1+\beta p}\right). \tag{11.19}$$

Inserting equation (11.17), we can compute the derivative of this cost with respect to the motor commands at a given p:

$$\begin{aligned} \frac{dJ}{d\mathbf{u}_h} &= 2\Gamma^T F^T C^T T C A^p \mathbf{x}^{(0)} - 2\Gamma^T F^T C^T T \mathbf{r}_h \\ &+ 2\Gamma^T F^T C^T T C F \Gamma \mathbf{u}_h + 2L\mathbf{u}_h \end{aligned}. \tag{11.20}$$

Setting equation (11.20) to zero and solving for $\mathbf{u}_h$, we have:

$$\mathbf{u}_h^*(p) = \left(L + \Gamma^T F^T C^T T C F \Gamma\right)^{-1} \Gamma^T F^T C^T T \left(\mathbf{r}_h - C A^p \mathbf{x}^{(0)}\right). \tag{11.21}$$

Figure 11.5 shows simulation results for a 50-deg saccade (thin line, marked "noise-free"), for which we assumed p = 160 ms. The simulation time step is 1 ms, $\mathbf{x}^{(0)} = [0 \quad 0 \quad 0]^T$, motor costs are $\lambda^{(i)} = 1$, and the target is at 50 degrees, $g = 50\pi/180$. The only variable in our system is tracking cost T, which has three parameters: $v = [5\times10^9 \quad 1\times10^6 \quad 80]$. We find that the simulated eye arrives at the target, but the velocity profile is symmetric in time (the line labeled noise-free in figure 11.5). In contrast to our simulation, the velocity profile of the eye during a real saccade (of 50 deg) is highly asymmetric, with the maximum speed attained near the start of the movement. That is, even though we picked a reasonable movement duration and found the optimal motor commands for that duration, we could not replicate the observed trajectory. Why did our simulation fail to produce a realistic movement?

The Importance of Signal-Dependent Noise

For a movement to show an asymmetric speed profile, it must be advantageous to produce large motor commands early into the movement rather than producing the same large motor commands near the end. Chris Harris and Daniel Wolpert (1998) noted that this advantage comes about when the system has a specific noise property: The relationship between motor commands and change in state depends on a noise that grows with the size of the motor commands, called signal-dependent

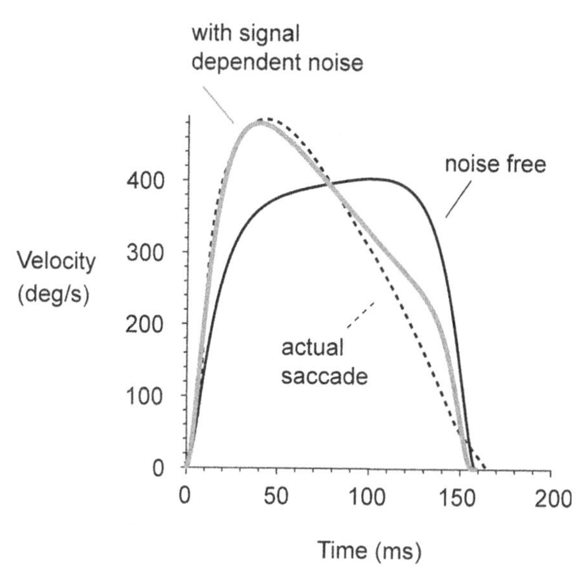

Figure 11.5
Saccades to a target at 50 degrees were generated for a noise-free system (equation 11.15) and for a system with signal-dependent noise (equation 11.22) under a policy that minimized endpoint variance and effort. The velocity profile is asymmetric in the case of a system with signal-dependent noise, showing a high initial velocity and then a slower tail. This is because endpoint accuracy depends more strongly on the motor commands that arrive later in the movement than those that arrive earlier. Asymmetric velocity profiles during large amplitude saccades arise in optimal control of a system that suffers from signal-dependent noise.

noise. As we will show, our simulation failed to produce a realistic movement because we imagined the system to be noise-free.

To consider the effect of noise in control of our system, suppose that motor commands u are affected by noise $\varepsilon \sim N(0,\kappa^2 u^2)$. That is, the noise is a Gaussian distribution with mean zero and variance that is proportional to u^2. We saw examples of this physiological noise in figure 10.12B. The fact that the variance of our noise is dependent on the motor commands is the key assumption. If the noise variance were independent of the motor commands, addition of noise would still have no effect on our solution in equation (11.23); this is because what matters is whether the noise has a nonzero derivative with respect to the motor commands. So let us now solve the optimal control problem by assuming that the dynamics of our eye is affected by signal-dependent noise:

$$\begin{aligned} \mathbf{x}^{(k+1)} &= A\mathbf{x}^{(k)} + \mathbf{b}\left(u^{(k)} + \varepsilon^{(k)}\right) \\ \varepsilon^{(k)} &\sim N\left(0, \kappa^2 \left(u^{(k)}\right)^2\right) \\ \mathbf{y}^{(k)} &= C\mathbf{x}^{(k)} \end{aligned} \quad . \tag{11.22}$$

Consider the endpoint variance of this system after a sequence of motor commands $u^{(0)},\cdots,u^{(p-1)}$:

$$\begin{aligned}\mathbf{x}^{(p)} &= A^p\mathbf{x}^{(0)} + \sum_{k=0}^{p-1} A^{p-1-k}\mathbf{b}\left(u^{(k)} + \varepsilon^{(k)}\right) \\ \text{var}\left[\mathbf{x}^{(p)}\right] &= \kappa^2 \sum_{k=0}^{p-1} \left(u^{(k)}\right)^2 A^{p-1-k}\mathbf{b}\mathbf{b}^T\left(A^{p-1-k}\right)^T\end{aligned}. \tag{11.23}$$

The variance at the end of the movement depends on the sum of squared motor commands. Why should we care about this variability? Because the variability affects our accuracy, which is part of the cost function that we are trying to minimize: when the state is a random variable, the expected value of the "squared error" (as in our cost function in equation 11.14) implicitly includes the sum of the squared bias and the variance of the state. If we choose large motor commands, they will produce a larger variance, and this will increase our cost. Interestingly, we see from equation (11.23) that the variance depends more strongly on the motor commands that arrive later: The power to which the matrix A is raised is smaller for larger k. Therefore, if we want reduced variance near the end of our movement, we should minimize the size of motor commands, particularly near the end of the movement. If we do this, we will reduce our endpoint variance, improve our accuracy, and as a result reduce our costs.

To express these ideas mathematically, let us see how the cost in equation (11.14) is affected by the addition of signal dependent noise. We define:

$$\boldsymbol{\varepsilon}_h = \begin{bmatrix} \varepsilon^{(0)} \\ \varepsilon^{(1)} \\ \vdots \\ \varepsilon^{(p-1)} \end{bmatrix} \quad U = \begin{bmatrix} u^{(0)} & 0 & \cdots & 0 \\ 0 & u^{(1)} & 0 & 0 \\ \vdots & 0 & \ddots & \vdots \\ 0 & 0 & 0 & u^{(p-1)} \end{bmatrix}. \tag{11.24}$$

The mean and variance of our noise vector are:

$$E[\boldsymbol{\varepsilon}_h] = \begin{bmatrix} 0 \\ \vdots \\ 0 \end{bmatrix} \qquad \text{var}[\boldsymbol{\varepsilon}_h] = \kappa^2 UU. \tag{11.25}$$

The state at the end of the movement is:

$$\mathbf{x}^{(p)} = A^p\mathbf{x}^{(0)} + F\Gamma(\mathbf{u}_h + \boldsymbol{\varepsilon}_h). \tag{11.26}$$

The expected value and variance of our state are:

$$\begin{aligned} E\left[\mathbf{x}^{(p)}\right] &= A^p\mathbf{x}^{(0)} + F\Gamma\mathbf{u}_h \\ \text{var}\left[\mathbf{x}^{(p)}\right] &= \kappa^2 F\Gamma UU\Gamma^T F^T \end{aligned}. \tag{11.27}$$

Because the cost in equation (11.14) is now a random variable, a reasonable thing to do is to minimize its expected value:

$$\begin{aligned} E[J] = & E\left[\mathbf{x}^{(p)}\right]^T C^T TCE\left[\mathbf{x}^{(p)}\right] + tr\left[C^T TC \operatorname{var}\left[\mathbf{x}^{(p)}\right]\right] \\ & -2E\left[\mathbf{x}^{(p)}\right]^T C^T T\mathbf{r} + \mathbf{r}^T T\mathbf{r} + \mathbf{u}_h^T L\mathbf{u}_h + \alpha\left(1 - \frac{1}{1+\beta p}\right) \end{aligned} \tag{11.28}$$

The trace operator in equation (11.28) comes about because of the identity:

$E\left[\mathbf{x}^T A\mathbf{x}\right] = E[\mathbf{x}]^T AE[\mathbf{x}] + tr[A \operatorname{var}[\mathbf{x}]]$.

The trace in equation (11.28) can be simplified:

$$\begin{aligned} tr\left[C^T TC \operatorname{var}\left[\mathbf{x}^{(p)}\right]\right] & = \kappa^2 tr\left[C^T TCF\Gamma UU\Gamma^T F^T\right] \\ & = \kappa^2 tr\left[U\Gamma^T F^T C^T TCF\Gamma U\right] \\ & = \kappa^2 \mathbf{u}_h^T diag\left[\Gamma^T F^T C^T TCF\Gamma\right]\mathbf{u}_h \end{aligned} \tag{11.29}$$

The term $diag[M]$ in equation (11.29) is the diagonal operator that generates a matrix with only the diagonal elements of the square matrix M. We see that the cost for a system with signal dependent noise (equation 11.28) is similar to the cost for a noiseless system (equation 11.14), with the crucial exception of an additional term (equation 11.29) that penalizes the "squared" motor commands by an amount proportional to the variance of the signal-dependent noise.

Let us now solve the optimal control problem for this system that has signal-dependent noise. We pick a movement duration p and find the optimal motor commands by setting the derivative of equation (11.28) with respect to $\mathbf{u}_h$ to zero and then solve for $\mathbf{u}_h$:

$$\begin{aligned} & S = diag\left[\Gamma^T F^T C^T TCF\Gamma\right] \\ & \mathbf{u}_h^*(p) = \left(L + \Gamma^T F^T C^T TCF\Gamma + \kappa^2 S\right)^{-1} \Gamma^T F^T C^T T\left(\mathbf{r} - CA^p\mathbf{x}^{(0)}\right) \end{aligned} \tag{11.30}$$

Figure 11.5 shows simulation results when the standard deviation of our signal dependent noise was set at $\kappa = 0.009$, with all other terms in the simulation unchanged. With this modest amount of noise the speed profile becomes asymmetric, particularly for large amplitude saccades. The simulated saccade now has its greatest speed early in the movement, closely resembling the trajectory of a real saccade (dashed line, figure 11.5). The reason for this is because for a system with signal dependent noise, it is better to produce large motor commands early rather than late, as the large noise in the early motor commands will be naturally dissipated by the viscous dynamics of the eye, resulting in little variability by the end of the movement.

Let us summarize the ideas thus far. Control of saccadic eye movements can be viewed as an open-loop process in the sense that motor commands do not rely on sensory feedback. Saccades exhibit characteristic timing and speed profiles. For example, large amplitude saccades tend to have an asymmetric speed profile, with the largest speeds attained early in the movement. If we assume a noise-free plant, and a cost function that penalizes endpoint inaccuracy (squared endpoint error), as well as effort (squared motor commands), then the motor commands that minimize such costs do not produce the asymmetric speed profiles. However, if we assume that the motor commands suffer from signal-dependent noise, then the expected value of the squared endpoint error implicitly incorporates a measure of endpoint variance. This variance depends more strongly on the motor commands that arrive later in the movement than those that arrive earlier. As a result, the policy (i.e., the sequence of motor commands) that minimizes the cost produces larger speeds earlier in the movement. The asymmetry in speed profiles of saccadic eye movements suggests that the brain is minimizing a cost that includes endpoint accuracy in the face of signal-dependent noise.

Step 2: Optimal Duration and the Influence of Reward

In step 1 we assumed that the movement should have a specific duration, and then found the optimal motor commands for that duration $\mathbf{u}_h^*(p)$. In our second step, the question that we wish to explore is as follows: if we see something interesting at 15 degrees, why should our brain produce a saccade that is about 70 ms in duration? If all that mattered were endpoint accuracy and motor costs, then we should make a very slow movement, something resembling smooth pursuit, not a saccade. However, if time to reward is itself a cost because time discounts reward, then this cost balances the effort and accuracy costs, encouraging us to move faster. The movement duration that we are looking for is one that minimizes the total cost J.

Figure 11.6A shows the effort, accuracy, and reward costs for a saccade to a target at 20 degrees. For a given movement period p, we generate the optimal policy $\mathbf{u}_h^*(p)$ in equation (11.30). Finally, we compute the costs in equation (11.28). The plot includes the total cost $E[J]$, accuracy costs $E\left[\left(\mathbf{x}^{(p)}-\mathbf{r}\right)^T T\left(\mathbf{x}^{(p)}-\mathbf{r}\right)\right]$, motor costs $\left(\mathbf{u}_h^*\right)^T L\mathbf{u}_h^*$, and costs of delaying the reward, J_p. On the left side of this figure we have plotted the costs when J_p is of a hyperbolic form, as in equation (11.1). Shorter-duration saccades have a large cost $E[J]$ because the costs associated with inaccuracy and effort increase as saccade duration decreases. With increasing saccade duration, the cost of delaying the reward increases. The function $E[J]$ has a minimum, and this is the optimum movement duration. At this minimum, the movement balances the need to be accurate and lazy versus the need to maximize reward.

Is there anything special about our hyperbolic cost, or would any function that penalizes time produce similar results? To answer this question, let us consider a quadratic temporal discounting of reward $J_p = \alpha p^2$. We see that for both hyperbolic and quadratic temporal discounting there exist parameter values such that a

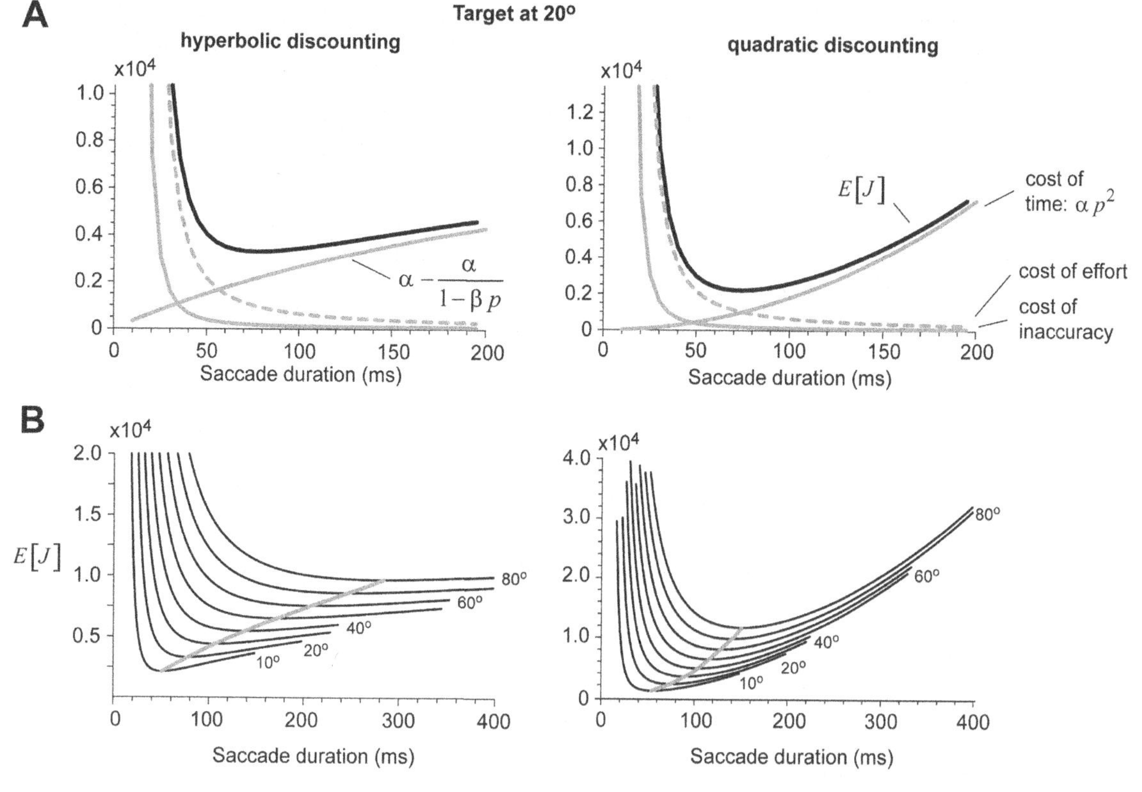

Figure 11.6
The effect of quadratic and hyperbolic discounting of reward on saccade durations. (A) Costs of a saccade to a target at 20 degrees. For the quadratic discount, $J_p = \alpha p^2$, $\alpha = 1.8 \times 10^5$. For the hyperbolic discount, $\alpha = 1.15 \times 10^4$ and $\beta = 3$. All other parameters are equal for the two simulations. Optimum saccade duration for both simulations is around 75 ms. (B) Expected value of the cost for various saccade amplitudes. The optimum saccade duration is noted by the gray line. Parameter values are unchanged from (A).

20-degree saccade will have its minimum total cost at around 75 ms. However, we see that a quadratic discounting of time implies that the loss as a function of movement duration increases rapidly. Therefore, in a quadratic regime there is little loss of reward when one compares two movements of 50 and 60 ms in duration, but much greater loss when one compares two movements of 200 and 210 ms in duration. In contrast, a hyperbolic discounting of time works exactly opposite. There is greater reward loss for short-duration saccades than for long-duration saccades. That is, as saccade durations increase, the sensitivity to passage of time decreases. As we will see, this is a crucial property of hyperbolic discounting, and one that allows us to account for the fact that saccade duration as a function of saccade amplitude grows faster than linearly (figure 10.3B).

We kept parameter values unchanged and computed optimal saccade durations for movements of various amplitudes in figure 11.6B. We see that for a quadratic temporal cost, increasing movement amplitudes accompany smaller and smaller changes in saccade durations. In contrast, for a hyperbolic temporal cost, increasing movement amplitudes accompany a faster than linear increase in saccade durations. The optimal movement duration is a balance between the desire to get to the goal state as soon as possible, while minimizing effort and inaccuracy. A hyperbolic cost of time produces a very different amplitude-duration relationship than a quadratic cost of time. To see why this is true, let us compute the condition that provides us with the optimal movement duration p^*. From equation (11.4), the derivative of our effort and accuracy costs must be equal to the derivative of our temporal cost:

$$-\left.\frac{d(J_x + J_u)}{dp}\right|_{p^*} = \left.\frac{dJ_p}{dp}\right|_{p^*}. \tag{11.31}$$

Figure 11.7 plots these two derivatives for hyperbolic, linear, and quadratic time costs. Because the effort and accuracy costs depend on stimulus location (saccade amplitude), there is a separate curve for each amplitude. The point at which the

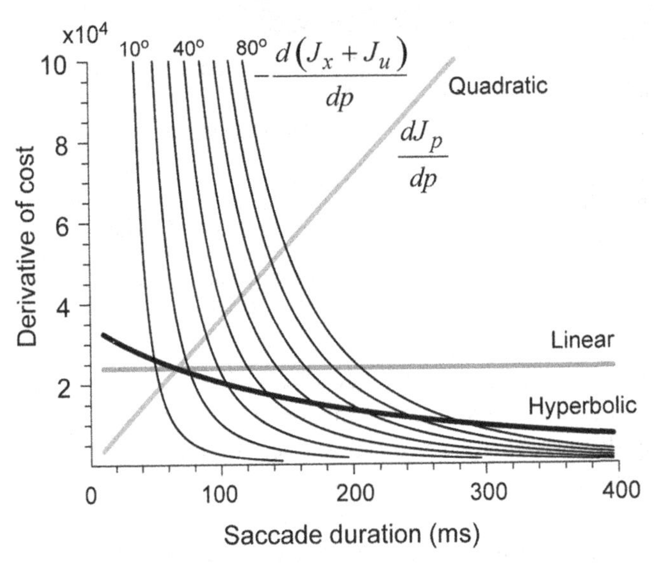

Figure 11.7
The optimal movement duration is one that satisfies equation (11.31), that is, a duration at which the derivative of the effort/accuracy cost is equal to the derivative of the temporal cost. The intersection between the derivative of the effort/accuracy cost and temporal cost is the optimal movement duration. When we set the quadratic, linear, and hyperbolic temporal costs to produce similar optimal movement times for one saccade amplitude (in this case 20 degrees), they produce very different duration-amplitude relationships for other amplitudes. Quadratic: $\alpha = 1.8 \times 10^5$. Linear: $\alpha = 2.4 \times 10^4$. Hyperbolic: $\alpha = 1.15 \times 10^4$, $\beta = 3$.

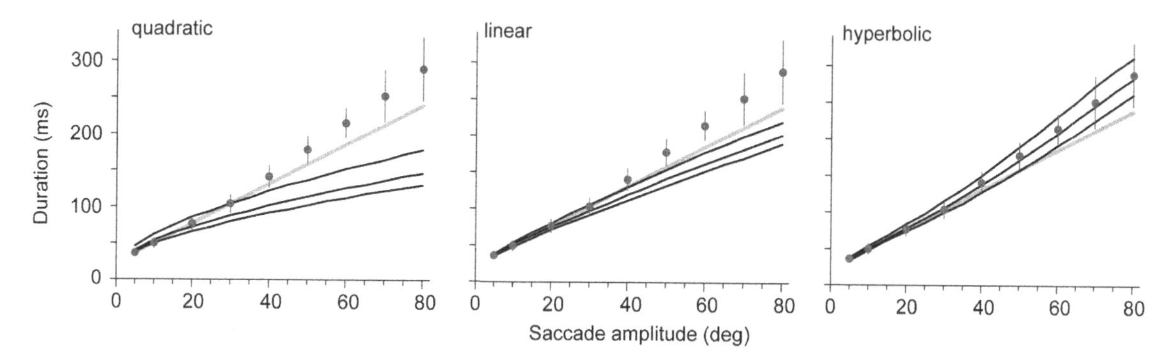

Figure 11.8
Optimal saccade durations as a function of amplitude for various temporal discount functions (quadratic, linear, and hyperbolic). The data points are mean and SD of saccade durations in healthy people (Collewijn, Erkelens, and Steinman, 1988). The gray line is for reference. The thin black lines are simulation results for various stimulus values, with the largest value producing shorter duration saccades. While linear and quadratic discounting schemes can account for durations of short amplitude saccades, they underestimate durations of long amplitude saccades. Simulation values are as follows. Quadratic: $\alpha = (1,2,3) \times 10^5$. Linear: $\alpha = (2,2.5,3.0) \times 10^4$. Hyperbolic: $\alpha = (8.5,9.5,11.5) \times 10^3$, $\beta = 3$.

derivative of the effort/accuracy cost crosses the temporal cost is the optimal movement duration. For example, for a 20-degree saccade all three temporal costs produce an optimal duration of around 75 ms. However, as saccade amplitudes increase, the three temporal costs predict very different saccade durations. When we compare the predicted pattern of saccade durations with actual data, we find that the hyperbolic function is an excellent fit. Figure 11.8 summarizes these results for three kinds of temporal cost functions: quadratic, linear, and hyperbolic. This figure also includes data from actual saccades (data from figure 10.3B). A quadratic cost of time produces reasonable estimates of saccade parameters for small amplitudes, but fails for larger amplitudes. The reason is that with a quadratic temporal cost, with passage of time the cost of time grows (derivative is increasing, as in figure 11.7). If we consider a linear cost of time, an approach that was suggested by Chris Harris and Daniel Wolpert (2006), once again we can produce reasonable estimates of saccade parameter for small amplitude saccades, but the simulations fail for larger amplitudes. However, in the case of a hyperbolic cost of time, we can account for durations of both small amplitude as well as large amplitude saccades. Indeed, the fact that movement durations increase faster than linearly as a function of movement amplitudes is consistent with a hyperbolic cost of time, but not a linear or quadratic one.

11.5 Cost of Time and Temporal Discounting of Reward

Why should the brain impose a hyperbolic cost on duration of movements? The answer, in our opinion, is that this cost expresses how the brain temporally discounts reward. We saw examples of temporal discounting of reward in the response of dopamine neurons (figure 11.2B), in the response of LIP neurons (figure 11.2C), and in the decisions that people made regarding money (figure 11.1A) or water (figure 11.1B). Suppose that the brain penalizes movement durations because passage of time delays the acquisition of reward. If this hypothesis is true, then it follows that movement kinematics should vary as a function of the amount of reward. For example, if we make a movement in response to a stimulus that promises little reward, α in equation (11.1) is small, and the motor and accuracy costs become relatively more important. As a consequence, when our brain assigns a low value to the stimulus, our movement toward that movement should be slow. To explore this idea, let us consider what happens to saccades when we alter the value of the stimulus α. Movement durations depend on the rate at which reward value is discounted in time (equation 11.31). That is, movement duration depends on the derivate of cost J_p. This derivative is:

$$\frac{dJ_p}{dp} = \frac{\alpha\beta}{(1+\beta p)^2}. \tag{11.32}$$

As α decreases, so does the derivative of the reward discount function. Figure 11.9A plots this derivative for various α. We see that the optimal movement duration increases as stimulus value α decreases. An increase in movement duration coincides with a decrease in peak movement speed.

For example, the opportunity to look at a face is a valued commodity, and physical attractiveness is a dimension along which value rises (Hayden et al., 2007). As α increases, durations of simulated saccades decrease, resulting in higher velocities. This potentially explains why people make faster saccades to look at faces (Xu-Wilson, Zee, and Shadmehr, 2009).

A hyperbolic function is a good fit to discharge of dopamine cells in the brain of monkeys that have been trained to associate visual stimuli with delayed reward (Kobayashi and Schultz, 2008). That is, the response of these cells to stimuli is a good predictor of the temporally discounted value of these stimuli (figure 11.2B). In Parkinson's disease (PD), many of the dopaminergic cells die. Let us hypothesize that this is reflected in a devaluation of the stimulus, that is, a smaller than normal α. In figure 11.9B we have plotted velocity-amplitude data from a number of studies that have examined saccades of people with moderate to severe PD. The saccades of PD patients exhibit an intriguing property: The peak speeds are normal for small amplitudes, but they become much slower than normal for large amplitudes. If we simply reduce stimulus value α, the model reproduces velocity-amplitude characteristics of PD patients (figure 11.9B).

If an abnormally small stimulus value can produce slow saccades, then an abnormally large value should produce fast saccades. In schizophrenia, saccade velocities

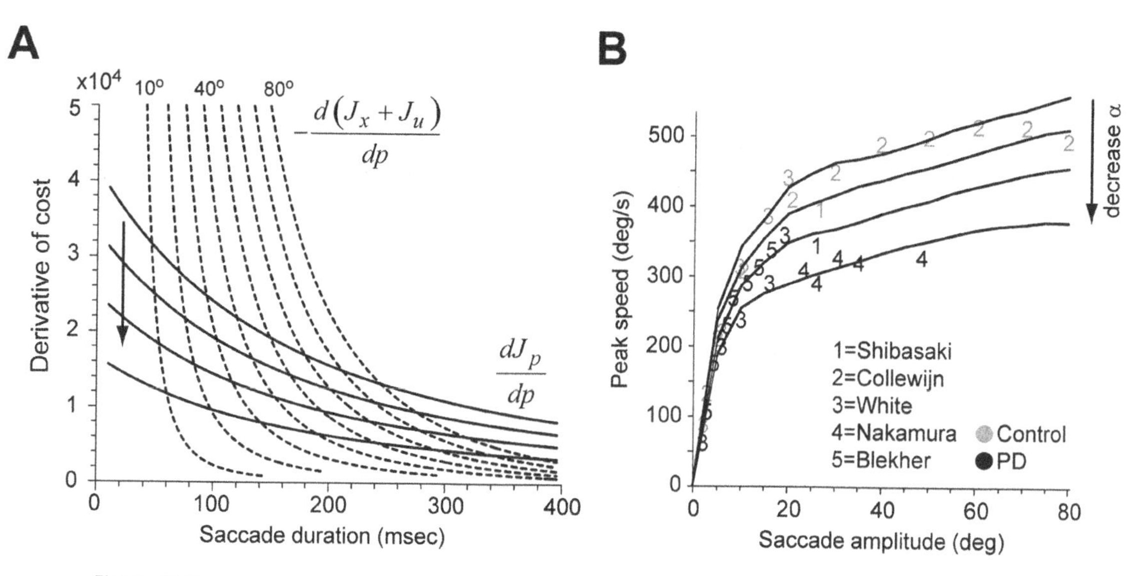

Figure 11.9
Effects of changing the stimulus value α on saccade durations and saccade peak velocities for a hyperbolic temporal discounting of reward. (A) Optimal saccade duration is the intersection of the derivative of the accuracy/effort cost (dashed lines) and the derivative of the temporal cost (solid lines). As α decreases (indicated by the arrow), the optimal saccade duration increases. Simulated values are: $\alpha = (8,10,12,14) \times 10^3$ and $\beta = 3$. (B) Peak saccade velocity as a function of amplitude for various stimulus values α. Simulation results are shown by the lines for $\alpha = (6 - 13.5) \times 10^3$ and $\beta = 3$. For each line, the stimulus value α was kept constant. Reducing stimulus value decreased saccade speeds, but the changes in speed were much bigger for large amplitude saccades than small amplitudes. Data from various papers are shown by the numbers in healthy controls and in patients with Parkinson's disease. These data are from Shibasaki, Tsuji, and Kuroiwa (1979); Collewijn, Erkelens, and Steinman (1988): White et al. (1983); Blekher et al. (2000); and Nakamura et al. (1991).

are faster than in healthy controls (Mahlberg et al., 2001). Schizophrenia is a complex disease that likely involves dysfunction of generation and uptake of many neurotransmitters, including dopamine, glutamate, and GABA. James Stone, Paul Morrison, and Lyn Pilowsky (2007) suggested that in the striatum of schizophrenic patients, there is greater than normal dopamine synthesis. Shitij Kapur (2003) noted that schizophrenics assign an unusually high salience to stimuli so that "every stimulus becomes loaded with significance and meaning." Indeed, currently available antipsychotic medications have one common feature: they block dopamine D2 receptors. The reward temporal discount function in schizophrenia has a higher slope with respect to controls (Heerey et al., 2007; Kloppel et al., 2008), implying a greater discount rate. In our framework, this produces a faster rise in the cost of time, increasing saccade speeds.

Consider another curious fact regarding saccades: As we age, the kinematics of our saccades change. Children produce faster saccades than young adults

(Fioravanti et al., 1995; Munoz et al., 2003). According to our theory, the differences in saccade kinematics should be a consequence of the way the child's brain temporally discounts reward. The availability of dopamine in the brain declines with age. For example, rhesus monkeys exhibit a 50 percent decline in dopamine concentrations in the caudate and putamen from youth to old age (Collier et al., 2007), and squirrel monkeys exhibit a 20 percent decline (McCormack et al., 2004). Leonard Green, Joel Myerson, and Pawel Ostaszewski (1999) measured the temporal discount rate of reward in both young children and adults and found that the initial slope of the discount function was 2–3 times larger in children than in adults. They would rather take a single cookie now than wait for a brief period in order to receive two cookies. Shadmehr et al. (2010) showed that by increase the slope of the temporal cost function (via parameter α) by a factor of two, the resulting saccades share the velocity-amplitude relationship found in children's saccades.

Let us now consider the fact that saccade velocities differ across species. For example, rhesus monkeys exhibit velocities that are about twice as fast as humans (Straube et al., 1997; Chen-Harris et al., 2008). Although there are small differences in the eye plants of monkeys and humans, such differences cannot account for the remarkably faster saccades in monkeys (Shadmehr et al., 2010). One possibility is that the differences in saccades are related to interspecies differences in valuation of stimuli and temporal discounting of reward. Indeed, some rhesus monkeys exhibit a much greater temporal discount rate than humans: When making a choice between stimuli that promise reward (juice) over a range of tens of seconds, thirsty adult rhesus monkeys (Kobayashi and Schultz, 2008; Hwang, Kim, and Lee, 2009) exhibit discount rates that are many times that of thirsty undergraduate students (Jimura et al., 2009) (compare figure 11.2A with figure 11.1B). If we take into account this higher slope of the temporal discount rate, the simulated monkey saccades will have velocities that were fairly consistent with the velocities that have been recorded from this species (Shadmehr et al., 2010).

11.6 State-Dependent Value of a Stimulus

Animals do not assign a value to a stimulus based on its inherent properties, but based on their own state when the stimulus was encountered. For example, birds that are initially trained to obtain equal rewards after either large or small effort, and that are then offered a choice between the two rewards without the effort, generally choose the reward previously associated with the greater effort. Tricia Clement and colleagues (Clement et al., 2000) trained pigeons to peck at circles.

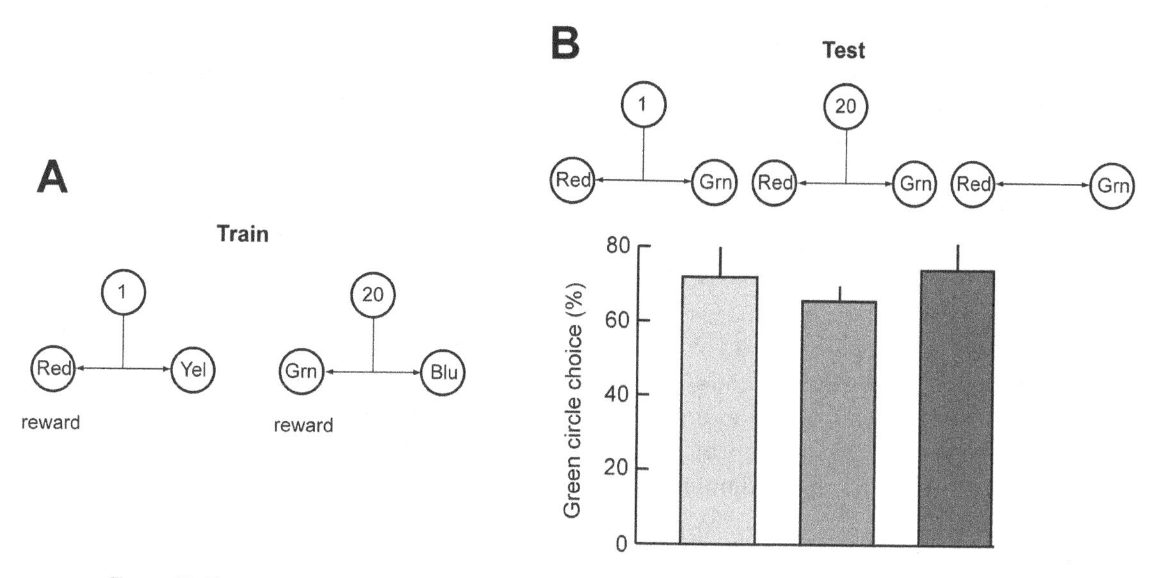

Figure 11.10
The value of a given reward depends on the state of the animal at the time that reward was attained. (A) Pigeons were trained to peck at circles. During training, on some trials a white circle would appear and after the pigeon pecked it once, two circles appeared (red and yellow). If the red circle was pecked, food was given. On other trials, the white circle had to be pecked twenty times before green and blue circles appeared. Pecking the green circle resulted in food. (B) On test trials, some pigeons saw a white circle, and after pecking it once they saw red and green circles. In about 70 percent of the trials, they pecked the green circle. Other pigeons saw a white circle, and after pecking it twenty times they saw red and green circles. They chose the green circle about 65 percent of the time. Pigeons that were only shown red and green circles during test trials also preferred the green circle. (Data from Clement et al., 2000.)

During training, on some trials a white circle would appear and after the pigeon pecked it once, two circles appeared (red and yellow), as shown in figure 11.10A. If the red circle was pecked, food was given. On other trials, the white circle had to be pecked twenty times before green and blue circles appeared. Pecking the green circle resulted in food. So the idea was to test whether after pecking twenty times the green circle would have greater value than the red, even though they both gave the same food. On test trials, some pigeons saw a white circle, and after pecking it once they saw red and green circles. In about 70 percent of the trials, they pecked the green circle (figure 11.10B). Other pigeons saw a white circle, and after pecking it twenty times they saw red and green circles. They chose the green circle about 65 percent of the time. Pigeons that were only shown red and green circles during test trials also preferred the green circle (figure 11.10B). This paradoxical result can be understood in terms of a greater utility (i.e., relative usefulness rather than absolute value) for the reward that was attained following a more effortful action. The green circle resulted in the same food as the red circle, but the food associated with the green circle was attained during training after twenty pecks. This phenomenon is called state-dependent valuation learning and is present in a wide variety of species from mammals to invertebrates (Pompilio and Kacelnik, 2010).

The state-dependent valuation of stimuli allows us to consider a curious fact: Kinematics of saccades to target of a reaching movement are affected by the load on the arm. For example, the peak speed of a saccade is higher when there is a load that resists the reach, and lower when the load assists the reach (van Donkelaar, Siu, and Walterschied, 2004). Why should varying the effort required to perform a reach to a target affect saccade velocities to that target? A reaching movement that is resisted by a load arrives at the target after a larger effort than one that is assisted. The more effortful state in which the reward is encountered favors assignment of a greater utility for that stimulus. This greater utility may contribute to a faster saccade.

11.7 Why Hyperbolic Discounting of Reward?

Why should passage of time discount reward in a hyperbolic fashion? Other than the fact that a hyperbolic function seems to fit the empirical data better (Myerson and Green, 1995), is there a rationale for this pattern? Alex Kacelnik (1997) has considered this question and made the following suggestion: Perhaps the choices that people make regarding time and reward are a reflection of a more fundamental normative law in which the objective is to maximize reward per unit of time. For example, suppose that you are given a choice between reward α_1 now and reward α_2 at delay t. Further suppose that you are asked to make these choices at intervals γ (this would represent an intertrial interval). If your objective is to maximize reward per unit of time, then you would compare two quantities: $J_1 = \frac{\alpha_1}{\gamma}$ and $J_2 = \frac{\alpha_2}{\gamma + t}$.

You should pick the choice that gives you the larger reward rate. Let us show that if you choose in this way, you are implicitly discounting hyperbolically. Suppose that given the choice between α_1 now and α_2 at t, you choose each at around 50 percent probability. This would imply that:

$$\frac{\alpha_1}{\gamma} = \frac{\alpha_2}{\gamma + t}.$$

Which implies that α_1 declines with time at a rate of γ^{-1}:

$$\alpha_1 = \frac{\alpha_2}{1 + \gamma^{-1} t}. \tag{11.33}$$

Therefore, a policy that maximizes the rate of reward would result in a hyperbolic temporal discounting of reward.

We can use this idea to reformulate the costs in our motor control problem. Suppose that we perform actions in such a way as to maximize reward per unit of time, while minimizing the effort expended per unit of time. That is, what matters is the difference between the reward that we hope to attain and the effort we expect to expend, per unit of time. If $\Pr(r = 1|\mathbf{u}(p))$ represents the probability of acquiring reward that has value α, given motor commands $\mathbf{u}$ that produce a movement of duration p, and J_u represents the effort costs associated with these commands, then the motor commands should be chosen to maximize the following:

$$J = \frac{\alpha \Pr(r = 1|\mathbf{u}(p)) - J_u}{1 + \gamma^{-1} p}. \tag{11.34}$$

In the numerator, the probability of success would generally increase as movement duration p increases. The term $(1 + \gamma^{1}p)^{-1}$ is a decreasing function of p, resulting in a cost J in equation (11.34) that will have a maximum at a particular movement duration. It remains to be seen whether maximizing the rate of reward is sufficient to explain movement patterns in biology, as well as the economic decision-making processes that are reflected in temporal discounting of reward.

Summary

Suppose that the objective of any voluntary movement is to place the body at a more valuable state. Further suppose that the value associated with this state is not static, but is discounted in time: We would rather receive the reward now than later. The value that one assigns a stimulus, and the rate at which this value declines, forms a reward temporal discount function. The temporal discounting of reward forms an implicit cost of time, that is, a penalty for the duration of the movement. This penalty is a hyperbolic function. If one assumes that motor commands are programmed to minimize effort while maximizing reward, and if one further assumes that reward loses value hyperbolically as a function of movement duration, then one can mathematically reproduce the relationship between movement duration, amplitude, and velocity in saccades.

Research over the last two decades indicates that the reward temporal discount function is affected by disease, is affected by development and aging, and is affected by evolution (interspecies differences in the brain). There appears to be a correlation between changes in the reward temporal discount function and changes in duration and velocities of saccadic eye movements. This correlation suggests that the motor commands that move the eyes reflect a specific cost of time, one in which passage of time discounts reward. Here, we used open-loop optimal control to describe a cost function that includes effort and reward, and then modeled control of eye movements to ask why saccades have their particular duration and velocities, and why these kinematics change as stimulus value or discounting of reward changes in the brain. The motor commands that move our body may be a reflection of an economic decision regarding reward and effort.

12 Optimal Feedback Control

Suppose you would like to retire wealthy and live in some tropical island. Currently, however, you are a young student taking classes at some university. What actions should you take today so that in thirty years, you will have reached your goal? Well, you have data that suggest a relationship between actions and consequences. These data come from what you have read about other people's lives, from what you have experienced about your own actions and their consequences, and from what you have observed in your friends and family. This is your forward model; it gives you a way to predict what might be the result of any potential action. Given your goal (retire in Bahamas), this forward model (actions and their predicted consequences), and your current state (young student taking classes), you compute an optimum set of actions that describe your plan to get from your current state to your goal. (Perhaps the fact that you are reading this book is part of this optimum sequence of actions!) However, your goal is very far away in time, and it would seem rational to reevaluate your plan occasionally. That is, as your state changes, it makes sense for you to reconsider what might be the best sequence of actions to get you to your goal.

Because the state you are in is influenced not only by your actions but also by unforeseen and possibly unforeseeable events, it is probably not very productive for you to compute a specific sequence of actions that you would want to do in the future. Instead, you need to devise a policy that specifies what actions should be performed for each possible state that you might find yourself in. Ideally, for each time point in the future, the policy specifies the action that you would perform for each possible state. This is called a *feedback control policy*, as the actions that are performed at any given time depend on your belief[1] about your state at that time. In effect, feedback control policies are instructions about how to get to the goal from any state, at any time in the future.

It may seem hard to believe, but the problem of generating motor commands, say to move your arm, is really not that different from the problem of retiring in Bahamas. The objectives and timeframes are different, but the basic problem is similar:

- Reaching the goal has a cost: the money you need to spend for retiring to the Bahamas or the effort that your muscles must make to reach a target.
- A forward model specifies what to expect from your actions: where you will end up retiring or the sensory consequences of your motor commands, and
- In both cases you have a sensory system that provides you with information about your state.

The basic framework is summarized in figure 12.1. Given the goal of the task and our belief about our current state, we generate some motor commands. These commands change the state of our body and our environment. Using our forward model, we predict what these state changes should be. We observe the state changes via our sensors. We combine what we predicted with what we observed to form a belief about the current state of the body/environment. Based on our current belief about our state, we apply our policy and form a new set of motor commands. The process continues until we reach our goal. Note that the sensory information may also lead to changes in the forward model. We may see, for example, that investing money on a particular stock may not be a good idea for retiring to the Bahamas, and this will lead us to revise our policy. This kind of change, however occurs over a slower time scale, compare to the scale of our instantaneous decisions.

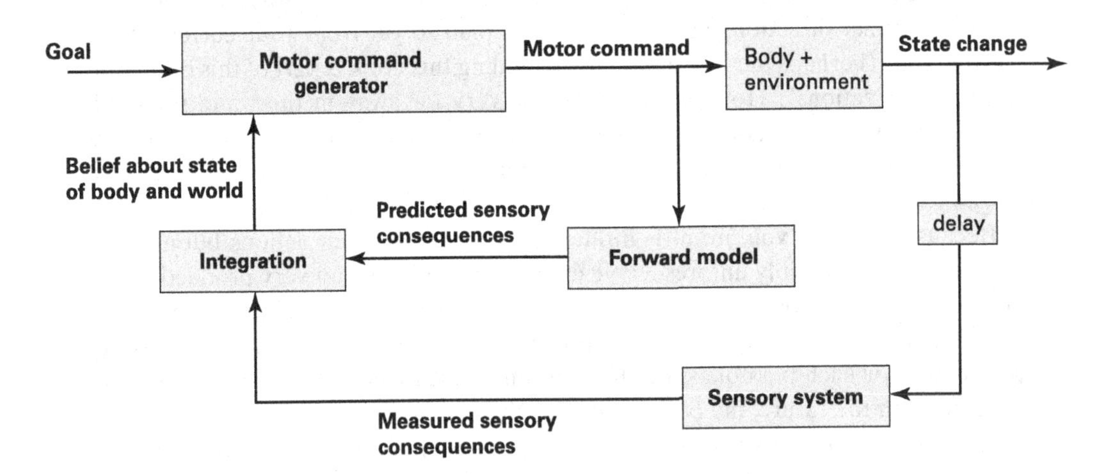

Figure 12.1
A general framework for goal directed behavior. Motor commands are generated in order to achieve a goal. The goal is expressed as a cost function. The motor command generator is a feedback controller that produces motor commands as a function of the current belief regarding the state of the body and the environment. This feedback controller is optimum in the sense that it produces motor commands that achieve the goal—that is, minimize the cost function. The state of the body and environment comes from a combination of sensory measurements (via sensory feedback) and predictions about the outcome of motor commands (via forward models).

Whereas in the previous chapter we computed a sequence of motor commands while ignoring the sensory feedback, here we want to consider forming feedback control policies that continuously adjust the motor commands in response to the sensory feedback. Our rationale is that our movements can be perturbed, and we want to have in place a policy that provides us with the best possible motor commands, regardless of where we end up during the movement. That is, we want to compute a time-varying sensory-motor transformation that, given an estimate of state—based on the integration of predictions with observations—we can compute the motor commands that are optimal in the sense that they will provide us with the best possible way to get to the goal.

12.1 Examples of Feedback-Dependent Motor Control

In the previous chapter, we focused on saccadic eye movements partly because these movements are so brief that sensory feedback appears to play no role in their control (Keller and Robinson, 1971; Guthrie, Porter, and Sparks, 1983). However, this does not mean that the state of the eye is not monitored during a saccade. Indeed, there are data suggesting that the motor commands that move the eyes during a saccade benefit from an internal feedback system, possibly a forward model that monitors the motor commands and predicts their sensory consequences. Let us consider some of the evidence regarding this state and goal-dependent process of generating motor commands.

Saccades are sometimes accompanied by blinks. In a blink, the eyelids close and reopen in a movement that takes about 100 ms (figure 12.2). However, during a blink the brain sends motor commands to not just the lids but the eyes as well: As the brain sends the commands to the lid muscles, it also sends commands to the extraocular muscles, causing a displacement of the eyes. (Indeed, the motion of the eyes during a blink is not due to a mechanical interaction between the lid and the eyes, but rather it is due to the specific motor commands to the eyes.) Thus, a blink during a saccade is a natural disturbance that affects the states of the eyes. Klaus Rottach, John Leigh, and colleagues (1998) used a "search coil" technique (which relies on a contact lens that is placed on the eyes) to measure eye and lid motion during a horizontal saccadic eye movement. They noted that a blink that occurred during a saccade significantly altered the kinematics of the saccade, slowing it initially and then producing an overshoot in the trajectory of the eyes (figure 12.2). Remarkably, they found that at saccade end the eyes were accurately stopped at the target: "Accuracy is almost unaffected" by the blink, they wrote. A more recent study in monkeys confirmed that despite the absence of visual feedback, blink-disturbed saccades were corrected in midflight, producing near normal endpoint accuracy (Goossens and Van Opstal, 2000). Importantly, the motion of the eyes during a

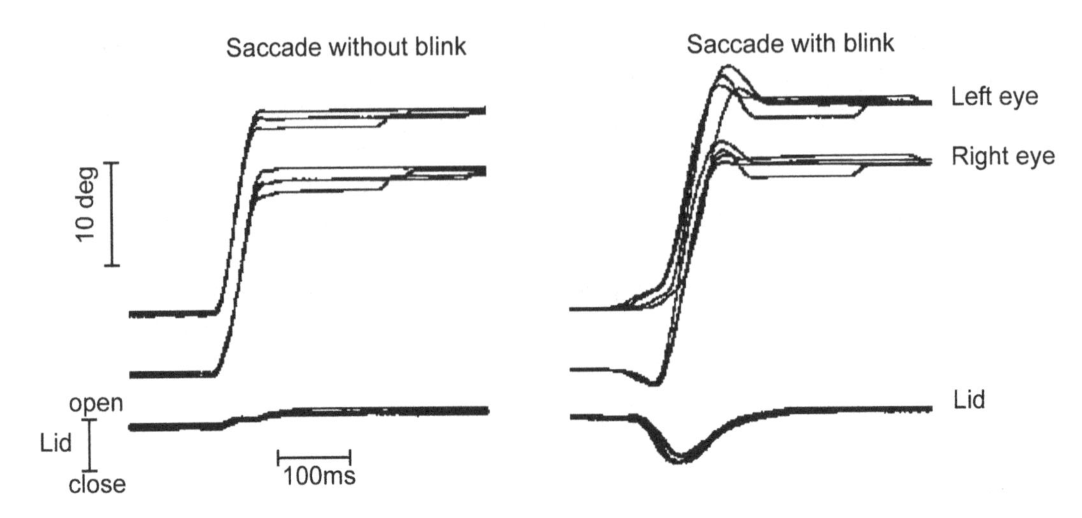

Figure 12.2
An eye blink that takes place during a saccadic eye movement disturbs the trajectory of the eyes, yet the eyes accurately arrive at the target. (From Rottach et al., 1998.)

blink-disturbed saccade was not simply a combination of commands that produced a blink and commands that produced a normal saccade. Rather, the data suggested that when the motor commands that initiated a saccade were corrupted by additional commands that were associated with a blink, the motor commands that followed responded to the state of the eye in a process resembling feedback control, effectively steering the eyes to the target.

A second example of this midflight correction of saccades comes from an experiment in which transcranial magnetic stimulation (TMS) was used to perturb an ongoing movement. TMS is usually placed over a specific part of the brain in order to briefly disturb the activity of a small region directly beneath the stimulating coil. However, Minnan Xu-Wilson, Jing Tian, Reza Shadmehr, and David Zee (2011) discovered that no matter where they placed the coil on the head, it always disturbed an ongoing saccade (figure 12.3). The discharge of the TMS coil appeared to engage a startle-like reflex in the brain that sent inhibitory commands to the saccadic system, resulting in a disturbance to the eye trajectory at a latency of around 60 ms (with respect to TMS pulse), and lasting around 25 ms. Interestingly, despite the fact that the task was performed in the dark and the target was extinguished after saccade onset, the TMS-induced disturbance was immediately corrected with additional motor commands that guided the eyes to the target. This is consistent with a view that as the saccadic motor commands are generated by one part of the brain, another part receives a copy and estimates the resulting state of the eye. When a TMS pulse is given to the head, it engages a startle-like neural system that inhibits

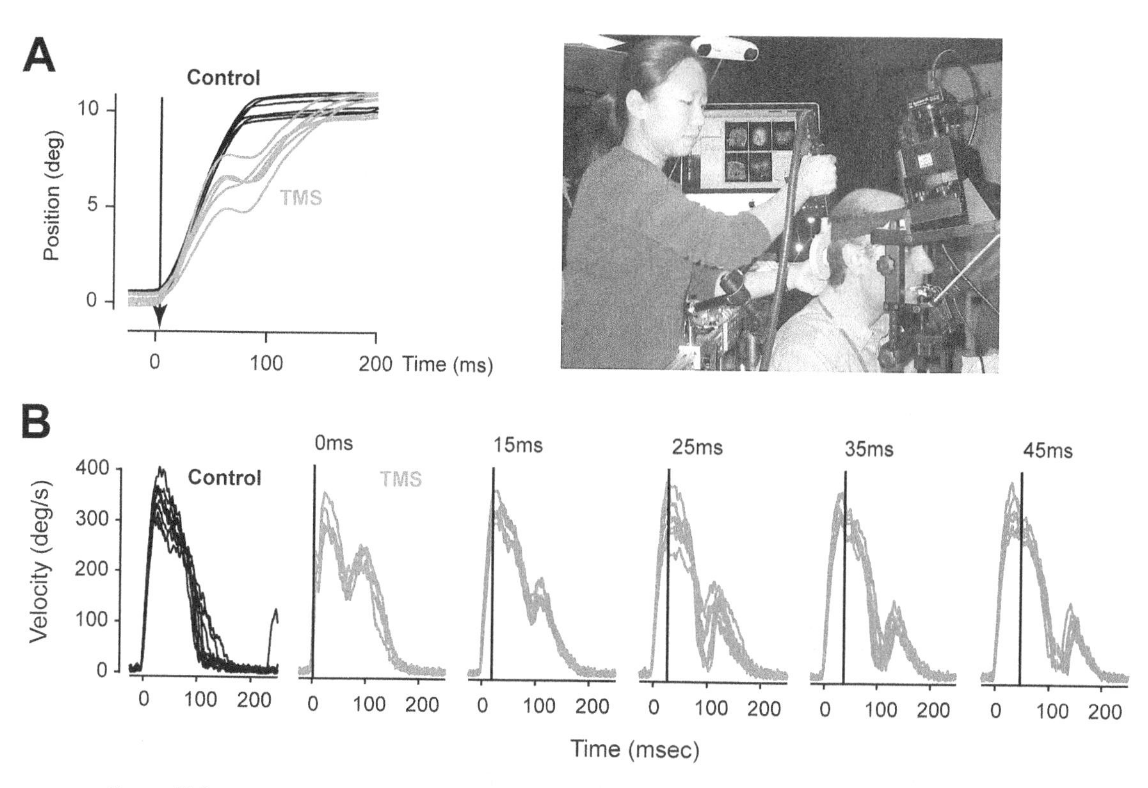

Figure 12.3
TMS applied to anywhere on the head inhibits the ongoing oculomotor commands, but the saccade is corrected midflight with subsequent motor commands that bring the eye to the target. (A) Experimental setup and example of TMS perturbed saccades. The arrow near saccade onset marks the time of a single TMS pulse. (B) TMS applied at various times after saccade onset inhibit the ongoing motor commands at a latency of around 60 ms, as reflected in the perturbation to saccade velocity.

the ongoing motor commands. The system that monitors the oculomotor commands responds to this perturbation and corrects for the motor commands so the eye will arrive near the target. Therefore, the motor commands to the eyes during a saccade are not preprogrammed in some open-loop form, but depend on internal monitoring and feedback.

A third example of this midflight correction of saccades comes from an experiment in which the target was predictably moved at saccade onset. Haiyin Chen-Harris, Wilsaan Joiner, Vincent Ethier, David Zee, and Reza Shadmehr (2008) performed an experiment in which people were shown a visual stimulus at 15 degrees on the horizontal meridian, that is, at (15,0). As soon as the saccade started, the target was removed and a new target appeared at 5-degree vertical displacement, that is, at (15,5) (figure 12.4). Therefore, the saccade completed with an endpoint

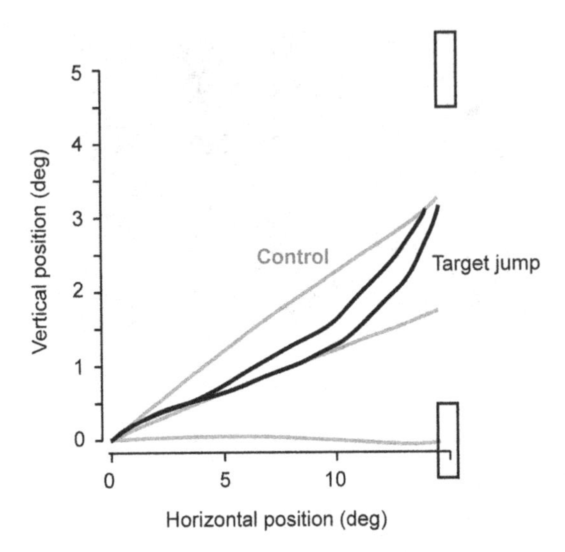

Figure 12.4
Volunteers were shown a target on the horizontal meridian at (15,0) deg. Upon saccade initiation, the target was jumped vertically to (15,5) deg. The resulting adaptation produced curved saccades, marked here with "target jump." In contrast, control saccades to targets that do not jump are generally straight. (From Chen-Harris et al., 2008.)

error. Trial after trial, the motor commands adapted in response to this endpoint error, moving the eyes away from the stimulus at (15,0) and toward (15,5). However, the result of adaptation was not a straight trajectory toward (15,5). (A straight trajectory is the normal response to a control target that appears at 15,5 or elsewhere, as shown by the trajectories labeled control in figure 12.4.) Rather, in this target-jump condition, saccades developed a curvature. (The saccades remained curved whether or not the target jumped on a particular trial, so we can be sure that the curvature was not due to visual input during the saccade.) The motor commands that moved the eyes appeared to be corrected as they were executed. The authors interpreted this data as evidence for a forward model that learned from endpoint errors: It learned to predict that a consequence of the motor commands was a displacement of the target. In effect, this internal feedback acted as a mechanism that steered the eyes toward the predicted position of the target.

In addition to internal feedback via a possible forward model, there are also instances in which sensory feedback affects the control of eye movements. A good example of this is in the case of natural eye movements in which both the eyes and head are free to move. Indeed, most eye movements are not done in isolation, but accompany head movements. An example of a natural (i.e., head-free) gaze shift is shown in figure 12.5A. The gaze shift begins with the eyes making a saccade, and

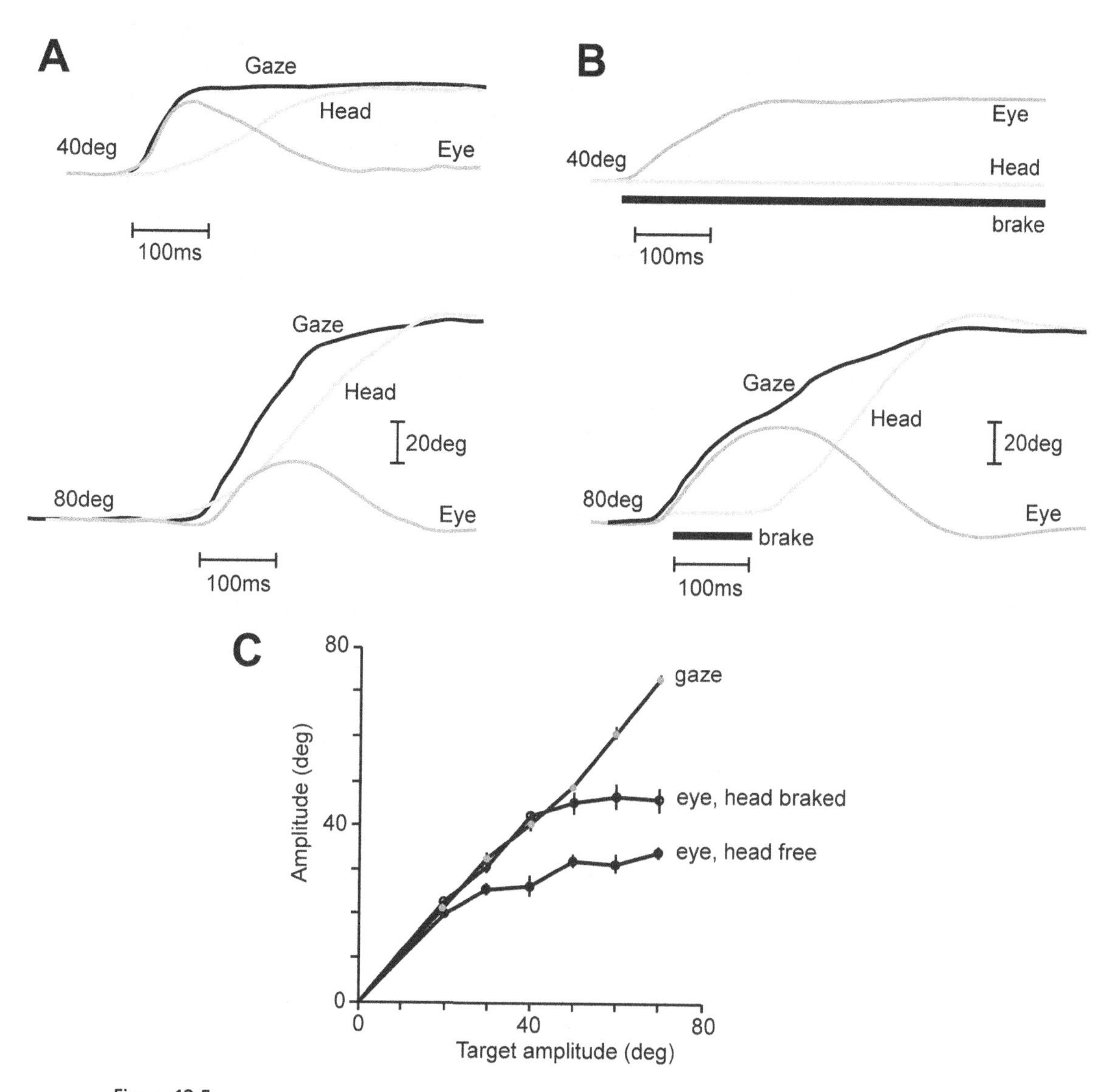

Figure 12.5
The coordinated response of the eye and head to visual targets. (A) Eye and head motion under head unrestrained condition. Volunteers were shown a target on the horizontal meridian, with amplitude marked at start of the traces. Head position is measured with respect to straight ahead. Eye position reflects the position of the globe with respect to a centered location in the head. Gaze is head position plus eye position. (B) Eye and head motion under head restrained condition. On random trials, a brake prevented motion of the head. The period of the break is indicated by the heavy black line. Motion of the eyes depends on the state of the head. (C) Amplitude of the eye saccade as a function of target amplitude in head free and head-braked condition. In the head-braked condition, the head was restrained for 100 ms at start of the gaze shift. Eye saccades are generally larger when head movement is prevented. (From Guitton and Volle, 1987.)

the head soon follows. For a target at 40 degrees, the gaze (sum of head and eye positions) is on the target by around 90 ms and maintains the target on the fovea, yet the head continues to rotate toward the target while the eyes rotate back to recenter with respect to the head. These natural gaze shifts are a good example of a coordinated motion in which multiple body parts cooperate in order to achieve a common goal: maintain the position of the target on the fovea. Emilio Bizzi, Ronald Kalil, and Vinenzo Tagliasco (1971) used this simple movement to answer a fundamental question: were the motor commands to the eyes preprogrammed and open-loop, or did these commands depend on the sensory feedback that measured the state of the head? To answer this question, they devised an apparatus that on random trials held the head stationary by applying a brake. They found that if the head was not allowed to move, the eyes made a saccade to the target, but did not rotate back. This was the case even if the visual stimulus was removed at saccade onset and the gaze shift took place in darkness. The experiment was later repeated by Daniel Guitton and Michel Volle (1987), whose data are shown in figure 12.5B. On a randomly selected trial the target was shown at 40 degrees, but the head was not allowed to rotate. The eyes made a saccade, but because the head was not allowed to move, the eyes did not rotate back. When the target was shown at 80 degrees, normally the eyes make a 30-degree saccade as the head rotates toward the target. However, when on a randomly selected trial the head was not allowed to rotate (brake condition, figure 12.5B), the eyes made a larger amplitude saccade as compared to when the head was free (figure 12.5A), and the eyes did not rotate back until the brake was released and the head was allowed to move. The fact that the eyes exhibited a larger displacement during head-braked trials is summarized in figure 12.5C. Together, these data demonstrate that during head-free eye movements, the motor commands to the eyes are not "open-loop" but depend on the state of the head.

These examples are consistent with the idea that the motor commands that move our body rely on two forms of feedback: internal predictions regarding state of the body/environment (figures 12.2–12.4), and sensory observations (figure 12.5). Let us now show that the feedback gains are also a reflection of available resources and expected rewards. That is, motor commands that are produced in response to sensory feedback are optimized with respect to some cost function. Jörn Diedrichsen (2007) considered a reaching task in which the two arms cooperated and shared a common goal. In one version of the task (one-cursor condition, figure 12.6A), there was a single cursor that reflected the average position of the left and the right hand. In this one-cursor condition, the goal was to move the cursor to the single target. There was also a two-cursor condition in which there was a cursor associated with each arm. In the two-cursor condition, the goal was to move each cursor to its own target. Diedrichsen's idea was that the feedback gains associated with how each arm

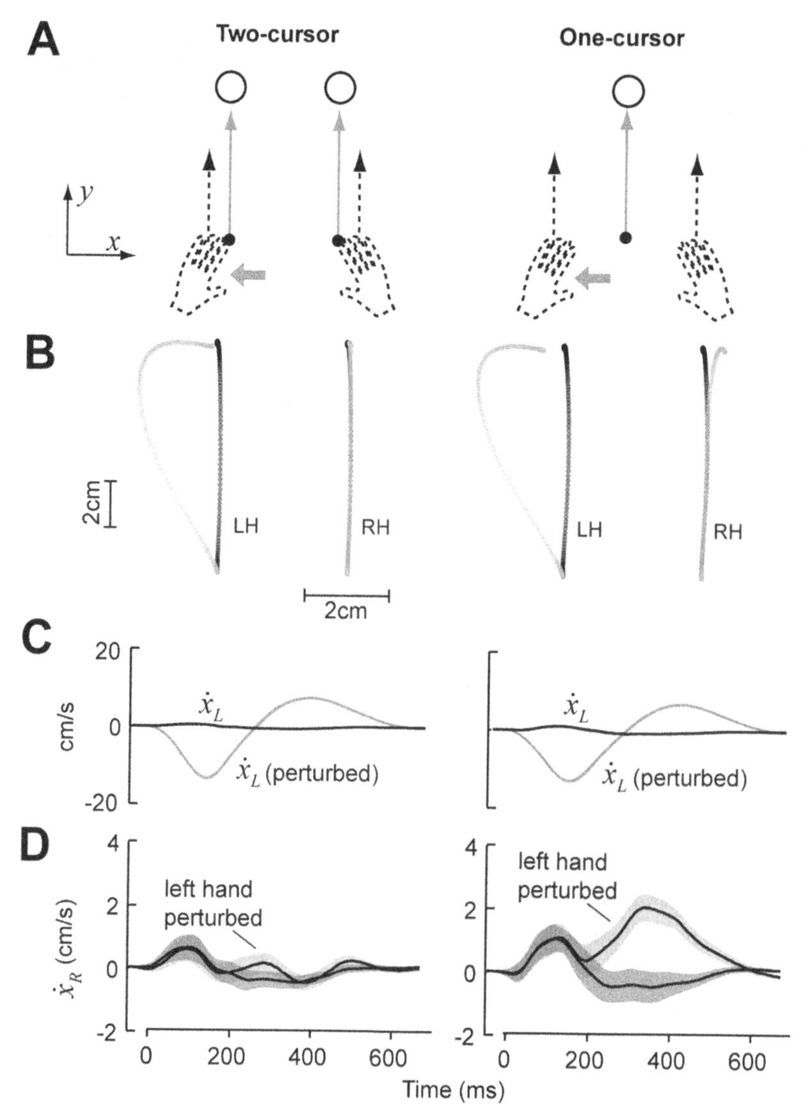

Figure 12.6
Response to a perturbation depends on the cost function. (A) In the two-cursor condition, each hand controlled a cursor and the goal was to place the cursors at the two targets. In the one-cursor condition, the cursor position represented the average position of the left and right hands and the goal was to place the cursor at the single target. Perturbations were applied to the left arm only. (B) Gray dots indicate left and right hand (LH, RH) positions in the condition in which the left hand was perturbed. The black dots reflect the unperturbed condition. In the two-cursor condition, the perturbation to the left hand produces no response in the right hand. In the one-cursor condition, the perturbation to the left hand produces a small response in the right hand. (C) Velocity of the left hand in the perturbed (gray line) and unperturbed conditions. (D) Velocity of the right hand. In the two-cursor condition, right hand velocity is nearly identical whether or not the left hand was perturbed. In the one-cursor condition, right hand velocity shows a corrective response when the left hand is perturbed. (From Diedrichsen, 2007.)

responded to a given perturbation should be different in these two conditions. In the two-cursor condition, the right arm should not respond to a perturbation that displaced the left arm, whereas in the one-cursor condition, the right arm should respond to this perturbation. The reason is that in the one-cursor condition (but not the two-cursor condition), the two arms have a common goal. This common goal should translate into cooperative behavior in which a perturbation that affects one arm should be handled by a reaction by both arms. In figure 12.6B we see that following a leftward perturbation to the left hand, the right hand moved slightly to the right in the one-cursor condition but not the two-cursor condition. This is illustrated better in the velocities of the left and the right arms in figure 12.6C and D: About 200 ms after move onset, the right arm responded to the perturbation to the left arm (figure 12.6D). That is, the right arm takes up some of the correction when the left arm is perturbed, but only if the two arms are working together in a task in which they share a common goal.

Suppose that in the one-cursor condition, a perturbation is given to the left arm. Suppose that this perturbation would require 2 N to compensate. Further suppose that motor commands cost us proportional to the squared magnitude. So if the left arm alone compensated for this perturbation, it would cost us 4 N^2. The optimum thing to do for this particular cost is to have the left and right arms each produce $1N$. Now the cost is 1 N^2 + 1 N^2, or 2 N^2. Cooperation leads to a smaller total cost. If we think of the force that each arm produces in response to the perturbation (a displacement) as a feedback gain, these results provide us with two ideas: (1) in the one-cursor task, the feedback gain of the left arm should be smaller than in the two-cursor task, and (2) in the one-cursor task, the feedback gain of the right arm should depend on the state of the left arm, whereas in the two-cursor task this feedback gain should depend only on the state of the right arm. What we need is a principled way to set these feedback gains so that the limbs cooperate to generate motor commands and bring the cursor to the goal in some efficient way.

In summary, during a movement the motor commands depend on the state of the body part that is being controlled, as well as the overall goal of the task. How does one generate motor commands that depend on the state of the body while simultaneously optimizing some long-term goal? We will consider this question in this chapter.

12.2 A Brief History of Ideas in Biological Control

Early theories of motor control stressed feedback mechanisms as the dominant circuits for voluntary movements. Sir Charles Sherrington (1923) looked at movements as "chains" of reflexes. He wrote: "Coordination, therefore, is in part the compounding of reflexes. . . . This compounding of reflexes with orderliness of coadjustment and of sequence constitutes coordination, and want of it incoordination." His idea was to explain all the richness of motor behavior as a combination of simpler and automatic transformations of senses into actions. As we discussed in chapter 3, developments in robotics shifted the focus of motor neuroscientists from feedback to preprogrammed, or "feedforward," control. The nonlinear dynamics of our limbs and the relatively long delays of sensory feedback would make it hard or impossible to generate stable movements without some form of anticipatory mechanism to compensate for inertial forces. Therefore the brain must be able to prepackage motor commands based on implicit knowledge of the body's mechanics. As it often happens in science, one extreme view replaced another. First, it was all reflexes, and then it became all open-loop control. But while there is evidence for the brain preprogramming movement patterns, there is also equally strong evidence for our ability to correct movements "on the fly" in response to incoming information. This is common sense and it has also been demonstrated by numerous studies of reaching movements, where the target is suddenly changed after the movement starts or the limb perturbed during the movement.

So, while movements are executed based on prior beliefs, our brain also pays attention to the incoming stream of sensory information. But how does it respond to sensory information? The idea of simple fixed reflexes is clearly inadequate. For example, recall that Cordo and Nashner (1982) showed how the muscles at the ankle became active when subjects are pulling on a handle but did not show activity with the same pulling action if a bar insured stability to the body (figure 4.1). The idea is that the response to sensory information is modulated by prior knowledge about the environment in which we move.

Optimal feedback control provides a framework to move beyond the antagonism of feedback vs. feedforward toward a view in which both prior beliefs and sensory-based actions coexist. In this view, a fundamental outcome of motor learning is to shape feedback, by establishing, on the basis of experience, how the brain must respond to incoming sensory input. Through learning, we acquire knowledge of the statistical properties of the environment; we learn how sensory inputs as well as motor commands are affected by uncertainty and how uncertainty itself has structure, for instance being larger in some directions than others. This knowledge is essential to tune future motor responses to sensory information. How can this tuning of feedback parameters be done optimally, so that we can reach our goals with minimal error and effort? Unfortunately the mathematical tools at our disposal are limited and the answer to this question can be given only for simple systems and simple forms of uncertainty. Nevertheless, answers in these simple cases can guide us toward the development of methods with broader range of applications. In the following sections we show how the combination of optimal control and optimal state estimation provide us with a way to relate feedback gains to the dynamical properties of a linear control system and to the statistical properties of signal dependent noise in motor commands and sensory signals.

12.3 Bellman Optimality Principle

Our aim is to produce motor commands so that they not only achieve a goal in some optimal sense (e.g., minimize a cumulative cost), but also respond to feedback—that is, we want the motor commands to be the best that they can be no matter which state we find ourselves in. A framework that is appropriate for solving this problem is optimal feedback control: We have a goal that we wish to achieve, and the goal is specified as a cost per unit of time (e.g., accuracy cost and effort cost). For example, suppose that we have a state specified by vector $\mathbf{x}$ and motor commands specified by vector $\mathbf{u}$. For simplicity, we assume that effort is a linear function of the command vector. Engineers use the term "control cost" because machinery is not yet endowed with a sense of effort. We have a cost per unit of time (i.e., cost per step) that depends on our state (which includes the goal state), and effort:

$$\alpha^{(k)} = \mathbf{u}^{(k)T} L \mathbf{u}^{(k)} + \mathbf{x}^{(k)T} T^{(k)} \mathbf{x}^{(k)}. \tag{12.1}$$

This cost is quadratic, but our discussion at this point does not require any specific form for the cost per step. So without loss of generality, suppose that the cost per step is of the form shown in equation (12.1). We will assume a finite horizon to our goal, which means that we wish to achieve the goal within some specified time period p. Our objective is to find a *policy* $\mathbf{u}^{(k)} = \pi(\mathbf{x}^{(k)})$ such that at each time step k, we can transform our state $\mathbf{x}^{(k)}$ into motor commands $\mathbf{u}^{(k)}$. If this policy is optimal, depicted by the term π^*, then it will minimize the sum total of costs $\sum_{k=0}^{p} \alpha^{(k)}$ from our initial time step $k = 0$ to the end step $k = p$.

To find this policy, we will rely on a fundamental observation of Richard Bellman (1957), who wrote:

> *The Principle of Optimality.* An optimal policy has the property that whatever the initial state and initial decision are, the remaining decisions must constitute an optimal policy with regard to the state resulting from the first decision.

To explain his idea, suppose that we start at the last time point p. We find ourselves at state $\mathbf{x}^{(p)}$. What is the optimum action $\mathbf{u}^{(p)}$ that we can perform? Because we are at the last time point, we have run out of time, and nothing that we can do now will have any bearing. Therefore, the optimum thing to do from a cost standpoint (equation 12.1) is nothing, $\mathbf{u}^{(p)} = 0$. So the optimal policy for the last time point is $\pi^*(\mathbf{x}^{(p)}) = 0$. Using equation (12.1), let us assign a value to the state that we find ourselves at time point p, given that we are using policy π^*:

$$v_{\pi^*}\left(\mathbf{x}^{(p)}\right) = \mathbf{x}^{(p)T} T^{(p)} \mathbf{x}^{(p)}. \tag{12.2}$$

The function $v_{\pi*}$ assigns a number to each state, typically implying that the closer we are to the goal state (which is a part of the vector $\mathbf{x}^{(p)}$), the better. [In a bit of confusing terminology, the smaller the value v for some state, the more valuable that state is for us. At this point, the concept of value seems identical to the concept of cost. The difference will become clearer at the next step.] Now let us move back one step to time point $p - 1$. If we find ourselves at state $\mathbf{x}^{(p-1)}$, what is the best action that we can perform? Suppose our policy $\pi(\mathbf{x}^{(p-1)})$ instructs us to perform action $\mathbf{u}^{(p-1)}$. How good is this policy? Well, given that we are at state $\mathbf{x}^{(p-1)}$ and have performed action $\mathbf{u}^{(p-1)}$, we will have incurred a cost specified by $\alpha^{(p-1)}$ for that time step. Furthermore, the action $\mathbf{u}^{(p-1)}$ will have taken us from state $\mathbf{x}^{(p-1)}$ to $\mathbf{x}^{(p)}$ with probability $p(\mathbf{x}^{(p)}|\mathbf{x}^{(p-1)},\mathbf{u}^{(p-1)})$. This probability is specifid by the dynamics of the system that we are acting on. The state $\mathbf{x}^{(p)}$ has a value. It seems rational that the goodness of our policy should be related to both the cost it incurs on the current time step $p - 1$, and the value of the state it takes us to in the next time step p. More precisely, according to Bellman, the value of any given state is the cost that we incurred in reaching that state (by whatever policy) plus the expected value produced by the optimal policy from that state to the goal. The reader may want to pause for a moment and see how this elegant concept is a reformulation of the optimality principle. More importantly, this idea can be directly translated into an equation. Thus, we assign a value to state $\mathbf{x}^{(p-1)}$ as follows:

$$v_{\pi}\left(\mathbf{x}^{(p-1)}\right)=\alpha^{(p-1)}+\int v_{\pi^*}\left(\mathbf{x}^{(p)}\right)p\left(\mathbf{x}^{(p)}\middle|\mathbf{x}^{(p-1)},\mathbf{u}^{(p-1)}\right)d\mathbf{x}^{(p)}. \tag{12.3}$$

The rightmost term in equation (12.3) is the mean value of the state that the optimal motor command takes us to:

$$v_{\pi}\left(\mathbf{x}^{(p-1)}\right)=\alpha^{(p-1)}+E\left[v_{\pi^*}\left(\mathbf{x}^{(p)}\right)\middle|\mathbf{x}^{(p-1)},\mathbf{u}^{(p-1)}\right]. \tag{12.4}$$

If our policy $\pi(\mathbf{x}^{(p-1)})$ were optimum, then it would produce motor commands $\mathbf{u}^{(p-1)}$ that minimize the sum of current cost $\alpha^{(p-1)}$, plus the value of the state that it takes us to, that is, minimize equation (12.4). (This is because the value of the next state is the smallest cost that we can incur if we were to perform the optimal policy at the next step.) Therefore, the optimum policy at time step $p - 1$ has the following property:

$$\pi^*\left(\mathbf{x}^{(p-1)}\right)=\underset{\mathbf{u}(p-1)}{\arg\min}\left\{\alpha^{(p-1)}+E\left[v_{\pi^*}\left(\mathbf{x}^{(p)}\right)\middle|\mathbf{x}^{(p-1)},\mathbf{u}^{(p-1)}\right]\right\}. \tag{12.5}$$

And so, if our policy was optimal, then the value of the state $\mathbf{x}^{(p-1)}$ would be related to the value of the state $\mathbf{x}^{(p)}$ as follows:

$$v_{\pi^*}\left(\mathbf{x}^{(p-1)}\right)=\min_{\mathbf{u}(p-1)}\left\{\alpha^{(p-1)}+E\left[v_{\pi^*}\left(\mathbf{x}^{(p)}\right)\middle|\mathbf{x}^{(p-1)},\mathbf{u}^{(p-1)}\right]\right\}. \tag{12.6}$$

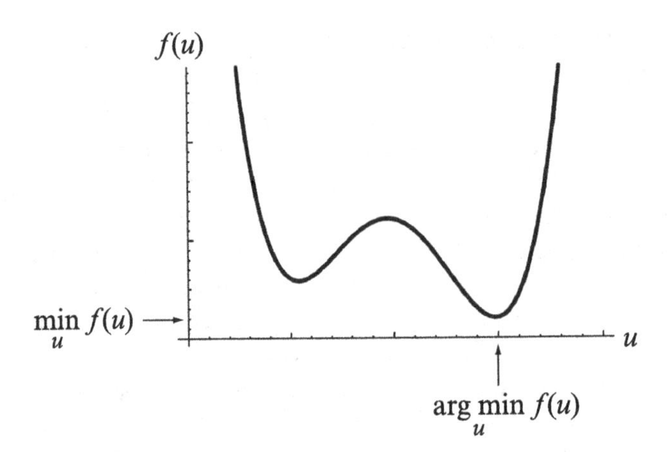

Figure 12.7
Schematic description of min $f(u)$ and arg min $f(u)$.

To help demystify the nomenclature, in figure 12.7 we have plotted what we mean by the terms *min* and *argmin*. The term *min* refers to the minimum value of a function, and the term *argmin* refers to the argument of a function for which the function has a minimum value. In general, the optimal policy $\pi^*(\mathbf{x}^{(k)})$ has the property that it produces the smallest value possible for state $\mathbf{x}^{(k)}$, where "value" refers to the cost that is accumulated if we were to apply the motor commands specified by this policy starting from time step k to the last time point p. If we write the expression in equation (12.6) for arbitrary time step k, we arrive at what is called the *Bellman equation*:

$$v_{\pi^*}\left(\mathbf{x}^{(k)}\right) = \min_{\mathbf{u}(k)} \left\{\alpha^{(k)} + E\left[v_{\pi^*}\left(\mathbf{x}^{(k+1)}\right) \middle| \mathbf{x}^{(k)}, \mathbf{u}^{(k)}\right]\right\}. \tag{12.7}$$

Equation (12.7) implies that if we knew the optimal value associated with states at time point $k+1$, that is, $v_{\pi^*}\left(\mathbf{x}^{(k+1)}\right)$, then we could find the optimum motor command $\mathbf{u}^{(k)}$, and this would provide us with the value associated with $v_{\pi^*}\left(\mathbf{x}^{(k)}\right)$. That is, we could recursively solve our problem by starting at some state (usually the end state $\mathbf{x}^{(p)}$), form an optimal value function at this state, and then work backward one step at a time. Using the Bellman equation (equation 12.7), we have a tool to break down the problem into smaller subproblems.

12.4 Control Policy

We are going to imagine that in order to make a movement, our brain formulates a goal in terms of a rewarding state that it wants to achieve. It has a model of how the motor commands influence the state of the body (i.e., a forward model), and if the movement is slow enough, the brain has access to sensory feedback during the movement. In reality, we do not know how the cost of achieving the goal is represented, or what costs may be involved in formulating the motor commands. However, there is some evidence that the rewarding state is discounted in time, i.e., it is better to get to the rewarding state sooner than later (Shadmehr et al., 2010). There is also some evidence that the efforts associated with the motor commands carry a cost, and the form of this effort cost is approximately quadratic (O'Sullivan, Burdet, and Diedrichsen, 2009). In the last chapter we showed that for control of very simple movements like saccades, motor commands that optimize the following cost function are fairly successful in accounting for movement kinematics:

$$J(p) = \mathbf{x}^{(p)T} T \mathbf{x}^{(p)} + \sum_{k=0}^{p} \mathbf{u}^{(k)T} L \mathbf{u}^{(k)} + \lambda \left(1 - \frac{1}{1+\beta p} \right). \tag{12.8}$$

In the preceding expression, the first term describes a cost for the distance to the goal state at the end of the movement, the second term describes a cost for the accumulated effort, and the last term describes a cost of time (a function that grows with movement duration). In our previous chapter we assumed that movements were open-loop (i.e., no feedback of any kind). For a given movement duration p we computed the sequence of motor commands that minimized the above cost. We then found the optimum movement duration by searching among the various total costs $J(p)$ for the one with the minimum cost. Here, we will assume that there is feedback during the movement. *We are no longer interested in computing the optimum motor commands. Rather, we wish to compute the optimum policy, something that transforms each state that we might encounter during the movement into the motor commands.* For a given movement duration p, our cost per step is:

$$\alpha^{(k)} = \mathbf{x}^{(k)T} T^{(k)} \mathbf{x}^{(k)} + \mathbf{u}^{(k)T} L \mathbf{u}^{(k)} + \frac{\lambda \beta}{1+\beta p}. \tag{12.9}$$

The last term in equation (12.9) is the average cost of time per step, i.e., the last term in equation (12.8) divided by p. (The term $T^{(k)}$ may be zero for all time steps except the last.) We will consider a system in which the motor commands produce signal-dependent noise, and the sensory observations also suffer from noise. We will estimate the state of the system using a Kalman filter. Our objective is to apply Bellman's theory in order to formulate a control policy in which the motor commands depend on our current estimate of the state of the system. Emanuel Todorov (2005) considered this problem and was first to describe a solution for the case in which the noise was signal dependent (which appears to be the case for biological systems). The derivations in this chapter are based on his work. Once we derive the solution, we will apply it to some of the movements that we presented in figures 12.2–12.6.

Suppose that we have a system of the form:

$$\begin{aligned} \mathbf{x}^{(k+1)} &= A\mathbf{x}^{(k)} + B\left(\mathbf{u}^{(k)} + \boldsymbol{\varepsilon}_u^{(k)}\right) + \boldsymbol{\varepsilon}_x \\ \mathbf{y}^{(k)} &= H\left(\mathbf{x}^{(k)} + \boldsymbol{\varepsilon}_s^{(k)}\right) + \boldsymbol{\varepsilon}_y \end{aligned}, \tag{12.10}$$

where $\boldsymbol{\varepsilon}_x$ and $\boldsymbol{\varepsilon}_y$ are zero mean Gaussian noise vectors with variance Q_x and Q_y:

$$\begin{aligned} \boldsymbol{\varepsilon}_x &\sim N(\mathbf{0}, Q_x) \\ \boldsymbol{\varepsilon}_y &\sim N(\mathbf{0}, Q_y) \end{aligned} \tag{12.11}$$

and $\boldsymbol{\varepsilon}_u$ and $\boldsymbol{\varepsilon}_s$ are zero-mean signal-dependent-noise terms, meaning that noise depends on the motor commands $\mathbf{u}$ and state $\mathbf{x}$, respectively:

$$\begin{array}{cc} \boldsymbol{\varepsilon}_u^{(k)} \equiv \begin{bmatrix} c_1 u_1^{(k)} \phi_1^{(k)} \\ c_2 u_2^{(k)} \phi_2^{(k)} \\ \vdots \\ c_m u_m^{(k)} \phi_m^{(k)} \end{bmatrix} & \boldsymbol{\varepsilon}_s^{(k)} \equiv \begin{bmatrix} d_1 x_1^{(k)} \mu_1^{(k)} \\ d_2 x_2^{(k)} \mu_2^{(k)} \\ \vdots \\ d_n x_n^{(k)} \mu_n^{(k)} \end{bmatrix} \\ \phi \sim N(0,1) & \mu \sim N(0,1) \\ c_i \geq 0 & d_i \geq 0 \end{array}. \tag{12.12}$$

The signal dependent motor noise $\boldsymbol{\varepsilon}_u$ affects the state $\mathbf{x}$, and the signal dependent sensory noise $\boldsymbol{\varepsilon}_s$ affects the observation $\mathbf{y}$. It is useful to express the signal dependent noise terms as a linear function of $\mathbf{u}$ and $\mathbf{x}$. To do so, we define:

$$\begin{array}{cc} C_1 \equiv \begin{bmatrix} c_1 & 0 & 0 \\ 0 & 0 & 0 \\ 0 & 0 & \ddots \end{bmatrix} & C_2 \equiv \begin{bmatrix} 0 & 0 & 0 \\ 0 & c_2 & 0 \\ 0 & 0 & \ddots \end{bmatrix} \\ D_1 \equiv \begin{bmatrix} d_1 & 0 & 0 \\ 0 & 0 & 0 \\ 0 & 0 & \ddots \end{bmatrix} & D_2 \equiv \begin{bmatrix} 0 & 0 & 0 \\ 0 & d_2 & 0 \\ 0 & 0 & \ddots \end{bmatrix} \end{array}, \tag{12.13}$$

and so we have:

$$\begin{aligned} \boldsymbol{\varepsilon}_u^{(k)} &= \sum_{i=1}^{m} C_i \mathbf{u}^{(k)} \phi_i^{(k)} \\ \boldsymbol{\varepsilon}_s^{(k)} &= \sum_{i=1}^{n} D_i \mathbf{x}^{(k)} \mu_i^{(k)} \end{aligned}. \tag{12.14}$$

In equation (12.14), m is the dimension of the vector $\mathbf{u}$ and n is the dimension of the vector $\mathbf{x}$. Because ϕ and μ are Gaussian random variables, $\boldsymbol{\varepsilon}_u$ and $\boldsymbol{\varepsilon}_s$ are also Gaussian with the following distribution:

$$\begin{aligned}\boldsymbol{\varepsilon}_u^{(k)} &\sim N\left(\mathbf{0}, \sum_{i=1}^{m} C_i \mathbf{u}^{(k)} \mathbf{u}^{(k)T} C_i\right) \\ \boldsymbol{\varepsilon}_s^{(k)} &\sim N\left(\mathbf{0}, \sum_{i=1}^{n} D_i \mathbf{x}^{(k)} \mathbf{x}^{(k)T} D_i\right)\end{aligned}. \tag{12.15}$$

And so our system has the following dynamics:

$$\begin{aligned}\mathbf{x}^{(k+1)} &= A\mathbf{x}^{(k)} + B\mathbf{u}^{(k)} + \boldsymbol{\varepsilon}_x + B\sum_i C_i \mathbf{u}^{(k)} \phi_i^{(k)} \\ \mathbf{y}^{(k)} &= H\mathbf{x}^{(k)} + \boldsymbol{\varepsilon}_y + H\sum_i D_i \mathbf{x}^{(k)} \mu_i^{(k)} \\ \boldsymbol{\varepsilon}_x &\sim N(\mathbf{0}, Q_x) \quad \boldsymbol{\varepsilon}_y \sim N(\mathbf{0}, Q_y) \\ \phi &\sim N(0,1) \qquad \mu \sim N(0,1)\end{aligned}. \tag{12.16}$$

We estimate of the state of our system by combining our predictions with our sensory observations using the Kalman framework:

$$\begin{aligned}\hat{\mathbf{x}}^{k|k} &= \hat{\mathbf{x}}^{k|k-1} + K^{(k)}\left(\mathbf{y}^{(k)} - H\hat{\mathbf{x}}^{k|k-1}\right) \\ \hat{\mathbf{x}}^{k+1|k} &= A\hat{\mathbf{x}}^{k|k} + B\mathbf{u}^{(k)}\end{aligned}. \tag{12.17}$$

Our estimate of state at time point $k + 1$ is simply our prior estimate at time point $k + 1, \hat{\mathbf{x}}^{(k+1)} = \hat{\mathbf{x}}^{k+1|k}$. In general, our estimate of state at any time point $k + 1$ is related to our observations on the previous trial $\mathbf{y}^{(k)}$, our prior beliefs $\hat{\mathbf{x}}^{(k)}$, and motor commands $\mathbf{u}^{(k)}$ as follows:

$$\hat{\mathbf{x}}^{(k+1)} = A\hat{\mathbf{x}}^{(k)} + AK^{(k)}\left(\mathbf{y}^{(k)} - H\hat{\mathbf{x}}^{(k)}\right) + B\mathbf{u}^{(k)}. \tag{12.18}$$

At this point we have described a procedure for estimating the state $\hat{\mathbf{x}}^{(k)}$. What motor command should we produce at this time point? We need to compute the policy $\pi^*\left(\hat{\mathbf{x}}^{(k)}\right)$ that transforms our estimate $\hat{\mathbf{x}}^{(k)}$ into motor command $\mathbf{u}^{(k)}$. Let us start at the last time point and consider what the value function $v_{\pi^*}\left(\mathbf{x}^{(p)}, \hat{\mathbf{x}}^{(p)}\right)$ might look like. Our cost at this last time point is specified by equation (12.9). At the last time point the best thing to do is nothing, that is, we will set $\mathbf{u}^{(p)} = \mathbf{0}$. When we do so, the cost at this last time point is a quadratic function of $\mathbf{x}^{(p)}$, and so the value function at this last time point is:

$$v_{\pi^*}\left(\mathbf{x}^{(p)}, \hat{\mathbf{x}}^{(p)}\right) = \mathbf{x}^{(p)T} T^{(p)} \mathbf{x}^{(p)} + \frac{\lambda\beta}{1+\beta p}. \tag{12.19}$$

In general, for linear systems for which the cost per step (equation 12.9) is quadratic in state, the value of the states under the optimal policy is also quadratic. For this reason, Todorov (2005) hypothesized that the value function for any given time step has the following form:

$$v_{\pi^*}\left(\mathbf{x}^{(k)}, \hat{\mathbf{x}}^{(k)}\right) = \mathbf{x}^{(k)T} W_x^{(k)} \mathbf{x}^{(k)} + \left(\mathbf{x}^{(k)} - \hat{\mathbf{x}}^{(k)}\right)^T W_e^{(k)} \left(\mathbf{x}^{(k)} - \hat{\mathbf{x}}^{(k)}\right) + w^{(k)}. \tag{12.20}$$

For the last time point, equation (12.20) certainly seems reasonable because if we set $W_x^{(p)} = T^{(p)}$, $W_e^{(p)} = 0$, and $w^{(p)} = \dfrac{\lambda\beta}{1+\beta p}$, we simply get the cost $\alpha^{(p)}$ under the optimal policy of $\mathbf{u}^{(p)} = \mathbf{0}$. If our policy were optimal, then the value function in one step would be related to the value function in the next step via the Bellman equation:

$$v_{\pi^*}\left(\mathbf{x}^{(k)}, \hat{\mathbf{x}}^{(k)}\right) = \min_{\mathbf{u}(k)} \left\{\alpha^{(k)} + E\left[v_{\pi^*}\left(\mathbf{x}^{(k+1)}, \hat{\mathbf{x}}^{(k+1)}\right) \middle| \mathbf{x}^{(k)}, \hat{\mathbf{x}}^{(k)}, \mathbf{u}^{(k)}\right]\right\}. \tag{12.21}$$

To find the optimal control policy we will proceed in the following three steps:

1. Starting at time step $k+1$, we will assume that the value function $v_{\pi^*}\left(\mathbf{x}^{(k+1)}, \hat{\mathbf{x}}^{(k+1)}\right)$ has the form specified by equation (12.20). We will apply the Bellman equation (equation 12.21) and compute $v_{\pi^*}\left(\mathbf{x}^{(k)}, \hat{\mathbf{x}}^{(k)}\right)$.
2. Next, we will find the motor commands $\mathbf{u}^{(k)}$ that minimize $v_{\pi^*}\left(\mathbf{x}^{(k)}, \hat{\mathbf{x}}^{(k)}\right)$.
3. By finding these motor commands, we will be able to check whether $v_{\pi^*}\left(\mathbf{x}^{(k)}, \hat{\mathbf{x}}^{(k)}\right)$ has the form specified by equation (12.20). That is, by having $\mathbf{u}^{(k)}$, we will check whether the value function at time step k is in fact a quadratic function of states. If it is, then we will have found the optimal motor commands for one time step.

By stepping back another step and so on, we will have found the optimal policy for all time steps. In the end, because we assume that we are controlling a linear system (equation 12.10), the value function will turn out to be quadratic both in the estimated state and in the motor command. Furthermore, the optimal policy will turn out to be linear in the state estimated at each step:

$$\mathbf{u}^{(k)} = -G^{(k)}\hat{\mathbf{x}}^{(k)}.$$

Our goal is to derive the gain matrix $G^{(k)}$ from the statistical properties of the motor and sensory processes, both of which are affected by signal dependent noise (equation 12.14). At the end of our story, our policy will be the sequence of gain matrices $G^{(k)}$ for time steps $k = 0, \cdots, p$, allowing us to produce motor commands for whatever state $\hat{\mathbf{x}}$ we happen to find ourselves at.

Step 1

To simplify the notation, we define the difference between our estimate of state and actual state as an estimation error:

$$\mathbf{e}^{(k)} \equiv \mathbf{x}^{(k)} - \hat{\mathbf{x}}^{(k)}. \tag{12.22}$$

By combining equations 12.16 and 12.18, we write the dynamics of error in our estimation of state as:

$$\mathbf{e}^{(k+1)} = \left(A - AK^{(k)}H\right)\mathbf{e}^{(k)} + \boldsymbol{\varepsilon}_x + B\sum_i C_i \mathbf{u}^{(k)}\phi_i^{(k)} \\ -AK^{(k)}\boldsymbol{\varepsilon}_y - AK^{(k)}H\sum_i D_i \mathbf{x}^{(k)}\mu_i^{(k)} \tag{12.23}$$

We rewrite equation (12.20) as:

$$v_{\pi^*}\left(\mathbf{x}^{(k+1)}, \hat{\mathbf{x}}^{(k+1)}\right) = \mathbf{x}^{(k+1)T} W_x^{(k+1)} \mathbf{x}^{(k+1)} + \mathbf{e}^{(k+1)T} W_e^{(k+1)} \mathbf{e}^{(k+1)} + w^{(k+1)}. \tag{12.24}$$

To minimize equation (12.21), given that we are at state $\mathbf{x}^{(k)}$ and $\hat{\mathbf{x}}^{(k)}$, and have produced motor command $\mathbf{u}^{(k)}$, we need to compute the expected value of the above value function. To do so, we will need the expected value and variance of $\mathbf{x}^{(k+1)}$:

$$E\left[\mathbf{x}^{(k+1)} \middle| \mathbf{x}^{(k)}, \hat{\mathbf{x}}^{(k)}, \mathbf{u}^{(k)}\right] = A\mathbf{x}^{(k)} + B\mathbf{u}^{(k)} \\ \mathrm{var}\left[\mathbf{x}^{(k+1)} \middle| \cdots\right] = Q_x + \sum_i BC_i \mathbf{u}^{(k)} \mathbf{u}^{(k)T} C_i^T B^T \tag{12.25}$$

We will also need the expected value and variance of $\mathbf{e}^{(k+1)}$:

$$E\left[\mathbf{e}^{(k+1)} \middle| \cdots\right] = \left(A - AK^{(k)}H\right)\mathbf{e}^{(k)} \\ \mathrm{var}\left[\mathbf{e}^{(k+1)} \middle| \cdots\right] = Q_x + \sum_i BC_i \mathbf{u}^{(k)} \mathbf{u}^{(k)T} C_i^T B^T + AK^{(k)} Q_y K^{(k)T} A^T \\ + \sum_i AK^{(k)} HD_i \mathbf{x}^{(k)} \mathbf{x}^{(k)T} D_i^T H^T K^{(k)T} A^T \tag{12.26}$$

With the preceding expressions, we can compute the expected value of equation (12.24):

$$E\left[v_{\pi^*}\left(\mathbf{x}^{(k+1)}, \hat{\mathbf{x}}^{(k+1)}\right) \middle| \mathbf{x}^{(k)}, \hat{\mathbf{x}}^{(k)}, \mathbf{u}^{(k)}\right] = E\left[\mathbf{x}^{(k+1)T} W_x^{(k+1)} \mathbf{x}^{(k+1)} \middle| \cdots\right] \\ + E\left[\mathbf{e}^{(k+1)T} W_e^{(k+1)} \mathbf{e}^{(k+1)} \middle| \cdots\right] + w^{(k+1)} \tag{12.27}$$

Recall that the expected value of a scalar quantity that depends on the quadratic form of a random variable $\mathbf{x}$ is:

$$E\left[\mathbf{x}^T A \mathbf{x}\right] = E[\mathbf{x}]^T A E[\mathbf{x}] + tr\left[A\,\mathrm{var}[\mathbf{x}]\right].$$

Therefore, the expected value terms on the right side of equation (12.27) are:

$$E\left[\mathbf{x}^{(k+1)T} W_x^{(k+1)} \mathbf{x}^{(k+1)} \middle| \cdots\right] = \left(A\mathbf{x}^{(k)} + B\mathbf{u}^{(k)}\right)^T W_x^{(k+1)} \left(A\mathbf{x}^{(k)} + B\mathbf{u}^{(k)}\right) \\ + tr\left[W_x^{(k+1)} Q_x\right] + \mathbf{u}^{(k)T}\left(\sum_i C_i^T B^T W_x^{(k+1)} BC_i\right)\mathbf{u}^{(k)} \tag{12.28}$$

$$E\left[\mathbf{e}^{(k+1)T} W_e^{(k+1)} \mathbf{e}^{(k+1)} \middle| \cdots\right] = \mathbf{e}^{(k)T}\left(A - AK^{(k)}H\right)^T W_e^{(k+1)} \left(A - AK^{(k)}H\right)\mathbf{e}^{(k)} \\ + tr\left[W_e^{(k+1)}\left(Q_x + AK^{(k)} Q_y K^{(k)T} A^T\right)\right] \\ + \mathbf{u}^{(k)T}\left(\sum_i C_i^T B^T W_e^{(k+1)} BC_i\right)\mathbf{u}^{(k)} \\ + \mathbf{x}^{(k)T}\left(\sum_i D_i^T H^T K^{(k)T} A^T W_e^{(k+1)} AK^{(k)} HD_i\right)\mathbf{x}^{(k)} \tag{12.29}$$

We can now write the Bellman equation in a form that we can use to find the optimal motor command at time point k:

$$v_{\pi^*}\left(\mathbf{x}^{(k)},\hat{\mathbf{x}}^{(k)}\right)=\min_{\mathbf{u}^{(k)}}\left\{\begin{array}{l}\mathbf{u}^{(k)T}L\mathbf{u}^{(k)}+\mathbf{x}^{(k)T}T^{(k)}\mathbf{x}^{(k)}+\dfrac{\lambda\beta}{1+\beta p}\\ +E\left[\mathbf{x}^{(k+1)T}W_x^{(k+1)}\mathbf{x}^{(k+1)}\middle|\mathbf{x}^{(k)},\hat{\mathbf{x}}^{(k)},\mathbf{u}^{(k)}\right]\\ +E\left[\mathbf{e}^{(k+1)T}W_e^{(k+1)}\mathbf{e}^{(k+1)}\middle|\cdots\right]+w^{(k+1)}\end{array}\right\}. \tag{12.30}$$

Step 2

To find the optimal motor command at time point k, we find the derivative of the sum on the right side of equation (12.30) with respect to $\mathbf{u}^{(k)}$ and set it equal to zero. Using equation (12.28) and equation (12.29), we can write equation (12.30) as:

$$v_{\pi^*}\left(\mathbf{x}^{(k)},\hat{\mathbf{x}}^{(k)}\right)=\min_{\mathbf{u}^{(k)}}\left\{\begin{array}{l}\mathbf{u}^{(k)T}\left(L+C_x^{(k+1)}+C_e^{(k+1)}+B^TW_x^{(k+1)}B\right)\mathbf{u}^{(k)}+\dfrac{\lambda\beta}{1+\beta p}\\ +\mathbf{x}^{(k)T}\left(A^TW_x^{(k+1)}A+T+D_e^{(k+1)}\right)\mathbf{x}^{(k)}+2\mathbf{x}^{(k)T}A^TW_x^{(k+1)}B\mathbf{u}^{(k)}\\ +\mathbf{e}^{(k)T}\left(A-AK^{(k)}H\right)^TW_e^{(k+1)}\left(A-AK^{(k)}H\right)\mathbf{e}^{(k)}\\ +tr\left[W_x^{(k+1)}Q_x\right]+tr\left[W_e^{(k+1)}\left(Q_x+AK^{(k)}Q_yK^{(k)T}A^T\right)\right]+w^{(k+1)}\end{array}\right\} \tag{12.31}$$

with the following shortcuts:

$$\begin{aligned}C_x^{(k+1)}&\equiv\sum\nolimits_i C_i^TB^TW_x^{(k+1)}BC_i\\ C_e^{(k+1)}&\equiv\sum\nolimits_i C_i^TB^TW_e^{(k+1)}BC_i\\ D_e^{(k+1)}&=\sum\nolimits_i D_i^TH^TK^{(k)T}A^TW_e^{(k+1)}AK^{(k)}HD_i\end{aligned}. \tag{12.32}$$

After we find the derivative of sum in equation (12.31) with respect to $\mathbf{u}^{(k)}$ and set it equal to zero, we have:

$$\begin{aligned}G^{(k)}&\equiv\left(L+C_x^{(k+1)}+C_e^{(k+1)}+B^TW_x^{(k+1)}B\right)^{-1}B^TW_x^{(k+1)}A\\ \mathbf{u}^{(k)}&=-G^{(k)}\mathbf{x}^{(k)}\end{aligned}. \tag{12.33}$$

This expression is our control policy for time step k. In practice, the state $\mathbf{x}^{(k)}$ is not observable and can only be estimated, and therefore the best that we can do is replace it with $\hat{\mathbf{x}}^{(k)}$:

$$\begin{aligned}\hat{\mathbf{x}}^{(k)}&=A\hat{\mathbf{x}}^{(k-1)}+AK^{(k-1)}\left(\mathbf{y}^{(k-1)}-H\hat{\mathbf{x}}^{(k-1)}\right)+B\mathbf{u}^{(k-1)}\\ \mathbf{u}^{(k)}&=-G^{(k)}\hat{\mathbf{x}}^{(k)}\end{aligned}. \tag{12.34}$$

Notice that the gain $G^{(k)}$ is inversely proportional to L, a variable that penalizes the effort expended on the task. Equation (12.33) implies that the greater the effort cost, the smaller the feedback gain. A second point to notice is that the feedback gain $G^{(k)}$ depends on $C_e^{(k+1)}$, which in turn depends on $W_e^{(k+1)}$. This matrix is a weight that penalizes the "squared" state estimation error $\mathbf{x}^{(k)}-\hat{\mathbf{x}}^{(k)}$ in equation (12.20). As we will see below, the estimation error depends on the Kalman gain $K^{(k+1)}$, implying that the feedback gain $G^{(k)}$ will also depend on the Kalman gain. The Kalman gain depends on the noise properties of the system, describing the uncertainty regarding our estimate of states. In summary, our policy, described as a time-dependent feedback gain, transforms our estimate of state into motor commands. This policy depends on the cost function that we are trying to minimize, as well as the uncertainties that we have regarding our estimate of state.

Step 3

As our final step, we need to show that when we apply the policy in equation (12.33), the resulting value function $v_{\pi^*}\left(\mathbf{x}^{(k)},\hat{\mathbf{x}}^{(k)}\right)$ remains in the quadratic form that we assumed for $v_{\pi^*}\left(\mathbf{x}^{(k+1)},\hat{\mathbf{x}}^{(k+1)}\right)$. We insert $\mathbf{u}^{(k)}=-G^{(k)}\hat{\mathbf{x}}^{(k)}$ from equation (12.33) into equation (12.31) and we have:

$$\begin{aligned} v_{\pi^*}\left(\mathbf{x}^{(k)},\hat{\mathbf{x}}^{(k)}\right) = & \hat{\mathbf{x}}^{(k)T}G^{(k)T}B^T W_x^{(k+1)}A\hat{\mathbf{x}}^{(k)} - 2\hat{\mathbf{x}}^{(k)T}G^{(k)T}B^T W_x^{(k+1)}A\mathbf{x}^{(k)} \\ & +\mathbf{e}^{(k)T}\left(A-AK^{(k)}H\right)^T W_e^{(k+1)}\left(A-AK^{(k)}H\right)\mathbf{e}^{(k)} \\ & +\mathbf{x}^{(k)T}\left(T^{(k)}+A^T W_x^{(k+1)}A+D_e^{(k+1)}\right)\mathbf{x}^{(k)} \\ & +tr\left[W_x^{(k+1)}Q_x+W_e^{(k+1)}\left(Q_x+AK^{(k)}Q_y K^{(k)T}A^T\right)\right]+\frac{\lambda\beta}{1+\beta p} \end{aligned} \qquad (12.35)$$

Using the identity $\hat{\mathbf{x}}^T Z\hat{\mathbf{x}}-2\hat{\mathbf{x}}^T Z\mathbf{x}=(\mathbf{x}-\hat{\mathbf{x}})^T Z(\mathbf{x}-\hat{\mathbf{x}})-\mathbf{x}^T Z\mathbf{x}$, we can simplify the first line of the above expression and remove the dependence on the interaction between $\hat{\mathbf{x}}^{(k)}$ and $\mathbf{x}^{(k)}$. When we do so, we arrive at the observation that the value function at time step k is quadratic:

$$\begin{aligned} v_{\pi^*}\left(\mathbf{x}^{(k)},\hat{\mathbf{x}}^{(k)}\right) &= \mathbf{x}^{(k)T}W_x^{(k)}\mathbf{x}^{(k)}+\mathbf{e}^{(k)T}W_e^{(k)}\mathbf{e}^{(k)}+w^{(k)} \\ W_e^{(k)} &\equiv \left(A-AK^{(k)}H\right)^T W_e^{(k+1)}\left(A-AK^{(k)}H\right)+G^{(k)T}B^T W_x^{(k+1)}A \\ W_x^{(k)} &\equiv T^{(k)}+A^T W_x^{(k+1)}A+D_e^{(k+1)}-G^{(k)T}B^T W_x^{(k+1)}A \\ w^{(k)} &\equiv tr\left[W_x^{(k+1)}Q_x+W_e^{(k+1)}\left(Q_x+AK^{(k)}Q_y K^{(k)T}A^T\right)\right]+\frac{\lambda\beta}{1+\beta p} \end{aligned} \qquad (12.36)$$

With equation (12.36), we have used induction to prove that our policy is optimal, since it satisfies the Bellman equation.

In summary, we start at time point p and set $W_x^{(p)}=T^{(p)}$, $W_e^{(p)}=0$, and $w^{(p)}=\dfrac{\lambda\beta}{1+\beta p}$. From this we compute $G^{(p-1)}$ (equation 12.33). We then use equation (12.36) to compute $W_x^{(p-1)}$, $W_e^{(p-1)}$, and $w^{(p-1)}$. From these weights we compute $G^{(p-2)}$, and so forth. As a result, we have a recipe to compute the feedback gains. Our feedback control policy is:

$$\pi^*\left(\hat{\mathbf{x}}^{(k)}\right)=-G^{(k)}\hat{\mathbf{x}}^{(k)}.$$

12.5 The Interplay between State Estimation and Control Policy

In chapter 4, we considered the problem of state estimation for a system that had signal-dependent noise, as in the system of equation (12.16). We found that the Kalman gain on step $k + 1$ was affected by the size of the motor commands on step k. Let us briefly review that result, as it has an impact on our ability to compute an optimal control policy. We found that if on step k our prior state uncertainty is $P^{(k|k-1)}$, then the Kalman gain $K^{(k)}$ has the following form:

$$K^{(k)} = P^{(k|k-1)}H^T\left(HP^{(k|k-1)}H^T + Q_y + \sum_i HD_i\hat{\mathbf{x}}^{(k)}\hat{\mathbf{x}}^{(k)T}D_i^T H^T\right)^{-1}. \tag{12.37}$$

The state uncertainty has the following form:

$$\begin{aligned} P^{(k|k)} &= P^{(k|k-1)}\left(I - H^T K^{(k)T}\right) \\ P^{(k+1|k)} &= AP^{(k|k)}A^T + Q_x + \sum_i BC_i\mathbf{u}^{(k)}\mathbf{u}^{(k)T}C_i^T B^T \end{aligned} \tag{12.38}$$

It is because of signal-dependent noise that the Kalman gain is a function of the state estimate $\hat{\mathbf{x}}$. Furthermore, because state uncertainty depends on the motor commands, as motor commands increase in size, so does state uncertainty. Therefore, the state uncertainty increases with the size of the motor commands, and the Kalman gain decreases with the size of the state. The implication being that if we are pushing a large mass (producing relatively large motor commands), then we will have a larger uncertainty regarding the consequences of these commands (as compared to pushing a small mass with a smaller amount of force). As a result, when we are producing large forces, the Kalman gain will be large, and we should rely more on the sensory system and our observations and less on our predictions.

Note that according to equation (12.37), the Kalman gain $K^{(k)}$ depends on $\hat{\mathbf{x}}^{(k)}$, which according to equation (12.34) depends on $\mathbf{u}^{(k-1)}$. When we minimized the value function in equation (12.31), we took the derivative of the sum with respect to $\mathbf{u}^{(k)}$. That sum had the term $K^{(k)}$ in it. Because $K^{(k)}$ does not depend on $\mathbf{u}^{(k)}$, our derivative is valid. However, we face a practical issue: in order to compute the sequence of feedback gains $G^{(0)}, G^{(1)}, \cdots, G^{(p)}$, we need to know the sequence of Kalman gains $K^{(0)}, K^{(1)}, \cdots, K^{(p)}$.

Emo Todorov (2005) considered this issue and suggested the following. Say that we compute a sequence of control gains $G^{(0)}, G^{(1)}, \cdots, G^{(p)}$, which are optimal for a sequence of *fixed* Kalman gains $K^{(0)}, K^{(1)}, \cdots, K^{(p)}$. These Kalman gains would not depend on the motor commands. Rather, $K^{(k)}$ will be computed so that for a given sequence of control gains, it will minimize the expected value of the future state $E\left[v_{\pi^*}\left(\mathbf{x}^{(k+1)}, \hat{\mathbf{x}}^{(k+1)}\right)\right]$. Once a sequence of fixed Kalman gains had been computed, we then recompute the sequence of control gains. He showed that this iterative approach was guaranteed to converge. Indeed, in practice the approach converges within a few iterations. The recipe for computing the Kalman gain is as follows. We begin at time step $k = 0$, set $S_e^{(0)}$ to be the prior uncertainty, and $S_x^{(0)} = \hat{\mathbf{x}}^{(0)}\hat{\mathbf{x}}^{(0)T}$, and compute the following sequence of Kalman gains:

$$\begin{aligned}
\hat{\mathbf{x}}^{(k+1)} &= A\hat{\mathbf{x}}^{(k)} + AK^{(k)}\left(\mathbf{y}^{(k)} - H\hat{\mathbf{x}}^{(k)}\right) - BG^{(k)}\hat{\mathbf{x}}^{(k)} \\
K^{(k)} &= S_e^{(k)}H^T\left(HS_e^{(k)}H^T + Q_y\right)^{-1} \\
S_e^{(k+1)} &= Q_x + \left(A - AK^{(k)}H\right)S_e^{(k)} + \sum_i C_i G^{(k)} S_x^{(k)} G^{(k)T} C_i^T \\
S_x^{(k+1)} &= AK^{(k)}HS_e^{(k)}A^T + \left(A + BG^{(k)}\right)S_x^{(k)}\left(A + BG^{(k)}\right)^T
\end{aligned} \quad (12.39)$$

To summarize, one begins by computing a set of Kalman gains $K^{(0)}, K^{(1)}, \cdots, K^{(p)}$. These gains can be computed from equations (12.37) and (12.38) with the signal-dependent noise component set to zero. We then compute a set of control gains $G^{(0)}, G^{(1)}, \cdots, G^{(p)}$ using equations (12.33) and (12.35). We then recompute the Kalman gains using equation (12.39), and then recompute the control gains. The Kalman and control gains converge within a few iterations.

12.6 Example: Control of Eye and Head During Head-Free Gaze Changes

In many laboratory experiments on eye movements, the head of the subject is kept at rest and only the eyes are allowed to move. But in more natural, unconstrained conditions the head participates by redirecting the gaze. When we look around a room, searching for our keys, our eyes and head move in fairly complex and well-coordinated patterns. As we shift the gaze from one point to another, the eyes tend to start the movement with a saccade (figure 12.5A). During the saccade, the head starts rotating. While the head is still moving, the saccade ends, and the eyes roll back in the head. If the motion of the head is perturbed, the saccade is altered in midflight (figure 12.5B). For example, if a brake is applied to the head for 100 ms at saccade onset, the saccade is longer in duration and amplitude as compared to when the head is free to rotate, implying that the motion of the eyes during the saccade is affected by the state of the head. Let us show that all of these behaviors are consistent with a very simple goal: keep the target on the fovea, and keep the eyes centered in the head. We will express this goal as a cost function, and then use optimal feedback control theory to produce movements that best achieve the goal, that is, minimize the cost. We will perturb the motion of the simulated system and test whether it produces movements that resemble the recorded data.

Our model of the eye dynamics is identical to the one that we used in the previous chapter. We have a third-order system where x_e represents eye position in the orbit, and f_e is the torque on the eye. We also assume that when $x_e = 0$ the eye is in the central neutral position in its orbit. The third-order system has three states $x_1 \equiv x_e$, $x_2 \equiv \dot{x}_e$, and $x_3 \equiv f_e$. We have

$$\begin{bmatrix} \dot{x}_1 \\ \dot{x}_2 \\ \dot{x}_3 \end{bmatrix} = \begin{bmatrix} 0 & 1 & 0 \\ -\frac{k_e}{m_e} & -\frac{b_e}{m_e} & \frac{1}{m_e} \\ 0 & 0 & -\frac{\alpha_2}{\alpha_1} \end{bmatrix} \begin{bmatrix} x_1 \\ x_2 \\ x_3 \end{bmatrix} + \begin{bmatrix} 0 \\ 0 \\ \frac{1}{\alpha_1} \end{bmatrix} u_e. \tag{12.40}$$

The parameters of the eye plant are set so that the resulting system has time constants of 224, 13, and 4 ms. So we set $k_e = 1$, $b_e = \tau_1 + \tau_2$, $m_e = \tau_1\tau_2$, $\alpha_2 = 1$, and $\alpha_1 = 0.004$, where $\tau_1 = 0.224$ and $\tau_2 = 0.013$. We represent equation (12.40) as:

$$\dot{\mathbf{x}}_e = A_e\mathbf{x}_e + \mathbf{b}_e u_e. \tag{12.41}$$

Our head model is similar to the eye model, but with time constants of 270, 15, and 10 ms. The head position, x_h, is an angle that we measure with respect to a stationary frame in the environment. Therefore, the direction of gaze in the same stationary frame is $x_e + x_h$. The noise-free version of our dynamical system in continuous time is:

$$\begin{aligned} A_c &\equiv \begin{bmatrix} A_e & \mathbf{0} \\ \mathbf{0} & A_h \end{bmatrix} \\ \begin{bmatrix} \dot{\mathbf{x}}_e \\ \dot{\mathbf{x}}_h \end{bmatrix} &= A_c \begin{bmatrix} \mathbf{x}_e \\ \mathbf{x}_h \end{bmatrix} + \begin{bmatrix} \mathbf{b}_e & \mathbf{0} \\ \mathbf{0} & \mathbf{b}_h \end{bmatrix} \begin{bmatrix} u_e \\ u_h \end{bmatrix} \end{aligned}. \tag{12.42}$$

Suppose that g represents the position of our target. We translate our continuous time model into discrete time with time step Δ as follows. Set the state of the system to be:

$$\mathbf{x} \equiv [\mathbf{x}_e \quad \mathbf{x}_h \quad g]^T. \tag{12.43}$$

In equation (12.43), g represents the goal location, described as an angle in the same coordinate system in which we measure position of our head. Define the following matrices using matrix exponentials:

$$\begin{aligned} A &\equiv \begin{bmatrix} \exp(A_c\Delta) & \mathbf{0}_{6\times1} \\ 0 & 1 \end{bmatrix} \\ B &\equiv \begin{bmatrix} \mathbf{b}_e & \mathbf{0}_{3\times1} \\ \mathbf{0}_{3\times1} & \mathbf{b}_h \\ 0 & 0 \end{bmatrix} \end{aligned}. \tag{12.44}$$

The discrete version of our model is:

$$\begin{aligned}
\mathbf{x}^{(k+1)} &= A\mathbf{x}^{(k)} + B\mathbf{u}^{(k)} + \boldsymbol{\varepsilon}_x^{(k)} + B\sum_i C_i\mathbf{u}^{(k)}\phi_i^{(k)} \\
\mathbf{y}^{(k)} &= H\mathbf{x}^{(k)} + \boldsymbol{\varepsilon}_y^{(k)} + H\sum_i D_i\mathbf{x}^{(k)}\mu_i^{(k)} \\
H &\equiv \begin{bmatrix} -1 & 0 & 0 & -1 & 0 & 0 & 1 \\ 1 & 0 & 0 & 0 & 0 & 0 & 0 \end{bmatrix} \\
\boldsymbol{\varepsilon}_x &\sim N(\mathbf{0}, Q_x) \quad \boldsymbol{\varepsilon}_y \sim N(\mathbf{0}, Q_y) \\
\phi &\sim N(0,1) \qquad \mu \sim N(0,1)
\end{aligned} \tag{12.45}$$

The H matrix in equation (12.45) implies that we can sense the position of the eye x_e, as well as the position of the target on the retina. The position of the target on the retina is the difference between the goal location and the sum of the eye and head positions: $g - (x_e + x_h)$. What we want is to place the target at the center of the fovea, so we assume a cost per step that penalizes the distance of the target to the fovea. We also want to keep the eyes centered in their orbit, and so we will penalize eccentricity of the eye. Finally, we want to minimize the motor commands to eye and head. Our cost per step is:

$$\begin{aligned}
\alpha^{(k)} &= \mathbf{y}^{(k)T}T^{(k)}\mathbf{y}^{(k)} + \mathbf{u}^{(k)T}L\mathbf{u}^{(k)} \\
&= \mathbf{x}^{(k)T}H^T T^{(k)}H\mathbf{x}^{(k)} + \mathbf{u}^{(k)T}L\mathbf{u}^{(k)}
\end{aligned} \tag{12.46}$$

Suppose that we want gaze position $x_e + x_h$ to arrive at target at time step k_1. The state cost matrix T is zero until time step k_1, and then is kept constant until end of simulation period p. The motor cost L is diagonal with equal costs for eye and head motor commands. We assumed nonzero signal-dependent motor noise $c_1 = c_2 = 0.01$ (as in equation 12.13), but zero signal-dependent sensory noise $d_i = 0$.

Figure 12.8A shows the behavior of our model for a target at 40 degrees. The gaze arrives at target at around 100 ms, as our cost function had implied, and this is accomplished through a cooperation of the eye and head systems. The gaze change begins with motion of the eye, making a 25-degree saccade, and is accompanied with motion of the head. The saccade amplitude is consistent with the data in figure 12.5C. Upon saccade completion the eyes roll back in the head as the head continues to move toward the target. The gaze change begins with motion of the eye because the dynamics of the eye present a lighter system to move, and therefore it costs less in terms of effort (squared motor commands) than moving the head. When the gaze is at the target—that is, the target is on the fovea—the eyes roll back. This is because we incur a cost for not having the eyes centered with respect to the head. These two costs, gaze on target, eyes centered, are sufficient to produce the coordinated motion of a natural gaze change.

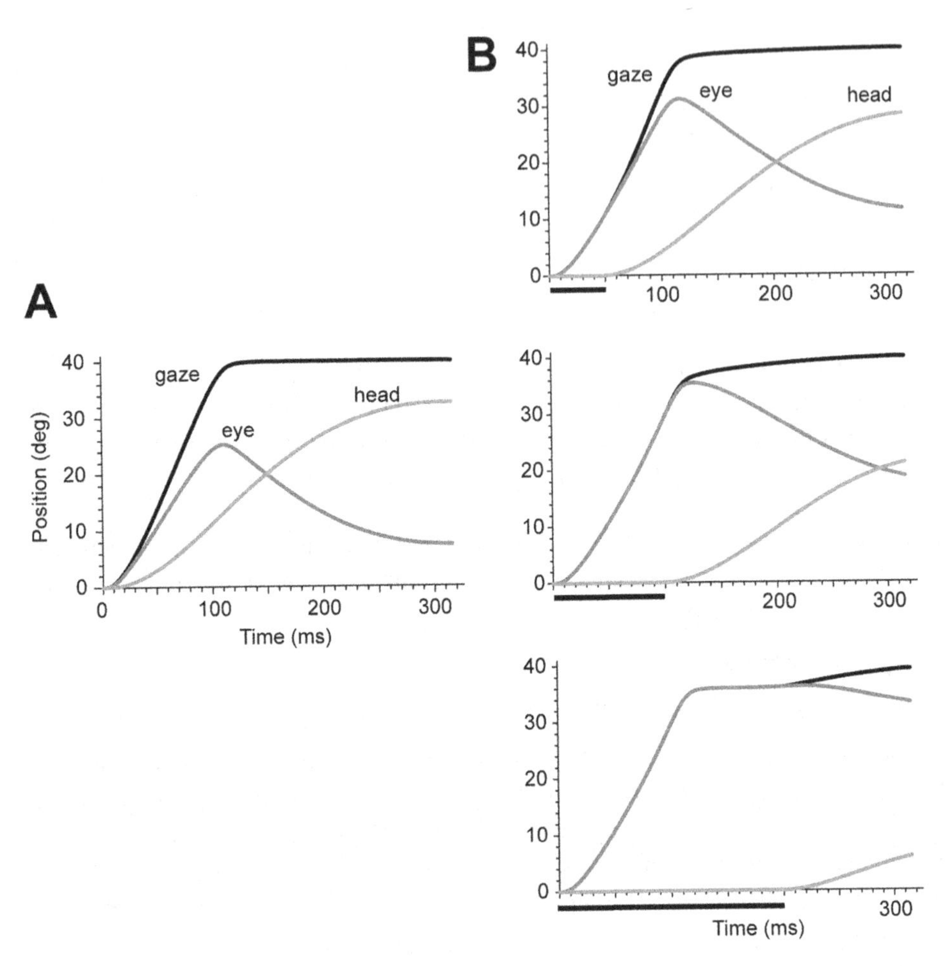

Figure 12.8
Simulation results for a head-free eye/head movement with target at 40 degrees. (A) An unperturbed movement. (B) Perturbed movements. The head was fixed and not allowed to move for 50, 100, and 200 ms. Parameter values: The state cost matrix T is zero until time step 110 ms, and then is kept constant at $T = [10^5 \ \ 0; \ \ 0 \ \ 300]$. Motor signal dependent noise standard deviation is $c_1 = c_2 = 0.01$. Motor costs are $L = [10 \ \ 0; \ \ 0 \ \ 10]$. State noise is $Q_x = 0.1 \times I_7$, where the term I_7 represents an identity matrix of size 7×7. Observation noise is $Q_y = I_2$.

Now let us consider a situation in which head motion is perturbed during the gaze change. To simulate this, we prevented the head from moving for 50, 100, or 200 ms at start of the gaze change (figure 12.8B). We find that the saccade made by the eye is lengthened so that the gaze still arrives near the target at around 100 ms, but that the eyes no longer roll back unless the head is allowed to move toward the target. The increased saccade amplitude is consistent with the data in figure 12.5C, in which we see that in the head-braked condition saccade amplitudes are increased. In the experimental data, the head-brake (figure 12.5B) also prevents the roll back of the eyes after saccade completion, consistent with our simulation results.

12.7 Limitations

The real advantage of using a cost function to describe the goal of a movement is that it removes the burden of having to describe the desired kinematics of the task in some a priori format. That is, we do not need to describe how we wish for the eyes and the head to move in order to produce a gaze change. We do not have a desired trajectory that specifies a sequence of positions and velocities that we wish for the eyes and the head to achieve. Our control system does not have feedback gains that are arbitrary attractors around this desired trajectory. Rather, we have a fairly simple goal (put the target on the fovea, keep the eye centered), and then we leave it up to the optimization process to find the controller that achieves this goal as well as it can be done.

However, there are potential problems with our simulations and the general framework. Let us highlight some of the limitations here.

In our present simulations we assumed a desired time at which gaze should arrive on target. Where does this desired time—that is, desired movement duration—come from? One possibility is that movement duration carries a cost because passage of time discounts rewards of the task. We used this idea in the previous chapter to show that head-fixed saccade durations reflect an optimization process in which passage of time discounts reward. A recent work showed that the duration of gaze changes in head-free movements can also be explained by a cost function like that of equation (12.9) in which movement durations carry a cost that has a hyperbolic form (Shadmehr et al., 2010). So in principle, a cost that involves three components (keep the target on the fovea, keep the eye centered, and get to the target as soon as possible) can account for the kinematics and timing of head-fixed and head-free gaze changes.

Following this line of thinking implies that before a movement begins, our brain considers all possible movement durations, and given the current cost and reward conditions per step, arrives at a control policy that carries a minimum total expected cost. Putting aside the fact that we have no idea how such an optimization process might biologically take place, our framework arrives at a specific movement duration for which we implement a control policy. However, if our movement is perturbed, does the perturbation alter our desired movement duration? It certainly seems as if it should. For example, say that during a movement a perturbation alters the position of the target. Our control policy has no trouble with this and will respond to the altered goal position. However, it will do so approximately in the same desired movement period as before. This is inconsistent with experimental data (Liu and Todorov, 2007). People allow more time to correct for a perturbation than is expected in our finite-horizon formulation of the problem. This hints that the framework of finite-horizon optimization is probably inappropriate and will need to be abandoned for a more general approach.

A second, more troubling problem is with the general theme of optimality devoid of evolutionary history. Our framework has proceeded with the assumption that biological behavior can be understood as minimization of a cost function in which we usually do not consider the constraints that are imposed by the history of the animal. To illustrate the problem, consider the task of trying to explain swimming behavior in seagoing mammals. Marine mammals swim by beating their horizontal tail flukes up and down. However, fishes swim by moving their tail flukes from side to side. How could there be two seemingly orthogonal but still "optimum" ways of swimming in water? In considering this question, Stephen Jay Gould (1995) writes: "the explanation probably resides in happenstances of history, rather than in abstract predictions based on universal optimality." In particular, marine mammals evolved from animals that used to run on land. Locomotion of these animals involved flexing their spinal column up and down. He writes:

> Thus, horizontal tail flukes may evolve in fully marine mammals because inherited spinal flexibility for movement up and down (rather than side to side) directed this pathway from a terrestrial past. The horizontal tail fluke evolved because whales carried their terrestrial system of spinal motion to the water. (p. 372)

Clearly, the mathematics can be expanded to represent constraints that are imposed by neuroanatomy and biomechanics. For example, we can incorporate models of local controllers as may be present in the spinal cord, and more detailed models of muscles and biomechanics. What is unclear, however, is whether our cost per step (equation 12.9), or our description of the system's dynamics (equation 12.45), need to incorporate some measure of the history of the species.

12.8 The Brain Finds a Better Way to Clear a Barrier

We have been spending the entirety of this book examining relatively simple movements like saccades and reaching, movements that probably have not changed much in the past few million years of evolution. During the last century, however, the

Figure 12.9
Techniques used in the high-jump competition. (A) Scissor technique, as illustrated by Platt Adams during the 1912 Summer Olympics. (B) Straddle techniques, as illustrated by Rolf Beilschmidt in 1977. (C) Fosbury technique. (Images from Wikimedia Commons.)

human brain has made fundamental breakthroughs in finding better ways to clear a barrier. The new ways of performing this task are real-world examples of better solutions to an optimal control problem.

In the high jump, you are asked to clear the highest hurdle possible. The task has been a part of track and field competition since the advent of such competitions in the modern Olympics. The earliest techniques were to hurdle over the bar: approach at an angle, scissoring the legs over the bar and land on the feet (figure 12.9A). In the late nineteenth century the technique was improved by taking off like the scissor but extending the back and flattening out over the bar. By the mid-twentieth century, the most successful approach was a straddle technique called the "western roll" or the "straddle," in which one begins by running toward the bar at a diagonal, and then kicks the outer leg over the bar and crosses the bar face down (figure 12.9B). In all these cases the feet cross the bar first, and there is at least a theoretical possibility of landing on your feet. Starting in the early twentieth century, all competitions included a landing area made of sand, sawdust, or woodchips. In the late 1960s, however, there was a revolutionary change in the way athletes jumped over the bar, and the change occurred because of the brain of one kid, Richard Fosbury, a high school sophomore in Medford, Oregon.[2]

Fosbury was a tall kid who in eight and ninth grade was still using the scissor jump to clear the bar. When he entered high school in the tenth grade, his coach insisted that he try the more modern western roll. However, Fosbury had trouble adapting his style. Rather than improve on his previous records, at his first meet as a sophomore he did not clear the opening height and failed on all three chances. The repeated failures continued that year until something amazing happened in a meet of a dozen schools near the end of the sophomore year in 1963 at Grants Pass, Oregon. Feeling desperate, Fosbury's coach had given him permission to revert back to the scissor technique. On his first jump that day he cleared 5'4", the height that he had achieved a year earlier. Richard Hoffer, a journalist who chronicled Fosbury's career, writes:

> The other jumpers were still warming up, waiting for the bar to be set at an age-appropriate height, while Fosbury continued to noodle around at his junior high elevations. If they, or anyone else, had been interested though, they might have seen an odd transformation taking place. Fosbury was now arching backward ever so slightly as he scissored the bar, his rear end now coming up, his shoulder going down. He cleared 5'6". He didn't even know what he was doing, his body reacting to desperation. His third round, his body reclined even more, and he made 5'8". On his fourth attempt Fosbury took a surprisingly leisurely approach to the bar and . . . completely flat on his back now, cleared 5'10"." The high jump was an event that measured advancement by fractions of an inch, sometimes over a year. Fosbury, conducting his own defiance, had just improved a half foot in one day.

Fosbury was crossing the bar head first, and landing on his neck and back on the sawdust and woodchips (figure 12.9C). Fortunately, in his junior year the school replaced the pit with foam and he reached 6'3" by the end of that year and 6'5.5" by the end of his senior year. After graduation, he enrolled in Oregon State University in the engineering program. By his second year, he had improved his record to 6'10". He was still not a world-class jumper, but he would become one by his third year, consistently crossing the 7-foot barrier. In the 1968 Olympics in Mexico City he was the only jumper to go over the bar head first. He won by crossing 7'4.25", an Olympic record.

In looking back at the change that Fosbury brought, we see two important ideas. First, the scissor technique had evolved into the western roll and the straddle, techniques in which you cross the bar face down. These are control policies that are now viewed as suboptimal in the sense that they are local minima. Fosbury, starting from the scissor evolved it into a jump in which you cross the bar facing the sky. By effectively restarting the search process for a better control policy, he was able to stumble upon a better solution. Second, his solution was made possible because there was a critical change in the cost function: kids were no longer jumping into piles of sawdust and woodchips, but had the luxury of landing on foam. The margin of safety and comfort was significantly increased if you happened to land on your neck. As a result of a healthy disrespect for history, and being present when technology altered the cost of the task, an engineering student found a fundamentally new solution to an old optimal control problem, and took home the Olympic gold for his genius.

Summary

We make movements in order to improve the state of our body and the environment. To make the movement, we must spent effort. We can express the purpose of the movement as obtaining a more rewarding state, while spending as little effort as possible. Mathematically, this is formulated as a cost function in which the state at the end of the movement is afforded a value (e.g., the closer we end up to the rewarding state, the better), discounted by the accumulated effort. Together this describes a cost function. The best movement is one in which the motor commands achieve the least cost. To produce these motor commands, we do not wish to preprogram them at movement start, because movements can be perturbed and we wish to be able to respond to sensory feedback. Rather, we wish to form a policy that produces motor commands based on our current belief about our state. This belief comes from our ability to make predictions about sensory consequences of our motor commands, and combine them to the actual sensory feedback. An optimal feedback controller describes a feedback dependent policy that produces motor commands that minimize a cost regardless of the state that we may find ourselves during the movement.

In this chapter we used Bellman's optimality principle to compute the policy that minimized a quadratic state and effort cost for a linear system with signal dependent motor and sensory noise. Our work followed the approach described by Todorov (2005). We applied the result to control of head-free gaze changes in which the eyes and the head cooperate to keep a target stimulus on the fovea. The motor commands to these two systems respond to sensory feedback, and we simulated conditions in which the head was perturbed, demonstrating how it affects the ongoing saccade of the eye.

Appendix

Derivative of Scalar Forms

$$\frac{d}{d\mathbf{x}}\mathbf{a}^T\mathbf{x} = \frac{d}{d\mathbf{x}}\mathbf{x}^T\mathbf{a} = \mathbf{a}$$

$$\frac{d}{d\mathbf{x}}\mathbf{x}^T A\mathbf{x} = \left(A + A^T\right)\mathbf{x}$$

$$\frac{d}{dX}\mathbf{a}^T X\mathbf{b} = \mathbf{a}\mathbf{b}^T$$

$$\frac{d}{dX}\mathbf{a}^T X^T\mathbf{b} = \mathbf{b}\mathbf{a}^T$$

$$\frac{d}{dX}\mathbf{a}^T X\mathbf{a} = \frac{d}{dX}\mathbf{a}^T X^T\mathbf{a} = \mathbf{a}\mathbf{a}^T$$

$$\frac{d}{dX}\mathbf{a}^T X^T C X\mathbf{b} = C^T X\mathbf{a}\mathbf{b}^T + CX\mathbf{b}\mathbf{a}^T$$

Derivative of Vector Products of the Form $\mathbf{w}^T\mathbf{x}\mathbf{x}$

$$\mathbf{w}^T\mathbf{x}\mathbf{x} = \begin{bmatrix} w_1x_1 & w_2x_2 & \cdots & w_mx_m \end{bmatrix} \begin{bmatrix} x_1 \\ x_2 \\ \vdots \\ x_m \end{bmatrix} = \begin{bmatrix} x_1(w_1x_1 + \cdots + w_mx_m) \\ x_2(w_1x_1 + \cdots + w_mx_m) \\ \vdots \\ x_m(w_1x_1 + \cdots + w_mx_m) \end{bmatrix}$$

$$\frac{d}{dw_i}\mathbf{w}^T\mathbf{x}\mathbf{x} = \begin{bmatrix} x_1x_i \\ x_2x_i \\ \vdots \\ x_mx_i \end{bmatrix}$$

$$\frac{d}{d\mathbf{w}}\mathbf{w}^T\mathbf{x}\mathbf{x} = \begin{bmatrix} x_1^2 & x_1x_2 & \cdots & x_1x_m \\ x_2x_1 & x_2^2 & \cdots & x_2x_m \\ \vdots & \vdots & \ddots & \vdots \\ x_mx_1 & x_mx_2 & \cdots & x_m^2 \end{bmatrix}$$

$$\frac{d}{d\mathbf{w}}\mathbf{w}^T\mathbf{x}\mathbf{x} = \mathbf{x}\mathbf{x}^T$$

Properties of the Trace Operator

$$\mathbf{r}^T\mathbf{r} = tr(\mathbf{r}\mathbf{r}^T)$$

$$tr(AB) = tr(A^TB^T) = tr(B^TA^T) = tr(BA)$$

$$tr(ABC) = tr(BCA)$$

Expected Value and Variance of Scalar and Vector Random Variables

$$\text{var}[\mathbf{x}] = E\left[(\mathbf{x}-\bar{\mathbf{x}})(\mathbf{x}-\bar{\mathbf{x}})^T\right] = E\left[\mathbf{x}\mathbf{x}^T\right] - \bar{\mathbf{x}}\,\bar{\mathbf{x}}^T$$

$$\text{cov}[\mathbf{x},\mathbf{y}] = E\left[(\mathbf{x}-\bar{\mathbf{x}})(\mathbf{y}-\bar{\mathbf{y}})^T\right] = E\left[\mathbf{x}\mathbf{y}^T\right] - \bar{\mathbf{x}}\,\bar{\mathbf{y}}^T$$

$$\begin{aligned} \text{cov}[A\mathbf{x},B\mathbf{y}] &= E\left[A(\mathbf{x}-\bar{\mathbf{x}})(B(\mathbf{y}-\bar{\mathbf{y}}))^T\right] \\ &= E\left[A(\mathbf{x}-\bar{\mathbf{x}})(\mathbf{y}-\bar{\mathbf{y}})^T B^T\right] = AE\left[(\mathbf{x}-\bar{\mathbf{x}})(\mathbf{y}-\bar{\mathbf{y}})^T\right]B^T \\ &= A\,\text{cov}[\mathbf{x},\mathbf{y}]B^T \end{aligned}$$

$$\text{cov}[\mathbf{x},\mathbf{y}] = \text{cov}[\mathbf{y},\mathbf{x}]^T$$

$$\text{var}[\mathbf{x}+\mathbf{y}] = \text{var}[\mathbf{x}] + \text{cov}[\mathbf{x},\mathbf{y}] + \text{cov}[\mathbf{y},\mathbf{x}] + \text{var}[\mathbf{y}]$$

$$\text{var}[\mathbf{a}^T\mathbf{x}] = \mathbf{a}^T\,\text{var}[\mathbf{x}]\mathbf{a}$$

$$\begin{aligned}\mathrm{var}[A\mathbf{x}] &= E\left[A(\mathbf{x}-\bar{\mathbf{x}})(A(\mathbf{x}-\bar{\mathbf{x}}))^T\right] = E\left[A(\mathbf{x}-\bar{\mathbf{x}})(\mathbf{x}-\bar{\mathbf{x}})^T A^T\right] \\ &= AE\left[(\mathbf{x}-\bar{\mathbf{x}})(\mathbf{x}-\bar{\mathbf{x}})^T\right]A^T = A\,\mathrm{var}[\mathbf{x}]A^T\end{aligned}$$

Expected Value of "Squared" Random Variables

$$E[x^2] = \mathrm{var}[x] + E[x]^2$$

$$\begin{aligned}E[\mathbf{x}^T\mathbf{x}] &= E[tr[\mathbf{x}\mathbf{x}^T]] = tr[E[\mathbf{x}\mathbf{x}^T]] \\ &= tr[\mathrm{var}[\mathbf{x}]] + tr\left[E[\mathbf{x}]E[\mathbf{x}]^T\right] \\ &= tr[\mathrm{var}[\mathbf{x}]] + E[\mathbf{x}]^T E[\mathbf{x}]\end{aligned}$$

$$E[\mathbf{x}^T A\mathbf{x}] = E[\mathbf{x}]^T AE[\mathbf{x}] + tr[A\,\mathrm{var}[\mathbf{x}]]$$

Notes

Chapter 1

1. The physiological basis for the "sense of north" is not well known and varies across species. Some are capable of detecting magnetic fields and orient to them. These include migratory birds, which travel to their destination for thousands of miles; cows, which reorient themselves while grazing under electric power lines; and certain bacteria, which are endowed with magnetic sensing organelles. Some rodents also have a physiological magnetic compass, as it was demonstrated in experiments on the African mole rat *Cryptomys hottentotus* (Burda et al., 1990): the presence of an artificial magnetic field deviated systematically the paths followed by the mole rats when building their nest inside a circular arena. In addition to the earth magnetic field, there are other subtle cues that are hard to suppress in the laboratory, like odors and small variations of colors and shape of the walls. Finally, there are navigation mechanisms by which the nervous system performs what sailors call "dead reckoning," the constant integration of visuomotor information that allows one to maintain a representation of one's position with respect to a fixed frame of reference.

2. Fourier first presented this idea in a paper that in 1807 he submitted to Institute de France, which appointed four noted mathematicians, including Laplace and LaGrange, to review the work. Unfortunately, LaGrange failed to see the importance of the work and objected to the idea that nonperiodic functions should be represented as sum of trigonometric functions. The paper was rejected. Discouraged, Fourier turned his attention to writing a series of books, *Description of Egypt*, for which he gained fame during his lifetime (remarkably, Fourier was less known as a mathematician during his lifetime than as an Egyptologist). Only fifteen years later did Fourier publish his mathematical results in his book *The Analytical Theory of Heat*. Lord Kelvin, a noted British mathematician, would later refer to Fourier's book as "a mathematical poem."

3. Here, we assume this number to be real. But, in general, vector spaces can be defined over complex numbers or any kind of scalar field. Scalar fields are structures where the fundamental four operations—addition, subtractions, multiplication, and division—are defined.

4. This statement can be demonstrated in more than one way. One is based on an insightful geometrical view of determinants. The determinant of a matrix is the signed volume of the parallelepiped included between the vectors that constitute the columns of the matrix. To see this, start with the simple case of a diagonal 3×3 matrix. Each column is a vector along the corresponding axis. The product of these vectors is the volume of the rectangular parallelepiped with three edges formed by the three vectors. This argument can be rapidly extended to more complex matrices with more rows and columns.

Each column of the matrix Φ is the representation of each basis vector in the frame established by the vectors themselves. The fact that the vectors of a base are linearly independent implies that span the full volume of their own space. Therefore the determinant of Φ cannot vanish.

5. We are still considering only real-valued matrices. To obtain the definition for complex-valued matrices, simply replace T (for transposed) with an asterisk (for complex conjugate).

6. There are several different types of integrals. The one that is most often used in function spaces is the Lebesgue integral, after Henri Lebesgue, another French mathematician. Another type of integral

operation is the Riemann integral, which is the one most commonly introduced in calculus classes. The distinction between Riemann and Lebesgue integrals is important but subtle, and beyond the scope of this text. In most practical cases the two methods give the same result.

Chapter 2

1. The reader may be familiar with one of the many version of a joke about extreme simplification. We found this one in the Wikipedia entry for "spherical cow":

Milk production at a dairy farm was low so the farmer wrote to the local university, asking help from academia. A multidisciplinary team of professors was assembled, headed by a theoretical physicist, and two weeks of intensive on-site investigation took place. The scholars then returned to the university, notebooks crammed with data, where the task of writing the report was left to the team leader. Shortly thereafter the physicist returns to the farm, saying to the farmer, "I have the solution, but it only works for spherical cows in a vacuum."

The joke reflects a common process in science, in which elements of reality are removed from a problem so as to render it tractable with the available mathematical means. Obviously a spherical cow model would not help much with understanding milk production. But it would not be too bad if you where to calculate the energy at impact of a cow falling from a cliff. Our planar gerbil with a one-dimensional circular retina is as extreme a simplification as the spherical cow. So, the reader should not think of it as a simplified model of the complex behavior of this marvelous little animal. However, this model highlights in accessible terms some of the extremely complex mathematical issues that the brain must deal to navigate within and localize itself within the environment.

2. In this book we adopt the convention, from linear algebra, to indicate the components of vectors as one-dimensional column arrays. This is useful to represent linear coordinate transformations as matrix-vector products and is readily extended to any number of dimensions. Thus we have

$$x = [x_1, x_2, \ldots, x_n]^T \equiv \begin{bmatrix} x_1 \\ x_2 \\ \cdots \\ x_n \end{bmatrix} \text{ and } Ax = \begin{bmatrix} a_{1,1} & a_{1,2} & \cdots & a_{1,n} \\ a_{2,1} & a_{2,2} & \cdots & a_{2,n} \\ \cdots & \cdots & \cdots & \cdots \\ a_{n,1} & a_{n,2} & \cdots & a_{n,n} \end{bmatrix} \cdot \begin{bmatrix} x_1 \\ x_2 \\ \cdots \\ x_n \end{bmatrix}.$$

3. This is also true in formal terms. An equivalence relation between elements of a set (indicated by the symbol ~) is a relation with three defining properties:

1. reflexive ($a \sim a$)
2. symmetric ($a \sim b \Leftrightarrow b \sim a$), and
3. transitive (if $a \sim b$ and $b \sim c$ then $a \sim c$)

The elements of a set that are equivalent to a given element define an *equivalence class*. The elements of a set are partitioned by an equivalence relation into a collection of nonoverlapping equivalence classes. One can easily see that the points (x,y) that map to the same projection ξ form an equivalence class and that all points of the 2D space external to the sensor circle are partitioned into such equivalence classes. The equivalence class constructed in this way from a function is also called a *fiber* of f at ξ.

Chapter 3

1. An isometric embedding is a transformation from a mathematical object to a space that contains the objects and preserves the object's metric properties. An example is the embedding of a sphere in the Euclidean 3D space. We can locate a point over the sphere by two coordinates, such as latitude and longitude. Alternatively, we can establish three Cartesian axes and describe the point by a triplet x,y,z. Importantly, the first type of description is non-Euclidean. Over the sphere, parallel lines may intersect and Pythagoras's theorem is violated. However, the second description is Euclidean. Looking at the earth

from outer space one would see that the meridians of longitude indeed meet at the poles, but they are not parallel lines. They are closed curves.

2. We adopt the standard notation in classical mechanics, where the lowercase q denotes a position and the uppercase Q represents a force in a system of "generalized coordinates." Generalized coordinates reflect the effective movement space of a mechanical system. Classical mechanics assumes that the essential law that governs the movement of any system is Newton's law, $F = ma$, applied to each of its constituents particles. However, the great majority of those particles are bound to stay at fixed distances from one another. While a simple pendulum contains billions of molecules, the motion of this immense set of particles is described by a single variable. This is called the "degree of freedom" of the pendulum. The idealized two-joint arm of figure 3.1 has only two degrees of freedom. The variables that describe each degree of freedom are called generalized coordinates. The generalized forces are the forces in that particular system of coordinates. For example, for an angular coordinate, the generalized force is a torque. For a linear coordinate, the generalized force is an ordinary force.

3. To see this, consider a simple second-order system, such as a spring and a mass or a pendulum. You may drive the system by applying an external force that will have the system reaching several times the same state, $s = [x, \dot{x}]^T$s (e.g., a given position at zero velocity) and leaving this state on different trajectories. Then, knowing only that the system is at s is no longer sufficient to know its future.

4. According to standard engineering terminology, equation (3.17) describes a PD control system, where P stands for "proportional" and D for "derivative."

Chapter 4

1.

$$\begin{aligned}
P^{(n|n)} &= \left(I - \mathbf{k}^{(n)}\mathbf{x}^{(n)T}\right)P^{(n|n-1)}\left(I - \mathbf{k}^{(n)}\mathbf{x}^{(n)T}\right)^T + \sigma^2\mathbf{k}^{(n)}\mathbf{k}^{(n)T} \\
&= P^{(n|n-1)} - 2\mathbf{k}^{(n)}\mathbf{x}^{(n)T}P^{(n|n-1)} + \mathbf{k}^{(n)}\mathbf{x}^{(n)T}P^{(n|n-1)}\mathbf{x}^{(n)}\mathbf{k}^{(n)T} + \mathbf{k}^{(n)}\sigma^2\mathbf{k}^{(n)T} \\
&= P^{(n|n-1)} - 2\mathbf{k}^{(n)}\mathbf{x}^{(n)T}P^{(n|n-1)} + \mathbf{k}^{(n)}\left(\mathbf{x}^{(n)T}P^{(n|n-1)}\mathbf{x}^{(n)} + \sigma^2\right)\mathbf{k}^{(n)T} \\
&= P^{(n|n-1)} - 2\mathbf{k}^{(n)}\mathbf{x}^{(n)T}P^{(n|n-1)} + \mathbf{k}^{(n)}\left(\mathbf{x}^{(n)T}P^{(n|n-1)}\mathbf{x}^{(n)} + \sigma^2\right)\mathbf{x}^{(n)T}\frac{P^{(n|n-1)}}{\left(\mathbf{x}^{(n)T}P^{(n|n-1)}\mathbf{x}^{(n)} + \sigma^2\right)}. \\
&= P^{(n|n-1)} - 2\mathbf{k}^{(n)}\mathbf{x}^{(n)T}P^{(n|n-1)} + \mathbf{k}^{(n)}\mathbf{x}^{(n)T}P^{(n|n-1)} \\
&= \left(I - \mathbf{k}^{(n)}\mathbf{x}^{(n)T}\right)P^{(n|n-1)}
\end{aligned}$$

Chapter 5

1. Here the term "belief" is not used with its ordinary meaning, but instead according to an accepted lexicon in statistical learning theory. In this more restricted sense, a belief is an expectation that the learning system has developed either from past experience or has encoded in its initial structural properties. There is not the assumption of conscious awareness associated with the more common use of the word.

Chapter 12

1. The idea that feedback depends upon a belief sounds like an oxymoron, because in the ordinary use of the word, beliefs are somewhat antagonistic to the concept of evidence or sensory information. One believes in a God regardless of any external evidence. Similarly, ideological belief is a construct that is not much affected by external inputs. However, here we use the term "belief" in the Bayesian sense: A belief is a combination of prior knowledge and current evidence.

2. Our source for this story is an article by Richard Hoffer, "The Revolutionary," in *Sports Illustrated*, September 14, 2009.

References

Anderson BDO, Moore JB. 1979. Optimal filtering. Englewood Cliffs, NJ: Prentice-Hall.

Bahill AT, Clark MR, Stark L. 1975. The main sequence: A tool for studying human eye movements. *Math Biosci* 24: 191–204.

Barbarulo AM, Grossi D, Merola S, Conson M, Trojano L. 2007. On the genesis of unilateral micrographia of the progressive type. *Neuropsychologia* 45: 1685–1696.

Bellman RE. 1957. Dynamic programming. Princeton, NJ: Princeton University Press.

Berniker M, Kording K. 2008. Estimating the sources of motor errors for adaptation and generalization. *Nat Neurosci* 11: 1454–1461.

Bizzi E, Kalil RE, Tagliasco V. 1971. Eye-head coordination in monkeys: Evidence for centrally patterned organization. *Science* 173: 452–454.

Blekher T, Siemers E, Abel LA, Yee RD. 2000. Eye movements in Parkinson's disease: Before and after pallidotomy. *Invest Ophthalmol Vis Sci* 41: 2177–2183.

Braun DA, Aertsen A, Wolpert DM, Mehring C. 2009. Motor task variation induces structural learning. *Curr Biol* 19: 352–357.

Bray S, O'Doherty J. 2007. Neural coding of reward-prediction error signals during classical conditioning with attractive faces. *J Neurophysiol* 97: 3036–3045.

Burda H, Marhold S, Westenberger T, Wiltschko R, Wiltschko W. 1990. Magnetic compass orientation in the subterranean rodent Cryptomys hottentotus (Bathyergidae). *Cell Mol Life Sci* 46: 528–530.

Burge J, Ernst MO, Banks MS. 2008. The statistical determinants of adaptation rate in human reaching. *J Vis* 8: 1–19.

Caithness G, Osu R, Bays P, Chase H, Klassen J, Kawato M, Wolpert DM, Flanagan JR. 2004. Failure to consolidate the consolidation theory of learning for sensorimotor adaptation tasks. *J Neurosci* 24: 8662–8671.

Carpenter RHS. 1988. Movements of the eyes. *J Neurophysiol* 103: 2275–2284.

Cerf M, Harel J, Einhasuer W, Koch C. 2008. Predicting human gaze using low-level saliency combined with face detection. In *Advances in Neural Information Processing Systems* (Platt JC; Koller D; Singer Y; Roweis S, eds.) pp. 241–248. Cambridge, MA: MIT Press.

Charpentier A. 1891. Analyse experimentale quelques elements de la sensation de poids. [Experimental study of some aspects of weight perception]. *Arch Physiol Normales Pthologiques* 3: 122–135.

Chen-Harris H, Joiner WM, Ethier V, Zee DS, Shadmehr R. 2008. Adaptive control of saccades via internal feedback. *J Neurosci* 28: 2804–2813.

Cheng S, Sabes PN. 2006. Modeling sensorimotor learning with linear dynamical systems. *Neural Comput* 18: 760–793.

Clement TS, Feltus JR, Kaiser DH, Zentall TR. 2000. Work ethic in pigeons: Reward value is directly related to the effort or time required to obtain the reward. *Psychon Bull Rev* 7: 100–106.

Collett T, Cartwright B, Smith B. 1986. Landmark learning and visuo-spatial memories in gerbils. *J Comp Physiol A* 158: 835–851.

Collewijn H, Erkelens CJ, Steinman RM. 1988. Binocular co-ordination of human horizontal saccadic eye movements. *J Physiol* 404: 157–182.

Collier TJ, Lipton J, Daley BF, Palfi S, Chu Y, Sortwell C, Bakay RA, Sladek JR, Jr, Kordower JH. 2007. Aging-related changes in the nigrostriatal dopamine system and the response to MPTP in nonhuman primates: Diminished compensatory mechanisms as a prelude to parkinsonism. *Neurobiol Dis* 26: 56–65.

Conditt MA, Gandolfo F, Mussa-Ivaldi FA. 1997. The motor system does not learn the dynamics of the arm by rote memorization of past experience. *J Neurophysiol* 78(1): 554–560.

Cordo PJ, Nashner LM. 1982. Properties of postural adjustments associated with rapid arm movements. *J Neurophysiol* 47: 287–302.

Criscimagna-Hemminger SE, Bastian AJ, Shadmehr R. 2010. Size of error affects cerebellar contributions to motor learning. *J Neurophysiol* 103: 2275–2284.

Criscimagna-Hemminger SE, Donchin O, Gazzaniga MS, Shadmehr R. 2003. Learned dynamics of reaching movements generalize from dominant to nondominant arm. *J Neurophysiol* 89: 168–176.

Criscimagna-Hemminger SE, Shadmehr R. 2008. Consolidation patterns of human motor memory. *J Neurosci* 28: 9610–9618.

Diedrichsen J. 2007. Optimal task-dependent changes of bimanual feedback control and adaptation. *Curr Biol* 17: 1675–1679.

Dissanayake M, Newman P, Clark S, Durrant-Whyte HF, Csorba M. 2001. A solution to the simultaneous localization and map building (SLAM) problem. *IEEE Trans Robot Autom* 17: 229–241.

Donchin O, Francis JT, Shadmehr R. 2003. Quantifying generalization from trial-by-trial behavior of adaptive systems that learn with basis functions: Theory and experiments in human motor control. *J Neurosci* 23(27): 9032.

Duhamel JR, Colby CL, Goldberg ME. 1992. The updating of the representation of visual space in parietal cortex by intended eye movements. *Science* 255: 90–92.

Eichenbaum H, Dudchenko P, Wood E, Shapiro M, Tanila H. 1999. The hippocampus, memory, review and place cells: Is it spatial memory or a memory space? *Neuron* 23: 209–226.

Epelboim J, Steinman RM, Kowler E, Pizlo Z, Erkelens CJ, Collewijn H. 1997. Gaze-shift dynamics in two kinds of sequential looking tasks. *Vision Res* 37: 2597–2607.

Ernst MO, Banks MS. 2002. Humans integrate visual and haptic information in a statistically optimal fashion. *Nature* 415: 429–433.

Ethier V, Zee DS, Shadmehr R. 2008. Spontaneous recovery of motor memory during saccade adaptation. *J Neurophysiol* 99: 2577–2583.

Fagg AH, Shah A, Barto AG. 2002. A computational model of muscle recruitment for wrist movements. *J Neurophysiol* 88: 3348–3358.

Fecteau JH, Munoz DP. 2006. Salience, relevance, and firing: A priority map for target selection. *Trends Cogn Sci* 10: 382–390.

Feldman A. 1966. Functional tuning of the nervous system during control of movement or maintenance of a steady posture. II. Controllable parameters of the muscles. III. Mechanographic analysis of the execution by man of the simplest motor task. *Biophysics* 11: 565–578, 766–755.

Fioravanti F, Inchingolo P, Pensiero S, Spanio M. 1995. Saccadic eye movement conjugation in children. *Vision Res* 35: 3217–3228.

Fiorillo CD, Tobler PN, Schultz W. 2003. Discrete coding of reward probability and uncertainty by dopamine neurons. *Science* 299: 1898–1902.

Flanagan JR, Beltzner MA. 2000. Independence of perceptual and sensorimotor predictions in the size-weight illusion. *Nat Neurosci* 3: 737–741.

Flanagan JR, Bittner JP, Johansson RS. 2008. Experience can change distinct size-weight priors engaged in lifting objects and judging their weights. *Curr Biol* 18: 1742–1747.

Flanagan JR, Rao AK. 1995. Trajectory adaptation to a nonlinear visuomotor transformation: Evidence of motion planning in visually perceived space. *J Neurophysiol* 74: 2174–2178.

Flash T, Hogan N. 1985. The coordination of arm movements: An experimentally confirmed mathematical model. *J Neurosci* 5: 1688–1703.

Frith C. 1996. Neuropsychology of schizophrenia: What are the implications of intellectual and experiential abnormalities for the neurobiology of schizophrenia? *Br Med Bull* 52: 618–626.

Frith CD, Blakemore S, Wolpert DM. 2000. Explaining the symptoms of schizophrenia: Abnormalities in the awareness of action. *Brain Res Brain Res Rev* 31: 357–363.

Fyhn M, Molden S, Witter MP, Moser EI, Moser MB. 2004. Spatial representation in the entorhinal cortex. *Science* 305: 1258.

Ganel T, Tanzer M, Goodale MA. 2008. A double dissociation between action and perception in the context of visual illusions: Opposite effects of real and illusory size. *Psychol Sci* 19: 221–225.

Gardner R, Hogan RE. 2005. Three-dimensional deformation-based hippocampal surface anatomy, projected on MRI images. *Clini Anat* 18: 481–487.

Georgopoulos AP, Schwartz AB, Kettner RE. 1986. Neural population coding of movement direction. *Science* 233: 1416–1419.

Ghahramani Z, Hinton GE. 1996. Parameter estimation for linear dynamical systems. Technical Report CRG-TR-96-2. Univ. Toronto.

Gilbert D. 2006. Stumbling on happiness. New York: Vintage Books.

Golla H, Tziridis K, Haarmeier T, Catz N, Barash S, Thier P. 2008. Reduced saccadic resilience and impaired saccadic adaptation due to cerebellar disease. *Eur J Neurosci* 27: 132–144.

Goodale MA, Milner AD. 1992. Separate visual pathways for perception and action. *Trends Neurosci* 15: 20–25.

Goossens HH, Van Opstal AJ. 2000. Blink-perturbed saccades in monkey. I. Behavioral analysis. *J Neurophysiol* 83: 3411–3429.

Gordon AM, Forssberg H, Johansson RS, Westling G. 1991. Visual size cues in the programming of manipulative forces during precision grip. *Exp Brain Res* 83: 477–482.

Gottlieb JP, Kusunoki M, Goldberg ME. 1998. The representation of visual salience in monkey parietal cortex. *Nature* 391: 481–484.

Gould SJ. 1995. Dinosaur in a haystack: Reflections in natural history. New York: Harmony Books.

Green DG. 1970. Regional variations in the visual acuity for interference fringes on the retina. *J Physiol* 207: 351–356.

Green L, Myerson J, Holt DD, Slevin JR, Estle SJ. 2004. Discounting of delayed food rewards in pigeons and rats: Is there a magnitude effect? *J Exp Anal Behav* 81: 39–50.

Green L, Myerson J, Ostaszewski P. 1999. Discounting of delayed rewards across the life span: Age differences in individual discounting functions. *Behav Processes* 46: 89–96.

Griffiths TL, Tenenbaum JB. 2006. Optimal predictions in everyday cognition. *Psychol Sci* 17: 767–773.

Guitton D, Volle M. 1987. Gaze control in humans: Eye-head coordination during orienting movements to targets within and beyond the oculomotor range. *J Neurophysiol* 58: 427–459.

Guthrie BL, Porter JD, Sparks DL. 1983. Corollary discharge provides accurate eye position information to the oculomotor system. *Science* 221: 1193–1195.

Haarmeier T, Thier P, Repnow M, Petersen D. 1997. False perception of motion in a patient who cannot compensate for eye movements. *Nature* 389: 849–852.

Hafting T, Fyhn M, Molden S, Moser M, Moser, EI. 2005. Microstructure of a spatial map in the entorhinal cortex. *Nature* 436: 801–806.

Haith A, Jackson C, Miall C, Vijayakumar S. 2008. Interactions between sensory and motor components of adaptation predicted by a Bayesian model. Advances in Computational Motor Control 7. Available at http://sites.google.com/site/acmcconference/.

Harris CM, Wolpert DM. 1998. Signal-dependent noise determines motor planning. *Nature* 394: 780–784.

Harris CM, Wolpert DM. 2006. The main sequence of saccades optimizes speed-accuracy trade-off. *Biol Cybern* 95: 21–29.

Hassabis D, Chu C, Rees G, Weiskopf N, Molyneux PD, Maguire EA. 2009. Decoding neuronal ensembles in the human hippocampus. *Curr Biol* 19: 546–554.

Hatada Y, Miall RC, Rossetti Y. 2006. Two waves of a long-lasting aftereffect of prism adaptation measured over 7 days. *Exp Brain Res* 169: 417–426.

Hayden BY, Parikh PC, Deaner RO, Platt ML. 2007. Economic principles motivating social attention in humans. *Proc Biol Sci* 274: 1751–1756.

Heerey EA, Robinson BM, McMahon RP, Gold JM. 2007. Delay discounting in schizophrenia. *Cogn Neuropsychiatry* 12: 213–221.

Held R, Freedman SJ. 1963. Plasticity in human sensorimotor control. *Science* 142: 455–462.

Hoffman DS, Strick PL. 1999. Step-tracking movements of the wrist. IV. Muscle activity associated with movements in different directions. *J Neurophysiol* 81: 319–333.

Hollerbach JM, Flash T. 1982. Dynamic interactions between limb segments during planar arm movements. *Biol Cybern* 44: 67–77.

Huang VS, Shadmehr R. 2009. Persistence of motor memories reflects statistics of the learning event. *J Neurophysiol* 102: 931–940.

Hull CL. 1930. Simple trial-and-error learning: A study in psychological theory. *Psychol Rev* 37: 241–256.

Hwang J, Kim S, Lee D. 2009. Temporal discounting and inter-temporal choice in rhesus monkeys. *Front Behav Neurosci* 3: 9.

Izawa J, Rane T, Donchin O, Shadmehr R. 2008. Motor adaptation as a process of reoptimization. *J Neurosci* 28: 2883–2891.

Izawa J, Shadmehr R. 2008. Online processing of uncertain information in visuomotor control. *J Neurosci* 28: 11360–11368.

Jacobs RJ. 1979. Visual resolution and contour interaction in the fovea and periphery. *Vision Res* 19: 1187–1195.

Jimura K, Myerson J, Hilgard J, Braver TS, Green L. 2009. Are people really more patient than other animals? Evidence from human discounting of real liquid rewards. *Psychon Bull Rev* 16: 1071–1075.

Jones KE, Hamilton AF, Wolpert DM. 2002. Sources of signal-dependent noise during isometric force production. *J Neurophysiol* 88: 1533–1544.

Kacelnik A. 1997. Normative and descriptive models of decision making: Time discounting and risk sensitivity. Ciba Foundation Symposium 208—*Characterizing Human Psychological Adaptations* (GR Bock and G Cardew, eds.), pp. 51–70. Chichester: Wiley.

Kagerer FA, Contreras-Vidal JL, Stelmach GE. 1997. Adaptation to gradual as compared with sudden visuo-motor distortions. *Exp Brain Res* 115: 557–561.

Kalman RE 1960. A new approach to linear filtering and prediction problems. *Trans ASME J Basic Engineering* 82 (series D): 35–45.

Kamin LJ. 1968. Attention-like processes in classical conditioning. Miami symposium on the prediction of behavior: Aversive stimulation (Jones MR, ed), pp. 9–33. Miami: University of Miami Press.

Kampe KKW, Frith CD, Dolan RJ, Frith U. 2001. Psychology: Reward value of attractiveness and gaze. *Nature* 413: 589.

Kapur S. 2003. Psychosis as a state of aberrant salience: A framework linking biology, phenomenology, and pharmacology in schizophrenia. *Am J Psychiatry* 160: 13–23.

Keller EL. 1973. Accommodative vergence in the alert monkey: Motor unit analysis. *Vision Res* 13: 1565–1575.

Keller EL, Robinson DA. 1971. Absence of a stretch reflex in extraocular muscles of the monkey. *J Neurophysiol* 34: 908–919.

Klassen J, Tong C, Flanagan JR. 2005. Learning and recall of incremental kinematic and dynamic sensorimotor transformations. *Exp Brain Res* 164: 250–259.

Kloppel S, Draganski B, Golding CV, Chu C, Nagy Z, Cook PA, Hicks SL, et al. 2008. White matter connections reflect changes in voluntary-guided saccades in pre-symptomatic Huntington's disease. *Brain* 131: 196–204.

Kluzik J, Diedrichsen J, Shadmehr R, Bastian AJ. 2008. Reach adaptation: What determines whether we learn an internal model of the tool or adapt the model of our arm? *J Neurophysiol* 100: 1455–1464.

Kobayashi S, Schultz W. 2008. Influence of reward delays on responses of dopamine neurons. *J Neurosci* 28: 7837–7846.

Kojima Y, Iwamoto Y, Yoshida K. 2004. Memory of learning facilitates saccadic adaptation in the monkey. *J Neurosci* 24: 7531–7539.

Kording KP, Beierholm U, Ma WJ, Quartz S, Tenenbaum JB, Shams L. 2007. Causal inference in multisensory perception. *PLoS ONE* 2: e943.

Kording KP, Tenenbaum JB, Shadmehr R. 2007. The dynamics of memory as a consequence of optimal adaptation to a changing body. *Nat Neurosci* 10: 779–786.

Kording KP, Wolpert DM. 2004. Bayesian integration in sensorimotor learning. *Nature* 427: 244–247.

Krakauer JW, Ghez C, Ghilardi MF. 2005. Adaptation to visuomotor transformations: Consolidation, interference, and forgetting. *J Neurosci* 25: 473–478.

Krakauer JW, Mazzoni P, Ghazizadeh A, Ravindran R, Shadmehr R. 2006. Generalization of motor learning depends on the history of prior action. *PLoS Biol* 4: e316.

Lee JY, Schweighofer N. 2009. Dual adaptation supports a parallel architecture of motor memory. *J Neurosci* 29: 10396–10404.

Lindner A, Thier P, Kircher TT, Haarmeier T, Leube DT. 2005. Disorders of agency in schizophrenia correlate with an inability to compensate for the sensory consequences of actions. *Curr Biol* 15: 1119–1124.

Liu D, Todorov E. 2007. Evidence for the flexible sensorimotor strategies predicted by optimal feedback control. *J Neurosci* 27: 9354–9368.

Ljungberg T, Apicella P, Schultz W. 1992. Responses of monkey dopamine neurons during learning of behavioral reactions. *J Neurophysiol* 67: 145–163.

Llinas RR. 2001. I of the vortex: From neurons to self. Cambridge, MA: MIT Press.

Louie K, Glimcher PW. 2010. Separating value from choice: Delay discounting activity in the lateral intraparietal area. *J Neurosci* 30: 5498–5507.

Maguire EA, Frackowiak RSJ, Frith CD. 1997. Recalling routes around London: Activation of the right hippocampus in taxi drivers. *J Neurosci* 17: 7103.

Maguire EA, Gadian DG, Johnsrude IS, Good CD, Ashburner J, Frackowiak RSJ, Frith CD. 2000. Navigation-related structural change in the hippocampi of taxi drivers. *Proc Natl Acad Sci USA* 97: 4398.

Mahlberg R, Steinacher B, Mackert A, Flechtner KM. 2001. Basic parameters of saccadic eye movements—differences between unmedicated schizophrenia and affective disorder patients. *Eur Arch Psychiatry Clin Neurosci* 251: 205–210.

Martin TA, Keating JG, Goodkin HP, Bastian AJ, Thach WT. 1996. Throwing while looking through prisms. I. Focal olivocerebellar lesions impair adaptation. *Brain* 119: 1183–1198.

Mazzoni P, Hristova A, Krakauer JW. 2007. Why don't we move faster? Parkinson's disease, movement vigor, and implicit motivation. *J Neurosci* 27: 7105–7116.

Mazzoni P, Krakauer JW. 2006. An implicit plan overrides an explicit strategy during visuomotor adaptation. *J Neurosci* 26: 3642–3645.

McCormack AL, Di Monte DA, Delfani K, Irwin I, DeLanney LE, Langston WJ, Janson AM. 2004. Aging of the nigrostriatal system in the squirrel monkey. *J Comp Neurol* 471: 387–395.

McCulloch W, Pitts W. 1943. A logical calculus of the ideas immanent in nervous activity. *Bull Math Biol* 5(4): 115–133.

McIntyre J, Zago M, Berthoz A, Lacquaniti F. 2001. Does the brain model Newton's laws? *Nat Neurosci* 4: 693–694.

McLaughlin S. 1967. Parametric adjustment in saccadic eye movements. *Percept Psychophys* 2: 359–362.

Medina JF, Garcia KS, Mauk MD. 2001. A mechanism for savings in the cerebellum. *J Neurosci* 21: 4081–4089.

Michel C, Pisella L, Prablanc C, Rode G, Rossetti Y. 2007. Enhancing visuomotor adaptation by reducing error signals: single-step (aware) versus multiple-step (unaware) exposure to wedge prisms. *J Cogn Neurosci* 19: 341–350.

Miller RR, Matute H. 1996. Biological significance in forward and backward blocking: Resolution of a discrepancy between animal conditioning and human causal judgment. *J Exp Psychol Gen* 125: 370–386.

Milner B. 1962. Les troubles de la memoire accompagnant des lesions hippocampiques bilaterales. In *Physiologie de l'hippocampe* (P Passouant, ed.), pp. 257–272. Paris: Centre National de la Recherche Scientifique.

Milner B, Corkin S, Teuber HL. 1968. Further analysis of the hippocampal amnesic syndrome: 14-year follow-up study of HM. *Neuropsychologia* 6: 215–234.

Montagnini A, Chelazzi L. 2005. The urgency to look: Prompt saccades to the benefit of perception. *Vision Res* 45: 3391–3401.

Morasso P. 1981. Spatial control of arm movements. *Exp Brain Res* 42: 223–227.

Mozer MC, Pashler H, Homaei H. 2008. Optimal predictions in everyday cognition: The wisdom of individuals or crowds? *Cogn Sci* 32: 1133–1147.

Munoz DP, Armstrong IT, Hampton KA, Moore KD. 2003. Altered control of visual fixation and saccadic eye movements in attention-deficit hyperactivity disorder. *J Neurophysiol* 90: 503–514.

Myerson J, Green L. 1995. Discounting of delayed rewards: Models of individual choice. *J Exp Anal Behav* 64: 263–276.

Nakamura T, Kanayama R, Sano R, Ohki M, Kimura Y, Aoyagi M, Koike Y. 1991. Quantitative analysis of ocular movements in Parkinson's disease. *Acta Otolaryngol Suppl* 481: 559–562.

Nash J. 1956. The imbedding problem for Riemannian manifolds. *Ann Math* 63(1): 20–63.

Ohyama T, Mauk MD. 2001. Latent acquisition of timed responses in cerebellar cortex. *J Neurosci* 21: 682–690.

O'Keefe J, Conway D. 1976. Sensory inputs to the hippocampal place units. *Neurosci Lett* 3: 103–104.

O'Keefe J, Dostrovsky J. 1971. The hippocampus as a spatial map: Preliminary evidence from unit activity in the freely-moving rat. *Brain Res* 34: 171–175.

O'Keefe J, Nadel L. 1978. The hippocampus as a cognitive map. Oxford: Clarendon Press.

O'Sullivan I, Burdet E, Diedrichsen J. 2009. Dissociating variability and effort as determinants of coordination. *PLOS Comput Biol* 5: e1000345.

Pekny SE, Criscimagna-Hemminger SE, Shadmehr R. 2011. Protection and expression of human motor memories. *J Neurosci*, in press.

Polit A, Bizzi E. 1978. Processes controlling arm movements in monkeys. *Science* 201(4362): 1235.

Pompilio L, Kacelnik A. 2010. Context-dependent utility overrides absolute memory as a determinant of choice. *Proc Natl Acad Sci USA* 107: 508–512.

Rabe K, Livne O, Gizewski ER, Aurich V, Beck A, Timmann D, Donchin O. 2009. Adaptation to visuomotor rotation and force field perturbation is correlated to different brain areas in patients with cerebellar degeneration. *J Neurophysiol* 101: 1961–1971.

Ramón y Cajal S. 1911. *Histologie du systeme nerveux de l'homme et des vertebretes*, Vols. 1 and 2. Paris: Maloine. [Reprinted in 1955 by Consejo Superior de Investigaciones Cientificas, Inst. Ramón y Cajal, Madrid.]

Reisman DS, Wityk R, Silver K, Bastian AJ. 2007. Locomotor adaptation on a split-belt treadmill can improve walking symmetry post-stroke. *Brain* 130: 1861–1872.

Rescorla RA, Wagner AR. 1972. A theory of Pavlovian conditioning: The effectiveness of reinforcement and non-reinforcement. In Classical Conditioning II: Current Research and Theory (Black AH, Prokasy WF, eds.), pp. 64–69. New York: Appleton Century Crofts.

Robinson DA, Gordon JL, Gordon SE. 1986. A model of the smooth pursuit eye movement system. *Biol Cybern* 55: 43–57.

Robinson FR, Soetedjo R, Noto C. 2006. Distinct short-term and long-term adaptation to reduce saccade size in monkey. *J Neurophysiol* 96: 1030–1041.

Rottach KG, Das VE, Wohlgemuth W, Zivotofsky AZ, Leigh RJ. 1998. Properties of horizontal saccades accompanied by blinks. *J Neurophysiol* 79: 2895–2902.

Samsonovich A, McNaughton BL. 1997. Path integration and cognitive mapping in a continuous attractor neural network model. *J Neurosci* 17: 5900–5920.

Scheidt RA, Reinkensmeyer DJ, Conditt MA, Rymer WZ, Mussa-Ivaldi FA. 2000. Persistence of motor adaptation during constrained, multi-joint, arm movements. *J Neurophysiol* 84: 853–862.

Schmidt RA. 1991. Motor learning and performance: From principles to practice. Champaign, IL: Human Kinetics Books.

Schmidt RA, Zelaznik H, Hawkins B, Frank JS, Quinn JT, Jr. 1979. Motor-output variability: A theory for the accuracy of rapid motor acts. *Psychol Rev* 47: 415–451.

Shadmehr R, Brandt J, Corkin S. 1998. Time-dependent motor memory processes in amnesic subjects. *J Neurophysiol* 80: 1590–1597.

Shadmehr R, Brashers-Krug T. 1997. Functional stages in the formation of human long-term motor memory. *J Neurosci* 17: 409–419.

Shadmehr R, Mussa-Ivaldi FA. 1994. Adaptive representation of dynamics during learning of a motor task. *J Neurosci* 14: 3208–3224.

Shadmehr R, Orban de Xivry JJ, Xu-Wilson M, Shih TY. 2010. Temporal discounting of reward and the cost of time in motor control. *J Neurosci* 30: 10507–10516.

Sherrington C. 1923. The integrative action of the nervous system. Cambridge: Cambridge University Press.

Shibasaki H, Tsuji S, Kuroiwa Y. 1979. Oculomotor abnormalities in Parkinson's disease. *Arch Neurol* 36: 360–364.

Shumway RH, Stoffer DS. 1982. An approach to time series smoothing and forecasting using the EM algorithm. *J Time Ser Anal* 3: 253–264.

Slijper H, Richter J, Over E, Smeets J, Frens M. 2009. Statistics predict kinematics of hand movements during everyday activity. *J Mot Behav* 41: 3–9.

Smith MA, Ghazizadeh A, Shadmehr R. 2006. Interacting adaptive processes with different timescales underlie short-term motor learning. *PLoS Biol* 4: e179.

Smith M, Shadmehr R. 2004. Modulation of the rate of error-dependent learning by the statistical properties of the task. Advances in Computational Motor Control 3. Available at http://sites.google.com/site/acmcconference/.

Snyder LH, Calton JL, Dickinson AR, Lawrence BM. 2002. Eye-hand coordination: Saccades are faster when accompanied by a coordinated arm movement. *J Neurophysiol* 87: 2279–2286.

Solstad T, Moser EI, Einevoll GT. 2006. From grid cells to place cells: A mathematical model. *Hippocampus* 16: 1026–1031.

Spong MW, Hutchinson S, Vidyasagar M. 2005. Robot modeling and control. Hoboken, NJ: John Wiley and Sons.

Stollhoff N, Menzel R, Eisenhardt D. 2005. Spontaneous recovery from extinction depends on the reconsolidation of the acquisition memory in an appetitive learning paradigm in the honeybee (Apis mellifera). *J Neurosci* 25: 4485–4492.

Stone JM, Morrison PD, Pilowsky LS. 2007. Glutamate and dopamine dysregulation in schizophrenia—a synthesis and selective review. *J Psychopharmacol* 21: 440–452.

Straube A, Deubel H, Ditterich J, Eggert T. 2001. Cerebellar lesions impair rapid saccade amplitude adaptation. *Neurology* 57: 2105–2108.

Straube A, Fuchs AF, Usher S, Robinson FR. 1997. Characteristics of saccadic gain adaptation in rhesus macaques. *J Neurophysiol* 77: 874–895.

Sutton R, Barto A. 1998. Reinforcement learning: An introduction. Cambridge, MA: MIT Press.

Takikawa Y, Kawagoe R, Itoh H, Nakahara H, Hikosaka O. 2002. Modulation of saccadic eye movements by predicted reward outcome. *Exp Brain Res* 142: 284–291.

Teyler TJ, DiScenna P. 1984. The topological anatomy of the hippocampus: A clue to its function. Brain Res Bull 12: 711–719.

Thrun S, Fox D, Burgard W, Dellaert F. 2001. Robust Monte Carlo localization for mobile robots. *Artif Intell* 128: 99–141.

Todorov E. 2005. Stochastic optimal control and estimation methods adapted to the noise characteristics of the sensorimotor system. *Neural Comput* 17: 1084–1108.

Todorov E, Jordan MI. 2002. Optimal feedback control as a theory of motor coordination. *Nat Neurosci* 5: 1226–1235.

Tolman EC. 1948. Cognitive maps in rats and men. *Psychol Rev* 55: 189–208.

Uno Y, Kawato M, Suzuki R. 1989. Formation and control of optimal trajectory in human multijoint arm movement: Minimum torque-change model. *Biol Cybern* 61: 89–101.

van Beers RJ, Wolpert DM, Haggard P. 2002. When feeling is more important than seeing in sensorimotor adaptation. *Curr Biol* 12: 834–837.

van Donkelaar P, Siu KC, Walterschied J. 2004. Saccadic output is influenced by limb kinetics during eye-hand coordination. *J Mot Behav* 36: 245–252.

Van Gemmert AW, Teulings HL, Stelmach GE. 2001. Parkinsonian patients reduce their stroke size with increased processing demands. *Brain Cogn* 47: 504–512.

van Overschee P, De Moor B. 1996. Subspace identification for linear systems. Boston: Kluwer Academic.

Vaziri S, Diedrichsen J, Shadmehr R. 2006. Why does the brain predict sensory consequences of oculomotor commands? Optimal integration of the predicted and the actual sensory feedback. *J Neurosci* 26: 4188–4197.

Wallace MT, Roberson GE, Hairston WD, Stein BE, Vaughan JW, Schirillo JA. 2004. Unifying multisensory signals across time and space. *Exp Brain Res* 158: 252–258.

Walls GL. 1962. The evolutionary history of eye movements. *Vision Res* 2: 69–80.

Wei K, Kording K. 2008. Uncertainty in state estimate and feedback determines the rate of motor adaptation. Advances in Computational Motor Control 7. Available at http://sites.google.com/site/acmcconference/.

White OB, Saint-Cyr JA, Tomlinson RD, Sharpe JA. 1983. Ocular motor deficits in Parkinson's disease. II. Control of the saccadic and smooth pursuit systems. *Brain* 106 (Pt 3): 571–587.

Widrow B, Hoff ME. 1960. Adaptive switching circuits. WESCON Convention Record Part IV, pp. 96–104.

Wilson MA, McNaughton BL. 1993. Dynamics of the hippocampal ensemble code for space. *Science* 261: 1055–1058.

Won J, Hogan N. 1995. Stability properties of human reaching movements. *Exp Brain Res* 107: 125–136.

Xu-Wilson M, Chen-Harris H, Zee DS, Shadmehr R. 2009. Cerebellar contributions to adaptive control of saccades in humans. *J Neurosci* 29: 12930–12939.

Xu-Wilson M, Tian J, Shadmehr R, Zee DS. 2011. TMS induced startle perturbs saccade trajectories and unmasks the internal feedback controller. *J Neurosci* 31: 11537–11546.

Xu-Wilson M, Zee DS, Shadmehr R. 2009. The intrinsic value of visual information affects saccade velocities. *Exp Brain Res* 196: 475–481.

Yarbus AL. 1967. Eye movements and vision. New York: Plenum Press.

Zarahn E, Weston GD, Liang J, Mazzoni P, Krakauer JW. 2008. Explaining savings for visuomotor adaptation: Linear time-invariant state-space models are not sufficient. *J Neurophysiol* 100: 2537–2548.

Zee D, Optican L, Cook JD, Robinson DA, Engel WK. 1976. Slow saccades in spinocerebellar degeneration. *Arch Neurol* 33(4): 243.

Index

Computational Neuroscience

Terence J. Sejnowski and Tomaso A. Poggio, editors

Neural Nets in Electric Fish, Walter Heiligenberg, 1991

The Computational Brain, Patricia S. Churchland and Terrence J. Sejnowski, 1992

Dynamic Biological Networks: The Stomatogastic Nervous System, edited by Ronald M. Harris-Warrick, Eve Marder, Allen I. Selverston, and Maurice Moulins, 1992

The Neurobiology of Neural Networks, edited by Daniel Gardner, 1993

Large-Scale Neuronal Theories of the Brain, edited by Christof Koch and Joel L. Davis, 1994

The Theoretical Foundations of Dendritic Function: Selected Papers of Wilfrid Rall with Commentaries, edited by Idan Segev, John Rinzel, and Gordon M. Shepherd, 1995

Models of Information Processing in the Basal Ganglia, edited by James C. Houk, Joel L. Davis, and David G. Beiser, 1995

Spikes: Exploring the Neural Code, Fred Rieke, David Warland, Rob de Ruyter van Steveninck, and William Bialek, 1997

Neurons, Networks, and Motor Behavior, edited by Paul S. Stein, Sten Grillner, Allen I. Selverston, and Douglas G. Stuart, 1997

Methods in Neuronal Modeling: From Ions to Networks, second edition, edited by Christof Koch and Idan Segev, 1998

Fundamentals of Neural Network Modeling: Neuropsychology and Cognitive Neuroscience, edited by Randolph W. Parks, Daniel S. Levine, and Debra L. Long, 1998

Neural Codes and Distributed Representations: Foundations of Neural Computation, edited by Laurence Abbott and Terrence J. Sejnowski, 1999

Unsupervised Learning: Foundations of Neural Computation, edited by Geoffrey Hinton and Terrence J. Sejnowski, 1999

Fast Oscillations in Cortical Circuits, Roger D. Traub, John G. R. Jeffreys, and Miles A. Whittington, 1999

Computational Vision: Information Processing in Perception and Visual Behavior, Hanspeter A. Mallot, 2000

Graphical Models: Foundations of Neural Computation, edited by Michael I. Jordan and Terrence J. Sejnowski, 2001

Self-Organizing Map Formation: Foundations of Neural Computation, edited by Klaus Obermayer and Terrence J. Sejnowski, 2001

Neural Engineering: Computation, Representation, and Dynamics in Neurobiological Systems, Chris Eliasmith and Charles H. Anderson, 2003

The Computational Neurobiology of Reaching and Pointing, edited by Reza Shadmehr and Steven P. Wise, 2005

Dynamical Systems in Neuroscience, Eugene M. Izhikevich, 2006

Bayesian Brain: Probabilistic Approaches to Neural Coding, edited by Kenji Doya, Shin Ishii, Alexandre Pouget, and Rajesh P. N. Rao, 2007

Computational Modeling Methods for Neuroscientists, edited by Erik De Schutter, 2009

Neural Control Engineering, Steven J. Schiff, 2011

Understanding Visual Population Codes: Toward a Common Multivariate Framework for Cell Recording and Functional Imaging, edited by Nikolaus Kriegeskorte and Gabriel Kreiman, 2011

Biological Learning and Control: How the Brain Builds Representations, Predicts Events, and Makes Decisions, Reza Shadmehr and Sandro Mussa-Ivaldi, 2012

Printed in the United States
by Baker & Taylor Publisher Services